Numerical Methods
in
OFFSHORE
ENGINEERING

Numerical Methods in OFFSHORE ENGINEERING

Edited by
O. C. Zienkiewicz
R. W. Lewis
K. G. Stagg
*Department of Civil Engineering,
University College Swansea*

A Wiley–Interscience Publication

JOHN WILEY & SONS
Chichester · New York · Brisbane · Toronto

Library of Congress Cataloging in Publication Data:
Main entry under title:

Numerical methods in offshore engineering.

(Wiley series in numerical methods in engineering)
'A Wiley–Interscience publication.'
Includes index.
1. Ocean engineering. 2. Numerical calculations.
I. Zienkiewicz, O. C. II. Lewis, Roland Wynne.
III. Stagg, Kenneth Geoffrey.
TC1650.N85   624'.1'09162   77-12565
ISBN 0 471 99591 6

Printed in Great Britain by J. W. Arrowsmith Ltd., Bristol.

Contributory Authors

K. H. Andersen — Engineer, Norwegian Geotechnical Institute, Oslo, Norway

Professor K. Bell — Norwegian Institute of Technology, Trondheim, Norway.

Dr. P. Bettess — Lecturer in Civil Engineering, University College of Swansea, Swansea, U.K.

Professor P. M. Byrne — Civil Engineering Department, University of British Columbia, Vancouver 8, Canada.

C. T. Chang — Research Assistant in Civil Engineering, University College of Swansea, Swansea, U.K.

D. J. Cronin — Engineer, Halcrow-Ewbank Petroleum and Offshore Engineering Company, Alliance House, 12 Caxton Street, London, U.K.

Dr. R. Dungar — Lecturer in Civil Engineering, University of Bristol, University Walk, Bristol, U.K.

Professor W. D. L. Finn — Civil Engineering Department, University of British Columbia, Vancouver 8, Canada.

Professor C. J. Garrison — Associate Professor of Mechanical Engineering, Naval Postgraduate School, Monterey, California, U.S.A.

P. J. George — Lloyd's Register of Shipping, Offshore Services Group, 71 Fenchurch Street, London, U.K.

Dr. P. S. Godfrey — Engineer, Halcrow-Ewbank Petroleum and Offshore Engineering Company, Alliance House, 12 Caxton Street, London, U.K.

O. E. HANSTEEN
Engineer, Norwegian Geotechnical Institute, Oslo, Norway.

DR. E. HINTON
Lecturer in Civil Engineering, University College of Swansea, Swansea, U.K.

R. HOBBS
Lloyd's Register of Shipping, Offshore Services Group, 71 Fenchurch Street, London, U.K.

DR. K. HOEG
Director, Norwegian Geotechnical Institute, Oslo, Norway.

PROFESSOR I. HOLAND
Norwegian Institute of Technology, Trondheim, Norway.

PROFESSOR P. HOLMES
Department of Civil Engineering, Liverpool University, Liverpool, U.K.

P. M. HOOK
Engineer, Halcrow-Ewbank Petroleum and Offshore Engineering Company, Alliance House, 12 Caxton Street, London, U.K.

DR. D. W. KELLY
Lecturer in Civil Engineering, University College of Swansea, Swansea, U.K.

K. W. LEE
Research Associate, Civil Engineering Dept., University of British Columbia, Vancouver 8, Canada.

DR. R. W. LEWIS
Lecturer in Civil Engineering, University College of Swansea, Swansea, U.K.

G. R. MARTIN
Senior Lecturer in Civil Engineering, University of Auckland, New Zealand.

C. G. W. MUSTOE
Lloyd's Register of Shipping, Offshore Services Group, 71 Fenchurch Street, London, U.K.

DR. D. J. NAYLOR
Lecturer in Civil Engineering, University College of Swansea, Swansea, U.K.

DR. R. NIELSEN
Pipe Line Technologists (Offshore) Limited, 17 Old Court Plaza, High Street, Kensington, London, U.K.

V. A. NORRIS — *Senior Research Assistant in Civil Engineering, University College of Swansea, Swansea, U.K.*

DR. J. W. PENDERED — *Pipe Line Technologists (Offshore) Limited, 17 Old Court Plaza, High Street, Kensington, London, U.K.*

PROFESSOR J. PENZIEN — *Civil Engineering Department, University of California, Berkeley, U.S.A.*

DR. J. H. PREVOST — *Engineer, Norwegian Geotechnical Institute, Oslo, Norway.*

PROFESSOR R. SIGBJÖRNSSON — *Norwegian Institute of Technology, Trondheim, Norway.*

DR. I. M. SMITH — *Lecturer in Civil Engineering, Simon Engineering Laboratories, University of Manchester, Manchester, U.K.*

R. G. TICKELL — *Lecturer in Civil Engineering, Liverpool University, Liverpool, U.K.*

W. S. TSENG — *Research Assistant, University of California, Berkeley, U.S.A.*

DR. B. J. WATT — *President Brian Watt Associates, Inc., P.O. Box 61361, Houston, Texas, U.S.A.*

PROFESSOR E. L. WILSON — *Civil Engineering Department, University of California, Berkeley, U.S.A.*

DR. L. A. WINNICKI — *Polish Academy of Science, Visiting Senior Research Assistant in Civil Engineering, University College of Swansea, Swansea, U.K.*

DR. T. A. WYATT — *Reader in Civil Engineering, Imperial College, London, U.K.*

PROFESSOR O. C. ZIENKIEWICZ — *Civil Engineering Department, University College of Swansea, Swansea, U.K.*

Preface

The need for exploitation of oil and gas resources in the 'offshore' area of seas and oceans has posed many technical problems in which a considerable extrapolation of knowledge beyond that available from land or coastal construction is needed. This challenge to the engineering profession necessitates the use of best *prediction* methods available for ensuring the safety and economy of the designs and here, naturally, computer based mathematical methods offer a unique and essential tool.

While the number of texts and symposia dealing with numerical methods and their application to engineering problems is large and covers well the background material, special techniques are applicable to each branch of engineering—and often new approaches are necessary. In this text we attempt to give a survey of both the problems and of methodologies applicable to 'offshore' structures. The procedures are not limited however to petroleum industry construction—although the major motivation comes from this activity. Most of the methods are equally applicable to all maritime structures, including harbour and shore installations.

With expertise of the subject being widely scattered between universities, research institutes and industry it appears that the best overall picture of the 'state of art' is available by an organized multi-author presentation. With this in mind the editors invited a series of contributions which were first presented orally at a Symposium held at Swansea (January 1977). Ensuing discussions and subsequent modification of chapters permits a reasonably coherent picture to be presented in this volume.

The reader will observe that the seventeen chapters fall into three major sections. Thus, after the basic review of Chapter 1 the first section of the book (Chapters 2–5) deals with problems of fluid loading, the second (Chapters 6–11) with general methods of computing dynamic response and finally, the third (Chapters 12–17) with foundation and sea bed problems.

Whilst in many of the areas intense research activity is at present conducted (such as for instance in the problems of cyclic soil response), the field is sufficiently well established for the book to form a basic reference for engineers engaged in it. This professional emphasis has given the motivation to the writers and editors alike—and we hope that the objectives will be achieved.

Acknowledgements

The editors would like to thank all the contributors who have unstintingly given their time in preparing and reviewing the manuscripts. Thanks are also due to the Computer Methods and Applications Group of the Institution of Civil Engineers for co-sponsoring the original Symposium held in Swansea in January 1977 and to the Institute of Numerical Methods in Engineering of University College of Swansea for acting as host on that occasion.

O. C. ZIENKIEWICZ
R. W. LEWIS
K. G. STAGG

Contents

Contents

Chapter 1

Basic Structural Systems– A Review of Their Design and Analysis Requirements

Brian J. Watt

1.1 INTRODUCTION

A casual observer at one of the many engineering symposia on numerical methods in the past decade would probably have been left with the impression that all of the interesting boundary and initial value problems had been solved, or at least that this was very close to happening. The power and versatility of numerical analysis methodology are such that he or she could readily be forgiven for believing this. While it is probably true for finite element methods that most of the major advances of a formulative kind have been made, much remains to be done before the developer of such systems can relegate all his programs to 'ordinary user' status. This is certainly the case for their application to offshore engineering.

The purpose of this introductory chapter is thus to identify some of the basic design problem areas and to help set the scene for the ensuing more detailed discussion on specific analysis topics.

There are many different activities and requirements offshore and hence many different structures used. The ensuing discussion will be limited to so-called 'fixed' structures where this is interpreted to mean structures installed and operated at one location throughout their design lives. This presentation is not intended as a comprehensive audit of all analysis problems in offshore engineering. It is hoped that it will at least provide a framework in which to view the more refined analytical discussions to follow in later chapters.

1.2 THE OCEAN ENVIRONMENT

1.2.1 Physical oceanography

It is appropriate to allocate a few words to the environment in which the offshore engineer works. The ocean can vary very widely at different places and times. In attempting to understand its behaviour, it is useful to refer to the analogy drawn by Vine[1] between physical oceanography and meteorology. Both embrace complex short term variations superimposed on predominantly seasonal cycles that are in turn affected by long-term climatic changes. The physical and chemical properties of seawater itself are also quite variable and thus whether one is interested on a macro or micro scale, the single dominant characteristic of the ocean environment is its changeability.

1.2.2   The meaning of 'offshore'

In its most general application, the term 'offshore' is usually taken to mean that part of the ocean where the present mudline is below the level of the lowest astronomical tide (L.A.T.) It is frequently also assumed to be restricted to the area of the continental shelves or a water depth of less than 200 m. This latter definition embraces an area comprising 8 per cent of the total wetted surface of the earth which is in turn equivalent to nearly 20 per cent of the total dry land surface.

1.2.3   Wind, waves and currents

The motions of the sea are the result of a superposition of many disturbing forces. Mathematical tools have been created to define these motions, but despite all efforts, any such model requires tremendous approximations which seldom represent the true facts with all their complexities.[2] The problem of wind offshore can be severe on parts of a structure and is especially important with regard to floating stability. Nevertheless, for most fixed structures in deeper waters the wind load represents less than 5 per cent of the total environmental loading. Current loads can be severe in some locations but it is usually wave loads which dominate the designer's thinking. Factors such as earthquakes and floating ice can change this completely but it is appropriate in this book to concentrate on waves, since this form of environmental loading dominates for most offshore installations currently being planned.

   The exact mechanism by which wind creates waves is not yet understood. Nevertheless, mathematical bases exist for predicting the wave profile and associated water particle kinematics for wind induced waves. The relative validity of the various theories was reviewed by Dean,[3] however, it does not need much observation of the sea to suggest that its surface is not characterized by regular, long-crested waves, whether Cnoidal, Airy, Stokes Fifth or otherwise. Its generally rather chaotic behaviour is better defined by the term 'sea state' and it is becoming increasingly recognized that realistic handling of the engineering problems requires some form of stochastic treatment.

1.2.4 Seabed geology

The geotechnical properties of a soil deposit are a complicated function of the origin, deposition, weathering and loading histories. While it is obvious that the morphological processes in the marine environment are different from those on land, it must be remembered that large parts of our present continental shelves have in recent geological time been raised above sea level, loaded by ice and sediment and subjected to subaerial weathering processes. Features such as buried river valleys, peat deposits, outwash channels and moraines have to be identified and avoided or quantified, along with other peculiarly marine features such as sandwaves and wave induced landslides.[4] Probably the most significant factor is the large percentage of the continental shelf area covered by recent deposits of marine clays and silts. These typically are under or normally consolidated with mudline shear strengths of about $4\,kN/m^2$, occurring in deposits up to many hundreds of feet thick. In summary, complex soil deposits covering if anything a greater range of strength and deformability than on land form the province of the offshore foundation engineer. Any numerical models of geotechnical problems are therefore likely to be at least as complex as those on land.

1.2.5 Earthquakes

The foci of severe earthquakes are located in the tectonic discontinuities between the plates which comprise the earth's mantle. Most of these discontinuities occur offshore and some are associated with major hydrocarbon deposits. Examples are the continental shelves off California and the Gulf of Alaska. The offshore engineer is faced with the usual land based seismic problems of predicting structural and foundation response as well as some unwelcome additional phenomena such as tsunamis and submarine landslides.

1.2.6 Floating ice

Exploration for mineral deposits has been actively pursued in polar regions for more than a decade. Although relatively little has been developed in a true offshore polar region, the first steps have been taken and planning is already underway for offshore structures to cope with the floating ice problem. A recent review of the information available on the genesis, morphology, physical properties and failure mechanisms of floating sea ice[5] indicated that the art of predicting ice loads on fixed structures is in its infancy. A great deal of research is needed to give the offshore engineer reliable numerate answers.

1.2.7 The ocean environment and design constraints

The above picture is brief and superficial but it serves to introduce some key characteristics of the design problem offshore. These are:

 (a) environmental loads originate from several sources
 (b) they are generally larger, more dynamic in character and more highly variable than on land
 (c) bottom conditions cover a wide range of soil types from very soft underconsolidated soils to rock

When one adds to these a fourth factor, namely the difficulty of underwater working, it is clear that the reliability aspect of design assumes a new perspective in the offshore environment.

1.3 THE CHOICE OF STRUCTURAL FORM

1.3.1 Historical perspective

The earliest engineered offshore structures were usually installed for either navigational or military purposes. The major share of offshore engineering experience, however, is associated with the exploitation of hydrocarbon deposits. The ensuing discussion is inevitably dominated by the requirements of typical oil and gas installations, but the engineering implications are relevant to several other offshore applications. It is entirely appropriate to view the problems through an oil industry perspective since this industry will continue to dominate the offshore scene for some time.

1.3.2 Foundation constraints

The structural options which are open to the offshore engineer are conditioned in large measure by foundation considerations. Figure 1.1 shows a highly simplified concept of an offshore structure. Fluid flow phenomena induce a

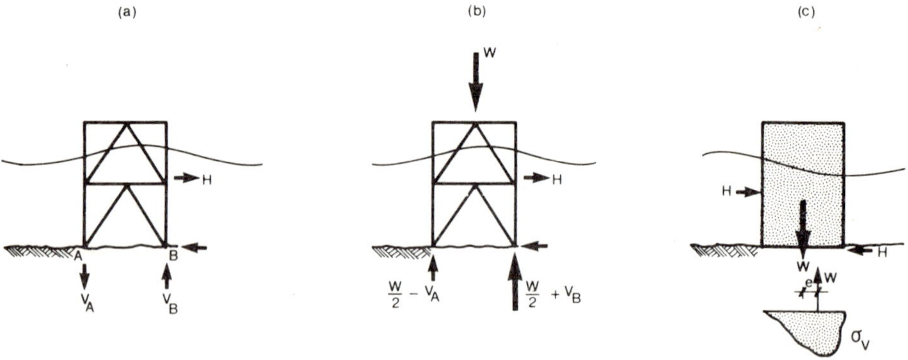

Figure 1.1 Piled and gravity foundations

high horizontal load and moment at the foundation level (mudline). For the two point support system shown in 1.1(a), a push-pull system of vertical reactions is required for moment equilibrium. Providing resistance to foundation uplift can sometimes be difficult and the requirement can be eliminated by providing additional vertical force as in 1.1(b). This is the concept of the so-called gravity structure, namely, the elimination of uplift problems by the provision of structural self weight. This is most often accomplished by using massive monolithic structures with continuous foundations as in 1.1(c). Framed gravity structures with discrete foundations as in 1.1(b) have been installed, though they are less common than the monolithic type.

It is evident from Figure 1.1 that:
System (a) implies a deep foundation (piles) to resist uplift
System (b) could use a deep or shallow foundation but downthrusts are larger than in (a)
System (c) implies much larger horizontal loads and requires strong soils close to the mudline.

1.3.3 Common structural systems

1.3.3.1 *The jacket or template*
It is proper to commence the discussion of structural systems with the so-called 'jacket' or template type of platform since it represents by far the largest number of offshore platforms installed to date. It developed to meet the needs of offshore drilling and production operations in the Gulf of Mexico and Lake Maracaibo. Both areas are characterized by large thicknesses of soft, recent marine sediments and foundation support has to be mobilized at some depth below the mudline. In order to reduce fluid loads and foundation movements a relatively transparent but stiff structure such as a space frame attached to long piles is an obvious solution.

A key difference between an offshore and onshore structure is the increased difficulty of construction offshore. A 'template' or 'jacket' structure is simply a space frame designed to make pile driving easier by obviating the need to provide temporary support for the piles during first driving. The principle is illustrated schematically in Figure 1.2. The jacket or template is placed in position, the piles are fed through the legs of the template and driven by means of a pile driver supported on a surface vessel. After driving to the design penetration depth, the piles are cut off at the head of the template and a prefabricated deck section is stabbed into the piles and field-weld connected. The deck weight is directly supported by the piles themselves. A typical eight pile jacket is shown in Figure 1.3.[6] The method of construction for small jackets is quite simple. The structure is prefabricated at some waterside facility, skidded onto a flat-topped barge and towed to location where it is lifted into position by a derrick barge. The only severe construction loads to be considered are those associated with the lifting operation.

As exploration and production proceeded further offshore and the location water depths increased, two things began to happen. First, the installation problems became more severe as the capacity of the seagoing derrick barges was exceeded. It thus became necessary to launch the larger jacket from its barge and to utilize ballasting and buoyancy procedures to bring about uprighting and sinking of the jacket, sometimes assisted by derrick barges. The second factor was the increase in foundation loads. It was no longer possible to make do solely with piling through the legs. Skirt piles driven in other plan positions and pile clusters around main legs began to be introduced.

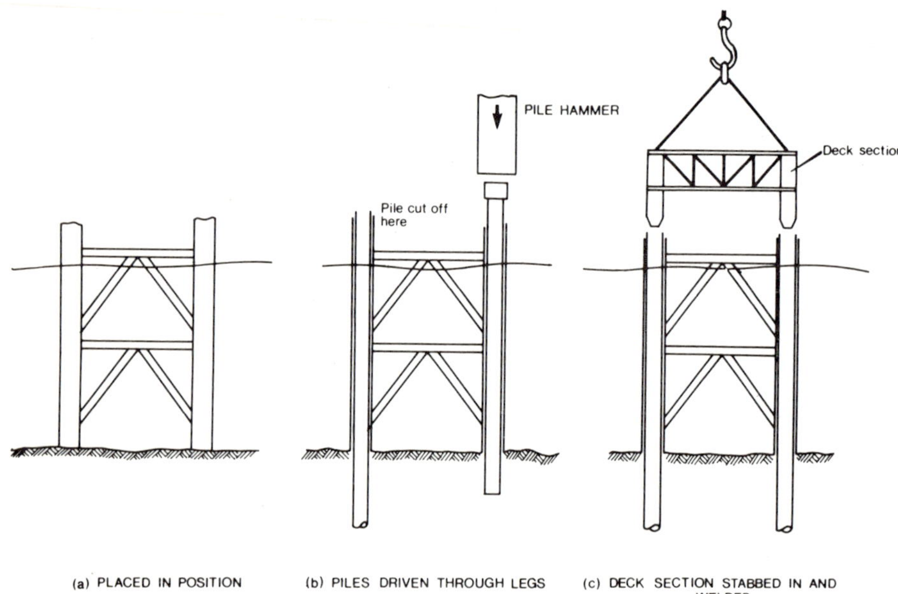

Figure 1.2 Principle of jacket or template

1.3.3.2 The tower

The tower is an extrapolation of the jacket to deep water. It is not only characterized by the use of pile clusters and skirt piles but the deck is now designed to be supported by the tower frame itself. Due to their size, towers are usually made self-buoyant either by enlarging several of the legs or by the use of a purpose-made buoyancy pontoon. Examples of the former type, which are sometimes called 'hybrids', have been installed on the Brent and Thistle fields

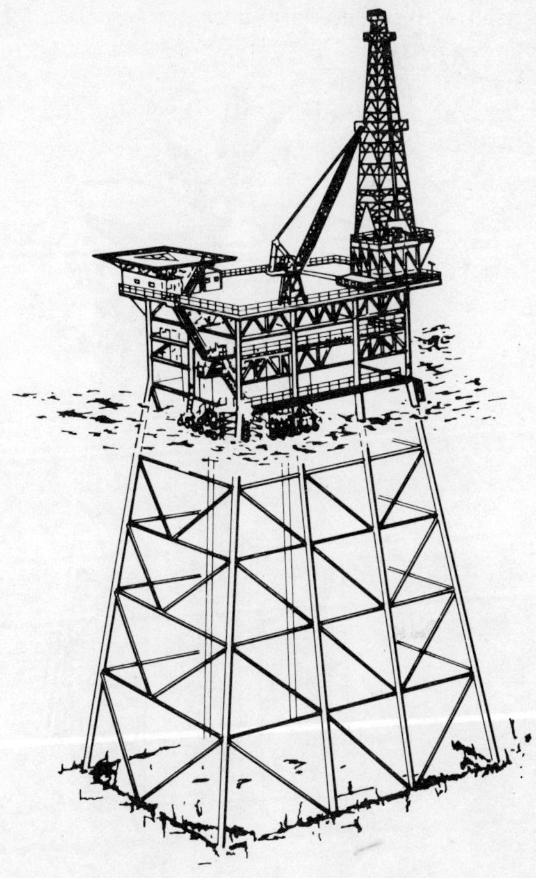

Figure 1.3 Eight pile jacket or template. (Reproduced by permission of Am. Soc. Civ. Eng.)

in the North Sea as shown in Figure 1.4. The Forties field towers used the supplemental buoyancy approach which has the advantage of reduced wave loads on the final structure but involves a more difficult installation operation.

The completed North Sea towers have been installed in depths up to 160 m and contain up to 40,000 tonnes of steel apiece, including deck and pilings. These do not however represent the largest framed offshore structures. The Exxon Corporation has recently installed a 260 m platform in southern California and Shell Oil Company are presently constructing a giant tower to be installed in more than 300 m of water in their Cognac field offshore Louisiana.

Figure 1.4 Hybrid tower with inbuilt buoyancy.
(Reproduced by permission of Inst. of Civil Eng.)

1.3.3.3 *Caissons*

Not all offshore installations are on a heroic scale. Many oil and gas fields are quite small and have minor deck load requirements. In addition, there are always step-out wells, flares and other installations which require a small but reasonably stable platform above sea level. A traditional and relatively inexpensive concept is the so-called caisson. This consists of a single tapered pile driven to a sufficient depth below the mudline to enable cantilever action to develop. This type of structure has been used in water depths up to 60 m.[7]

1.3.3.4 *Concrete gravity platforms*

The pleistocene glaciation of the North Sea region created areas with heavily overconsolidated soils at or very close to the present mudline. These can support large loads at the soil surface and are hence well suited to gravity type foundations. A number of factors combined to create a suitable market for concrete gravity structures and several have already been installed. There are some differences between the designs built to date but the most common type comprises a large cellular base supporting three or four concrete towers which in turn support a steel deck. This type is shown in Figure 1.5(a) while 1.5(b) shows another type which is of more monolithic construction and hence much more rigid.

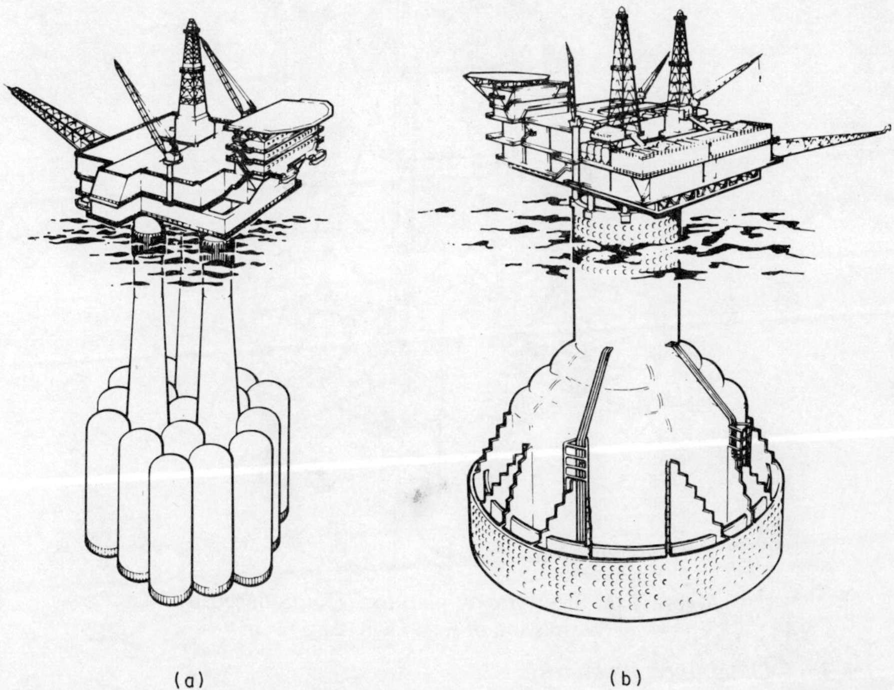

Figure 1.5 Concrete gravity platforms. (Reproduced by permission of Inst. Civil Eng.)

1.3.3.5 Steel gravity platforms and hybrids
The gravity structure limelight has undoubtedly been stolen by the concrete platform. Steel gravity platforms have been installed on the Loanga field offshore Nigeria where the presence of rock close to the mudline ruled out the possibility of tension piles (see Figure 1.6). Other steel gravity concepts are being actively marketed as are so-called concrete-steel 'hybrid platforms' which consist of a steel space frame supported on a large concrete gravity base. To the author's knowledge, no platforms of the latter type have as yet been ordered.

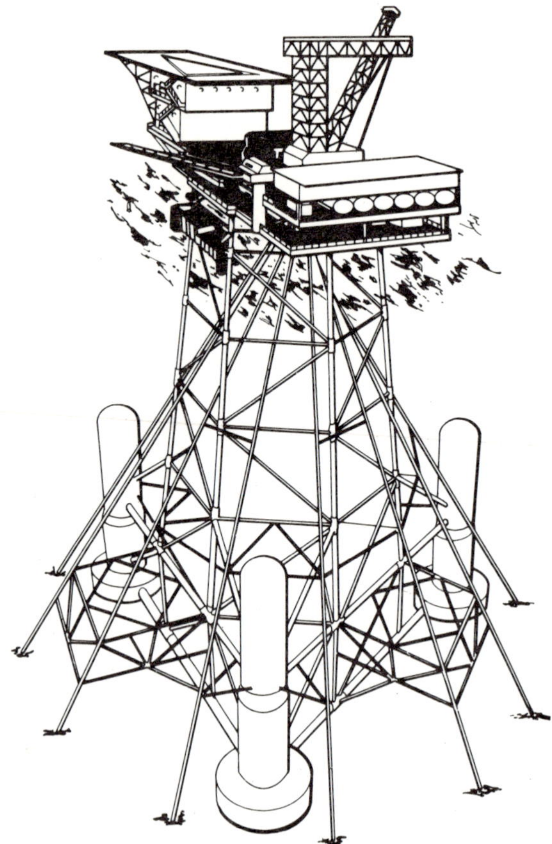

Figure 1.6   Steel gravity platform. (Reproduced by
permission of Inst. Civil. Eng.)

1.3.3.6   *Other fixed structures*

There are many other fixed structures used offshore such as ballasted down
concrete barges in the swamps of Louisiana, the piled oil storage Khazzans off
Dubai[8] and the Ekofisk storage tank.[9] One could perhaps include the man-
made islands in southern California and the Canadian Arctic. One of the most
interesting classes of structure is that employed in Cook Inlet, Alaska; an area
with large tides, high current velocities and floating ice. Figure 1.7 shows
solutions which were adopted to meet the high horizontal ice loads in the Inlet.

1.3.3.7   *Compliant structures*

All structures deform under load. A compliant one, as its name implies, is
designed to move so that the effects of the environmental load are mitigated.
The analogy is a ship riding out a storm at anchor. As the scope of the anchor
chain is reduced, the stiffness of the anchoring system increases. The ship

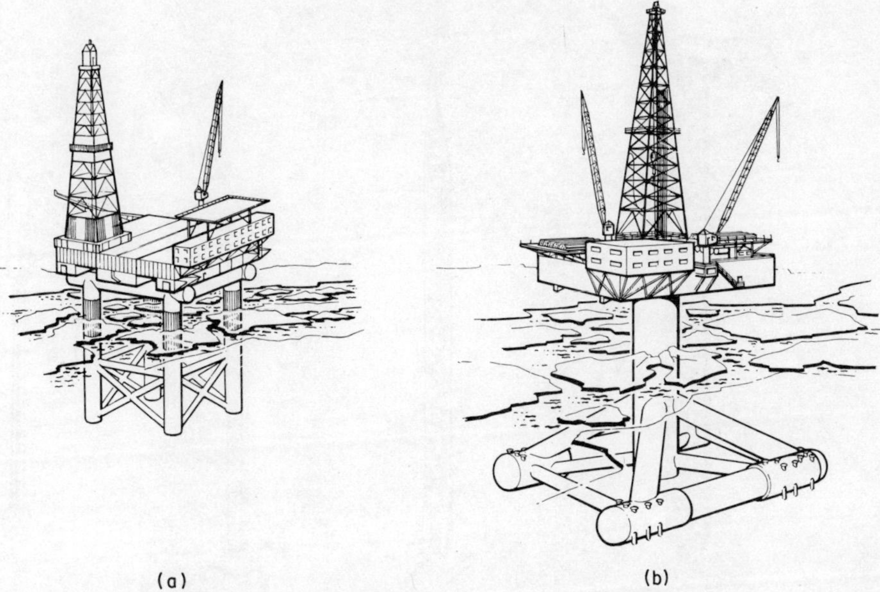

Figure 1.7 Ice-resistant structures in Cook Inlet, Alaska

movements are reduced but at the price of anchor forces increasing. The trade-off in a compliant structure is between excursion amplitude and restraining force. The first crude oil from the North Sea was produced from a compliant 'platform'. This is a modified semisubmersible drilling rig, permanently moored on location in the Argyll field.[10] Purpose made tethered or tension leg structures of various kinds are being developed and deployed as flares, single point moorings, and drilling and production platforms. Some examples are shown in Figure 1.8.

1.3.4 Factors governing form selection

Many factors must be considered in selecting the most desirable offshore hardware. Among the obvious ones anticipated by an experienced 'on-shore' engineer would be:

    capital cost
    cash call
    maintenance cost
    deliverability
    extreme environmental loads
    foundation conditions
    available construction infrastructure and labour
    materials availability

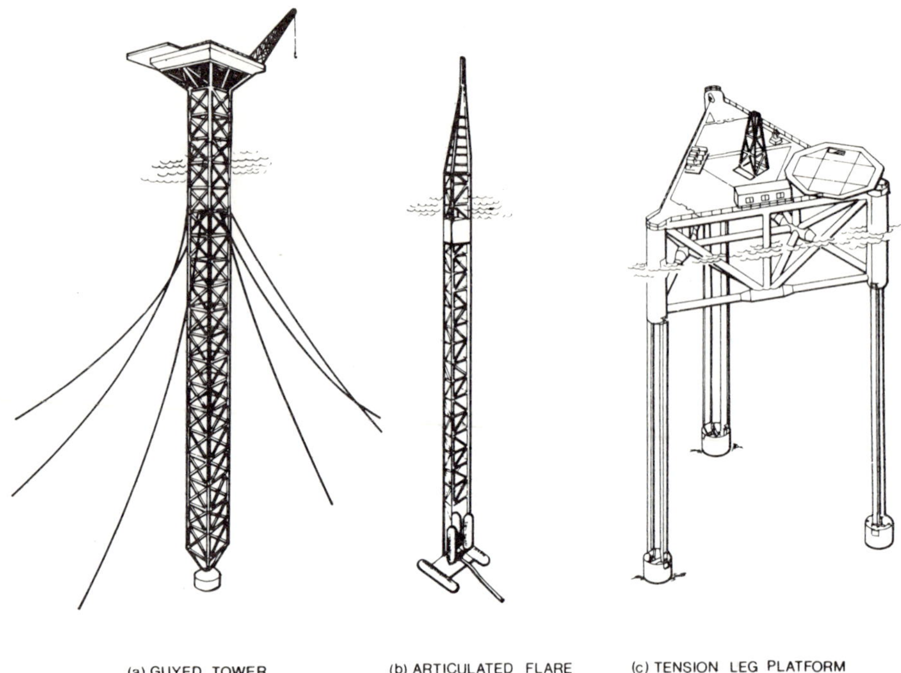

Figure 1.8 Compliant systems. (Reproduced by permission of Inst. Civil Eng.)

The offshore engineer would in addition be preoccupied with:

accessibility (by sea and air)
offshore construction weather window
weather persistence characteristics
availability of offshore plant (derrick barges)
reliability

One of the biggest factors is the last, namely, reliability. It is not difficult to see why when one considers the consequences of the failure of a large offshore installation in human, financial and environmental terms.

1.4   DESIGN AND ANALYSIS

1.4.1   Selection of design criteria

Bea[11,12] described a system for the development of platform design criteria and showed how it has been applied to an installation area in the Gulf of Alaska with both earthquakes and waves. The system relies on many years experience, primarily in the Gulf of Mexico with one category of structure, namely, framed, piled steel towers or jackets. It would be highly desirable if this approach were used generally but it is probably fair comment that a rigorous criteria selection analysis of the kind described by Bea is only practised by a handful of companies.

The application of this method to new structural types is difficult in view of the absence of statistics on operational performance. It nevertheless represents the direction in which sound engineering practice must go. A major implication of the approach, is that the so-called 'design-wave' is not a physical event associated with a given annual probability of exceedence or return interval for the site. Rather it is a platform design criterion which is a function not only of the sea conditions but also typical platform resistances. The analyst must hence exercise discretion in using the design wave concept and interpreting the results.

1.4.3 Statutory requirements

These vary from country to country but the internationalism of the industry has resulted in a broad consensus on basic principles. In the U.K., the public interest is protected by the Offshore Installations (Construction and Survey) Regulations Act 1974 which imposes requirements for the certification of offshore installations. This is the only statutory requirement and the act specifies little of a truly technical nature. Advisory information is contained in the notes issued by the U.K. Department of Energy.[13] These are currently being revised and give advice on the selection of environmental criteria, design practice etc. Other bodies such as the American Petroleum Institute, FIP and the Certification Authorities, Lloyds Register, Det norske Veritas, Bureau Veritas and American Bureau of Shipping have issued design rules or guides. Some of these give a reasonably specific definition of the type of problem to be analysed, but there is considerable freedom in selecting the actual analysis method.

1.4.4 Analysis inputs to design decisions

Structures are built as a result of decisions made, not analyses being conducted. The role of analysis is to enable the decisions to be made with an adequate level of confidence. Design is a dynamic process and the degree of analytical refinement changes markedly from the initial feasibility stage through to the final design checks. The methods discussed in the following sections are generally more applicable to the more advanced stages of design. In the interests of a reasonably organized presentation, the analysis requirements will be discussed initially under the headings of the two major classes of structure, namely piled framed towers and concrete gravity structures. The problems of aseismic design and ice loads are then discussed followed by fatigue and analysis of compliant systems.

1.5 ANALYSIS REQUIREMENTS FOR FRAMED, FIXED STRUCTURES

1.5.1 Fluid loadings

The basic deterministic approach to analysing wave loads on a framed structure comprising linear prismatic elements is to use the well known Morison equation.[14] This states that the fluid force is the vector sum of an inertial and a drag component. These are out of phase due to the phase shift between particle velocity and acceleration. For waves in which the ratio of characteristic structural diameter to wave length is less than 0.2, drag effects dominate. This is the case for most framed towers in fully arisen seas. The calculation of the fluid loads on the entire structure is relatively straightforward, if tedious. Assumptions are required concerning the inertial and drag coefficients.C_m and C_D. The water particle kinematics are determined from the incident wave, provided its profile is assumed (e.g. Stokes Fifth) and its height and period are defined. The fluid load on the overall structure is obtained by summing the component forces on each prismatic element, making due allowance for shift in relation to wave crest and angle of attack. This is quite a straightforward mechanistic approach which has been used for many years. It requires no elaborate fluid flow analysis and the validity of the basic Morison equation has been checked many times in the laboratory in the context of frame structures.

Despite all this, there is sufficient doubt as to the accuracy of its application to complete structures for the oil industry to undertake a multimillion dollar, large scale test project in which an Ocean Test Structure[15] is being installed and instrumented to measure fluid loads. The reasons behind some of the suspected inaccuracies concern factors such as energy spreading and directionality characteristics of real seas as opposed to those which are assumed in analysis. In summary, the primary need right now is not for an analysis method which gives a more accurate measure of fluid flow around submerged prismatic members. It is rather a matter of being able to deal with the water particle kinematics in a random three-dimensional sea.

1.5.2 Static frame analysis

There is little that one needs to teach the offshore engineering world about the use of matrix methods for analysing space frames. Systems involving ordinary frame analysis with preprocessors to calculate wave loads have been marketed and used routinely for a number of years. As an industry, this helped to reduce the analysis and design costs to a very low level, so much so that some engineering managers found it difficult at first to adjust to the significantly higher costs of engineering on the major North Sea structures. A minor problem relates to the treatment of the foundation supports. The pile-soil combination exhibits non-linear stiffness characteristics.[16] This is usually allowed for by solving the problem iteratively using a set of non-linear springs at the mudline support points.

The vast majority of steel framed towers and jackets have been installed in less than 100 m of water. Their fundamental structural periods are low compared with the wave periods and dynamic amplification is minimal. Most of these analyses are therefore of a quasistatic type.

1.5.3 Dynamic analysis

Malhotra and Penzien[17] discussed the non-deterministic dynamic analysis of wave excited structures. While analysis capabilities are being developed on a number of fronts there are still severe problems. The coupled soil-structure-water system has associated non-linearities, due to the velocity squared damping from structure-fluid interaction and the force-deflection and damping non-linearities at the foundation level.

The random nature of the loading suggests the use of a stochastic approach combined with modal decomposition to reduce the number of freedoms. This requires the use of a linear transfer function which means eliminating the non-linearities described above. Malhotra and Penzien[17] suggested a method which enabled one to obtain a quasi-linear response operator. However, Tickell *et al.*[18] have shown recently that the use of a linearized spectral approach can lead to large errors when predicting extreme loads on drag-driven structures.

This problem can be eliminated by using a deterministic approach and solving the equations of motion by direct integration. The difficulty one faces is the large number of degrees of freedom and the obvious temptation to use modal analysis or some other form of condensation technique. Furthermore, there is still the need to make allowance for the random characteristics of the loading and such a deterministic approach requires to be interpreted in some statistical fashion in order to provide a meaningful representation of the problem.

Whichever method one adopts, there are two broad areas of uncertainty, the one relates to the modelling of the dynamic system and the other, the characterization of the load.

Ruhl[19] analysed the results of a measurement program on an instrumented platform in the Gulf of Mexico. Predictions of the dynamic response were made on both deterministic and spectral bases and resulted in errors between 13 and 25 per cent, despite good correlation between measured and predicted structural periods. This implies that attention should be focused on the assumptions regarding loads. The work of Marshall[20,21] regarding directionality and spreading effects may hold the clue.

It is difficult to give a precise definition of the state of the art as practised. Informal enquiries by the author among oil companies and certification authorities in the U.K. and U.S.A. suggest that to date little has usually been done beyond the stage of a first eigenanalysis to determine modal response. This has usually been considered justifiable for installation water depths up to about 150 m since the frequency ratios have suggested low dynamic amplification for extreme events. (This is not the case for the low strain range, high cycle fatigue problems which are discussed in Section 1.8.)

In summary, methodologies for both deterministic and stochastic analysis have been developed, but not widely used. This situation is now changing with the move into deeper waters. Probably the greatest need is for the development of reliable means of simplifying the approaches but still giving meaningful design answers.

There is insufficient published information regarding parametric sensitivity of structural response to factors such as soil-pile stiffness, linearization of drag effects, etc. In short, the analysis methods exist but we lack experience in applying them.

It would be remiss of any discussion on dynamics to ignore factors such as wave slap and slam on horizontal members near the free surface or the hydroelastic oscillations associated with vortex shedding. These dynamic problems are of great importance in the design of individual members but their treatment at the present time is largely empirical.

1.5.4 Joint analysis

Failures in welded tubular frames occur predominantly at the nodes or joints of the structure. These are the most highly stressed regions and the nominal member stresses from the frame analysis do not reflect the true stresses which will affect factors such as fatigue strength and brittle fracture. A detailed evaluation of the stresses at the joints requires a separate analysis of each joint. In actual practice the large number of joints and the existence of a limited number of joint types has led to the use of 'stress concentration factors'. The nominal member axial and bending stresses are multiplied by these factors to give the estimated stresses at the so-called 'hot spots'. The determination of hot spot stresses such as the punching shears shown in Figure 1.9 and derivation of the stress concentration factors is ideally performed by the application of three-dimensional finite element analyses using isoparametric elements.[22] Marshall[20] has described the basic considerations with regard to tubular joint design.

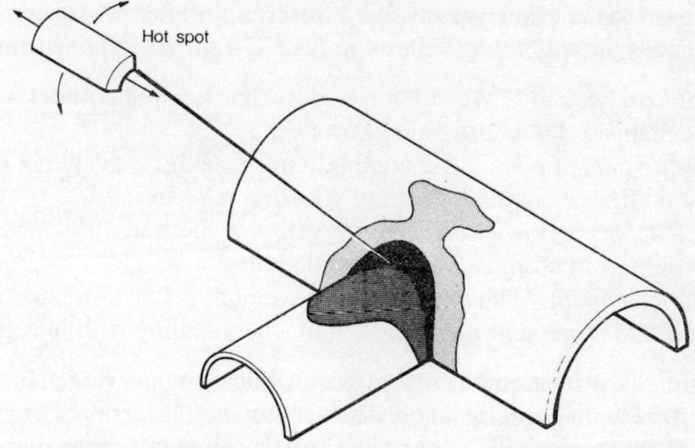

Figure 1.9 Hot spot stress analysis

1.5.5 Foundation problems

The state of the art of pile design for jacket structures was reviewed by McClelland.[23] Soil is a highly complicated material which exhibits among others, plastic, viscous, consolidation, hysteretic and cyclic degradation characteristics. Numerical models of soil behaviour have been used for a number of years. Unfortunately the state of the art of estimating soil deformability properties is so limited that most practical geotechnical analyses, where deformations are of interest, are virtually curve-fitting exercises. This is not meant as an unkind criticism of geotechnical engineers or numerical analysis methods. It is a fact of life that it is extremely difficult to predict soil deformations accurately even on the basis of carefully measured unit soil properties and refined analysis methods. However a combination of field measured full-scale behaviour and numerical analysis can lead to reliable backfigured parameters for use in subsequent analyses.

The three key problems of pile design are:

pile driveability
axial load transfer and stiffness
lateral and rotational pile head stiffness

Most pile driving analyses are based on the one-dimensional wave equation approach using a discretized model developed by Smith.[24] McClelland *et al.*[25] and Fox *et al.*,[26] have reported on its application to actual offshore foundation problems. The analysis is sensitive to a number of inputs involving ill-defined soil properties. The method is widely used but its application is treated with some scepticism in view of some of the assumptions which are required regarding soil parameters.

With regard to the other aspects, it is entirely appropriate to quote verbatim the conclusions drawn by McClelland in his 1974 state of the art lecture.[23]

1. *Axial load capacity.* 'We still have a less than adequate understanding of the soil mechanism that controls load transfer'.
2. *Laterally loaded piles.* 'Less complete information is available for establishing p-y (stiffness) curves for laterally loaded piles in sand.'
3. *Axial deformations.* 'Less fully developed methods are available for numerical definition of an equivalent axial spring'.
4. *Group behaviour.* 'The problem of movement prediction becomes more complicated and at present more uncertain when dealing with pile groups'.

These comments are based on current geotechnical engineering practice which relies extensively on numerical analysis techniques. The array of higher-order elements, dampers, non-linear, gap and slip-stick elements etc available should in theory be quite up to the task of modelling soil-structure interaction problems. It is evident that we have some way to go before the gap between principle and practice is bridged.

1.5.6   Construction and installation problems

The deployment and installation of platform structures has several interesting analysis problems. The first of these concerns the dynamic response during towout and upending or launching. Much of this work is done using model tests. However numerical analysis techniques have several advantages, not only in determining the probable behaviour during the critical stages of upending and sinking, but also in simulation exercises for launch operator training and in real time control systems.

The other aspect concerns pressure resistant design. The buoyancy legs of a large hybrid tower can be easily 5 m in diameter. It is evident that buckling analysis is very important considering the nature of the stiffened shells used, construction tolerances, complicated joint geometry and ambient pressures of $1.5\,\text{N/mm}^2$. The required buckling analysis of a 3-D structure is available in many of the large commercial finite element analysis packages.

1.5.7   Summary

The above brief presentation is summarized schematically in Figure 1.10. The picture is a varied one. At the level of overall frame analysis there is high confidence and everyday usage of numerical methods. Below the foundation level, methods have been developed to model soil-pile interaction but there is as yet insufficient confidence in their application. Fluid loadings on framed structures can be incorporated readily in a regular frame analysis package with the aid of a suitable preprocessor. However there are difficulties in defining a

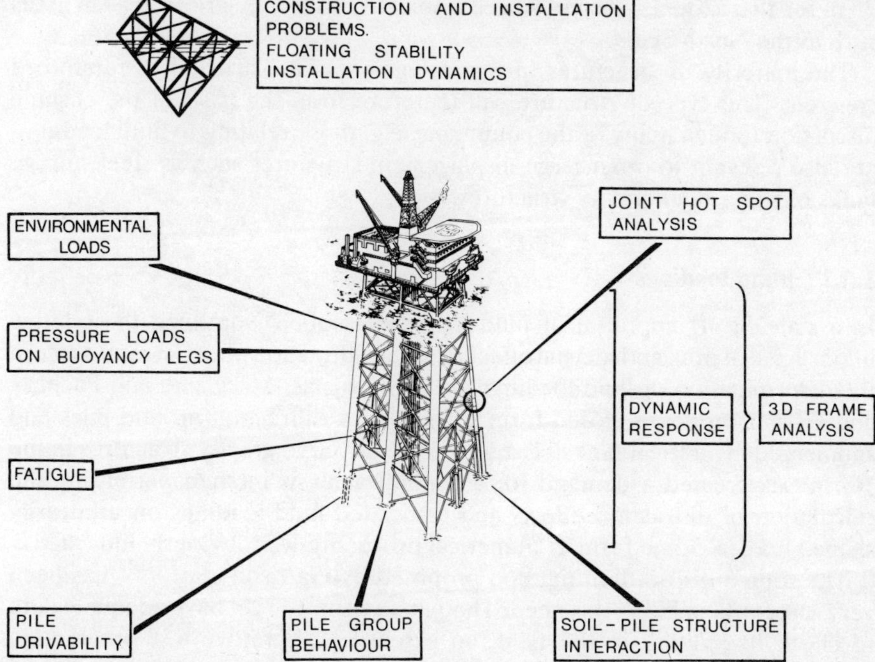

Figure 1.10 Current analytical problem areas for steel towers

suitable three-dimensional load input. Detailed stress and buckling analyses of 3-D problems are required and used. Dynamic analysis fundamentals have been developed but relatively few dynamic response analyses to predict extreme loads appear to have been carried out in practice.

1.6 ANALYSIS REQUIREMENTS FOR LARGE DISPLACEMENT FIXED STRUCTURES

Keulegan and Carpenter[27] showed that the ratio (N_{KC}) of water particle orbit width to characteristic structural dimension defines the relative importance of wave induced drag and inertial effects. As one might expect, the smaller this number becomes, the greater is the relative importance of inertial effects. Where the Keulegan–Carpenter number N_{KC} is less than 5 (equivalent to a diameter/wave length ratio of about 0·2) not only do inertial effects dominate, but the diffraction of the incident wave by the structure must be accounted for in calculating the fluid loads. As an example, for a low sea state with a characteristic zero crossing period of 7 seconds, diffraction becomes important for structural diameters of more than 15 m. This figure would increase to about

70 m for the 12 to 15 second waves associated with fully arisen seas in areas such as the North Sea.

The majority of structures in this category are constructed of reinforced concrete. This type of structure will therefore form the basis of the ensuing discussion though many of the comments, e.g. those relating to fluid loadings, are also relevant to other large displacement structures such as steel storage tanks or some steel gravity structures.

1.6.1 Fluid loadings

In a state of art appraisal of fluid loadings Hogben[28] outlined the relative importance of drag and inertial effects and the diffraction phenomenon's role in the determination of fluid loadings on large objects. MacCamy and Fuchs,[29] and Gran[30] proposed closed form solutions for diffraction around piles and submerged cylindrical tanks. The advent of the large gravity structures in the North Sea created a demand for a general analysis method permitting the calculation of diffraction effects and associated fluid loadings on arbitrarily shaped bodies. Some form of numerical procedure was obviously indicated.

The source distribution method proposed by Garrison *et al.*[31–33] has been very successful in filling this need. Though we do not as yet have adequate data to check its validity in the field, an extensive laboratory test programme conducted by Hogben and Standing[34] indicated high confidence in the method for regular long-crested waves. Finite element methods can also be used to predict fluid loads on fixed and floating structures but successful techniques have only recently been developed and have not yet had much impact in practice. It is the author's opinion that this is one area in which the powerful finite element method may be the less desirable alternative. The real power of the F.E. method is demonstrated in coping with continuum problems charac-terized by marked changes in the medium properties, e.g. anisotropy, inhomogeneity, complicated geometry, etc. In the present instance, the analyst is primarily interested in modelling the conditions at the boundaries while assuming a predictably smooth variation of the field variables within the fluid continuum. A boundary solution procedure such as the source distribution technique is ideally suited to this situation since one can usually comfortably neglect factors such as inhomogeneity of the fluid medium. The F.E. method however has to pay the penalty of modelling both the boundary and the continuum itself although banded matrix properties reduce the computational effort. A further disadvantage of the finite element method is the difficulty of simulating distant boundary conditions.

One of the most interesting recent developments is the use of solution techniques which combine the flexibility of the finite element method in discretizing the problem in the region of maximum interest with the efficiency of a boundary solution procedure to characterize the remote boundary conditions. Chen and Mei[35] were among the first to use this type of approach in solving a wave refraction-diffraction problem. Zienkiewicz[36,37] showed how the finite element method and boundary solution procedures could be effectively combined in solving a range of field problems. A related development for characterizing remote boundaries is the creation of so-called 'infinite elements' for evaluating fluid-structure interaction problems.[38] These approaches represent a very promising development for other applications also.

In this book recent developments in both techniques are presented (viz. Ch. 3–5) which promise to have an impact on future computation processes.

One of the major criticisms levelled in Section 1.5.1 dealing with framed structures concerned the inadequate modelling of the three-dimensional sea behaviour. Similar criticisms will be made of two-dimensional diffraction analyses, once we have obtained reliable field data. Even in the case of a two-dimensional approach, the wave force on submerged objects is quite sensitive to wave period.

There are no ready guidelines for selecting an appropriate period for the design extreme wave. If one adds to this the mitigating effects of energy spreading, it is not difficult to foresee a move towards stochastic analyses with directional and spreading effects accounted for. The diffraction analysis, whether by source distribution or finite element method will still be required however to generate the load transfer functions used in such a stochastic procedure.

1.6.2 Structural analysis

The popular technical press has been more than generous in its use of superlatives to describe the construction of the present generation of North Sea concrete platforms. While the problems of scale undeniably create some interesting challenges for the designer, the emphasis on their sheer size and cost has tended to mask some of the novel design difficulties posed by this new breed of structures. Unlike most land-based installations, many of these problems arise before the structure is finally commissioned to perform its planned function as a drilling and production facility.

Consider the typical construction scenario presented in Figure 1.11. The first stage consists of part construction of the base caisson in a dry dock. When construction has advanced sufficiently for the embryo structure to have adequate strength and floating stability, the dock is flooded and the structure towed to a protected deepwater construction site where it is moored. The remainder of the concrete structure is then completed, with both caisson walls and legs usually being slipformed. The prefabricated deck structure which may weigh up to 10,000 tonnes is then loaded onto the platform legs. This is

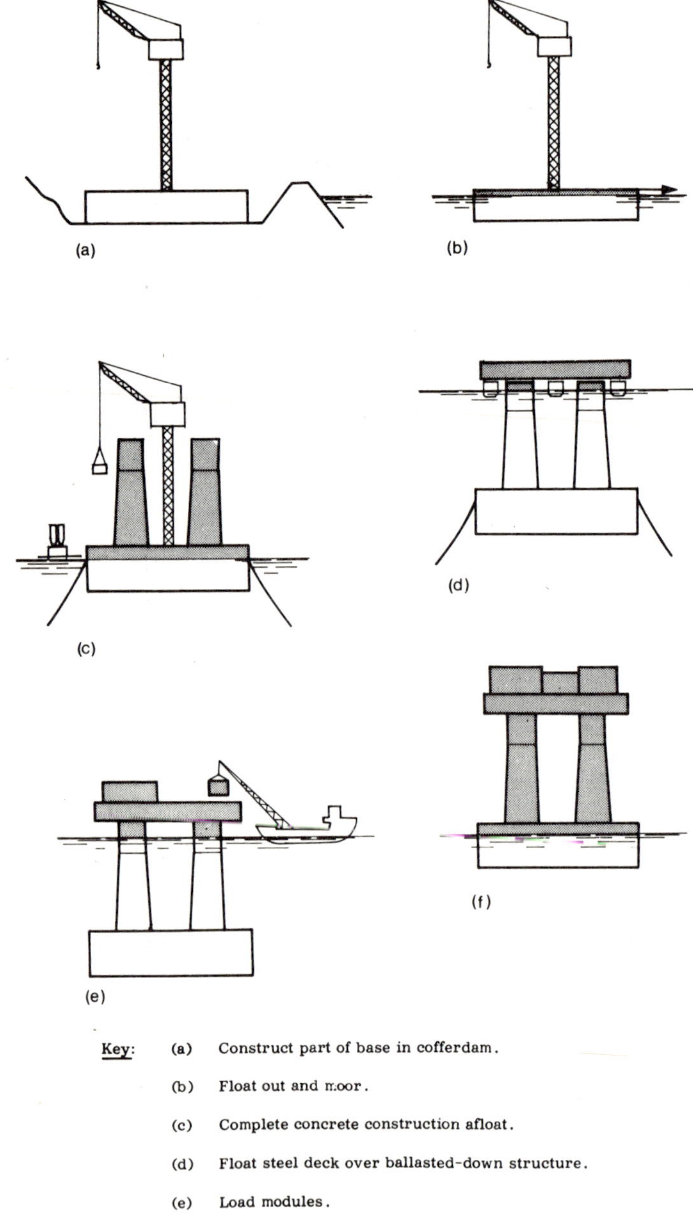

Key: (a)   Construct part of base in cofferdam.

         (b)   Float out and moor.

         (c)   Complete concrete construction afloat.

         (d)   Float steel deck over ballasted-down structure.

         (e)   Load modules.

         (f)   Refloat to towing mode.

Figure 1.11   Typical construction scenario for concrete gravity platforms

conveniently accomplished by ballasting down the concrete structure so that the deck can be floated into position over the legs. At this stage, the concrete caisson is only partly flooded and is thus subjected to very large net ambient pressures. Considering that the reserve buoyancy is very small at this juncture, it is evident that the deck loading operation represents a critical stage in the construction. Deballasting of the structure enables the deck to be lifted off its delivery pontoons and the structure is then raised to an adequate level for loading of the modules of topside equipment.

The structure complete with its large payload is then refloated to the planned towing draft and the tow to location carried out when suitable weather conditions are forecast. On reaching its installation location, the structure is ballasted until touchdown is effected. Damage due to impact on touchdown is avoided by a slow descent and the use of downstand dowels to arrest structural oscillations during the final stage of descent. Virtually all of the structures to date have been equipped with steel or concrete skirts. These are driven to the required penetration by further ballasting. The water filled voids under the structure may be grouted before final flooding of the structure is carried out.

Once it is in operation, the storage of hot crude or the presence of the hot well conductors and marine risers inside the structure will result in differential heating of the structure and associated thermally induced loads. Thus in addition to the much vaunted need to resist a 100 year wave of the order of 30 m the structure must be designed to be satisfactory in respect of other characteristics such as:

watertightness
resistance to high pressures
floating stability
towout draft limitations
payload carrying capacity
placement on an unprepared site
resistance to thermal loads

This is by no means an exhaustive list but it suffices to indicate that the designer's options are constrained in several respects at all stages from construction through operation. It is clear therefore that while the concept of a concrete gravity structure is simple, its behaviour will not be. An exhaustive discussion of these problems is clearly outside the scope of this presentation but certain problem areas are worth highlighting.

1.6.2.1 *Pressure loads*

During the critical deck mounting operation, parts of the caisson are subjected to differential pressures typically in the range 1·3 to 1·6 N/mm² or at least three hundred times the design distributed live load in a typical building. Two basic structural systems have been used for the majority of structures to date. These are shown diagrammatically in Figure 1.12. In the first of these, the caisson consists of an assembly of cylindrical cells as shown in Figure 1.12(a). The alternative system shown in Figure 1.12(b) comprises a cellular box with straight walls which act as a system of struts to support a pressure resistant skin.

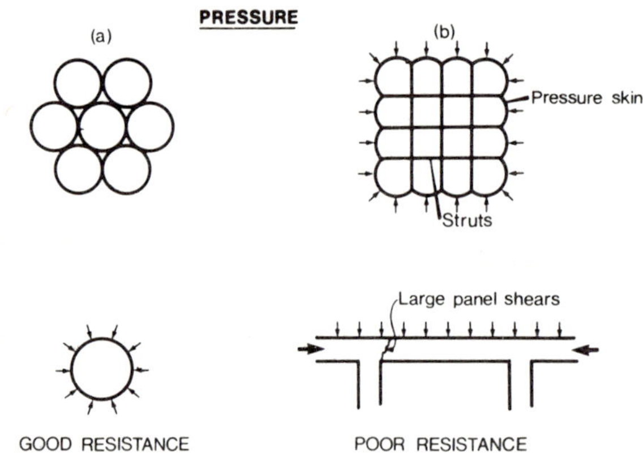

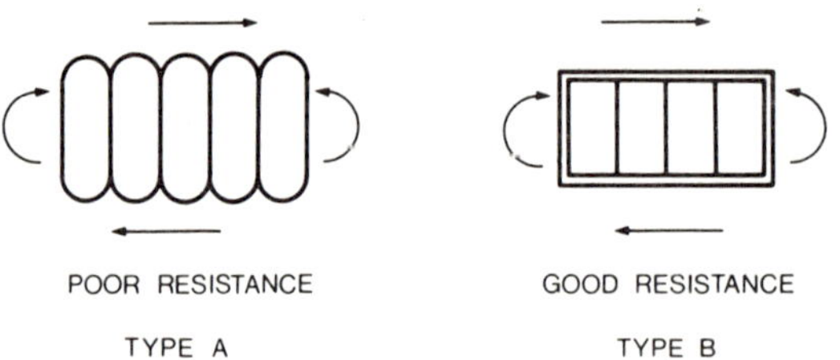

Figure 1.12   Typical caisson structural systems

In both cases, the ambient pressure is accepted by cylindrical surfaces which ensure that the pressure loads are resisted by direct membrane compression, at least in plan. When one looks at the method for capping top and bottom of each of these systems, a different picture emerges. The first system can be conveniently capped with hemispherical domes which will permit the simple membrane action of the concrete to be retained. In some cases, a larger radius of curvature dome has been used to permit the elimination of the top shutter during construction. While this has construction advantages, it introduces large thrusts and associated bending near the springing points. The closing of the square cellular base has been most frequently accomplished by the means of flat slabs although in certain instances polyhedral capping surfaces have been used.

The first question to be considered is that of overall stability. The thickness to diameter ratios of the typical concrete cylinders shown in Figure 1.12(a) are usually high enough to ensure thick shell type behaviour and a fairly small risk of elastic instability. Similarly, for the rectangular configuration shown in Figure 1.12(b), cell sizes can usually be selected which will ensure that individual cell walls are unlikely to give buckling problems. However, for this type of system, it is possible that overall instability may be associated with the behaviour of the complete cellular structure. This can obviously be analysed using an appropriate buckling analysis and there is nothing in the boundary conditions or element configurations which appear to be particularly exacting.

The biggest problem appears to be associated more with the question of the triaxial stress state and the exceedence of the appropriate rupture criterion for reinforced concrete. This has been described as 'implosion' by Haynes,[39] who has carried out a series of experimental investigations on the failure of concrete pressure vessels under high ambient pressures. Although no general theory exists to date, the evidence available would suggest that the problem lies not so much in determining the appropriate stress conditions as in the selection of the correct failure criterion. It still remains to be proved how important factors such as strength loss of the concrete due to the build-up of pore pressures are in influencing the resistance to high pressures.

Since the boundary conditions are so simple, the overall analysis of pressure loads is usually quite trivial and the only difficult analytical aspect relates to the effect of the restraints imposed by adjacent parts on the structure. For example, the influence of a fixed connection between the cylinders in Figure 1.12(a) on shear stress and bending has to be investigated by means of a stiffness analysis. The suite of elements available in most of the major analysis packages is quite adequate for the purposes of this type of investigation. A slight level of sophistication may be introduced by the investigation of the effect of creep when considering that some of these pressure loads may be acting for many months. As the stress distribution will remain sensibly constant, the problem is clearly one of finding the right constitutive law rather than of conducting a detailed time-dependent analysis to determine deformations.

By far the biggest problem with regard to pressure loads is one of design rather than analysis, in that there are few guidelines to assist the engineer in designing for the very large panel shears and in particular, the allowance which should be made for the assistance given by axial prestress. The British Code CP110 reduces the allowable shear stress in slabs to half of that in beams, without stating why this is the case although it is suspected that this has to do with the non-uniform distribution of shear forces around the perimeter of non-circular plates. The ACI Code 318/71 allows the designer some freedom to use the influence of the axial prestress but it is still a far from satisfactory situation facing the designer. Clearly, there is considerable scope for investigating the behaviour in shear of the heavily loaded thick slabs.

1.6.2.2 Thermal loads

The problems with thermally induced loads in nuclear pressure vessels have been studied for many years and it is tempting to believe that a direct transplant of technology would suit the problems of thermally induced loads in concrete platforms. The pressure vessel consists essentially of a single structural element which, apart from start up conditions is associated with a steady state heat flow regime. The gradient of temperature and pressure is outward and apart from the problems created by local penetrations, the thermal regime is fairly uniform throughout. This is fundamentally not the case for a concrete gravity platform, particularly where there is storage of hot crude. In the latter event, parts of the structure are heated while others are quite cold. A hot/cold fluid interface exists in one or more parts of the structure and its position moves with time. Load cycling is a characteristic feature, despite operational attempts to maintain a sensibly constant thermal regime.

Clearly, the problem is a fairly complicated one in several respects. In the first instance, it is necessary to establish a typical range of thermal regimes, including transient states to be designed for. Coupled convective and conductive heat transfer complicates the problem of analysis and most practising engineers have thus far been prepared to make the convenient assumption that steady state conditions are obtained quite smoothly and rapidly and that problems of thermal shock need not be assessed. Analysis techniques for carrying out such investigations were developed by Hsu and Nickell[40] and applied to a class of initial-boundary-value problems that are commonly referred to as diffusion-convection problems. The author is unaware of any published results based on the use of methods such as these to investigate typical thermal regimes in concrete storage vessels.

Even given a sensibly constant thermal regime within the structure, the thermoviscoelastic nature of concrete results in several time dependent phenomena of significance to the designer. An analysis of typical loads associated with assumed elastic behaviour of the structure shows local loads at an unacceptably high level. Richmond[41] has shown how creep can assist to provide a marked reduction in the thermoelastic stresses. On the other hand, he has also shown that particular combinations of prestress and thermal loading can result in higher stresses when the thermal differential is removed than are incurred under the initial elastic regime. Most important of all, creep effects in redundant structures can cause locked-in stresses after the removal of the initial load regime. Reversal of the load will then create a far worse condition than occurred on first loading. The initial evidence would suggest therefore that as in the case of many other structures, creep has an ameliorating effect on stresses unless cycling of the load occurs.

1.6.2.3 Environmental loads
Section 1.6.1 described the type of diffraction analysis which can be used to predict the distribution of pressure over the structure resulting from wave loading. Pressure variation over the structure is quite smooth and the analysis of structural load distribution under environmental forces is relatively straightforward insofar as fluid loadings are concerned. The question of seismic loads is considered later.

1.6.2.4 Foundation loads
The local distribution of forces applied by the foundation to the structure is less reliably predicted than the loads caused by environmental factors such as waves and currents. Starting at the time of installation, the structure is installed on a totally unprepared site which has its own topographic and stratigraphic characteristics. Not only is it difficult to sample these adequately in advance, but it is also a fairly demanding exercise to emplace a structure with a displacement of over half a million tonnes in exactly the planned location.

Bjerrum[42] described some of the problems to be anticipated. Apart from the obviously expected moments, torsions and shears induced by non-uniform resistances over the site, local pressures of the order of $2 \cdot 0 \, \text{N/mm}^2$ can be applied to the underside of the structure by high spots. Such pressures, when superimposed on the already very high hydrostatic differential pressures can create an almost intolerable situation for the structural designer. Very often, the problem is turned around and the question becomes one of designing a structure-foundation interface such that these difficulties do not occur.

The entire philosophy of foundation design is different from that of a typically landbased structure due to the uncertainties regarding the site properties as well as the fact of placing the structure on an unprepared bed. The geotechnical engineer must therefore provide the structural designer with a range of terrain models which will embrace the expected range of conditions anticipated at the installation location. These can then be used to generate support conditions for analysis of the forces in the structure resulting from installation as well as the response to environmental loads.

Both the short and long term reactions must be considered and some of the special influences due to effects such as cyclic loading must be considered as discussed in a later section.

1.6.2.5 Other structural analyses
There are a number of other problems as one might anticipate with regard to the design of the structure. Among the current preoccupations of the industry is

the question of fatigue. This relates not only to the problems in the concrete part of the structure where tension crack propagation, fatigue of the reinforcement, bond degradation etc are less than perfectly understood, but also with regard to the design of the deck. Most of the decks which have been used to date have fully fixed moment connections at the top of the concrete legs. Under the action of wave loads it is evident therefore that cycling of the stresses in a steel deck will occur, requiring an investigation of the fatigue problems associated with these stress reversals. The question of high cycle low amplitude fatigue is discussed in a later section.

1.6.3 Geotechnical analyses for gravity platforms

Excessive conservatism in the design of a gravity structure foundation is very costly. If one adds to this the problems associated with installing a complete structure on an unprepared site with somewhat uncertain foundation conditions, it is clear that the geotechnical engineer's role in the design of the gravity structure assumes a greater relative importance than it does for most conventional landbased projects.

1.6.3.1 Installation problems

Among the most interesting of the geotechnical analyses required for the successful installation of a gravity platform is the prediction of skirt penetration resistance. At first sight, this would appear to be a relatively straightforward matter of applying limit analysis techniques such as are employed in conventional bearing capacity problems. An examination of the theories for predicting deep foundation bearing capacity by Vesic[43] showed an order of magnitude difference between the various theories for predicting bearing capacities in sand. Work by Kerisel[44] had indicated practical upper limits on resistance which do not agree with classical strength theories and penetration tests results from North Sea site investigations showed far higher penetration resistance than could reasonably be expected from the properties of typical soils.[42] Current practice tends to follow the pile design type of approach with due adjustment to allow for an upper limit on penetration resistance as revealed by Kerisel. The problem is not simply resolved as in the case of bearing capacity analyses in which one takes a conservative approach and picks a low value. In the present circumstances, the platform cannot be left sitting on partially penetrated skirts and a high premium is therefore placed on obtaining a reliable estimate of the maximum resistance likely to be encountered. The present state of affairs is unsatisfactory in analysis terms and there is obviously scope for a general investigation of the problem. Considering the inhomogeneity of typical installation sites, and variation in skirt profiles, a general numerical analysis technique probably based on a Lagrangian formulation with a general dilatant constitutive law could perhaps provide the answer.

1.6.3.2   Stability analyses

For the extreme environmental load case, the force resultant at the mudline is characterized by much larger values of obliquity and eccentricity than are typical for the majority of foundation problems. It is therefore necessary to use a generalized bearing capacity theory which takes full account of the three-dimensional nature of the problem. The most general theory of bearing capacity for such situations was developed by Hansen[45] and his procedure has been fairly widely used, despite its obvious drawbacks. Among the most serious of these is the assumption of homogeneous foundation conditions, a condition which is unlikely at best for most gravity structure sites. Accordingly, other limit analysis techniques such as the method of slices are being increasingly used. Vaughan *et al.*[46] applied finite element techniques to the analysis of the stability of large gravity structures, taking into account the variation in shear strength with depth. A primary drawback with this approach is that for reasons of economy, the analysis is usually limited to the assumption of plane strain conditions and the results therefore are only likely to be representative of the conditions near the plane of rotation. While they are useful in identifying the initiation of yielding and the development of failure modes, the prediction of overall bearing capacity is likely to be somewhat suspect for the truly three-dimensional nature of the problem. It is the author's view that a combination of such analysis methods together with model testing approaches such as those used by Rowe *et al.*[47] will provide the most useful means of attacking the problem of predicting both bearing capacity and deformation of the foundation.

1.6.3.3   Hydraulic instability

Henkel[4] identified the role of waves in inducing bottom instability and it is generally recognized that any foundation stability analysis must take into account the effective change in surcharge caused by the transient pressure wave. However, this phenomenon and the obvious question of scour around the structure are not the only hydraulic factors to be investigated in the design. The presence of virtually incompressible ambient fluid during the application of overturning forces to the base of the structure results in the development of a pressure differential between the underside of the structure and the surrounding free water. In addition to the obvious question of eliminating potential piping problems around skirts, it is in some cases necessary to evaluate the rate of dissipation of this pressure differential in order to check the validity of the assumptions with regard to the overturning environmental loads. For example, it is fairly common practice in predicting the wave loads on a gravity structure to assume a so-called 'closed gap' under the structure. This assumption is only valid if the water pressures under the structure are not dictated by the hydraulic boundary conditions outside the skirts. The investigation of this requires one to look at the response of the foundation as a consolidation phenomenon. Approximate analyses carried out by Chan[48] using a simplified finite difference model indicated that this is a reasonable assumption for typical skirt configurations and soil conditions in the North Sea.

1.6.3.4 *Cyclic degradation effects*

The questions of liquefaction and cyclic softening of clays have received considerable attention during recent years, particularly the latter in view of the predominantly clay profiles underlying the gravity structures installed to date. (Note: The Ekofisk Tank[9] is the major exception.) A joint industry project on the repeated loading on clay has been carried out and the results published by Andersen *et al.*[49] The programme showed quite clearly that cyclic loading effects could result in a significant change in both stiffness and shear strength of normally and overconsolidated clays. Despite the substantial nature of the investigation, and the advances which have been made in understanding the effects of cyclic loading on soil specimens, no generally accepted methodology has yet been developed for conducting both stability and deformation analyses which incorporate the findings of this programme.

1.6.4 Dynamic response to waves

The prediction of the dynamic response of a gravity platform to environmental load transients such as waves or earthquakes represents a very challenging set of demands largely because of the coupling between the fluid-structure-soil systems. General methods have been proposed by Eatock-Taylor[50] and Moan *et al.*[51] for the dynamic response of structures to wave inputs. The application of such methods is still very much in its infancy and there are many fundamental questions to be answered, from the validity of two versus three-dimensional models to the selection of realistic sea state spectra. The response of coupled dynamic systems forms a major part of the proposed discussion for this conference. The following paragraphs are intended to provide a brief introduction to some of the factors involved and their relative importance as seen in the context of the present state of the art of actual design.

1.6.4.1 *Structure - water coupling*

Both finite element and source distribution techniques have already been used to investigate dynamic coupling between submerged structures and their ambient fluids. Liaw and Chopra[52] used a finite element characterization of both structure and fluid to assess the effects for a typical submerged intake tower responding to seismic excitations. A source distribution technique was used by Garrison and Berklite[31] to assess the dynamic response of other bluff bodies. The work of these investigators has shown that in most cases, the dynamic system can be adequately characterized by using the so-called 'added mass' approach to allow for the participation of the ambient fluid in the response. Added mass factors are relatively insensitive to frequency except near the free surface where gravity waves generated by the structural motions attenuate the added mass effect, but there are no definitive guidelines yet as to how this and other forms of hydraulic damping should be incorporated in a dynamic response model. Among the other factors still to be thoroughly explored is the degree of coupling between the added mass contributions in the different modes of motion.

1.6.4.2 Foundation-structure coupling

The aseismic design of nuclear reactors has resulted in an intensive study of the influence of foundation-structure coupling in relation to earthquakes. Despite the difference in the frequency contents of earthquake versus wave excitations, the basic physical phenomena are still similar. A full evaluation of the methods for analysing these effects was carried out by the American Society of Civil Engineers and the results are due to be published shortly (see ASCE 1976).[53] The key conclusion from this work is that the soil-structure interaction effects have a very significant influence on the response of both structure and foundation. Most of the experience which has been gained has been in relation to nuclear power plants and much of the work currently being done on developing suitable structure-foundation models for gravity platforms is based directly on its findings.

1.6.4.3 Selection of analysis model

Provided that remote boundary conditions can be adequately characterized, it is possible to idealize both structure and foundation using a finite element model. The primary advantage of this approach is its suitability for inhomogeneous foundation conditions and non-linear response of both structure and foundation. Unfortunately, it has some severe disadvantages, not only in terms of overall cost but also because the technique frequently requires one to use large numbers of freedoms to discretize the foundation at the expense of a less accurate characterization of the structure itself. Consequently, there is a strong tendency towards the use of closed form impedance functions to define the foundation support conditions. Standard solutions are available for half-space conditions,[54] but this is considered a doubtful assumption for structures on the scale of a typical gravity platform where a significant change in soil stiffness with depth is likely to occur.

A two stage procedure is sometimes used, the first comprising a dynamic finite element analysis of the foundation in order to determine the frequency dependent foundation impedances, followed by the use of these in a dynamic response analysis of the structure itself.

1.6.4.4 Deterministic analyses for waves

Relatively few response history analyses of the dynamic response of a gravity structure to a wave train have been published. There are several reasons for this, only one of which is the cost. The only attempts of which the author is aware have been based on the direct integration approach which has several theoretical difficulties associated with it. In the first instance, there is no simple way of incorporating frequency dependent foundation behaviour in a time domain analysis. It can be argued with some justification that the foundation impedance is relatively insensitive to the range of frequencies involved in wave excitations, but this is only likely to be true in the lowest modes. The second and more compelling reason is the fact that the forcing function itself is dependent on the frequency content and amplitudes of the wave train and therefore it is not possible to generate a simple correlation between forcing function intensity at various parts of the structure and sea surface elevation with time.

The author and his colleagues have formulated an approach[55] which may be used to obtain a time history response to waves, incorporating frequency dependent effects in both foundation and wave input. This is based on the use of discrete Fourier Transforms in a similar fashion to the approach used for the analysis of earthquake response problems.[54]

1.6.4.5 Stochastic analyses

Even if one is successful in carrying out a response history analysis of the kind mentioned above, the validity of the wave train itself is bound to be suspect. In the first instance, considering the response to extreme events, there is an obvious lack of wave train data and one is forced to generate a synthetic record or to scale up from measured wave trains corresponding to a higher probability event. Secondly, the random nature of the sea implies that a number of such analyses would have to be carried out in order to obtain an adequate statistical interpretation of the problem. It is not surprising therefore that most investigators appear to favour the use of a stochastic approach where the sea state is characterized by a power density spectrum. This approach is also subject to limitations, the first and most obvious being the use of a linear transfer function in order to obtain the response spectrum. Fortunately, concrete gravity structures are better favoured in this respect than steel templates since the wave driving on a gravity structure is predominantly inertial and the response operator can be treated as linear with some justification. Also, the foundation response is sensibly linear for the usual range of foundation safety factors for the 100 year event. There are inevitable arguments as to the suitability of various standard wave input spectra for these analyses and attention to this is given in several chapters of this book.

1.6.4.6 Designing for dynamic effect

As stated earlier, the analysis of dynamic response of gravity structures is at an early stage of development and it is not surprising therefore that there are as yet no generally agreed design criteria to be used in relation to dynamic response calculations. Several sources of uncertainty have been mentioned above, probably the most important of these is the foundation compliance. As in the case of piled structures this is of fundamental importance and moreover is difficult to predict in advance. It should thus be treated in a parametric fashion.[51]

1.6.5 Review of analysis requirements

The above areas of current analysis interest are summarized in Figure 1.13. Although the list is not an exhaustive one, it contains a considerable variety of problem types ranging from three-dimensional stress analysis, through steady and transient flow problems to viscoelastic and elastoplastic deformations in structure and foundation. The engineering analyst would be hard pressed to find an equivalent set of challenges in any other single project.

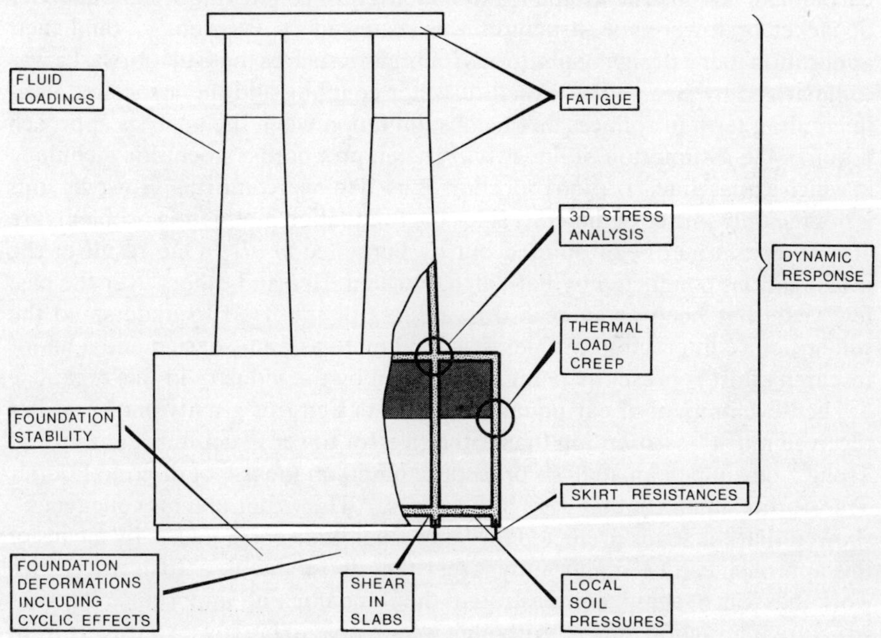

Figure 1.13 Current analytical problem areas for gravity platforms

1.7 OTHER ENVIRONMENTAL LOADS

The introduction to this chapter identified several forms of environmental loading which are significant in the development of a safe design for an offshore structure. Among these are winds, currents, floating ice and earthquakes, but the first two are usually of secondary importance. It is only relatively recently that the question of aseismic design offshore has attracted much interest; however the development of fields offshore southern California and New Zealand and the lease sale earlier this year in the Gulf of Alaska have resulted in an apparent gearing up by the industry to tackle these problems. Another problem area which has attracted some interest is that of loads due to floating ice. These two issues will be briefly considered.

1.7.1　Seismic loads

Evidence of an increased awareness of the seismic design problem is the fact that the most recent issue of the American Petroleum Institute's API RP2A[56] (Recommended Practice for Planning, Designing and Constructing Fixed Offshore Platforms) in January 1976 introduced a substantial section on earthquake design. The available methodology for conducting seismic analyses of jacket or tower type structures was reviewed by Penzien,[57,58] and their application to the design of platforms for regions such as the Gulf of Alaska was summarized by Bea.[12] The structure-water coupling and the associated non-linear drag term introduces an obvious limitation when the analysis approach requires the assumption of linearity. Penzien proposed a stochastic technique in which a quasi linearization procedure is used to overcome this. However, this is not the only source of difficulty since the foundation interaction problems are also severe as had been pointed out by Parmalee *et al.*[59] One result of the investigations conducted by Parmalee, Penzien, Bea and others over the past few years has been an increased awareness of the need to understand the soil-structure interaction problems as the key to aseismic design and a major research effort is presently being undertaken by the industry in this regard.

The development of earthquake resistant designs for gravity platforms has received much less attention than is the case for tower structures. Penzien and Tseng[60] developed an analysis procedure based on the use of discrete Fourier Transforms similar to that used by Veletsos.[54] The writer and his colleagues[61] showed that the loads predicted for large magnitude earthquakes by means of this approach can be significantly larger than those due to a 30 m wave. This work has once again demonstrated the fundamental importance of soil-structure interaction effects. Although we are able to borrow extensively from the methodology developed for nuclear reactor design, the mass and stiffness characteristics of a typical gravity platform differ from those of a nuclear power plant and much remains to be done in understanding the importance of these differences. For example, Figure 1.14 shows the seismic response of a typical gravity platform to a natural earthquake (TAFT accelerogram). The structure-foundation system has a fundamental period of about 4·5 seconds and the

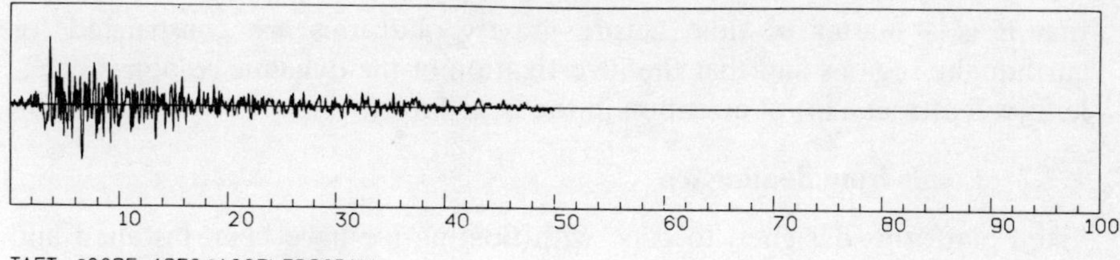

TAFT "S6BE–1952" ACCELEROGRAM

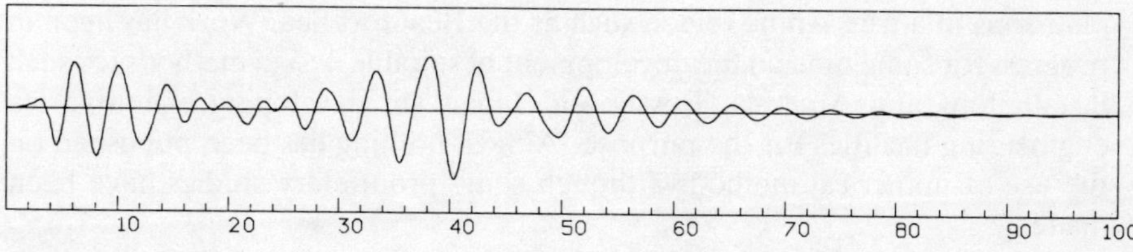

DECK DISPLACEMENT

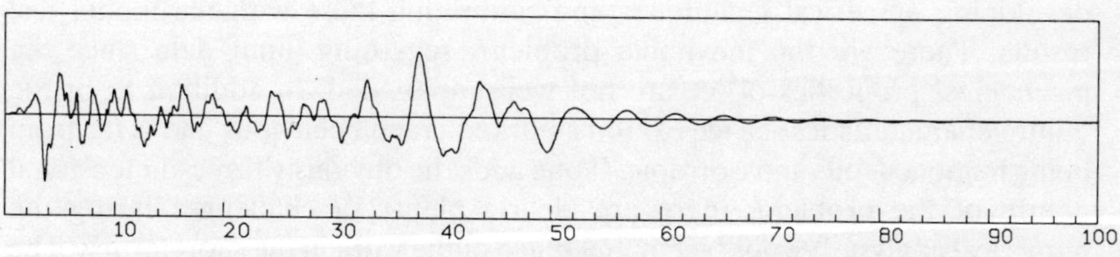

BASE DISPLACEMENT

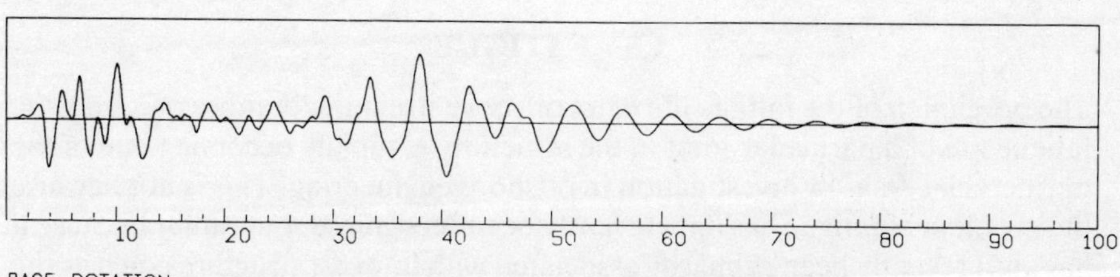

BASE ROTATION

Figure 1.14　Seismic response of gravity platform

response at deck level at a late stage in the duration of the record clearly demonstrates the importance of the low amplitude, long period components in the earthquake accelerogram. In making this point, it is worth remembering that the biggest uncertainty in earthquake engineering is the earthquake itself.[62] Other investigations are in hand to evaluate the non-linear response of such systems to extreme magnitude events in order to assess their survivability characteristics and develop appropriate design criteria. The writer is confident that it is a matter of time before gravity platforms are constructed for earthquake regions and that the investigation of the dynamic response problems will attract a lot of attention in the near future.

1.7.2 Loads from floating ice

Fixed platforms designed to cope with floating ice have been installed and successfully operated in Cook Inlet, Alaska for about a decade, but apart from some man made islands in the Canadian Arctic, there are as yet no offshore platforms in a true Arctic region such as the Beaufort Sea. Work has been in progress for some time on the development of suitable design methods for such installations and America, Russia and Japan are developing extensive ice engineering facilities for the purpose. Almost nothing has been published on the use of numerical methods although some proprietary studies have been made.

The emphasis for fixed structures must be on horizontal loads. Most attempts to predict these have been based on model tests but the time is ripe for developing numerical techniques and calibrating these with the model test results. There are the inevitable problems regarding input data since the mechanical properties of ice are not well understood. In addition to brittle failure characteristics sea ice exhibits marked creep behaviour and is far from being homogeneous and isotropic. If one adds the obviously three-dimensional nature of the problem, there are clearly plenty of challenges facing the numerical analyst. It would seem that much of the work in the short term will be of a conceptual nature.

1.8 FATIGUE

The prediction of the fatigue life of an offshore structure, or more precisely the fatigue life of a particular joint in the structure, is rapidly becoming one of the most crucial areas of investigation in offshore engineering. This is not because the problem of fatigue has only recently been recognized, but rather because it has until recently been primarily associated with low cost structures such as the caisson described by Hong and Brooks.[67] In a review of the problems of tubular joint design, Marshall[21] stated that at that time, no fixed template type platforms had collapsed as a result of fatigue failure, even though there had been problems with waterline braces. The primary reason for this increased interest is the focus on high cycle, low strain amplitude fatigue where experience of cumulative damage problems has been rather limited. The importance of fatigue analysis in the design of the current generation of steel offshore platforms for the North Sea was recently reviewed by Williams and Rinne.[63]

A basic fatigue analysis consists of three elements, namely, the determination of the relationship between 'hot spot' stress and wave amplitude, the prediction of the number of load cycles at each strain or stress amplitude and the substitution of the stress amplitude-load cycle data in some cumulative damage criterion to determine the fatigue life. Virtually all of the published work on fatigue has been based on the well known Palmgren–Miner cumulative fatigue damage law.[64,65] This linear damage criterion underlies the so-called design codes which have been published by authorities such as Lloyds, the British Standards Institution and the American Welding Society. Credibility of the linearity assumption was defended by Marshall[20] as being justifiable within the current range of design uncertainties, but this assumption still continues to attract the criticism of fatigue researchers such as Kirk,[66] who maintain that the linear damage theory results in a significant overestimate of fatigue life.

The major differences of opinion relating to fatigue design however seem to originate in the matter of treating the wave loads, and two basic schools appear to have emerged. The first of these is basically a deterministic approach, as suggested by Maddox,[67] who opted for a deterministic, rather than a stochastic treatment of the dynamic amplification effects in view of the non-linear drag problems which were discussed earlier. Maddox used modal analysis and a time history integration of the equations of motion in order to determine stress responses at the hot spots and a statistical treatment of the long term wave data in order to assess fatigue life in terms of the Palmgren–Miner rule. Other investigators such as Vugts and Kinra,[68] took a probabilistic approach to the fatigue analysis problem and considered that the linearity assumption could be justified since the low amplitude, short period waves associated with the fatigue problem implied predominantly inertial, rather than drag driving of the structures. Williams and Rinne[63] compared both deterministic and spectral methods considering both quasistatic and dynamic analyses. Not surprisingly, they pointed out that one of the biggest problems associated with either method is the selection of appropriate input data regarding sea state and wave force calculation.

Among the factors emerging as being of primary importance are the directionality and energy spreading characteristics of the sea. Some of the implications of these factors were discussed by McDowell and Holmes,[69] and Marshall[21] who developed a dynamic and fatigue analysis procedure using directional spectra and applied this to a 1000 ft platform currently being constructed for the Gulf of Mexico. This work has shown that the three-dimensional nature of the sea must be included in any fatigue analysis in order to obtain meaningful results.

1.9 OTHER STRUCTURAL TYPES

Based on our experience to date, the cost of a fixed platform would appear to increase exponentially with the water depth on location. It is inevitable therefore that the offshore engineer should be seeking new structural systems to meet the demands of the frontier areas. An obvious solution is to provide systems which can function entirely on the seabed, such as the wet or dry subsea completion systems which have been developed and are being actively marketed at present. The only systems of this kind which have been installed to date are being used in association with other more conventional oil field hardware and there are strong grounds for believing that it will be some considerable time before a totally self-contained system can be developed which will eliminate the need for a permanently emplaced surface vessel or platform. There is an obvious interest in developing so-called compliant systems which represent a reasonable compromise between the need to maintain such a link and the cost of providing it by means of a rigid structure. Some of the design problems associated with compliant systems were reviewed by Godfrey[70] who pointed out that there are severe design disadvantages for certain types of compliant structure. Nevertheless, the industry is confident that such systems have a future and large scale test programmes have been carried out on the guyed tower and the so-called tension leg structure shown in Figure 1.8 and several articulated flares have been installed. One of the main reasons behind carrying out such test programmes is the difficulty of obtaining realistic predictions of the motions of such systems in different sea states. The pursuit of suitable compliant systems will inevitably lead the analyst into some intriguing non-linear dynamic regimes.

1.10   CONCLUSION

This paper has attempted to provide an overview of the technical problems which are of primary interest to the civil engineer operating in the offshore environment at present. The range of problems is so large that the treatment has necessarily been superficial. However, it is hoped that it has served to illustrate those areas in which numerical analysis techniques are currently being applied in practice.

Several common threads run through most of these problem areas, one of the most obvious being the coupled nature of the problems. Not only does this make the problems themselves more interesting, but it has the added advantage that it demands the closest collaboration between engineers working in the different fields of structural, soil and fluid mechanics. The offshore engineer must address problems in all of these areas simultaneously if reasonable solutions are to be found.

A second obvious factor is the difficulty of characterizing the random environmental forces, whether these are due to winds, ice, earthquakes or waves. In conclusion, it is appropriate to quote directly from the excellent text by Blair Kinsman[71] on wind waves. 'It is very easy to forget the chaotic state of the sea when one is engaged in theoretical work. Our means are seldom commensurate with our ends, and the only way to make progress is to simplify and regularize. When one has given a great deal of effort and thought to a problem, one has an emotional investment in any results obtained. It is very difficult for any man to evaluate justly the distance his work lies from 'physical reality', or even the distance it lies from the problem he would like to have solved. Most of us need to be reminded. Therefore, go wave watching. If you will watch waves with a seeing eye, you will never confuse the regularity of any simplified approximation, like the sinusoid, with ocean waves as they really are'.

Acknowledgement

The author is indebted to Ove Arup & Partners for permission to publish this survey.

REFERENCES

1. Vine, A. C. (1969). 'Physical Oceanography', *Handbook of Ocean and Underwater Engineering*, McGraw-Hill.
2. Bretschneider, C. L. (1969). 'Wind generated waves', *Handbook of Ocean and Underwater Engineering*. Ed. John J. Myers. McGraw-Hill.
3. Dean, R. G. (1967). 'Relative validities of water wave theories', *Proc. ASCE Conf. on Civil Engineering in the Oceans*, San Francisco, 1–30.
4. Henkel, D. J. (1970). 'The role of waves in causing submarine landslides', *Geotechnique*, **20**, 75–80.
5. Boeing/Arup (1976). Unpublished internal report on joint review of Arctic Engineering problems by Boeing Engineering and Construction and Ove Arup and Partners.
6. Mashburn, M. K. and Hubbard, J. L. (1967). 'An Ocean Structure', *Proc. Conf. on Civil Engineering in the Oceans*, San Francisco, Sept. 1967.
7. Hong, S. T. and Brooks, J. C. (1976). 'Dynamic behaviour and design of offshore caissons', *OTC 2555, Offshore Technology Conference*, Houston, 363–374.
8. Wees, T. A. and Chamberlain, R. S. (1971). 'Khazzan Dubai No. 1 pile design and installation', *ASCE, Jour. SMFD SM10. Proc. Paper No. 8439*, 1415–1429.
9. Gerwick, B. C. and Hognestad, E. (1973). 'Concrete Oil Storage Tank to rest in North Sea', Civil Engineering, *ASCE*, **43**, No. 8, Aug. 1973, 81–85.
10. *Offshore Engineer*, 'The Argyll Field', January 1975.
11. Bea, R. G. (1976). 'Earthquake criteria for platforms in the Gulf of Alaska', *OTC 2675, Offshore Technology Conference*, Houston.
12. Bea, R. G. (1974). 'Development of safe environmental criteria for offshore structures', *CE-7 Note*, Shell Oil Co., Houston, Texas.
13. U.K. Dept. of Energy (1974). *Guidance on the Design and Construction of Offshore Installations*, London, HMSO.
14. Morison, J. R., *et al.* (1950). 'The force exerted by surface waves on piles', *Petroleum Transactions AIME*, **189**.
15. Exxon Production Research Co. (1975). *Ocean Test Structure Technical Plan*, Exxon Production Research Co., October 1975.

16. Matlock, H. (1970). 'Correlations for design of laterally loaded piles in soft clay', *OTC 1204, Offshore Technology Conference*, Houston.
17. Malhotra, A. K. and Penzien, J. (1910). 'Nondeterministic analysis of offshore structure', *Jour. EMD, ASCE*, **12**, 1970.
18. Tickell, R. G., Burrows, R., and Holmes, P. (1976). 'Long-term wave loading on offshore structures', *Proc. Instn. Civil Engineers*, Part 2, 1976, N61, 145–162.
19. Ruhl, J. A. (1976). 'Offshore platforms: Observed behaviour and comparisons with theory', *OTC 2553, Offshore Technology Conference*, 333–352, Houston.
20. Marshall, P. W. (1974). 'Basic considerations for tubular joint design in offshore construction', *W.R.C. Bulletin 193*, April 1974.
21. Marshall, P. W. (1976). 'Dynamic and fatigue analysis using directional spectra', *OTC 2537, Offshore Technology Conference*, 133–157, Houston.
22. Zienkiewicz, O. C. (1977). *The Finite Element Method in Engineering Science*, 3rd Edn., McGraw-Hill.
23. McClelland, B. (1974). 'Design of deep penetration piles for ocean structures', *ASCE Jour. Geot. Eng. Div.*, **100**, GT7, 1974.
24. Smith, E. A. L. (1960). 'Pile driving analysis by the wave equation', *Proc. ASCE*, Aug. 1960, 35.
25. McClelland, B., Focht, J., and Emrich, W. J. (1969). 'Problems in design and installation of offshore piles', *Jour. Soil Mechanics and Foundation Division, ASCE*, **95**, SM 6, 1491–1513, November 1969.
26. Fox, D. A. *et al.* (1976). 'North Sea platform piling–development of the Forties Field piles from West Sole and Nigg Bay experience and tests', *Proc. Instn. of Civil Engineers Conference on Design and Construction of Offshore Structures*, London.
27. Keulegan, G. H. and Carpenter, L. H. (1958). 'Forces on cylinders and plates in an oscillating fluid', *Jour. of the National Bureau of Standards*, **60**, No. 5, May 1958.
28. Hogben, N. (1974). 'Fluid loading on offshore structures. A State-of-Art Appraisal—Wave Loads', *National Physical Laboratory Ship T.M. 381*, Teddington, Feb. 1974.
29. MacCamy, R. C. and Fuchs, R. A. (1954). 'Wave forces on piles—a diffraction theory', *Beach Erosion Board Technical Memorandum No. 69*.
30. Gran, S. (1973). 'Wave forces on submerged cylinders', OTC 1817, *Proc. Offshore Technology Conference*, Houston.
31. Garrison, C. J. and Berklite, R. N. (1972). 'Hydrodynamic loads induced by earthquakes', *Proc. Offshore Technology Conference*, Houston, Texas, O.T.C. 1554.
32. Garrison, C. J. and Chow, P. Y. (1972). 'Wave Forces on submerged bodies', *ASCE Waterways*, Harbors and Coastal Engineering Div., 375–392.
33. Garrison, C. J. *et al.* (1974). 'Wave forces on large volume structures', *OTC 2137*, Houston.
34. Hogben, N. and Standing, R. (1974). 'Wave loads on large bodies', *International Symposium on the Dynamics of Marine Vehicles and Structures in Waves*, University College, London.
35. Chen, H. S. and Mei, C. C. (1974). *Oscillations and Wave Forces in a Man-made Harbor in the Open Sea*, Presented at 10th Naval Hydrodynamics Symposium, June 1974.
36. Zienkiewicz, O. C. (1975). 'The finite element method and boundary solution procedures as general approximation methods for field problems', *Proc. World Conf. on Finite Element methods in Structural Mechanics*, Bournemouth, England, October 1975.

37. Zienkiewicz, O. C., Kelly, D. W., and Bettess, P. (1977). 'The coupling of the final element method and boundary solution procedures', *In. J. Num. Math. Eng.*, **11**, 355–375.
38. Zienkiewicz, O. C. and Bettess, P. (1975). 'Infinite elements in the study of fluid structure interaction problems', *Proc. 2nd Int'l. Symposium on Computing Methods in Applied Science and Engineering*, Versailles, France, December 1975.
39. Haynes, H. N. (1976). 'Collapse behaviour of pressurised concrete shells', *Proc. BOSS Conf. on Behaviour of Offshore Structures*, Trondheim, Norway.
40. Hsu, M. B. and Nickell, R. E. (1974). 'Coupled convective and conductive heat transfer by finite element methods', *Proc. Conf. in Finite Element Methods in Flow Problems*, 427, Swansea, Wiley.
41. Richmond, B. (1976). 'The time–temperature dependence of stresses in offshore concrete structures', *Proc. Conf. on Design and Construction of Offshore Structures*, Instn. of Civil Engineers, London.
42. Bjerrum, L. (1973). 'Geotechnical problems involved in foundations of structures in the North Sea', *Geotechnique*, **23**, 319–358.
43. Vesic, A. S. (1967). 'Ultimate loads and settlements of deep foundations in sand', *Proc. Symposium on Bearing Capacity and Settlement of Foundations*, Duke Univ., Durham, N.C. 53.
44. Kerisel, J. (1961). 'Fondations profondes en milieux sableux: Variation de la force portante limite en fonction de la densite, de la profondeur, du diametre et de la vitesse d'enforcement', *Proc. 5th Int. Conf. on Soil Mech. and Found. Eng.*, **II**, 73–84.
45. Hansen, J. Brinch (1970). 'A revised and extended formula for bearing capacity', *Geotekniske Institut København Bulletin 28*, 5–11.
46. Vaughan, P., *et al.* (1976). 'Stability analysis of large gravity structures', Preprint— *BOSS Conference on Behaviour of Offshore Structures*, Trondheim (1916).
47. Rowe, P. W. and Craig, W. H. (1976). 'Studies of offshore caissons founded on Oosterschelde sand', *Proc. Conf. on Design and Construction of Offshore Structures*, Instn. of Civil Engineers, London.
48. Chan, K. C. (1976). 'A preliminary study of one-dimensional flow through saturated soils, caused by wave loadings', Ove Arup and Partners, *Offshore Projects Group internal report*, May 1976.
49. Andersen, K. H., Brown, S. F., Foss, I., Pool, J. H., and Rosenbrand, W. F. (1976). 'Effect of cyclic loading on clay behaviour', *Proc. Conf. on Design and construction of offshore structures*, Instn. of Civil Engineers, London.
50. Eatock-Taylor, R. (1975). 'Structural dynamics of offshore platforms', *Proc. offshore structures Conf. I.C.E. London*, 1975, 125–132.
51. Moan, T., Haver, S., and Vinje, T. 'Stochastic dynamic response analysis of offshore platforms, with particular reference to gravity-type platforms', *Offshore Technology Conference, Houston*, 1975, OTC 2407.
52. Liaw, C. Y. and Chopra, A. K. (1973). 'Earthquake response of axisymmetric tower structures surrounded by water', *Report No. EERC 73–25,* University of California, Berkeley.
53. ASCE (1976). 'Analyses for soil structure interaction effects for nuclear power plants'. Report by the Ad Hoc Group on Soil-structure Interaction, Nuclear Structures and Materials Committee of the Structural Division of ASCE. *Final Draft, April 1976.*

54. Veletsos, A. S. (1975). *Dynamics of Structure-foundation Systems*, Presented at the symposium on structural and geotechnical mechanics honoring Nathan M. Newmark, Univ. of Illinois, October 1975.

55. Ove Arup and Partners (1976). 'DAFT—Dynamic analysis using Fourier transforms for soil-structure systems onshore and offshore', *Interim Report to U.K. Dept of Energy, Contract No. ESA/Con/505*, June 1976.

56. API (1976). 'Recommended practice for planning, designing and constructing fixed offshore platforms', *API RP2A American Petroleum Institute Division of Production*, Dallas, Texas.

57. Penzien, J. (1975). 'Seismic analysis of platform structure—foundation systems', *OTC 2352, Proc. Offshore Technology Conference, Houston, Texas.*

58. Clough, R. W. and Penzien, J. (1975). *Dynamics of Structures*, McGraw-Hill.

59. Parmalee, R., *et al.* (1964). 'Seismic effects on structures supported on piles extending through deep sensitive clays', *EERC Report 64–2,* Univ. of California, Berkeley.

60. Penzien, J. and Tseng, W. S. (1976). 'Seismic analysis of gravity platforms including soil—structure interaction effects', *O.T.C. Paper 2674, Proc. Offshore Technology Conference*, Houston, Texas, May 1976.

61. Watt, B. J., Boaz, I. B., and Dowrick, D. J. (1976). 'Response of concrete gravity platforms to earthquake excitations', *OTC Paper 2673, Offshore Technology Conference*, Houston, Texas.

62. Page, R. A. (1975). 'Evaluation of seismicity and earthquake shaking at offshore site', *OTC 2354.*

63. Williams, A. K. and Rinne, J. E. (1976). 'Fatigue analysis of steel offshore structures', *Proc. Instn. Civil Engineers*, Part 1, 1976, **60**, Nov. 635–654.

64. Palmgren, A. (1924). 'Die Lebendsdauen von Kugelhagern'. *VDIZ*, 1924, **68**, 14, 339.

65. Miner, M. A. (1945). 'Cumulative damage in fatigue', *Jour. Appl. Mech.*, **12**, A-159.

66. Kirk, C. L. (1975). Discussion, *Proc. Conf. on Offshore Structures, ICE*, London 1975, 204.

67. Maddox, N. R. (1974). 'Fatigue analysis for deep water fixed bottom platforms', *OTC 2051, Offshore Technology Conference*, Houston.

68. Vugts, J. H. and Kinra, R. K. (1976). 'Probabilistic fatigue analysis of fixed offshore structures', *OTC 2608*, 889–906.

69. McDowell, D. M. and Holmes, P. 'General research problems in the design of offshore structures', *Proc. Offshore Structures Conf. ICE*, London, 1975, 165–170.

70. Godfrey, P. S. (1976). 'Compliant drilling and production platforms', *Proc. Conf. on Design and Construction of Offshore Structures*, Instn. of Civil Engineers, London, October 1976.

71. Kinsman, B. (1965). *Wind Waves—Their Generation and Propagation on the Ocean Surface*, Prentice-Hall.

Chapter 2

Approaches to Fluid Loading, Probabilistic and Deterministic Analyses

R. G. Tickell and P. Holmes

2.1 INTRODUCTION

The loading imposed on members of an offshore structure subject to wave and current action represents one of the major design considerations in offshore engineering and has been the subject of intensive experimental and theoretical research over the past thirty years, which is some indication of the complexity of the problem. A knowledge of wave loads is essential in ensuring an acceptable probability of survival under some extreme conditions (first excursion failure) and in designing for an adequate fatigue life.

Wave loading is due to the relative motion between the fluid and structure, and consideration of the problem must logically begin with the specification of the wave or wave-current field. Although swell, from some distant storm, may be described by regular and deterministic wave theories, the wave environment within the generation zone is strictly a random process. Thus a fundamental bifurcation is encountered in the treatment of the storm-wave field as a random process or in quasi-deterministic form. This chapter discusses some of the numerical procedures related to both approaches.

Once the appropriate wave model is established a load mechanism is employed which relates the fluid motions to the resulting forces experienced by a structural element. The in-line component of loading is normally evaluated using the well-known Morison[1] equation for a member with dimensions such that the presence of the member does not significantly disturb the incident wave field. In addition to the in-line component, the member may experience a transverse or lift force as a result of changes in the boundary layer and the formation of vortices within the flow. When the member is large, relative to some characteristic wave-length of the wave field, diffraction of the incident wave must be considered in the treatment.[2] This chapter is concerned with situations in which Morison's equation is appropriate and diffraction solutions are discussed elsewhere.

The calculation of structural response, in terms of displacements or stresses, is reasonably straightforward once the loading is established and provided that the excitation is quasi-static relative to the dynamic properties of the structure-foundation system. However, a coupling between the fluid and structural motions may result from the dynamic response of the system to wave excitation. This represents a feedback element in the loading mechanism requiring the calculation of the dynamic response of the system and this problem is the subject of later chapters.

2.2   THE WAVE ENVIRONMENT

The wave activity of major concern in the design of offshore structures results from wind action. The physics of this process is not yet fully understood but there are recognized to be several mechanisms involved; initial generation by a resonance mechanism between turbulence-induced pressure fields and the resulting water-surface deformation,[3] deformation of the wind field by waves of significant height resulting in energy transfer,[4] wave-wave interaction in which energy is transferred from short to long wave-lengths,[5] and ultimately, the breaking of waves so as to maintain a balanced or equilibrium state between energy input to a given frequency of wave motion and that lost to turbulence in the breaking process.

The result of these various physical processes is an assumed superposition of wave of different heights, periods and directions of propagation which may be regarded as a three-dimensional random process. Due to the incomplete understanding of the generation process descriptions of wave conditions for design are formulated, in part, on quasi-empirical relationships, as discussed below.

The three elements of this formulation are:
(1) Deterministic wave theories;
(2) Probabilistic properties of a random sea; and
(3) Wave-energy spectral-density methods.

The following discussion is based on the simplified problem of a long-crested sea, in which the waves propagate in one direction only. The extension of the numerical methods to short-crested, three-dimensional seas involves few additional considerations of a fundamental nature but computations are more tedious and experimental verification is relatively sparse.[6] In addition the variation of mean wave direction can be taken account of in the design computations.

2.2.1   Deterministic wave theories

Most water-wave theories essentially obtain a solution for a 'velocity-potential', ϕ or its orthogonal, the stream-function, ψ, which must satisfy the continuity (Laplace) equation:

$$\frac{\partial^2 \phi}{\partial x^2} + \frac{\partial^2 \phi}{\partial y^2} = 0 \tag{2.1}$$

and the energy (dynamic) equation:

$$p/\rho = -gy - \frac{\partial \phi}{\partial t} - \frac{1}{2}\left[\left(\frac{\partial \phi}{\partial x}\right)^2 + \left(\frac{\partial \phi}{\partial y}\right)^2\right] \tag{2.2}$$

where p is the pressure

ρ is the density of the fluid
g is the gravitational acceleration
(x, y) are Cartesian co-ordinates
and incompressible, irrotational motion is assumed.

The solution for ϕ is governed by boundary conditions:

(1) At the air-water interface, where:
(a) the velocity of a particle must be tangential to the surface slope; the kinematic boundary condition, and
(b) $p/\rho = 0$ and the energy equation must be satisfied; the dynamic boundary condition, and
(2) At the sea floor where the vertical velocity is zero.

The velocity potential (or stream-function) describes the flow within the fluid and at its boundaries, such that the horizontal and vertical components of water particle velocities at a point x, y are:

$$u = \frac{\partial \phi}{\partial x} = -\frac{\partial \psi}{\partial y}$$

$$v = \frac{\partial \phi}{\partial y} = \frac{\partial \psi}{\partial x} \tag{2.3}$$

The major problem in solving for ψ or ϕ arises from the boundary conditions to be applied at the air-water interface, $\eta(t)$, which is itself part of the solution sought. There are therefore several solutions in common use: Linear Wave Theory, Stokes' III and V Order in 'deep' water; Cnoidal theories to first and second order and Solitary wave theory in 'shallow' water, and numerical solutions of which the most popular are Streamfunction.[7,8]

Linear wave theory is summarized below, followed by a brief discussion of a numerical solution. Information on other wave theories may readily be obtained from the literature, e.g., Wiegel.[9]

Linear (Airy) wave theory is so termed following a linearization of the boundary condition at the air-water interface, $y = 0$. Note that y is measured positive upwards from mean-water-level.

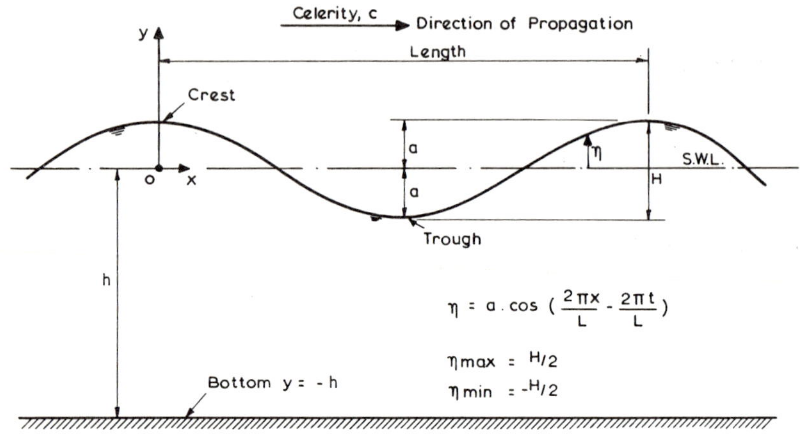

Figure 2.1   Definition of terms—linear wave theory

The relevant equations from this solution are as follows, reference Figure 2.1:

$$\eta(t) = a\,\cos\,(kx - \omega t) = \frac{H}{2}\cos 2\pi\left(\frac{x}{L} - \frac{t}{T}\right) \tag{2.4}$$

where H = wave height = $2a$

       k = wave number = $2\pi/L$ (radians/metre)

       L = wave length

       ω = wave frequency = $2\pi/T$ (radians/second)

       T = wave period = L/c

The phase-velocity, c, and wave length, L, are given by:

$$L = L_0 \tanh kh \qquad c^2 = \frac{g}{k}\tanh(kh) \tag{2.5}$$

and the total energy per unit area, E, is $\rho g a^2/2$, subscript 0 denoting 'deep' water conditions ($kh > \pi$).

Forces exerted by waves on structures are related to horizontal and vertical water-particle velocities, u and v respectively and to horizontal and vertical accelerations, $\dot{u}$ and $\dot{v}$. From the velocity potential these are given by:

$$u = a\omega \left[\frac{\cosh k(y+h)}{\sinh kd} \right] \cos (kx - \omega t) \qquad (2.6)$$

$$v = a\omega \left[\frac{\sinh k(y+h)}{\sinh kd} \right] \sin (kx - \omega t) \qquad (2.7)$$

and local particle accelerations follow by differentiation with respect to time. (For a further derivation of these equations see Ch. 3.4.)

The higher-order wave theories of Stokes' involve series formulations for $\eta(t)$ in which the coefficients of the series are functions of wave height. Although the classical theories are not numerical by nature, their practical application involves a major use of computers in evaluating fluid motions and programs for these theories are readily available.

Dean's Stream Function theory[7,10] is numerical, and as the name implies, it is developed in terms of the stream function, ψ, which is related to fluid velocities as defined above.

Lines on which ψ is a constant are streamlines, parallel at every point to the local velocity vector. By using a stream function rather than a velocity potential to describe the wave, it is therefore possible to satisfy the kinematic boundary condition exactly, by making the free surface a line of constant ψ.

A suitable stream function, which satisfies exactly the requirements of continuity, and of zero vertical velocity on the sea bed is given by:

$$\psi = A_1 \sinh ky \cdot \cos kx + A_2 \sinh 2ky \cdot \cos 2kx + \cdots A_n \sinh nky \cdot \cos nkx \qquad (2.8)$$

The series may be extended to any order without extensive algebraic manipulation since the coefficients $A_1 \ldots A_n$ are computed numerically for each wave. In addition to the coefficients, the wave length and the value of ψ on the surface must be treated as unknowns. The procedure, given H, T and h, is to find values for all of the unknowns such that the error in the dynamic boundary condition at a large number of points on the free surface is minimized. Several minimization techniques for such non-linear functions exist and are available in computer libraries. Because it may easily be extended to any order, the stream function theory is applicable over a wide range of conditions. However, some care is needed in ensuring that a numerically satisfactory solution is physically meaningful.

There are considerable difficulties in measuring particle velocities in waves, and few reliable experimental results have been obtained. Consequently, studies of the relative validities of different wave theories in given conditions have concentrated on the respective errors in those boundary conditions which are not satisfied exactly. Comparisons between the boundary condition errors in different theories have been carried out notably by Dean,[11] Le Méhauté[12] and Silvester.[13]

A modification to the stream function theory described above permits an additional boundary condition to be introduced: that of requiring that the free surface to agree with an observed profile. With this additional restraint the coefficients are adjusted as before to minimize all boundary condition errors. Once this is achieved all characteristics of the wave can be computed.

Recent work, Chaplin,[14] enables the orbit of a particle on the free surface to be determined from a known wave profile. From the geometry of the orbit all free surface velocities and particle accelerations can be obtained. This method has been used successfully to reduce the scatter in wave slamming coefficients calculated from force measurements in the laboratory.

Direct finite difference or finite element procedures are today available for the solution of many fluid mechanics problems including wave motion (e.g. marker–cell technique). Such solutions are however seldom called for in the periodic wave problem.

2.2.2 Probabilistic properties of a random sea

At some position offshore, a wave gauge might well record a variation of surface elevation, η, similar to that shown in Figure 2.2. The subsequent analysis of such a time series $\eta(x, t)$ could proceed as follows. If the surface elevation is assumed to be statistically stationary during the period of observation and to be ergodic, then estimates of the cumulative distribution function, $P(\eta)$, and the probability density function, $p(\eta)$, could be obtained from the digitized record of values n_i ($i = 1, 2, \cdots N$, where $N = T_R/\Delta t$; T_R is the time for which the record was taken and Δt is the digitizing interval).

If N' is the number of observations for which $\eta' < \eta \leq \eta' + d\eta$ then the probability density function is:

$$p(\eta')\,d\eta = \lim_{N \to \infty} N'/N \tag{2.9}$$

and the cumulative distribution function is:

$$P(\eta') = \text{Prob}\,\{\eta \leq \eta'\} = \int_{-\infty}^{\eta'} p(\eta)\,d\eta \tag{2.10}$$

Similarly, the jth order statistical moment or expectation of η is given by:

$$E(\eta^j) = \lim_{N \to \infty} \left(\sum_{i=1}^{N} n_i^j \right)/N \tag{2.11}$$

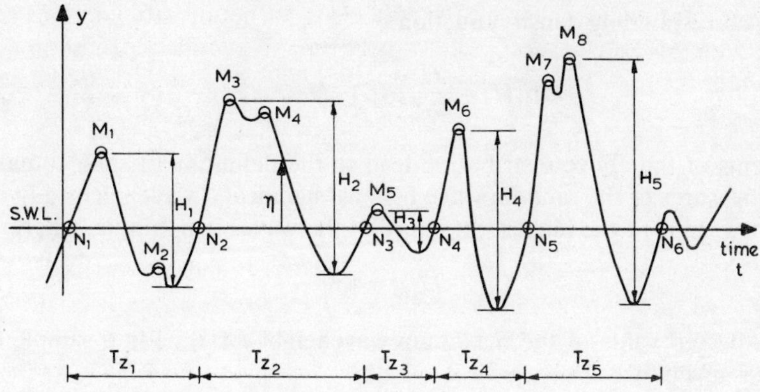

d—Water depth

y—+ve upwards from SWL

$\eta(x, t)$—The surface elevation at (x, t) measured from SWL (+ve upwards)

N_i—Upcrossing of $y = 0$ by $\eta(x, t)$

M_i—Maxima of $\eta(x, t)$

T_{zi}—time between successive upcrossings (zero crossing period of ith wave)

H_i—Waveheight of ith wave-range of $\eta(x, t)$ from highest crest to lowest trough between successive upcrossings

Figure 2.2   Definition sketch of random wave variables

To a first order, surface elevation can be well represented by a zero mean Gaussian density function:

$$p(\eta) = \frac{1}{\sqrt{(2\pi)}\sigma_\eta} \exp\{-\eta^2/2\sigma_\eta^2\} \tag{2.12}$$

where σ_η is the standard deviation of $\eta(t)$.

Further details of the probabilistic concepts involved are given in the Appendix to this Chapter.

If a suitable convention is adopted, the random sea shown in Figure 2.2 can be described in terms of wave height (H) and period (T). In engineering applications the usual definition of waveheight is the maximum surface elevation range between successive upcrossings of still-water-level and this leads to the definition of zero-crossing period T_z. Longuet-Higgins[15] has shown that, for certain conditions the cumulative distribution of waveheight is approximated by the Rayleigh distribution:

$$P(H) = 1 \cdot 0 - \exp\{-H^2/8\sigma_\eta^2\} \tag{2.13}$$

or the corresponding density function:

$$p(H) = \frac{H}{4\sigma_\eta^2} \exp\{-H^2/8\sigma_\eta^2\} \tag{2.14}$$

Moments of the above distribution lead to the definition of such commonly used measures of the random wave field as 'significant' waveheight, $H_{1/3}$, the expected value of the highest one-third of the waves in a sample, where:

$$H_{1/3} = 4\sigma_\eta \tag{2.15}$$

The expected value of the maximum waveheight $E(H_{max})$ in a sample of N waves is given by:

$$E(H_{max}) \simeq [H_{1/3}\sqrt{(2 \log_e N)}]/2 \qquad for\ N > 100 \tag{2.16}$$

At this point it is common practice for the designer to switch from a probabilistic to a deterministic mode by applying some regular wave theory with an appropriate waveheight (e.g., $H_{1/3}$, $H_{1/10}$ or H_{max}). A problem is immediately encountered in the selection of a wave period to be associated with the waveheight parameter. Some progress has been made in this area and Longuet-Higgins[16] and Cavanie[17] have proposed forms for the bivariate distribution of H and T which still require further experimental verification.

The probabilistic models developed thus far relate to short-term stationary conditions. For the purpose of the first excursion failure problem models must be developed for the prediction of the expected maximum waveheight, the 'design wave', over a long-term period such as 50 or 100 years. The long-term wave climate at a particular site is commonly summarized by a bivariate histogram of observed $H_{1/3}$ and T_z, see Figure 2.3. For each $H_{1/3}$, T_z, pair, a short-term conditional Rayleigh distribution may be written:

$$P(H|H_{1/3_i}, T_{z_j}) = 1\cdot0 - \exp\{-2H^2/H_{1/3_i}^2\} \tag{2.17}$$

and a convolution of Equation (2.17) with the bivariate histogram gives the long-term individual waveheight distribution, $P_{LT}(H)$ as:

$$P_{LT}(H') = \text{Prob}\ \{H \le H'\} = 1\cdot0 - \left\{\sum_i \sum_j \exp\{-2(H'/H_{1/3_i})^2\}n_{ij}/T_{z_j}\right\}/\bar{N}_z \tag{2.18}$$

where n_{ij} is the proportion of time for which:

$$(H_{1/3})_i - \Delta H_{1/3}/2 \le H_{1/3} \le (H_{1/3})_i + \Delta H_{1/3}/2$$

$$(T_z)_j - \Delta T_z/2 \le T_z \le (T_z)_j + \Delta T_z/2$$

PERIOD, T_z (SECONDS)

H 1/3(m)	4·5	5·5	6·5	7·5	8·5	9·5	10·5	11·5	12·5	13·5
0·3	14	40	34	8						
0·9	64	159	135	40	4					
1·5	18	103	164	78	24	2				
2·1	6	53	126	95	33	7	1			
2·7		19	103	72	41	8	2			
3·3		9	46	71	31	3	1			
3·9		1	23	63	38	6	1			
4·5			6	20	31	10	2	1		
5·1			5	13	15	12	1			
5·7			1	9	4	6	3			
6·3				2	2	4	2			2
6·9					1	3	7			
7·5				1	1		4		2	
8·1					2					
8·7						2	2			
9·3										2

PARTS PER 1924·0

Figure 2.3 Typical bivariate histogram of H_s
and T_z

$\bar{N}_z$ is the long-term average number of waves per unit time:

$$\bar{N}_z = \sum_i \sum_j n_{ij}/T_{zj} \tag{2.19}$$

The derived long-term individual waveheight distribution is, typically, based on one-year's data and its extrapolation to a return-period of 50 or 100 years is based on fitting the data to a Weibull or Gumbel distribution function given, respectively, by:

$$P(H) = 1 - \exp\left[-\left(\frac{H-\alpha}{\beta}\right)^{\gamma}\right] \tag{2.20}$$

$$P(H) = \exp\left\{-\exp\left[-\left(\frac{H-\delta}{\xi}\right)\right]\right\} \tag{2.21}$$

where α, β, γ, δ and ξ are parameters of the distribution functions determined from the data.

The long-term distribution of H resulting from the data of Figure 2.3 is shown in Figure 2.4, plotted, in this case, on a Weibull probability scale, the distribution being extrapolated to give the waveheight with a required probability of exceedance or to estimate the height with a given average return period (for extensions to this treatment see Nolte[18]).

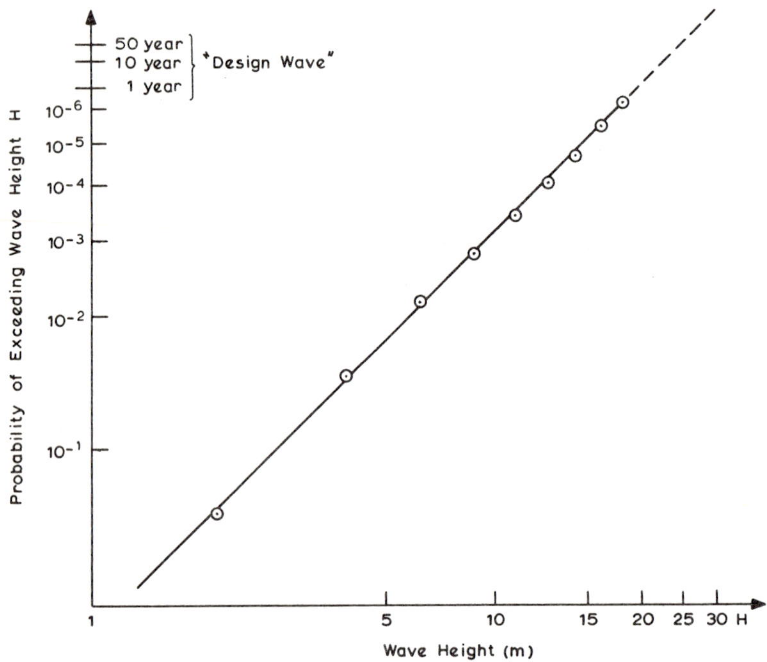

Figure 2.4   Long-term distribution of individual waveheight

2.2.3   Wave-energy spectral-density methods

The probabilistic description presented above is deficient in that the frequency content of the wave field and the relative distribution of energies at different frequencies are not retained in the formulation. Such information is given by the wave energy spectral density function. The formal derivation and properties of the spectral density function are given in the Appendix.

The surface-elevation spectrum, $S_\eta(\omega)$, is crucial to a wide range of design situations in the offshore environment. In design one of many standard forms may be adopted, specified by a small number of properties or characteristic measures of the random sea, e.g. Pierson–Moskowitz[19]:

$$S_\eta(\omega) = \frac{\alpha g^2}{\omega^5} \exp\left\{-\frac{\beta \omega_0^4}{\omega^4}\right\} \tag{2.22}$$

where

$$\alpha = 0\cdot0081$$

$$\beta = 0\cdot74$$

$$\omega_0 = g/U$$

U is the characteristic wind speed at 19·5 m above still-water-level, and Equation (2.22) corresponds to the fully-developed sea condition. Many other forms have been used and recent work on the development of the spectrum in fetch-limited conditions has resulted in the JONSWAP spectrum.[20] The JONSWAP spectrum may be written in the following form:

$$S_\eta(\omega) = \frac{\alpha g^2}{\omega^5} \exp\left\{-1\cdot25\left(\frac{\omega_m}{\omega}\right)^4\right\} \cdot \gamma^{\exp\{-(\omega-\omega_m)^2/2\sigma^2\omega_m^2\}} \tag{2.23}$$

and

$$\sigma = \sigma_a \text{ for } \omega \le \omega_m$$

$$= \sigma_b \text{ for } \omega > \omega_m$$

This is of similar form to Equation (2.22) except that α and ω_m, the frequency of the spectral peak, are functions of wind velocity, U, and fetch. The additional parameters γ, σ_a and σ_b showed no functional relationships to fetch and the experimentally-derived mean values for the original study were $\gamma = 3\cdot3$, $\sigma_a = 0\cdot07$ and $\sigma_b = 0\cdot09$. The JONSWAP spectrum was not intended by its originators as a formulation for engineering use—it has so been used but caution in its general application is essential. This raises a general point concerning the specification of a design spectrum. For example, given a pair of $H_{1/3}$ and T_z values two spectral forms result even for the Pierson–Moskowitz spectrum depending on whether the spectrum is derived from $H_{1/3}$ or from T_z.

Although the concept of the spectral density function is derived from the Fourier transform of the auto-correlation (see Appendix), measured spectra are more efficiently calculated from the direct sequence of observed values of surface elevation by Fast Fourier Transform methods.[21,22] Extremely efficient algorithms are available for the calculation of the basic discrete Fourier coefficients η_k, especially when the numbr of data points, N, is 2^q.

$$\eta_k = \frac{1}{N} \sum_{j=0}^{N-1} \eta_j \exp\{-i(2\pi jk/N)\} \tag{2.24}$$

The 'raw' continuous spectrum is obtained from Equation (2.24) as:

$$S'(\omega_k) = \frac{T_R}{2\pi} \eta_k^* \eta_k \tag{2.25}$$

where η_k^* denotes the complex conjugate of η_k and $\omega_k = 2\pi k/T_R$. The variances of the spectral estimates is high in the raw state given by Equation (2.25). It is common practice to smooth the spectrum by applying a running average of estimates at adjacent frequency points. Further smoothing may be performed by taking the FFT of segments of the total record and averaging the resulting spectra.

Several other practical points should be mentioned. The digitizing interval Δt must be chosen such that no significant energy exists at frequencies above $\pi/\Delta t$ rads/sec, the Nyquist frequency, irrespective of one's interest in such high frequencies. If this step is not taken or the record is not subject to an initial filtering, energy above $\pi/\Delta t$ will be 'folded' into the lower frequency estimates. It is also necessary to remove any trend from the data to avoid large estimates at zero frequency and to taper the record at the beginning and end by a cosine or other appropriate function.[23]

2.2.4 Simulation of a random sea

The spectrum of η describes the distribution of variance in the frequency domain and from linear wave theory, Section 2.2.1, the variance is related to the energy of the wave field.

$$\sigma_\eta^2 = \int_0^\infty S_\eta(\omega)\,d\omega \tag{2.26}$$

A numerical model of a long-crested random sea can be based on the summation of a finite number of first order waves with amplitudes derived from the spectrum $S_\eta(\omega)$ in the following manner:

$$\eta(x, t) = \sum_{j=1}^{M} a_j \cos\left(k_j x - \omega_j t + \theta_j\right) \tag{2.27}$$

where

$$a_j = \sqrt{2 S_\eta(\omega_j)\,\Delta\omega} \tag{2.28}$$

and the wave number k_j is related to the frequency and water depth, d, by:

$$\omega^2 = gk \tanh kh \tag{2.29}$$

If the phase angle θ_j is selected from an independent and uniform probability distribution of θ between 0 and 2π then as $M \to \infty$, the Central Limit Theorem may be invoked to establish the limiting form of $p(\eta)$ as Gaussian, which is consistent with Equation (2.12) above.

Borgman[24] and Shinozuka[25] have discussed the use of Equation (2.27) and similar models for the digital simulation of ocean waves and other random processes. Borgman suggests that the frequencies of ω_j be chosen from the cumulative spectrum to yield equal amplitude (energy) components and Shinozuka shows that simulation may be performed by taking ω as a random frequency drawn from a distribution related to the spectral density function. Provided that M is sufficiently large, such models appear to be most successful in simulating η with the required target spectrum and probability distribution. Borgman[24] also develops the simulation of surface elevation using a linear filter in the form:

$$\eta_j = \sum_{i=-N}^{N} a_i b_{j-i} \qquad j = N+1, N+2 \ldots \tag{2.30}$$

where b is a random Gaussian sequence with a known spectrum and the filter constants a_i are calculated from the Fourier transform of the frequency response function relating b and η. This approach is the basis for many of the random wave synthesizers used to control wave generators in laboratory tanks. The sequence b is drawn from the pseudo-random binary output of a shift register. The filter constants are in this instance weighting resistances whose values are chosen from a knowledge of the target spectrum $S_\eta(\omega)$, and the transfer function between the generator motions and the resulting waves.

The primary objective of this discussion is not the simulation of the sea surface but to develop the models for fluid motion required in the calculation of wave loading. Thus, further application of linear wave theory will yield expressions for the horizontal and vertical fluid velocities and accelerations, u, v, $\mathrm{d}u/\mathrm{d}t = \dot{u}$, $\mathrm{d}v/\mathrm{d}t = \dot{v}$, respectively. For example:

$$u(x, y, t) = \sum_{j=1}^{M} \sqrt{(2S_\eta(\omega_j)\,\Delta\omega)}\,\frac{\cosh k_j(y+d)}{\sinh k_j d} \cos(k_j x - \omega_j t + \theta_j) \tag{2.31}$$

from which it follows that:

$$S_u(\omega) = \left[\omega^2 \frac{\cosh^2 k(y+d)}{\sinh^2 kd}\right] S_\eta(\omega) \tag{2.32}$$

etc. Thus the bracketed term in Equation (2.32) is a transfer function relating the particle velocity spectrum to the surface elevation spectrum.

Such linear transforms imply Gaussian distributions of the fluid motions, which is reasonable, except within the free surface zone where the variation of η, above and below the point considered, leads to an intermittent or discontinuous random process for fluid motions. This aspect is discussed further when considering the loads on members in the free surface zone.

2.3 WAVE-INDUCED LOADS

2.3.1 The loading mechanism

Morison's[1] equation for the in-line horizontal force on a vertical member represents the total force per unit length as the sum of a linear inertial and a non-linear drag component.

$$F = K_i \dot{u} + K_d u |u| \tag{2.33}$$

For tubular members with diameter D:

$$K_i = C_m \rho \pi D^2 / 4$$

$$K_d = C_d \rho D / 2 \tag{2.34}$$

where ρ is the density of sea water and C_m, C_d are the much debated inertia and drag coefficients. It is recognized that C_m and C_d are functions of all the variables which are not adequately represented in Equation (2.33), such as depth of immersion, wave conditions, diameter and surface roughness. Thus the problem of wave loading falls into two parts. Firstly, ways must be found to evaluate the coefficients appropriate to the particular conditions under consideration and only then can the second stage proceed in which the wave loads on the structure are calculated from Equation (2.33). Numerous studies have been undertaken to provide guides to the selection of C_m and C_d. A common approach has been to measure forces on sections of a member and to measure the properties of the incident wave. Some wave theory is then employed to calculate the velocities and accelerations through the wave cycle and this data, together with the observed force time-history, is combined in Equation (2.33) to yield estimates of C_m and C_d. The values may be fitted in a least square error sense throughout the wave cycle. They may be simply derived for the phase angles corresponding to zero velocity or acceleration or they may be chosen to provide a best fit to the force maxima. Aagard and Dean[26] made use of the ability of the stream function theory to fit observed profiles in calculating the fluid motions corresponding to the measured forces and in a further work, Dean[27] discusses criteria for the suitability of wave and wave force data in determining the inertia and drag coefficients. In probabilistic mode Pierson and Holmes[28] and Borgman[29] developed methods for evaluating C_m and C_d based on moments of the observed force distributions or by spectral relationships. A criterial review[30] of the extensive data relating to Morison coefficients has recently been made in which the methods described above are considered.

Some useful conclusions may be drawn from the various studies undertaken. The coefficients are related to parameters of the fluid motion such as Reynold's number and Keulegan-Carpenter number. The scale at which the experiments are performed is important and much of the laboratory data is unsuitable for use in prototype, highly-turbulent conditions. However, there are important exceptions to this, notably the work undertaken by Sarpkaya.[31] Finally, it is apparent that the subsequent use of the derived Morison coefficients should be consistent with the wave theory or method of analysis used in their derivation from force data. For example, values of C_m and C_d resulting from a statistical analysis may represent expected values for the total period of observation. It would be inappropriate to use such values together with a deterministic wave theory, to calculate the load on a member at some particular phase angle.

Before leaving the discussion of Morison's equation it is necessary to note that in the presence of steady currents, it is present practice to use the total wave and current velocity in the non-linear drag term. Furthermore, in calculating the wave motion it must be recognized that there is an interaction between the current and wave field leading to a modification of the wave properties. This problem has been considered by Dalrymple[32] and Dalrymple and Dean[33] in terms of the stream function wave theory. Huang *et al.*[34] discussed the interaction between currents and random waves which can lead to a significant modification of the wave spectrum.

It is common in steady flow problems to express the transverse or lift force in a form similar to that of the in-line drag component of force.

$$F_T = C_L \rho \frac{D}{2} u^2 \qquad (2.35)$$

However, numerous experimental studies in oscillating flow have shown that F_T is irregular and weakly correlated with the in-line flow but may be of the same order of magnitude as the in-line force. One approach to the problem, used in several studies[35,36] is to represent the force in a series form:

$$F_T = \rho \frac{D}{2} u^2_{\max} \sum_{j=1}^{N} C_{L_j} \cos \{ j\omega t + \theta j \} \qquad (2.36)$$

where $u_{\max}$ is the maximum horizontal velocity
        ω the basic wave frequency
        C_{L_j} and θ_j are the lift coefficients and phase angles related to the j-th harmonic force.

Sarpkaya's work[31] indicates that the frequency content of the transverse force may not always be composed of exact harmonics of the fundamental wave frequency. In principle, the lift coefficients may be evaluated from experimental data in much the same way as previously described for the Morison coefficients. Little data is available in this area for prototype conditions and more work is required to develop a complete in-line and transverse loading mechanism for use in analyses of wave loads on offshore structures.

2.3.2   Spectral description of wave loads

Borgman[37] derived an approximate spectral representation of force per unit length based on the Morison equation by expanding the non-linear drag term in series form. The velocity and acceleration are considered as independent Gaussian processes—by virtue of their linear relationship to surface-elevation.

If only the first term of the series is retained, the solution is given:

$$S_F(\omega) = \frac{8K_d^2\sigma_u^2}{\pi} \cdot S_u(\omega) + K_i^2 S_{\ddot{u}}(\omega) \tag{2.37}$$

$S_F(\omega)$ is thus approximated by the sum of two linear transformations of independent inputs $S_u(\omega)$ and $S_{\ddot{u}}(\omega)$ and it therefore follows consistently that the probability density function of force is Gaussian with zero mean (in the absence of currents) and with variance σ_F^2 given by:

$$\sigma_F^2 = \int_0^\infty S_F(\omega)\, \mathrm{d}\omega \tag{2.38}$$

Thus, for a narrow-band spectrum, the probability density of peak force and force range would be closely represented by a Rayleigh density analogous to the development of the statistics of waveheight given earlier.

Retaining further terms in the series representation of the drag component of force involves spectral self-convolutions which must be evaluated numerically. The most efficient method would be to compute the solution in the τ-domain with subsequent Fourier Transformation to the frequency domain. The effect of spectral convolution is to produce energy in the force spectrum at frequencies higher than the input wave frequencies. The energy levels are low and despite the low structural damping coefficients of members of offshore platforms, typically 3–5 per cent of critical, it is unlikely that this excitation causes the significant stress fluctuations experienced in offshore structures at frequencies in the region of 5 to 20 Hz. It may be that the latter result from vortex shedding processes which, as yet, are inadequately understood.

2.3.3   Probabilistic description of wave loads

Wave loads may be described[38] in the probability domain by the standard method of transformation of variables applied to the multi-variate Gaussian density function for fluid motions. Let the wave loads at n positions on a structure be represented by the vector $\{F\} = \{F_1 \cdots F_n\}$ and the vector $\{u\}$ represent the corresponding velocities and accelerations of the fluid $\{u_1, u_2 \cdots u_n, \dot{u}_1$ etc$\}$. The joint pdf. of $\{u\}$ is in the form:

$$p(u_1, u_2 \cdots \dot{u}_n) = \exp\left(-\{u\}^T[\mu]^{-1}\{u\}/2\right)/[(2\pi)^n\sqrt{(\mathrm{Det}\,\mu)}] \tag{2.39}$$

where $[\mu]$ is a matrix of second moments of $\{u\}$. These terms can be found from integration of the auto and cross-spectral density functions such as $S_{u_1u_1}(\omega)$ and $S_{u_1\dot{u}_2}(\omega)$, which may be developed from linear random wave theory, Equations (2.6), (2.7), (2.32) and the Appendix. In mapping between the $\{u\}$ and $\{F\}$ probability spaces it is necessary to introduce the auxiliary variables F_{n+1} to F_{2n}, which may in this case be taken as the velocities variables, i.e., $F_{n+1} = u_1$, etc. The joint pdf. of forces may then be written as:

$$p(F_1, F_2 \cdots F_{2n}) = \frac{p(u_1, u_2 \cdots \dot{u}_n)}{|\text{Det}\,[J]|} \tag{2.40}$$

where $[J]$ is the Jacobian transformation matrix, derived from the differentials of the Morison Equation (2.33) written for each of the n positions. The auxiliary variables may be removed by n-fold integration to give:

$$p(F_1 \cdots F_n) = \left((2\pi)^n \sqrt{(\text{Det}\,\mu)} \prod_{j=1}^{n} K_{ij} \right)^{-1}$$

$$* \underset{n\text{-fold}}{\int_{-\infty}^{\infty} \cdots \int_{-\infty}^{\infty}} \exp\left(-\{u\}^T [\mu]^{-1} \{u\}/2\right) \mathrm{d}u_1 \cdots \mathrm{d}u_n \tag{2.41}$$

where K_{ij} is the coefficient in Equation (2.33) associated with the inertial component of load at the j-th position. For the quasi-static case of wave loading on a structure, Equation (2.40) could be subject to a further transformation, in terms of the stiffness matrix for the structure, to yield the joint density function for stress or displacement response variables. However, this approach is extremely lengthy by virtue of the n-fold integrations involved in evaluating expressions such a those in Equation (2.41). The numerical computations become too lengthy for more than a few degrees of freedom, despite the use of special quadrature procedures. The value of this approach is that no linearization of the drag term is used and this can be vital to the statistical description of wave loads on small diameter members, for which the drag component is significant.

It can be shown[28] that the univariate probability density of force (for zero current) can be expressed in the form:

$$p(F) = (2\pi\sigma_{\psi_1}\sigma_{\psi_2})^{-1} \int_{-\infty}^{\infty} \exp\left[-\frac{1}{2}\left(\frac{\psi_1^2}{\sigma_{\psi_1}^2} + \frac{\psi_2^2}{\sigma_{\psi_2}^2} \right) \right] \mathrm{d}\psi_1 \tag{2.42}$$

where $\psi_1 = u\sqrt{(K_d)}$ and $\psi_2 = \dot{u}K_i$. The parameters σ_{ψ_1} and σ_{ψ_2} may be found in terms of the second and fourth order moments of the force, i.e. $E(F^2)$ and $E(F^4)$. Analyses of prototype strain records for a North Sea platform,[39] and from laboratory experiments,[38] indicate that the probability distribution for structural response variables will follow a form similar to Equation (2.42). An example is given in Figure 2.5 of the comparison between the observed strain distribution and the theoretical distribution, fitted by the observed second and fourth moments of response.

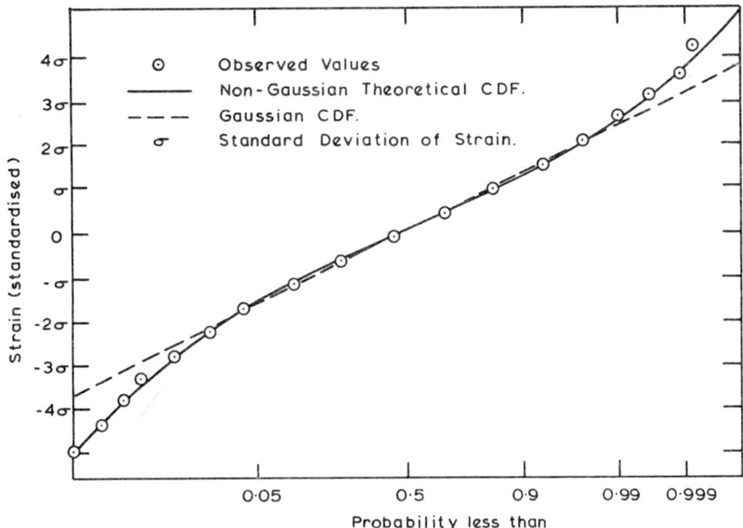

Figure 2.5   A comparison of observed and theoretical strain distributions

2.3.4   Peak forces and force ranges

In designing for the extreme load and for fatigue it is necessary to derive expressions for the density functions of peak force and force range. Several forms of solution are available depending on the linearization of the drag term, or its retention, and assumptions regarding the band-width of the force spectrum.

For the linearized solution the basic force variate is Gaussian and, for narrow-band spectra, the peak forces and force ranges have a Rayleigh distribution. This is analogous to the probabilistic properties of water-surface-elevation, its maxima and waveheights.

Retention of the non-linear drag term results in a non-Gaussian distribution of force and the corresponding solutions for maxima and force ranges are more complex.

Consideration of the joint statistics of F, $\dot{F}$ and $\ddot{F}$ leads to a general wide-band distribution of peak force,[38,40] but some useful simplifications are possible if a narrow-band force spectrum is assumed such that the occurrence of a positive maximum is implied by an upcrossing of a given level of force. The first form of this distribution may be written as:

$$P_{\text{nbl}}(F) = 1 \cdot 0 - \int_0^\infty \dot{F} p(F, \dot{F}) \, \mathrm{d}\dot{F} \Big/ \int_0^\infty \dot{F} p(F = 0, \dot{F}) \, \mathrm{d}\dot{F} \qquad (2.43)$$

and a further simplification can be achieved if the assumption of independence between F and $\dot{F}$ is reasonable.

$$P_{nb2}(F) = 1\cdot0 - p(F)/p(F=0) \qquad (2.44)$$

where subscript nb denotes narrow-band.

Although several important assumptions have been made to reach the peak distribution $P_{nb2}(F)$, given in Equation (2.44), the value of this expression is appreciated when it is recalled that the basic pdf. of force, $p(F)$, may be defined in terms of a few simple statistical moments. The three models for peak force distribution are compared with Figure 2.6 with the Rayleigh distribution resulting from the linearized approach. A measure of the relative importance of inertial and drag components is given by the kurtosis, the ratio between the fourth and the square of the second moment of force (i.e. $E(F^4)/E^2(F^2)$). The example shown is broadly representative of the situation for loads on a bracing member of a jacket platform. Calculations for stress range required for fatigue design show a similar trend. The effect of the non-linearity is clearly indicated but it remains to be seen whether this factor is significant in terms of the fatigue problem. However, in the calculation of extreme loads, it is certainly important that the non-linearity be properly considered.[41]

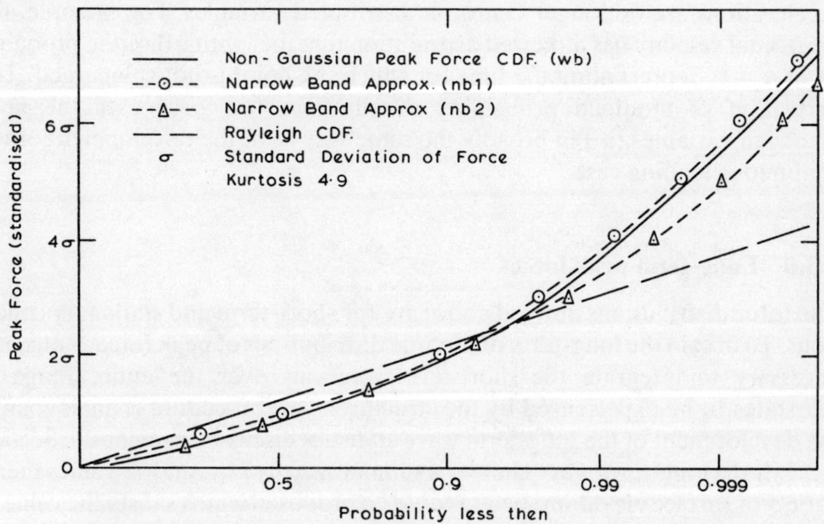

Figure 2.6 A comparison of theoretical distributions of peak force

The basic and peak probability density functions for force can be modified,[42] to include the presence of steady currents by setting the mean of the velocity variables in $\{u\}$ to the values of the current profile, and the elements in matrix

$[\mu]$ are calculated from the auto and cross-spectral density functions derived from the modified surface elevation spectrum $S_\eta(\omega)$. It must be recognized that the above modifications represent only a first order attempt at dealing with the interaction between waves and currents by direct superposition of velocity fields.

2.3.5 Intermittent loading

Further modifications must be made to the probabilistic model of wave loads when members lie within the free surface zone. The fluid motion at some particular level y relative to still-water-level represents a discontinuous time history as the point lies above and below the free surface during the passage of the waves. One approach to this problem[42] is to define modified velocity and accelerations $(u', v', \dot{u}', \dot{v}')$ in the following manner.

$$u' = \xi[\eta - y]u$$
$$\dot{u}' = \xi[\eta - y]\dot{u}, \text{ etc.}$$

(2.45)

where $\xi[x] = 1$ for $x \geq 0$ and $\xi[x] = 0$ for $x < 0$. The new set of velocities and accelerations are no longer Gaussian distributed variables. For example, the horizontal velocity has a skewed distribution together with a discrete probability at $u' = 0$, representing the time for which the point is not submerged. The derivation of modified probability distributions for force uses the non-Gaussian variables $\{u'\}$ in broadly the same way as in the development of the continuous loading case.

2.3.6 Long-term peak forces

The force distributions derived above are for short-term and stationary conditions. To obtain the long-term or lifetime distributions of peak force,[43] etc., it is necessary to integrate the short-term solutions over the entire range of sea-states to be experienced by the structure. This procedure is analogous to the development of the long-term wave statistics discussed previously, Section 2.2.2. If the long-term wave climate is summarized by $H_{1/3}$ and T_z values then a variety of surface elevation spectra could be proposed which satisfy the value of $H_{1/3}$ or T_z, or both. It is only necessary to consider the few most severe sea-states in estimating the extreme load statistics. Fatigue analysis could be based on either the long-term stress-range history or on computing the short-term cumulative damage for a number of sea-states and then convoluting the expected damage for each sea-state by the proportion of time for which that condition exists.

2.4 DISCUSSION AND CONCLUSIONS

Storm waves are a complex three-dimensional random process. However, significant progress has been made in modelling the wave field by deterministic and by probabilistic methods. The numerical forms of deterministic wave theories are very flexible and can be extended to a high order although verification of their validity at full scale is required. By comparison, random wave theory has only been developed to the level of a linear solution, based on a first order spectrum. This may be sufficient for engineering design purposes, but further investigations are warranted into wave-current interaction and the problem of flow near the free surface.

Despite many criticisms, Morison's equation remains the most appropriate loading mechanism for the calculation of in-line wave forces for frame type structures. Suitable values for the Morison coefficients can be obtained from experiments provided that attention is paid to the dependence of C_m and C_d upon the characteristic flow parameters. It would appear that further work is warranted on the combination of in-line and transverse loading mechanisms for both rigid and flexible members.

Linearized spectral analysis is appropriate to the situations in which wave loading is dominated by the linear inertial component. When the drag component of loading is significant, it is important that the non-linearity is adequately represented. The short-term non-Gaussian models which result from the above, have shown reasonable agreement with measured data. Work is now in progress to assess the importance of the non-linearity to the prediction of fatigue-life and first excursion failure.

Extensions of these solutions are possible in several respects. More realistic descriptions of the long-term wave environment should be developed, including the directionality of the incident wave field, and spectra appropriate to partially-developed sea-states. The solutions should now be extended to consider the effect of dynamic response of the structure and, in particular, the importance of the non-linearity in the hydrodynamic damping requires both experimental and theoretical investigations.

REFERENCES

1. Morison, J. R., O'Brien, M. P., Johnson, J. W., and Schaaf, S. A. (1950). 'The forces exerted by surface waves on piles', *Pet. Trans.*, **189**, TP2846.
2. Garrison, C. J. and Chow, P. Y. (1972). 'Wave forces on submerged bodies', J. Waterways, Harbors & Coastal Eng. Div., *ASCE*, **98**, WW3, August 1972.
3. Phillips, O. M. (1957). 'On the generation of waves by turbulent winds', *J. Fluid Mech.*, **2**.
4. Miles, J. W. (1957–62). 'On the generation of surface waves by shear flows', *J. Fluid Mech.*

5. Longuet-Higgins, M. S. 'A non-linear mechanism for the generation of sea waves', *Proc. Royal Soc*, Series A, 311.
6. Marshall, P. W. (1976). 'Dynamic and fatigue analysis using directional spectra', Paper No. 2537, p. 143, *Offshore Technology Conference*, Houston, Texas.
7. Dean, R. G. (1974). 'Evaluation and development of water wave theories for engineering application'. *Special Report No. 1*, Vol. 1, U.S. Army Coastal Eng. Res. Centre, November 1974.
8. Chan, R. K. and Street, R. L. (1970). 'A computer study of finite amplitude water waves', *J. Computational Physics*, **6**.
9. Wiegel, R. L. (1963). *Oceanographical Engineering*, Prentice-Hall.
10. Chaplin, J. R. (1976). 'Mathematical models for gravity waves', *Mathematical and Physical Modelling*, Universities of Liverpool and Manchester, April 1976.
11. Dean, R. G. (1970). 'Relative validities of water wave theories', *J. Waterways, Harbors & Coastal Eng. Div. ASCE*, WW1.
12. Le Méhauté, B. (1969). *Shore Protection Manual*, U.S. Army Coastal Eng. Res. Centre, 1973.
13. Silvester, R. (1974). *Coastal Engineering*, Vol. 1, Elsevier.
14. Chaplin, J. R. (1977). 'Surface particle orbits in uniform waves', *Waterways Harbors & Coastal Eng. Div. ASCE*, May 1977.
15. Longuet-Higgins, M. S. (1952). 'On the statistical distribution of the heights of sea waves', *J. Mar. Res.*, **XI**.
16. Longuet-Higgins, M. S. (1975). 'On the joint distribution of the periods and amplitudes of sea waves', *J. Geophys. Res.*, **80**, June 1975.
17. Cavanié, Arham, M., and Ezraty, R. (1976). 'A statistical relationship between individual heights and periods of storm waves', *Int. Conf. on Behaviour of Offshore Structures, BOSS*, Trondheim.
18. Nolte, K. (1973). 'Statistical methods for determining extreme sea states', *2nd Conf. Port & Ocean Eng. Under Arctic Conditions*, Univ. of Iceland.
19. Pierson, W. J. and Moskowitz, L. (1964). 'A proposed spectral form for fully developed wind seas based on the similarity theory of S. A. Kitaigordskii', *J. Geophys. Res.*, **69**, 5181.
20. Hasselman, K. *et al.* (1973). *Measurements of Wind-wave Growth and Swell Decay During the Joint North Sea Wave Project (JONSWAP)*, Deutches Hydrographisches Institut, Hamburg.
21. Newland, D. E. (1975). *An Introduction to Random Vibrations and Spectral Analysis*, Longman.
22. Bendat, J. S. and Piersol, A. G. (1971). *Random Data: Measurement and Analysis Procedures*, 2nd Edn, Wiley, New York.
23. Bingham, C., Godfrey, M. D., and Tukey, J. W. (1967). 'Modern techniques of power spectrum estimation', *IEEE Trans. Audio and Electroacoustics*, Vol. AU-15.
24. Borgman, L. E. (1969). 'Ocean wave simulation for engineering design', *J. Waterways Harbors & Coastal Eng. Div. ASCE*, **95**, WW4, November 1969.
25. Shinozuka, M. and Jan, C. M. (1972). 'Digital simulation of random processes and its applications', *J. Sound & Vibn.*, **25**.
26. Aagard, R. and Dean, R. G. (1974). 'Wave forces: data analysis and engineering calculation method', *OTC, 1969, Paper 1008*—see also *OTC, 1974, Paper 2073*.
27. Dean, R. G. (1976). 'Methodology for evaluating suitability of wave and wave force data for determining drag and inertia coefficients', *BOSS*, **1**, Trondheim.

28. Pierson, W. J. and Holmes, P. (1965). 'Irregular wave forces on a pile', *J. Waterways, Harbors & Coastal Eng. Div. ASCE*, **91**.
29. Borgmann, L. E. (1972). 'Statistical models for ocean waves and wave forces', *Recent Advances in Hydroscience*, Vol. 8, Academic Press.
30. *A Critical Evaluation of the Data on Wave Force Coefficients*, British Ship Research Association, August 1976.
31. Sarpkaya, T. (1976). 'Vortex shedding and resistance in harmonic flow about smooth and rough circular cylinders', *BOSS*, **1**, Trondheim.
32. Dalrymple, R. A. (1973). 'Water-wave models and wave forces with shear currents', *Coastal & Offshore Eng. Lab., Univ. of Florida*, August 1973.
33. Dalrymple, R. A. and Dean, R. G. (1975). 'Waves of maximum height on uniform currents', *J. Waterways, Harbors & Coastal Eng. Div. ASCE*, **101**, August 1975.
34. Huang, N. E. *et al.* (1972). 'Interaction between steady non-uniform currents and gravity waves with applications for current measurement', *J. Physical Oceanography*, **2**.
35. Isaacson, M. St. J. and Maull, D. (1976). 'Transverse forces on vertical cylinders in waves', *J. Waterways, Harbors & Coastal Eng. Div. ASCE*, **102**, February 1976.
36. Chakrabarti, S. K., Wolbert, A. L., and Tam, W. A. (1976). 'Wave forces on vertical circular cylinders', *J. Waterways, Harbors & Coastal Eng. Div. ASCE*, **102**, May 1976.
37. Borgman, L. E. (1967). 'Spectral analysis of ocean wave forces on piling', *J. Waterways, Harbors & Coastal Eng. Div. ASCE*, **93**.
38. Tickell, R. G. (1977). 'Continuous random wave loading on structural members', *Structural Engineer*, May 1977.
39. Holmes, P. *Analysis of Wave Forces on a North Sea Structure*, unpublished.
40. Tung, C. C. (1974). 'Peak distribution of random wave-current force', *J. Eng. Mech. Div. ASCE*, **100**.
41. Tickell, R. G. (1976). Discussion on 'Prediction of extreme wave-induced loads on ocean structures', *Ochi & Wang, BOSS*, **2**, Trondheim.
42. Pajouhi, K. and Tung, C. C. (1975). 'Statistics of random wave field', *J. Waterways, Harbors & Coastal Eng. Div. ASCE*, November 1975.
43. Tickell, R. G., Burrows, R., and Holmes, P. (1976). 'Long-term wave loading on offshore structures', *Proc. ICE*, **61**.

Appendix

PROBABILITY THEORY AND RANDOM PROCESSES

Most environmental loads encountered in offshore engineering are random in nature, which implies random structural responses in terms of stress or deformation. Deterministic design procedures have been developed in the past to ease the designers' problems when dealing with waves, winds or earthquakes. However, there is now a justifiable increase in the use of probabilistic techniques or analyses. This Appendix briefly reviews the fundamental concepts involved in such analyses in five sections:

1. The Random Variable
2. The Random Process
3. The Spectrum
4. The Response of Linear Systems to Random Excitation
5. The Statistics of Maxima and Extreme Values

The general references given at the end of this Appendix should be consulted for further or more detailed development of the concepts.

1 THE RANDOM VARIABLE

Probability theory is the mathematics of uncertainty and it is concerned with the possible outcomes or results of some situation in which the actual outcome cannot be predetermined. If $\{R_1, R_2 \cdots R_n\}$ represents the complete set of possible results, then the probability of R_j (Prob $\{R_j\}$) can be defined as the chance that R_j will be the result on some particular occasion. Probability is a numerical, non-negative quantity and by convention, it lies in the range zero to one, which represent respectively the limits of impossibility and certainty.

A random variable is a variable which can take all the numerical values of the possible outcomes in a situation and hence levels of probability can be associated with individual values of a random variable. A random variable X can be described by its cumulative distribution function, $P(X)$:

$$P(x) = \text{Prob}\,\{X \le x\} \tag{2A.1}$$

where, by definition:

$$P(x = -\infty) = \text{Prob}\,\{X \le -\infty\} = 0\cdot0$$

$$P(x = +\infty) = \text{Prob}\,\{X \le +\infty\} = 1\cdot0 \tag{2A.2}$$

and

$$\text{Prob}\{x_1 < X \le x_2\} = P(x_2) - P(x_1) \tag{2A.3}$$

A typical cumulative distribution function (cdf) is shown in Figure 2A.1(a). The c.d.f may be used to describe random variables which can take either discrete or continuously variable numerical values. However, a difficulty is encountered with continuous random variables in that Prob $\{X = x\}$ is generally zero since there are an infinite number of possible values of X, when taken to an infinite number of significant figures. The concept of the probability that X lies within

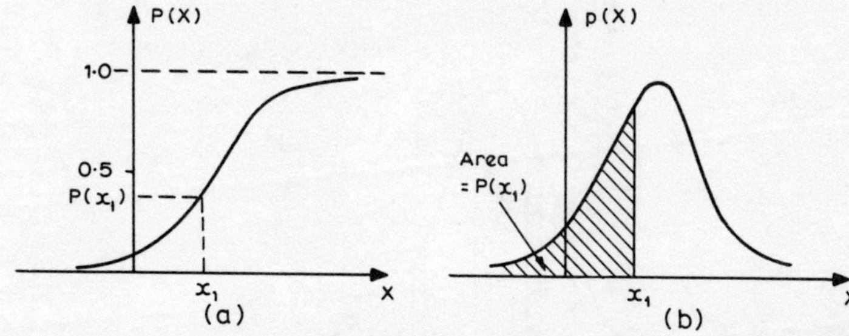

Figure 2A.1 Cumulative distribution function and probability density function of X

a narrow range of x is introduced to avoid the difficulty and this leads to a definition of the probability density function, $p(X)$:

$$\text{Prob}\{x < X \le x + \Delta x\} = P(x + \Delta x) - P(x)$$
$$= \Delta P(x) \tag{2A.4}$$

If $P(x)$ is differentiable then:

$$p(x) = \lim_{\Delta x \to 0} \frac{\Delta P(x)}{\Delta x} = \frac{dP(x)}{dx} \tag{2A.5}$$

and it follows that:

$$\int_{-\infty}^{x_1} p(x)\,dx = P(x_1) - P(-\infty) = P(x_1)$$

$$\int_{-\infty}^{+\infty} p(x)\,dx = P(\infty) - P(-\infty) = 1\cdot0 \tag{2A.6}$$

A typical probability density function (p.d.f) is shown in Figure 2A.1(b).

If Y is a known function, $f(X)$, of X then Y is also a random variable and the mean, $\bar{Y}$, or expected value, $E\{Y\}$, of Y may be defined in terms of $f(X)$ and the p.d.f of X.

$$E\{Y\} = \bar{Y} = \int_{-\infty}^{\infty} f(x)p(x)\,\mathrm{d}x \qquad (2\text{A}.7)$$

A particularly useful set of functions $f(X)$ is given by the moments of X, $(m_j, j = 0, 1, 2 \cdots n)$, which are the moments of the curve $p(x)$ taken about the vertical axis through the origin $x = 0$, see Figure 2A.2(a). Thus:

$$m_j = E\{X^j\} = \overline{X^j} = \int_{-\infty}^{\infty} x^j p(x)\,\mathrm{d}x \qquad (2\text{A}.8)$$

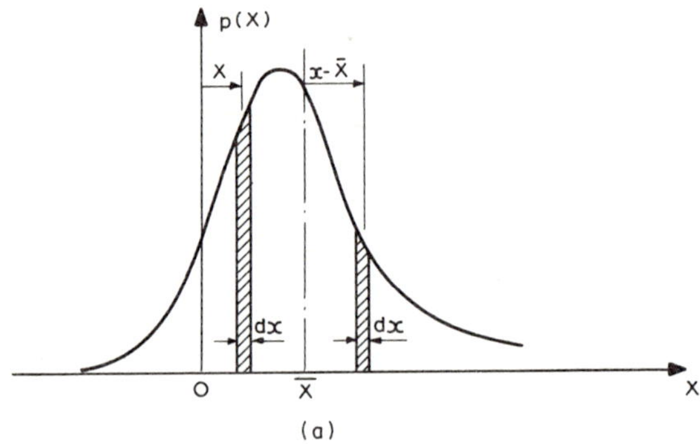

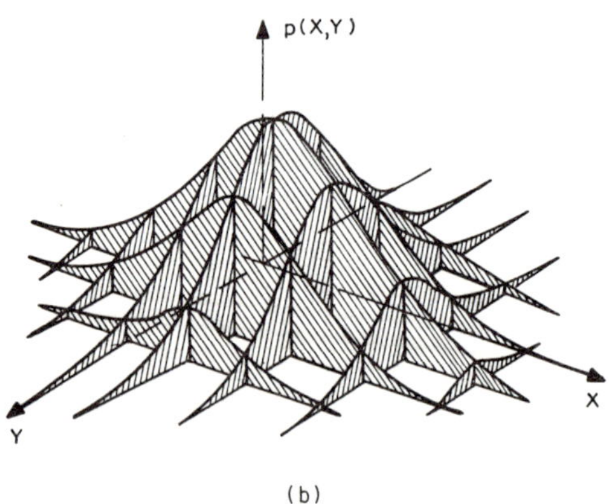

Figure 2A.2 (a) Moments of a probability density function of X. (b) Joint probability density function

In particular,

$$m_0 = 1{\cdot}0 \tag{2A.9}$$

the mean, average or expected value of X is:

$$m_1 = E\{X\} = \bar{X} = \int_{-\infty}^{\infty} xp(x)\,\mathrm{d}x \tag{2A.10}$$

and the second moment or mean square value of X is:

$$m_2 = E\{X^2\} = \overline{X^2} = \int_{-\infty}^{\infty} x^2 p(x)\,\mathrm{d}x \tag{2A.11}$$

Moments may be taken about any axis. For example, the variance of X, σ_X^2, is the second moment of X about an axis through the mean value of X, and σ_X is the standard deviation of X.

$$\sigma_X^2 = E\{(X - \bar{X})^2\} = \int_{-\infty}^{\infty} (x - \bar{X})^2 p(x)\,\mathrm{d}x$$
$$= m_2 - m_1^2 \tag{2A.12}$$

The following familiar parameters are used to describe a probability distribution.

$$\text{Mean} = m_1$$
$$\text{Standard Deviation} = \sigma_X = (\mu_2)^{1/2}$$
$$\text{Skewness} = \mu_3/\sigma_X^3$$
$$\text{Kurtosis} = \mu_4/\sigma_X^4$$
$$\text{Mode} = \text{value of } X \text{ for } p(x) = \text{maximum}$$
$$\text{Median} = \text{value of } X \text{ for } P(X) = 0{\cdot}5$$

where μ_n is the nth moment about the mean. These parameters are often used in fitting hypothesized density functions to data by the method of moments.

The probability measures developed above may be extended in a straightforward manner to deal jointly with several random variables. For example, the joint p.d.f of X and Y is $p(X, Y)$, as illustrated in Figure 2A.2(b).

$$p(x, y) = \lim_{\substack{\Delta x \to 0 \\ \Delta y \to 0}} \frac{\text{Prob}\,\{x < X \le x + \Delta x, \, y < Y \le y + \Delta y\}}{\Delta x\,\Delta y} \tag{2A.13}$$
$$= \frac{d^2 P(x, y)}{\mathrm{d}x\,\mathrm{d}y}$$

$$\int\int_{-\infty}^{\infty} p(x, y)\,\mathrm{d}x\,\mathrm{d}y = 1{\cdot}0 \tag{2A.14}$$

The individual or marginal p.d.f of X or Y may be obtained from the joint p.d.f, but the converse does not hold in general:

$$p(x) = \int_{-\infty}^{\infty} p(x, y)\, dy \qquad (2A.15)$$

The conditional probability density function $p(x|y)$ describes the probability for X given that Y equals some value y, thus:

$$p(x|y) = \lim_{\Delta x \to 0} \frac{\text{Prob}\,\{x < X \le x + \Delta x \,|\, Y = y\}}{\Delta x} \qquad (2A.16)$$

An important relationship exists between the joint p.d.f and conditional p.d.f.

$$p(x, y) = p(x|y)p(y) \qquad (2A.17)$$

Note that if X is statistically independent of Y then $p(x|y) \equiv p(x)$ and in this special case

$$p(x, y) = p(x)p(y) \qquad (2A.18)$$

Multi-variate moments are defined in the same manner as for uni-variate or single random variables

$$E\{f(X, Y)\} = \int\!\!\int_{-\infty}^{\infty} f(x, y)p(x, y)\, dx\, dy \qquad (2A.19)$$

For example:

$$E\{XY\} = \overline{XY} = \int\!\!\int_{-\infty}^{\infty} xyp(x, y)\, dx\, dy \qquad (2A.20)$$

The joint p.d.f of two dependent random variables (V, W) may be derived from the joint p.d.f of X, Y and the relationship between the variables.

$$V = f(X, Y), \qquad W = g(X, Y) \qquad (2A.21)$$

If the inverse relationships have the roots x_i, y_i where $v = f(x_i, y_i)$ and $w = g(x_i, y_i)$ then:

$$p(v, w) = \sum_i \frac{p(x_i, y_i)}{|\text{Det}\, J(x_i, y_i)|} \qquad (2A.22)$$

Where the modulus $|\quad|$ ensures that $p(V, W)$ is never negative and Det $J(\quad)$ is the determinant of the Jacobian transformation matrix, defined as:

$$J(x, y) = \begin{vmatrix} \dfrac{\partial f}{\partial x} & \dfrac{\partial f}{\partial y} \\[2mm] \dfrac{\partial g}{\partial x} & \dfrac{\partial g}{\partial y} \end{vmatrix} \qquad (2A.23)$$

The simplest example of transformation of variables is that of Y being a function of X with a complete one-to-one correspondence.

$$Y = f(X), \quad \text{then } p(y) = p(x) / \left| \frac{df}{dx} \right| \qquad (2A.24)$$

Dummy or auxiliary variables may be introduced if the number of independent variables exceeds the number of dependent variables in a transformation. The dummy variables may be removed by integration after making the transformation.

$$W = f(X, Y) \qquad (2A.25)$$

Introducing V as the dummy variable such that $V = X$

$$p(w) = \int_{-\infty}^{\infty} p(v, w) \, dv = \int_{-\infty}^{\infty} \frac{p(x, y)}{|\text{Det } J(x, y)|} \, dx \qquad (2A.26)$$

2 THE RANDOM PROCESS

The concept of a random process is a development from that of a random variable in which consideration is given to the variation of the random variable X with time, $X(t)$, or space, $X(s)$, or both, $X(s, t)$. A simple example is the variation of the sea surface elevation in both time and space.

If a set of observations are taken, the values obtained are no longer random but are simply whatever they have been measured to be. However, considering the value of $X(t)$ at some time $t = t_1$, but prior to measurement, then $X(t_1)$ is a random variable which could take any one of an infinite number of values within the permissible range of X, with a probability specified by the p.d.f $p(x : t_1)$. Note that $p(x : t_1)$ is not a joint p.d.f of X and t_1 but rather the p.d.f of X at time t_1. The same argument can be advanced for $X(t_2)$, $X(t_3) \cdots$ and each combination of values could be regarded as a permissible realization of the time series $X(t)$. The infinite collection of these realizatons constitutes the ensemble of the random process, represented in Figure 2A.3.

The expected value of $X(t)$ at $t = t_1$ is defined as the average value of $X(t_1)$ taken across the whole ensemble and as such it is a function of t.

$$E\{X(t_1)\} = \int_{-\infty}^{\infty} x p(x : t_1) \, dx \qquad (2A.27)$$

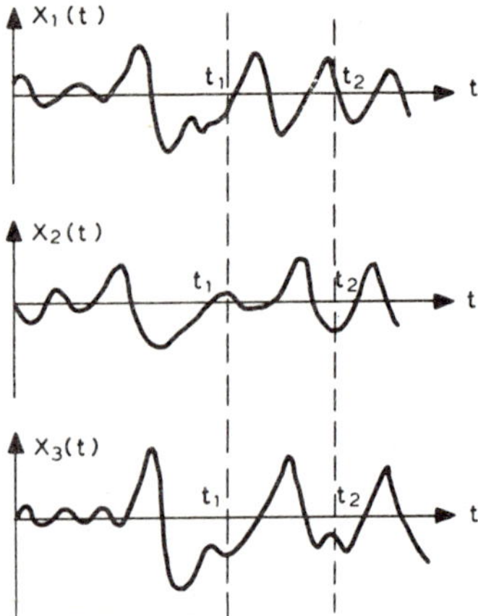

Figure 2A.3 Part of the infinite ensemble of $X(t)$

Second and higher moments may similarly be defined:

$$E\{X(t_1)X(t_2)\} = R_{xx}(t_1, t_2)$$

$$= \int\int_{-\infty}^{\infty} x_1 x_2 p(x_1: t_1, x_2: t_2)\, dx_1\, dx_2$$

(2A.28)

$R_{xx}(t_1, t_2)$ is the auto-correlation function which represents the expected value of the product of $X(t_1)$ and $X(t_2)$ taken on each realization in turn. The second order p.d.f, $p(x_1: t_1, x_2: t_2)$, gives the joint probability of $X(t_1)$ in the range $x_1 \pm \Delta x/2$ and $X(t_2)$ in the range $x_2 \pm \Delta x/2$.

The random process is termed stationary, in a statistical sense, if the various probability measures of the process (e.g. moments and p.d.f's) do not vary with absolute time and therefore are independent of the origin of time measurement. The property of stationarity permits simplifactions in describing the random process:

$$E\{X(t_1)\} = E\{X(t_2)\} = \bar{X}$$

(2A.29)

and the auto-correlation function is solely dependent on the difference in time or lag $\tau = t_2 - t_1$.

$$R_{xx}(\tau) = E\{X(t)X(t+\tau)\}$$

(2A.30)

The sea surface may well be represented adequately by a stationary random process for some short period of time (e.g. 30 minutes). But the sea is

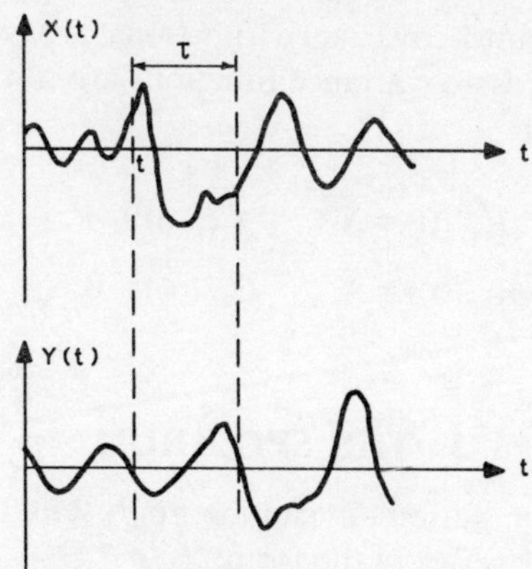

Figure 2A.4   Ergodic time series of
$X(t)$ and $Y(t)$

non-stationary as storms develop or decay in intensity over a time scale of
hours or days.

Only one realization of the ensemble can be obtained in practice, preventing
the calculation of moments across the ensemble. However, if the realization is
considered to be typical of all the ensemble then estimates of moments or other
probability measures can be made by averaging with time along that realization
rather across the ensemble, Figure 2A.4. Under these circumstances the
stationary random process is termed ergodic. The moments or temporal
averages for an ergodic process are:

$$E\{X(t)\} = \bar{X} = \lim_{T \to \infty} \frac{1}{T} \int_{-T/2}^{T/2} X(t)\,dt \tag{2A.31}$$

$$R_{xx}(\tau) = \lim_{T \to \infty} \frac{1}{T} \int_{-T/2}^{T/2} X(t)X(t+\tau)\,dt \tag{2A.32}$$

Auto-correlation functions may be generalized to deal with two variables $X(t)$
and $Y(t)$ giving the cross-correlation function, $R_{xy}(\tau)$.

$$R_{xy}(\tau) = \lim_{T \to \infty} \frac{1}{T} \int_{-T/2}^{T/2} X(t)Y(t+\tau)\,dt \tag{2A.33}$$

Auto-covariance and cross-covariance functions are derived from the basic
time series by subtraction of the respective means:

$$C_{xx}(\tau) = \lim_{T \to \infty} \frac{1}{T} \int_{-T/2}^{T/2} (X(t)-\bar{X})(X(t+\tau)-\bar{X})\,dt \tag{2A.34}$$

$$C_{xy}(\tau) = \lim_{T \to \infty} \frac{1}{T} \int_{-T/2}^{T/2} (X(t)-\bar{X})(Y(t+\tau)-\bar{Y})\,dt \tag{2A.35}$$

Auto-correlation and auto-covariance functions are symmetrical in τ but cross-correlation and cross-covariance functions are not.

Inspection shows that:

$$R_{xx}(0) = \overline{X^2} \qquad C_{xx}(0) = \sigma_X^2$$
$$R_{xx}(\infty) = \bar{X}^2 \qquad C_{xx}(\infty) = 0 \tag{2A.36}$$

3   THE SPECTRUM

A function $f(t)$ with a period T can be represented, subject to certain conditions, by a Fourier series in the form:

$$f(t) = \sum_{n=-\infty}^{\infty} F(n\,\Delta\omega)\exp(in\,\Delta\omega t) \tag{2A.37}$$

where the Fourier coefficients are:

$$F(n\,\Delta\omega) = \frac{1}{T}\int_{-T/2}^{T/2} f(t)\exp(-in\,\Delta\omega t)\,dt \tag{2A.38}$$

with $\Delta\omega = 2\pi/T$ and $\exp(i\theta) = \cos\theta + i\sin\theta$. Combine Equations (2A.37) and (2A.38):

$$f(t) = \sum_{n=-\infty}^{\infty}\left\{\frac{1}{2\pi}\int_{-T/2}^{T/2} f(t)\exp(-in\,\Delta\omega\,t)\,dt\right\}\exp(in\,\Delta\omega t)\,\Delta\omega \tag{2A.39}$$

A non-periodic function may be represented by Equation (2A.39) above by considering an infinite period of repetition $(T \to \infty)$. In the limit, the summation becomes an integral, $\Delta\omega \to d\omega$ and $n\,\Delta\omega$ becomes the continuous function ω, thus:

$$f(t) = \int_{-\infty}^{\infty}\left\{\frac{1}{2\pi}\int_{-\infty}^{\infty} f(t)\exp(-i\omega t)\,dt\right\}\exp(i\omega t)\,d\omega \tag{2A.40}$$

or

$$f(t) = \int_{-\infty}^{\infty} F(\omega)\exp(i\omega t)\,d\omega \tag{2A.41}$$

$$F(\omega) = \frac{1}{2\pi}\int_{-\infty}^{\infty} f(t)\exp(-i\omega t)\,dt \tag{2A.42}$$

Equations (2A.41) and (2A.42) constitute a Fourier transform pair which permits problems formulated in the time domain to be solved in the frequency domain and vice-versa. A rigorous derivation of the relationships above requires that:

$$\text{(a)} \qquad \int_{-\infty}^{\infty} |f(t)| \, dt;$$

and

$$\text{(b)} \qquad \int_{-\infty}^{\infty} (f(t))^2 \, dt$$

(2A.43)

exist and are finite.

A stationary random process, $X(t)$, violates the first of the above conditions and the integrals must, in practice, have finite limits by virtue of the finite sample sizes available for analyses. However, the introduction of an additional function $u(t)$ permits the Fourier transform of a truncated process to be taken:

$$f(t) = u(t)X(t) \tag{2A.44}$$

where

$$u(t) = 1 \cdot 0 \quad \text{for } -T/2 < t < T/2$$

$$= 0 \cdot 0 \quad \text{elsewhere}$$

and T is the duration of the sample.

The effect on the Fourier transform of introducing the data sampling window $u(t)$ is discussed in texts on spectral analysis.

Consider the auto-correlation function for $f(t)$.

$$R_f(\tau) = E\{f(t)f(t+\tau)\}$$

$$= \frac{1}{T} \int_{-\infty}^{\infty} f(t)f(t+\tau) \, dt$$

$$= \frac{1}{T} \int_{-\infty}^{\infty} f(t)\left\{ \int_{-\infty}^{\infty} F(\omega) \exp(i\omega t) \, d\omega \, \exp(i\omega\tau) \right\} dt \tag{2A.45}$$

$$= \frac{1}{T} \int_{-\infty}^{\infty} F(\omega) \int_{-\infty}^{\infty} f(t) \exp(i\omega t) \, dt \, \exp(i\omega\tau) \, d\omega$$

$$= \frac{2\pi}{T} \int_{-\infty}^{\infty} F(\omega)F^*(\omega) \exp(i\omega\tau) \, d\omega$$

where $F^*(\omega)$ denotes the complex conjugate of $F(\omega)$.

A new function is introduced in Equation (2A.45):

$$S_f'(\omega) = \frac{2\pi}{T} F(\omega)F^*(\omega) \tag{2A.46}$$

and so

$$R_f(\tau) = \int_{-\infty}^{\infty} S_f'(\omega) \exp(i\omega\tau)\, d\omega \qquad (2A.47)$$

where $S_f'(\omega)$ is the spectral density function of $f(t)$. It follows that the other transform in the pair is:

$$S_f'(\omega) = \frac{1}{2\pi} \int_{-\infty}^{\infty} R_f(\tau) \exp(-i\omega\tau)\, d\tau \qquad (2A.48)$$

The spectral density function of $X(t)$ is, in the limit:

$$S_X'(\omega) = \lim_{T\to\infty} \frac{2\pi}{T} F(\omega)F^*(\omega) \qquad (2A.49)$$

and then

$$S_X'(\omega) = \frac{1}{2\pi} \int_{-\infty}^{\infty} R_X(\tau) \exp(-i\omega\tau)\, d\tau \qquad (2A.50)$$

$$R_X(\tau) = \int_{-\infty}^{\infty} S_X'(\omega) \exp(i\omega\tau)\, d\omega \qquad (2A.51)$$

The auto-spectral density function $S_X'(\omega)$ is symmetrical in ω since the auto-correlation function is real and symmetrical in τ.

It will often be more convenient to deal with a one-sided spectrum $S_X(\omega)$ rather than the two-sided form $S_X'(\omega)$.

$$\begin{aligned} S_X(\omega) &= 2S_X'(\omega) \qquad & \omega \geq 0 \\ &= 0 \qquad & \omega < 0 \end{aligned} \qquad (2A.52)$$

Now:

$$R_X(\tau) = \int_{0}^{\infty} S_X(\omega) \exp(i\omega\tau)\, d\omega \qquad (2A.53)$$

if $\tau = 0$ in Equation (2A.53).

$$E\{X^2\} = R_X(0) = \int_{0}^{\infty} S_X(\omega)\, d\omega \qquad (2A.54)$$

and therefore $S_X(\omega)$ may be interpreted as describing the way in which the mean square of the process $E\{X^2\}$ is distributed in the frequency domain. In particular, within the limits of linear random wave theory, wave energy is proportional to mean square surface elevation. Therefore, the surface elevation spectral density function describes the distribution of wave energy between the frequency components present within the random sea.

The components contributing to the total mean square value of the process may be distributed over a wide range of frequencies (wide-band process) or concentrated about some particular frequency (narrow-band process). Typical wide-band and narrow-band processes are shown in Figure 2A.5 together with the limiting case of a narrow-band process, namely a sinusoid which must by definition have an infinite value of spectral density at its own frequency, Figure 2A.5(c). The ith order moment of the spectral density function (m_i') may be written as:

$$m_i' = \int_0^\infty \omega^i S_X(\omega)\, \mathrm{d}\omega \qquad (2A.55)$$

The spectral moments must not be confused with the moments or expectations of the probability density function, Equation (2A.8).

For the particular case of a mean-zero Gaussian process, the moments define certain useful parameters of the process and its spectrum.

$$E\{T_z\} = 2\pi\{m_0'/m_2'\}^{1/2} \qquad (2A.56)$$

$$E\{T_c\} = 2\pi\{m_2'/m_4'\}^{1/2} \qquad (2A.57)$$

$$\varepsilon = \left\{ 1 - \frac{(m_2')^2}{m_0' m_4'} \right\}^{1/2} \qquad (2A.58)$$

T_z and T_c are the zero up-crossing and crest periods respectively and ε is the spectral bandwidth. For a narrow-band process ε will tend to zero and for a wide-band process ε will approach unity.

A Fourier transform may be taken of the cross-correlation function $(R_{xy}(\tau))$ in a similar manner to (2A.50) which then defines the cross-spectral density function $S_{xy}'(\omega)$.

$$S_{xy}'(\omega) = \frac{1}{2\pi} \int_{-\infty}^\infty R_{xy}(\tau) \exp(-i\omega\tau)\, \mathrm{d}\tau \qquad (2A.59)$$

$$R_{xy}(\tau) = \int_{-\infty}^\infty S_{xy}'(\omega) \exp(i\omega\tau)\, \mathrm{d}\omega \qquad (2A.60)$$

The cross-correlation function for two random processes is not generally symmetrical about $\tau = 0$ and therefore the cross-spectral density function in a complex function which must describe both the amplitude and phase relationships between the processes.

$$S_{xy}'(\omega) = Co_{xy}(\omega) - iQu_{xy}(\omega) \qquad (2A.61)$$

The real part of the cross-spectral density function is an even function of frequency and is termed the co-spectrum, $Co_{xy}(\omega)$. The imaginary part of $S_{xy}'(\omega)$ is an odd function of frequency and is called the quadrature spectrum,

$Qu_{xy}(\omega)$. As the names imply, the two parts of the spectrum give the contributions to the mean square value formed from the product of the processes derived from the in-phase and out-of-phase components respectively.

In practice, estimates of auto- and cross-spectral density functions are not derived from the formal definitions given in terms of the appropriate correlation functions. The calculation of lagged products of the correlation functions is a very lengthy process and it is more efficient to estimate spectra, $S'_x(\omega)$, from the Fourier transforms, $F_x(\omega)$ and $F_y(\omega)$, as in Equation (2A.46).

$$S'_x(\omega) \simeq \frac{2\pi}{T} F^*_x(\omega) F_x(\omega) \tag{2A.62}$$

and

$$S'_{xy}(\omega) \simeq \frac{2\pi}{T} F^*_x(\omega) F_y(\omega) \tag{2A.63}$$

Fast Fourier Transform (FFT) sub-routines are widely available for making the raw spectral estimates in Equations (2A.62) and (2A.63) which must then be smoothed to yield stable spectral density functions.

4 THE RESPONSE OF LINEAR SYSTEMS TO RANDOM EXCITATION

A framework for the description of a random process has now been established which can be used in studying the response of a system to some random excitation. In the present context, the excitation could well be wave, wind or earthquake-induced loads and the system be a fixed or floating structure.

This section will be confined to the analysis of linear systems since the treatment of general non-linearity is beyond the scope of a brief review. If X_i represents the ith excitation or input and Y represents the output of the simple linear system under consideration, then:

$$Y = L(X_1, X_2 \cdots) = L(X_1) + L(X_2) + \cdots \tag{2A.64}$$

and

$$Y = L(cX_1) = cL(X_1) \tag{2A.65}$$

where the operator $L(\ \ )$ represents the system and c is a constant multiplier.

If the system possesses no memory then the statistics of response may be derived, in terms of the excitation, via the transformation of variables method described earlier. For example, the stress response of a stiff structure may be dictated by the instantaneous wave loads during an extreme storm and therefore the analysis is essentially quasi-static.

If the system has a memory then the response is dependent on the magnitude and order of loads applied prior to the point in time under consideration. The many and varied methods of analysis may be carried out in either the time or frequency domain. The majority of analyses are based on the use of an impulse response function, $h(t)$, or a frequency response function, $H(\omega)$, to describe the behaviour of the system, $L(\ )$.

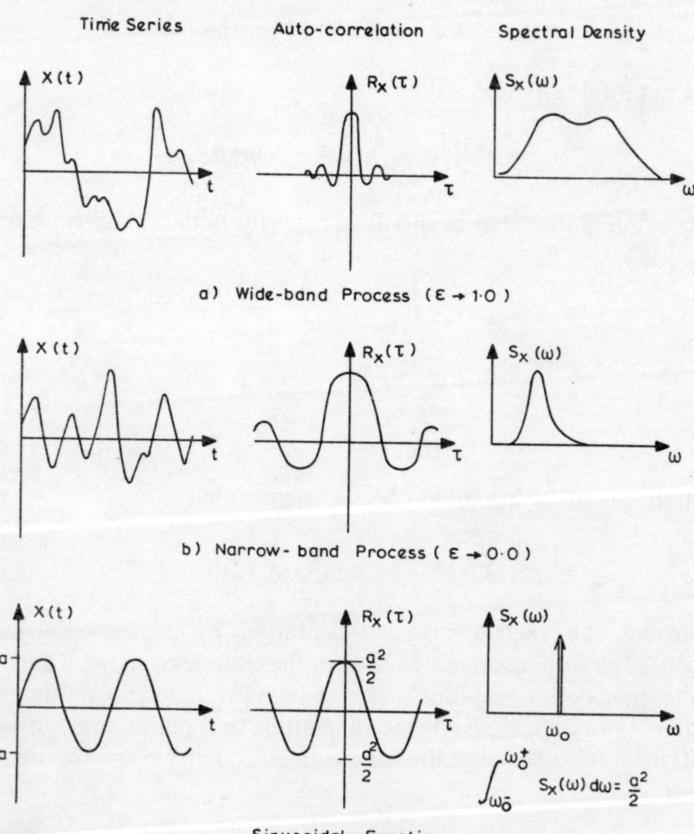

Figure 2A.5 Spectral density and correlation functions for typical time series

The response of the system $Y(t_1)$, shown in Figure 2A.6(a), at $t = t_1$ is due to an impulsive excitation $X(\tau)$ applied to the system at $t = \tau$, for a very short interval $d\tau$. In terms of the impulse response function for the system:

$$Y(t_1) = h(t_1 - \tau)X(\tau)\,d\tau \tag{2A.66}$$

Any general excitation could be treated as a series of such impulses [Figure

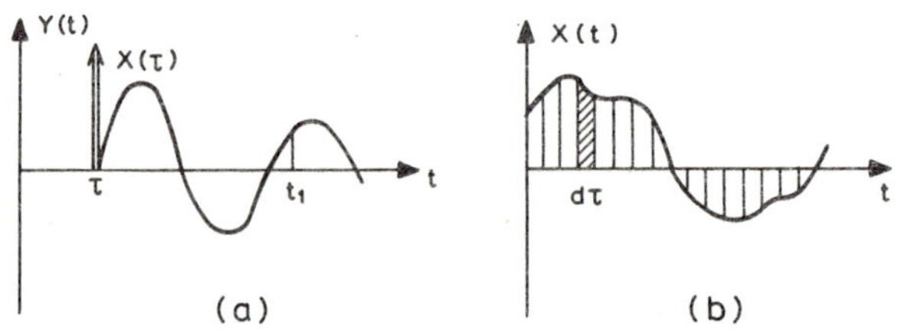

Figure 2A.6   Impulse response function

6(b)] and the total response is then:

$$Y(t_1) = \int_0^{t_1} h(t_1 - \tau)X(\tau)\,d\tau \tag{2A.67}$$

The limits of integration may be modified by virtue of the causality principle.

$$Y(t) = \int_{-\infty}^{\infty} h(t - \tau)X(\tau)\,d\tau \tag{2A.68}$$

since

$$h(t - \tau) = 0 \qquad t < \tau$$

and

$$X(\tau) = 0 \qquad \tau < 0$$

An alternative form of Equation (2A.68) is given by:

$$Y(t) = \int_{-\infty}^{\infty} h(\tau)X(t - \tau)\,d\tau \tag{2A.69}$$

Now consider the special case of excitation by a sinusoidal function, $X_0 \exp(i\omega t)$. The response some time after the commencement of excitation is termed the steady-state response and may be represented by another sinusoidal function, $Y_0 \exp(i\omega t)$. Both the magnitude and phase angle of response may be derived from the excitation using the frequency response function for the system.

$$Y(t) = Y_0 \exp(i\omega t) = H(\omega)X_0 \exp(i\omega t) \tag{2A.70}$$

Substitution of the expression for sinusoidal excitation in Equation (2A.69) shows, by comparison with Equation (2A.70), that the impulse and frequency response functions constitute a Fourier transform pair, though the factor $1/2\pi$ has changed positions.

$$H(\omega) = \int_{-\infty}^{\infty} h(t) \exp(-i\omega t)\,dt \tag{2A.71}$$

$$h(t) = \frac{1}{2\pi} \int_{-\infty}^{\infty} H(\omega) \exp(i\omega t)\, d\omega \tag{2A.72}$$

If the system is subject in general to a vector $\{X(t)\}$ of m stationary random excitations then the vector of the n random responses $\{Y(t)\}$ may be described in terms of their first and second moments using the $n \times m$ matrix $[h(t)]$ or $[H(\omega)]$ relating, in turn, the jth response variable $Y_j(t)$ to the ith excitation $X_i(t)$. From Equation (2A.69), the response vector is:

$$\{Y(t)\} = \int_{-\infty}^{\infty} [h(\tau)]\{X(t-\tau)\}\, d\tau \tag{2A.73}$$

and the expected value of response for stationary processes follows as:

$$E(\{Y(t)\}) = E\left(\int_{-\infty}^{\infty} [h(\tau)]\{X(t-\tau)\}\, d\tau \right)$$

$$= \int_{-\infty}^{\infty} [h(\tau)] E(\{X(t-\tau)\})\, d\tau \tag{2A.74}$$

$$= \int_{-\infty}^{\infty} [h(\tau)]\, d\tau\, E(\{X(t)\})$$

Similarly, if $[R_{xx}(\tau)]$ is the $m \times m$ matrix of correlation functions of $X_i(t)$ and $[R_{yy}(\tau)]$ is the corresponding $n \times n$ matrix of response correlation functions, then:

$$[R_{yy}(\tau)] = E([\{Y(t)\}\{Y(t+\tau)\}^T) \tag{2A.75}$$

where $\{\ \}^T$ represents the vector transpose. Substituting Equation (2A.73) in Equation (2A.75) and introducing the variables of integration τ_1 and τ_2 gives, after taking the expectation within the integrals:

$$[R_{yy}(\tau)] = \int_{-\infty}^{\infty} [h(\tau_1)] \int_{-\infty}^{\infty} [R_{xx}(\tau+\tau_1-\tau_2)][h(\tau_2)]^T\, d\tau_2\, d\tau_1$$

$$\tag{2A.76}$$

The two-sided spectral density matrix of response, $[S'_{yy}(\omega)]$ is defined by the term-by-term Fourier transform of the correlation matrix, $[R_{yy}(\tau)]$.

$$[S'_{yy}(\omega)] = \frac{1}{2\pi} \int_{-\infty}^{\infty} [R_{yy}(\tau)] \exp(-i\omega\tau)\, d\tau \tag{2A.77}$$

Substituting Equation (2A.76) in Equation (2A.77) and using Equation (2A.71) it may be shown that the spectral densities of response can be obtained in terms of the corresponding spectral densities of excitation.

$$[S'_{yy}(\omega)] = [H^*(\omega)][S'_{xx}(\omega)][H(\omega)]^T \qquad (2A.78)$$

where $H^*(\omega)$ is the complex conjugate of $H(\omega)$,

$$[S'_{yy}(\omega)] = \begin{vmatrix} S'_{y_1y_1}(\omega) & S'_{y_1y_2}(\omega) & \cdots & S'_{y_1y_n}(\omega) \\ S'_{y_2y_1}(\omega) & S'_{y_2y_2}(\omega) & \cdots & \cdot \\ \cdot & \cdot & \cdots & \cdot \\ S'_{y_ny_1}(\omega) & \cdot & \cdots & S'_{y_ny_n}(\omega) \end{vmatrix} \qquad (2A.79)$$

and

$$[H(\omega)] = \begin{vmatrix} H_{11}(\omega) & H_{12}(\omega) & \cdots & H_{1m}(\omega) \\ H_{21}(\omega) & H_{22}(\omega) & \cdots & \cdot \\ \cdot & \cdot & \cdots & \cdot \\ H_{n1}(\omega) & & \cdots & H_{nm}(\omega) \end{vmatrix} \qquad (2A.80)$$

The response vector $\{Y(t)\}$ will have a multivariate Gaussian probability distribution provided that the excitation applied to the linear system is also Gaussian distributed. Relationships such as Equations (2A.74), (2A.76) and (2A.78) give the necessary first and second moments for a complete specification of the multivariate Gaussian response.

The detailed form of the impulse and frequency response functions depend on the nature of the excitation and response variables chosen to characterize the problem. For example, the excitation could take the form of forces, ground acceleration or surface elevation variations and the response variables could be structural deformations, internal stresses or the motions of a floating body. Further details are given in the general references quoted.

The treatment of non-Gaussian excitation or the effect of system non-linearities is more difficult and must be the subject of separate and special consideration beyond the scope of this review.

5 THE STATISTICS OF MAXIMA AND EXTREME VALUES

In any design application it is usually necessary to have a probabilistic description of the peak response of a system subject to random excitation. The estimation of the statistics for the extreme peak response for some duration or a knowledge of the response range are common requirements. Such properties of a general random process may be found from the joint distribution of the basic process and its time differentials.

Two important relationships may be derived for a stationary random process $X(t)$. Considering Figure 2A.7(a), the expected number of crossings per unit time of a threshold level α is given by $E\{N(\alpha)\}$, which includes up and down crossings.

$$E\{N(\alpha)\} = \int_{-\infty}^{\infty} |\dot{X}| p(X = \alpha, \dot{X}) \, \mathrm{d}\dot{X} \qquad (2A.81)$$

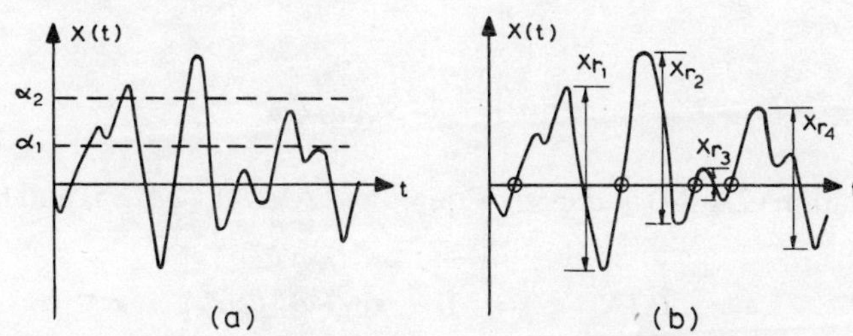

Figure 2A.7 Threshold crossings and ranges

The expected number of peaks per unit time above a level α is given by $E\{M(\alpha)\}$, where a peak is defined by $\dot{X} = 0$ and $\ddot{X} < 0$:

$$E\{M(\alpha)\} = -\int_{\alpha}^{\infty} \int_{-\infty}^{0} \ddot{X} p(X, \dot{X} = 0, \ddot{X}) \, \mathrm{d}\ddot{X} \, \mathrm{d}X \qquad (2A.82)$$

where $\dot{X} = \mathrm{d}X/\mathrm{d}t$ and $\ddot{X} = \mathrm{d}^2 X/\mathrm{d}t^2$. A reasonable model for the cdf. of peaks is given by $P_p(X)$:

$$P_p(X = \alpha) = 1 \cdot 0 - E\{M(\alpha)\}/E\{M(-\infty)\} \qquad (2A.83)$$

where $E\{M(-\infty)\}$ is the total peak occurrence rate and Equation (2A.83) includes peaks above and below the mean value (i.e. the full wide-band treatment). Differentiation of Equation (2A.83) gives the peak pdf. For many applications it is reasonable to assume that $X(t)$ is a narrow-band process and in such situations $E\{M(\alpha)\}$ is approximated by $E\{N_+(\alpha)\}$, where $E\{N_+(\alpha)\}$ is the threshold crossing rate for level α in the upwards direction. Thus for a mean-zero stationary process, the narrow-band approximation for the peak cdf. is given by:

$$P_p(X = \alpha) = 1 \cdot 0 - E\{N_+(\alpha)\}/E\{N_+(0)\} \qquad (2A.84)$$

where, from Equation (2A.81):

$$E\{N_+(\alpha)\} = E\{N(\alpha)\}/2 \qquad (2A.85)$$

In the particular case of a Gaussian distributed random process $X(t)$, it is found that $dX/dt = \dot{X}(t)$ is also Gaussian distributed. In addition, $X(t)$ and $\dot{X}(t)$ are statistically independent and therefore:

$$p(x, \dot{x}) = (2\pi\sigma_X\sigma_{\dot{X}})^{-1} \exp\left\{-\frac{1}{2}\left(\frac{x^2}{\sigma_X^2} + \frac{\dot{x}^2}{\sigma_{\dot{X}}^2}\right)\right\} \qquad (2A.86)$$

where

$$\sigma_X^2 = \int_{-\infty}^{\infty} S_X'(\omega)\,d\omega \qquad (2A.87)$$

and

$$\sigma_{\dot{X}}^2 = \int_{-\infty}^{\infty} \omega^2 S_X'(\omega)\,d\omega \qquad (2A.88)$$

From Equation (2A.84) and using Equations (2A.81), (2A.85) and (2A.86), it follows that:

$$P_p(X = \alpha) = 1{\cdot}0 - \exp\{-\alpha^2/2\sigma_X^2\} \qquad (2A.89)$$

which is the standard form of the Rayleigh cdf. If the range X_r of the random process is defined in terms of the upwards zero-crossing convention, see Figure 2A.7(b), then:

$$X_r = 2\alpha \qquad (2A.90)$$

and therefore the cdf. of range is:

$$P(X_r) = 1{\cdot}0 - \exp\{-X_r^2/8\sigma_X^2\} \qquad (2A.91)$$

and the range pdf. is:

$$p(X_r) = \frac{X_r}{4\sigma_X^2} \exp\{-X_r^2/8\sigma_X^2\} \qquad (2A.92)$$

Equations (2A.91) and (2A.92) are familiar forms for the cdf. and pdf. of waveheight (H) when $X_r = H$ and $X(t) = \eta(t)$, the surface elevation.

If $X(t)$ is not a Gaussian process then the full wide-band or narrow-band relationships Equations (2A.83) and (2A.84) must be used. However, it will be found that even for relatively wide-band spectra, the narrow-band approximations are generally acceptable, particularly for the higher values of maxima or for the estimation of range statistics.

If $X_{\max}$ is the largest peak value of $X(t)$ in a sample of N peaks, then the probability that a particular peak is equal to or less than $x_{\max}$ is, by definition, $P_p(x_{\max})$. If the individual peaks are statistically independent then the probability that each peak is equal to or less than $x_{\max}$ is $\{P_p(x_{\max})\}^N$ and defining

$P_{ep}(x_{\max})$ as the cdf. of the extreme peak, it follows that:

$$P_{ep}(x_{\max}) = \{P_p(x_{\max})\}^N \qquad (2\text{A}.93)$$

and the extreme peak cdf. is:

$$p_{ep}(x_{\max}) = N\{P_p(x_{\max})\}^{N-1} p_p(x_{\max}) \qquad (2\text{A}.94)$$

The expected value of $x_{\max}$ may be obtained from taking the first moment of $p_{ep}(x_{\max})$ and the most probable peak value corresponds to the mode of $p_{ep}(x_{\max})$.

GENERAL REFERENCES

Bendat, J. S. and Piersol, A. G. (1971). *Random Data: Analysis and Measurement Procedures*, 2nd Edn., J. Wiley, New York.
Clough, R. W. and Penzien, J. (1975). *Dynamics of Structures*, 1st Edn., McGraw-Hill, New York.
Newland, D. E. (1975). *Random Vibration and Spectral Analysis*, 1st Edn., Longman, London.
Papoulis, A. (1965). *Probability Random Variables and Stochastic Processes*, 1st Edn., McGraw-Hill, New York.
Price, W. G. and Bishop, R. E. D. (1974). *Probabilistic Theory of Ship Dynamics*, 1st Edn., Chapman and Hall, London.

Chapter 3

Hydrodynamic Loading of Large Offshore Structures: Three-Dimensional Source Distribution Methods

C. J. Garrison

3.1 INTRODUCTION

Large gravity structures as well as some steel space-frame structures are constructed in dry-dock and are towed as floating structures to their placement site. During towing conditions it is necessary to assess the stability, seaworthiness and loading of the structure under storm conditions which could develop during the sometimes extended exposure of the structure in the floating condition. When the structure is finally in place, it is necessary to assess with a high degree of certainty the magnitude of the loads induced by the extreme waves associated with the design storm conditions as well as the loads associated with the long term periodic loading for purposes of fatigue calculations. Moreover, if the structure is to be located in an earthquake zone the dynamic earthquake-induced response of the structure which is strongly affected by the inertia of the surrounding fluid is important. Thus, even though the numerical procedures in free-surface hydrodynamics tend to be rather involved, very practical needs exist for dealing with the hydrodynamics of large offshore structures.

This chapter presents some of the concepts, techniques and numerical procedures used in the evaluation of hydrodynamic forces on large offshore structures subject to wave motion. The Morison formulation of the hydrodynamic force acting on a body is discussed in relation to the diffraction theory and regions of applicability of the two procedures are proposed. The numerical procedures for calculating loads on fixed and floating structures are treated in detail. Attention is primarily directed toward (a) wave induced loading of fixed structures which are composed of a large displacement caisson with superstructure constructed of small diameter members, and (b) large displacement floating caissons subjected to waves. The analysis is based primarily on the inviscid theory of hydrodynamics, solutions being constructed using the method of distributed sources with allowances made for viscous effects when small diameter members, which give rise to gross flow separation, are involved.

3.2 THE RELATIONSHIP BETWEEN THE MORISON EQUATION AND DIFFRACTION THEORY

There are basically two different approaches in use today for the computation of fluid-structure interaction associated with fixed or floating structures: (a) application of the so-called Morison equation and (b) diffraction (or potential flow) theory. However, there is often confusion regarding the relationship between these two procedures, and their respective limits of applicability. Therefore, even though the present emphasis is placed on the inviscid theory, it is appropriate to preface this with a discussion of the more familiar Morison equation procedure and its relationship to the inviscid diffraction theory.

Let us begin with the most general procedure, recognizing that diffraction theory, as it has come to be known, refers to the inviscid, incompressible and irrotational (potential flow) solution of fluid-structure interaction. In the linear diffraction theory the solution to the fluid-structure problem is sought such that the linearized free-surface boundary condition is satisfied as well as the kinematic boundary condition on the surface of the body and on the sea floor. Moreover, the waves caused by the presence of the body and/or its motion satisfy a radiation condition at some large distance away from the body. Thus, notwithstanding the possibility of certain numerical limitations in application, the diffraction theory is based on the exact solution to the interaction of either a fixed or floating body with a fluid, but because of its underlying assumptions admits of two fundamental limitations: (1) that arising from the assumption of zero viscosity of the fluid and (2) that arising from the assumption of small amplitude motion as implied by the application of the linearized free-surface boundary condition. While it is impossible to precisely delineate the limits of the theory as imposed by these two complex effects, there exists certain experimental and theoretical results which shed some light on such limits. Thus, in the following, an attempt is made to indicate which parameters are important with respect to the limits of applicability and to establish approximate limit criterion for use in application.

Consider first the limitation of diffraction theory, i.e. that due to linearization. Because of the velocity-squared term occurring in Bernoulli's equation, among other things, the free surface boundary condition is non-linear, and the boundary-value problem associated with diffraction theory is made tractable by assuming the motion (waveheight) to be of small amplitude so that this boundary condition may be linearized. An indication of the effect of this linearization may be obtained by considering results of extensions of the diffraction theory to the second-order which have been made by Raman and Venkatanarasaiah,[1] and Garrison[2] for the case of wave interaction with fixed bodies and by Lee[3] and Potash[4] for the case of a two-dimensional body oscillating in still water. The theory is highly complex and not easily applied so extensive results have not yet been generated. However, as a general assessment, it appears that the nonlinear effects become significant only in the case of shallow-water, large-amplitude waves. Results indicate that in practice non-linear diffraction theory will clearly be of great value in the calculation of wave forces acting on caisson-type coastal structures such as berthing caissons for ship loading structures (see, for example, Apelt and Macknight[5]) where highly non-linear waves prevail and the caisson is surface piercing. But for most applications where the caisson is placed in fairly deep water (i.e. where the water depth to wave length is not small) non-linear free surface effects in diffraction mechanics is generally negligible.

Raman and Venkatanarasaiah[1] indicate that for the case of a vertical surface-piercing circular cylinder (pile) the non-linear effect is less than 5 per cent when the water depth, h, to wave length, L, ratio is greater than $0 \cdot 25$ for wave steepness up to $0 \cdot 09$. For $h/L < 0 \cdot 25$ non-linear effects become more pronounced increasing more or less proportionally to wave steepness. However, the higher-order components in non-linear waves die out very rapidly with depth so in cases where a caisson-type structure is fairly deeply submerged as opposed to surface piercing, the non-linear components have much less effect than in the case of surface piercing structures.

Consider now the second limitation of the diffraction theory: that arising from the neglect of the effect of viscosity. At high Reynolds number it is well-known that the effect of viscosity is concentrated in the boundary layer region of the flow. In the case of bluff bodies such as circular cylinders, boundary layer development results in flow separation and wake development and for this reason the local pressures and resulting forces acting on the body are different than calculated on the basis of the inviscid flow assumption. The questions which must be addressed, in this connection then, are: what are the parameters which control separation and wake development and what are the practical limits on the diffraction theory as established by the neglect of viscous effects?

These questions are obviously highly complex and depend on the body shape, type of oscillatory motion, the amplitude of the fluid motion relative to the dimensions of the object and the Reynolds number. However, it is possible to obtain approximate answers by considering basic experimental results obtained from measuring forces due to oscillatory flow past a circular cylinder. In Figure 1 the added mass coefficient,* C_m, and drag coefficient, C_d, associated

---

* The inertia coefficient used in the Morison equation[6] is defined as $C_M = (1 + C_m)$ where C_m denotes the added mass coefficient. *Vide* Ch. 2, p. 56 and Ch. 4, p. 142.

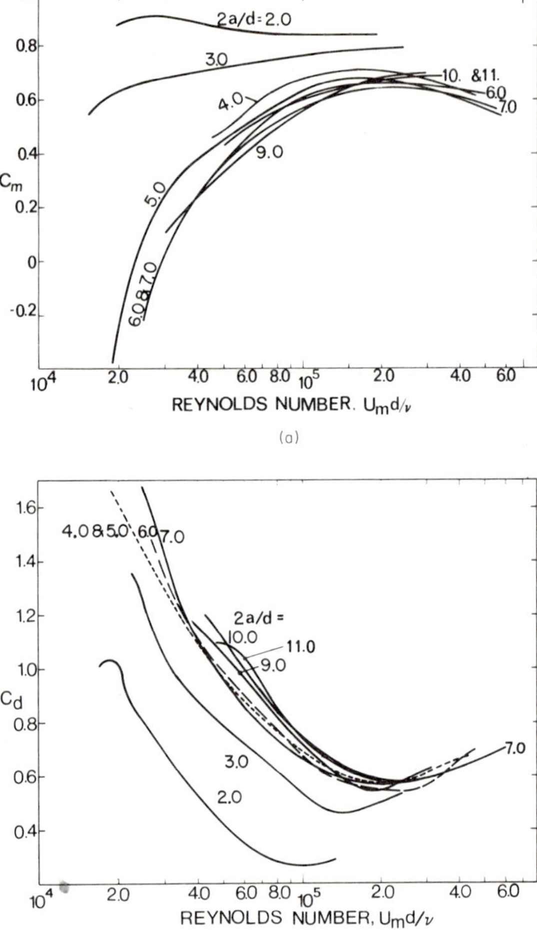

Figure 3.1 Added mass and drag coefficients for periodic flow about a smooth circular cylinder of diameter d

U_m—Maximum cycle velocity
ν—Kinematic viscosity
a—Amplitude of motion

with the oscillation of a smooth circular cylinder in still water* are plotted as a function of the Reynolds number based on the maximum velocity in the cycle,

---

* The force acting on a fixed cylinder in oscillatory flow differs from that associated with oscillation of the cylinder in still water. For a fixed body in oscillatory flow $C_M = 1 + C_m$ whereas $C_M = C_m$ when the body oscillates in still water.[7]

U_m, and the relative amplitude of the motion, a. It may be recalled that in inviscid flow the drag coefficient is zero and the added mass coefficient is unity for the circular cylinder. Accordingly, the extent to which the experimental results differ from the inviscid flow results is a good indicator of the magnitude of viscous effects and can be used to establish an approximate limit to the region of applicability of the inviscid theory in so far as it is dependent on this effect. The results at high Reynolds number corresponding to values of $2a/d$ of $\leq 1\cdot0$ indicate that the added mass coefficient is essentially equal to its inviscid flow value of $1\cdot0$. As $2a/d$ is increased beyond this value separation and wake development resulting from viscous effects increase and C_m tends to decrease from unity considerably.

The drag coefficient plotted in Figure 3.1(b) also shows corresponding expected trends at small value of $2a/d$. When $2a/d$ is small the degree of separation and wake development is small and, accordingly, the drag coefficient is small. Taking the ratio of the two components of force in the Morison equation and using the high Reynolds number values for C_m and C_d from Figure 1 shows that the drag force would be less than $5\cdot0$ per cent of the inertia force up to $2a/d = 1\cdot0$. Although the drag force is $5\cdot0$ per cent of the inertia force for this case it would have yet less net effect on the maximum force since the drag and inertia components in the Morison equation are 90 degrees out of phase. Thus, the value of $2a/d = 1\cdot0$ appears to represent a rather conservative limit for the inviscid theory; it is likely that this limit may be somewhat greater.

To relate this discussion to the wave force problem it is necessary to remember that the orbital motion in a wave is directly proportional to the wave height. Thus, the double amplitude of the fluid particles in wave motion is equal to the waveheight, H, adjusted downward due to the attenuation with depth beneath the free surface and adjusted upward to account for the effect of finite depth. Figure 3.2 which has been prepared to show the respective limits of difffraction theory and the Morison equation for a vertical cylinder indicates the limit on diffraction theory at $H/d = 1\cdot0$ due to viscous effects. This limit is, of course, strictly valid only for the case of deep water where the double amplitude of the particle orbits are approximately equal to the wave height.

Having now established that the general diffraction theory is not limited to any specific body shape, although it admits of the two limitations discussed, it is of interest to consider its relationship to the Morison equation. Historically the Morison equation was introduced as a semi-intuitive formula for the computation of wave forces on objects which were small in relation to the wave length such as a small diameter pile. Although this approximation is accepted as intuitively valid it is possible to show rigorously that the basic assumptions underlying its application, namely that the body is small in relation to the wave length, leads to considering the flow past the body in question with velocity and acceleration such as would occur at its centre if it were not present in the flow.

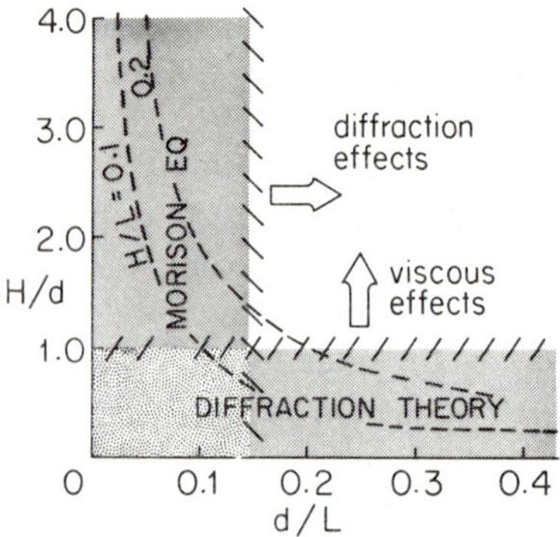

Figure 3.2 Regions of validity—
wave interaction with a pile
H—Wave height = 2
d—Pile diameter
L—Wave length

The rigorous proof follows along the lines of singular perturbation theory (see for example, Van Dyke[8]) wherein the outer expansion of the inner solution is matched with the inner expansion of the outer solution. In the case of wave-structure interaction the outer solution is represented by some wave theory, either linear or non-linear. The inner problem is represented by the viscous time-dependent flow past the body in question and the velocity field encountered by the body is considered to extend to infinity. The inner solution must match at infinity with the velocity induced by the wave at the centre of the body if it were not present in the flow at all. This procedure is common in fluid mechanics and is used, for example, in the rigorous approach to boundary layer theory, Prandtl lifting line theory and slender body theory.

The Morison equation as such is not suggested by the formal perturbation theory, only the flow situation which must be considered. The form of the force equation which is commonly used to represent the inner flow involves both a drag and inertia term involving the cross-flow velocity and acceleration, respectively (i.e., the Morison equation), and is semi-intuitive in that the inertia term is exactly of the form which results from inviscid flow theory while the drag term is of the form used in steady flow. It can be reconciled with the general dimensional analysis of the problem only if C_m and C_d are considered to be functions of the Reynolds number, relative displacement and phase angle of the periodic motion. The assumption of constant values of C_m and C_d throughout a complete cycle of the motion is convenient and appears to be valid for smooth cylinders, but recent results of May[9] indicates that it is invalid in the case of rough cylinders at least at intermediate values of $2a/d$. It appears that there remains a considerable amount to be learned about the calculation of wave loads by use of this simple force equation.

The Morison equation approach to the calculation of wave forces on a body represents the asymptotic form of the diffraction theory in the limit as the size of the body (or diameter in the case of elongated shapes such as a cylinder) relative to the wave length approaches zero. Thus, if viscous effects are disregarded the results of diffraction theory approach those based on the inertia term in the Morison equation as the diameter to wave length approaches zero. The results of MacCamy and Fuchs[10] which represent the exact solution to the diffraction problem for a vertical pile are particularly convenient for demonstrating this behaviour. MacCamy and Fuchs' diffraction theory for the cylinder gives

$$C_M = 1 + C_m = \frac{4}{\pi} \frac{1}{(\pi d/L)^2} \frac{1}{\sqrt{J_1'(\pi d/L)^2 + Y_1'(\pi d/L)^2}} \tag{3.1}$$

where J_1' and Y_1' denote derivatives of Bessel functions of the first and second kind of order unity with respect to their arguments.

Equation (3.1) is plotted against the diameter to wave length ratio, d/L, in Figure 3.3 where it is evident that the inertia coefficient, $1 + C_m$, approaches $2 \cdot 0$ as d/L tends to zero. This plot indicates the reduction of the diffraction theory to the Morison equation as $d/L \to 0$ and indicates that $d/L = 0 \cdot 1$ to $0 \cdot 2$ might represent a practical limit on the validity of the Morison equation insofar as diffraction effects are concerned. Above this value of d/L scattering of the incident wave appears to be significant and, accordingly, the Morison equation,

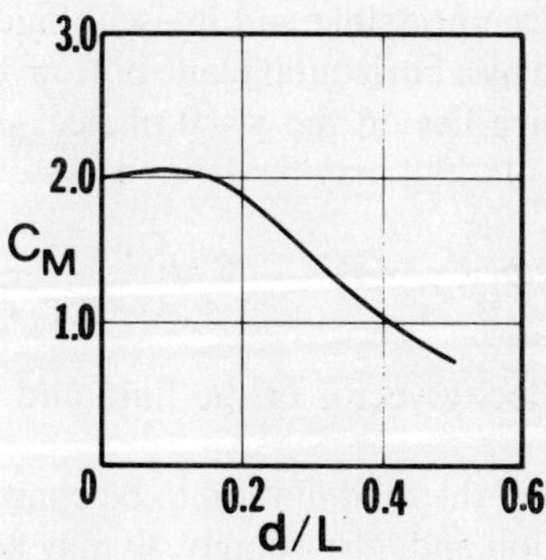

Figure 3.3　Inertia coefficient for a pile computed from diffraction theory

d—Pile diameter
L—Wave length

which represents the asymptotic limit for small d/L, becomes invalid. This limit on the Morison formulation of the hydrodynamic force is indicated in Figure 3.2.

Figure 3.2, which was introduced several years ago[11] but without dimensions, represents a rather convenient method of presentation of the regions of applicability of the inviscid diffraction theory and the Morsion equation for the case of a pile. It shows the overlap region bounded by $H/d = 1 \cdot 0$ and $d/L = 0 \cdot 15$ where both theories are valid and a region at both large H/d and d/L where both viscous effects and diffraction effects are important and, consequently, neither theory is valid. However, contours of constant values of wave steepness, H/L, plotted on Figure 3.2 indicate that the region of large H/d and d/L is of little practical importance since the breaking limit for deep water is at about $H/L = 0 \cdot 14$. Thus, Figure 3.2 suggests that conveniently viscous effects and diffraction effects should never be important at the same time. It appears that viscous effects should become important only in regions where the Morison equation is valid.

3.3   GREEN'S FUNCTION METHOD

As a preliminary to addressing particular problems from practice, the general concepts and numerical procedures pertinent to the application of the Green's function method to the solution of problems in hydrodynamics is first discussed. Application to specific problems regarding offshore structures is taken up subsequently.

Consider for the present purposes an object with surface denoted by $S(x, y, z) = 0$ which has characteristic dimensions a to be either immersed or semi-immersed in an incompressible and inviscid fluid as indicated in Figure 3.4. The fluid is bounded by a horizontal plane bottom boundary at $y = -h$ and a mean free surface which lies on the $y = 0$ plane. Assuming the fluid to be irrotational, a velocity potential may be defined as

$$\mathbf{q} = \nabla \Phi(x, y, z, t), \qquad \nabla \equiv \left[\frac{\partial}{\partial x}, \frac{\partial}{\partial y}, \frac{\partial}{\partial z} \right]^T \tag{3.2}$$

where $\mathbf{q}$ denotes the velocity vector of the fluid and Φ denotes the velocity potential.

The time dependence of the fluid motion to be considered here is restricted to simple harmonic motion and, accordingly, Φ may be expressed as

$$\Phi = \mathrm{Re}\left[\phi(x, y, z)\, \mathrm{e}^{-i\omega t} \right] \tag{3.3}$$

where $\omega = 2\pi/T$ denotes the frequency of the motion, T being the period, and ϕ denotes the complex potential.

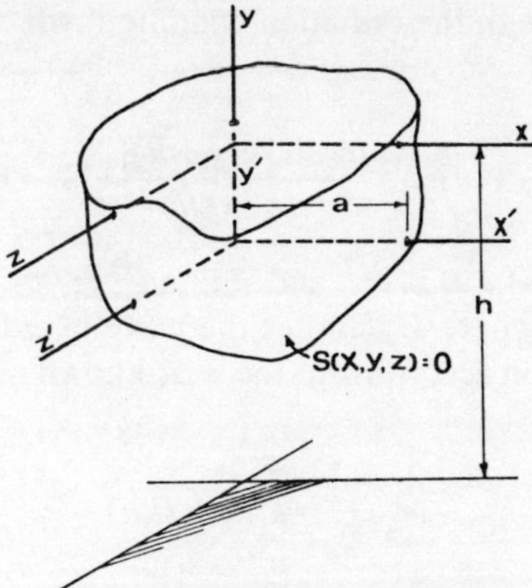

Figure 3.4   Immersed surface

The differential equation governing the fluid motion follows from application of the continuity equation, $\nabla^T . \mathbf{q} = 0$, to Equation (3.2), yielding the Laplacian,

$$\nabla^2 \phi = 0 \tag{3.4}$$

Boundary conditions for the geometry specified include the kinematic condition on the ocean floor,

$$\frac{\partial \phi}{\partial y} = 0 \quad on \quad y = -h \tag{3.5}$$

The boundary condition to be applied on the immersed surface in problems considered here will be of the form,

$$\frac{\partial \phi}{\partial n} - v_n \quad on \quad S(x, y, z) = 0 \tag{3.6}$$

where v_n denotes the specified complex function which represents the magnitude of the normal component of velocity on the immersed surface given by $V_n = \mathrm{Re}\,[v_n(x, y, z)\,e^{-i\omega t}]$.

On the mean free surface both a kinematic and dynamic condition are imposed resulting in the well-known linearized boundary condition valid for small amplitude motion,

$$\frac{\partial \phi}{\partial y} - \frac{\omega^2}{g}\phi = 0 \quad on \quad y = 0 \tag{3.7}$$

Furthermore, in order to insure that the velocity potential has the correct behaviour in the far field, the radiation condition, which may be written (see Ch. 4, p. 000)

$$\phi(r, \theta, y) - \lambda(\theta) r_1^{-1/2} \frac{\cosh\left[k(y+h)\right]}{\cosh(kh)} e^{ikr_1} \to 0, \quad r_1 \to \infty \tag{3.8}$$

is imposed where $r_1 = [x^2 + z^2]^{1/2}$ and $\theta = \tan^{-1}(z/x)$. The wave number is defined as $k = 2\pi/L$, where L denotes the wave length and is related to the frequency of the motion according to the well-known relationship from linear theory,

$$\nu \equiv \frac{\omega^2}{g} = k \tanh(kh) \tag{3.9}$$

Lamb[12] has shown through application of the Green's theorem that for the infinite fluid case the velocity potential associated with the flow about a body can be described by either a simple source or doublet distribution over the surface of the body. This representation of the potential is rather easily extended to the case of a fluid of finite depth with free surface and, accordingly, either singularity may be used to represent free surface flows although additional terms are needed to satisfy the free surface and bottom conditions. The fundamental difference between these two approaches is that in the case of the doublet distribution the normal component of velocity is continuous across the surface $S(x, y, z) = 0$ while the value of ϕ undergoes a jump. Because of the continuity in normal velocity the doublet distribution is convenient for representation of a thin shell with fluid on either side (see, for example, Chakrabarti[13]).

By contrast, the source distribution produces normal velocities which are discontinuous across the surface S but if the flow on one side only is of interest, and this is usually the case, the source distribution appears to be the more convenient method and is applied in the following.

The potential at some point (x, y, z) in the fluid region may be expressed in terms of a surface distribution of sources

$$\phi = \frac{1}{4\pi} \iint_S f(\xi, \eta, \zeta)\, G(x, y, z; \xi, \eta, \zeta)\, dS \tag{3.10}$$

where (ξ, η, ζ) denotes a point on S and $f(\xi, \eta, \zeta)$ denotes the unknown source distribution. The integral is to be carried out over the complete immersed surface of the object. The Green's function, (source potential) $G(x, y, z; \xi, \eta, \zeta)$ must, in order for the representation in Equation (3.10) to be valid, satisfy all the boundary conditions of the problem, Equations (3.5), (3.7) and (3.8), with the exception of the kinematic condition, Equation (3.6), and have a source-like behaviour. In short, G must be of the form

$$G = \frac{1}{R} + G^* \tag{3.11}$$

where

$$R = [(x-\xi)^2 + (y-\eta)^2 + (z-\zeta)^2]^{1/2} \tag{3.12}$$

and G^* represents a regular function satisfying $\nabla^2 G^* = 0$ throughout the fluid region.

The particular expression for G appropriate to the boundary-value problem posed is given in two different forms by Wehausen and Laitone.[14] One of these is the integral form,

$$G = \frac{1}{R} + \frac{1}{R'} + 2\text{P.V.} \int_0^\infty \frac{(\mu+\nu)\,e^{-\mu h}\cosh\left[\mu(\eta+h)\right]\cosh\left[\mu(y+h)\right]}{\mu\sinh(\mu h) - \nu\cosh(\mu h)} J_0(\mu r)\,d\mu$$

$$+ i\frac{2\pi(k^2-\nu^2)\cosh\left[k(\eta+h)\right]\cosh\left[k(y+h)\right]}{k^2 h - \nu^2 h + \nu} J_0(kr) \tag{3.13}$$

where R is defined in Equation (3.11), ν by Equation (3.8) and

$$R' = [(x-\xi)^2 + (y+2h+\eta)^2 + (z-\zeta)^2]^{1/2}$$

$$r = [(x-\xi)^2 + (z-\zeta)^2]^{1/2}$$

P.V. denotes the principal value of the integral. An alternate series form of G is given by[15]

$$G = \frac{2\pi(\nu^2-k^2)}{k^2 h - \nu^2 h + \nu}\cosh\left[k(\eta+h)\right]\cosh\left[k(y+h)\right][Y_0(kr) - iJ_0(kr)]$$

$$+ 4\sum_{k=1}^{\infty}\frac{(\mu_k^2+\nu^2)}{\mu_k^2 h + \nu^2 h - \nu}\cos\left[\mu_k(y+h)\right]\cos\left[\mu_k(\eta+h)\right]K_0(\mu_k r) \tag{3.14}$$

in which J_0 and Y_0 denote, respectively, Bessel functions of the first and second kind of order zero and K_0 denotes the modified Bessel function of the second kind of order zero. The quantities μ_k denote the real positive roots of the equation

$$\mu_k\tan(\mu_k h) + \nu = 0 \tag{3.15}$$

Equations (3.13) and (3.14) represent Green's functions satisfying the complete free surface boundary condition (3.7) and are valid at all frequencies. However, for motion at high-frequency as induced, for example, by earthquakes, Equation (3.7) can be approximated by taking $\phi = 0$ on $y = 0$ and

Green's function satisfying this boundary condition may be written,[16]

$$G = \frac{1}{R} + \frac{1}{R_1} + \sum_{n=1}^{\infty} \left\{ (-1)^n \left[\frac{1}{R_{2n}} + \frac{1}{R_{4n}} \right] + (-1)^{(n+1)} \left[\frac{1}{R_{3n}} + \frac{1}{R_{5n}} \right] \right\}$$

(3.16a)

in which

$$r = [(x - \xi)^2 + (z - \zeta)^2]^{1/2}$$
$$R = [r^2 + (y - \eta)^2]^{1/2}$$
$$R_1 = [r^2 + (y + \eta)^2]^{1/2}$$
$$R_{2n} = [r^2 + (y - 2nh - \eta)^2]^{1/2}$$
$$R_{3n} = [r^2 + (y + 2nh + \eta)^2]^{1/2} \qquad (3.16b)$$
$$R_{4n} = [r^2 + (y + 2nh - \eta)^2]^{1/2}$$
$$R_{5n} = [r^2 + (y - 2nh + \eta)^2]^{1/2}$$

On the other hand, if the period of the motion is large as might occur in the interaction of waves of large period with a body, the free surface boundary condition takes on the asymptotic form $\partial\phi/\partial y = 0$ on $y = 0$, i.e. the rigid wall condition. The Green's function appropriate to this case is

$$G = \frac{1}{R} + \frac{1}{R_1} + \sum_{n=1}^{\infty} \left[\frac{1}{R_{2_n}} + \frac{1}{R_{3_n}} + \frac{1}{R_{4_n}} + \frac{1}{R_{5_n}} \right] \qquad (3.16c)$$

The asymptotic forms of G given in Equation (3.16) are generated by a series which represents the summation of image sources with respect to both the bottom and free surface. Solutions based on Equation (3.16) tend to converge very rapidly with n and, therefore, are very efficient with respect to computer evaluation. They have the added advantage of being real which, interpreted physically, means that in asymptotically low or high-frequency motion no waves are generated or scattered by an immersed surface.

The solution to the boundary-value problem as given by Equation (3.10) satisfies Equations (3.4), (3.5), (3.7) and (3.8) since G itself satisfies these conditions. When the kinematic boundary condition on the immersed surface [Equation (3.6)] is applied the following integral equation is obtained:*

$$-f(x, y, z) + \frac{1}{2\pi} \iint_S f(\xi, \eta, \zeta) \frac{\partial G}{\partial n}(x, y, z; \xi, \eta, \zeta) \, dS = 2v_n(x, y, z) \qquad (3.17)$$

---

* The validity of this equation comes from noting the result of integration around a source to obtain the first term.

where $\partial G/\partial n$ denotes the derivative of the Green's function in the outward normal direction and (x, y, z) denotes points on $S(x, y, z) = 0$. This derivative of G may be evaluated from

$$\frac{\partial G}{\partial n} = \nabla G \cdot \mathbf{n} = \frac{\partial G}{\partial x} n_x + \frac{\partial G}{\partial y} n_y + \frac{\partial G}{\partial z} n_z \tag{3.18}$$

where $\partial G/\partial x$, $\partial G/\partial y$ and $\partial G/\partial z$ are determined through a straightforward differentiation of G given in Equations (3.13), (3.14) or (3.16). The vector $\mathbf{n}$ represents the unit normal vector, $\mathbf{n} = i n_x + j n_y + k n_z$, where n_x, n_y and n_z denote the x, y and z components, respectively, such that $n_x^2 + n_y^2 + n_z^2 = 1$.

3.3.1 Numerical solution

The integral equation (3.17) may be solved numerically beginning with the subdivision of S into N quadrilateral patches or panels of area $\Delta S_j (j = 1, 2, \ldots, N)$ and identifying as node points the centroid of each panel. The continuous formulation of the solution indicates that Equation (3.17) is to be satisfied at all points (x, y, z) on the immersed surface defined by $S(x, y, z) = 0$ but in order to obtain a discretized numerical solution it is necessary to relax this requirement and apply the condition at only N control points. The location of these control points may, in principle, be chosen arbitrarily but for convenience the N node points at the panel centroids are used. Thus, in the discretization process Equation (3.17) is replaced by the N equations,

$$-f_i + \frac{1}{2\pi} \iint\limits_{S} f(\xi, \eta, \zeta) \frac{\partial G}{\partial n}(x_i, y_i, z_i; \xi, \eta, \zeta)\, \mathrm{d}S = 2v_{n_i} \tag{3.19}$$

Furthermore, the surface integral in Equation (3.19) may be written as the sum of the integrals over the N panels of area ΔS_j and, as an approximation, the source strength function $f(\xi, \eta, \zeta)$ may be taken as constant over each panel so that Equation (3.19) becomes,

$$-f_i + \alpha_{ij} f_j = 2v_{n_i}, \qquad i, j = 1, 2, \ldots, N \tag{3.20}$$

where the repeated index denotes summation and where

$$\alpha_{ij} = \frac{1}{2\pi} \iint\limits_{\Delta S_j} \frac{\partial G}{\partial n}(x_i, y_i, z_i; \xi, \eta, \zeta)\, \mathrm{d}S \tag{3.21}$$

In physical terms, α_{ij} denotes the velocity induced at the ith node point in the direction normal to the surface by a source distribution of unit strength distributed uniformly over the jth panel.

In Equation (3.20) the function v_n is considered specified and the elements of the α-matrix are defined through the Green's function according to Equation (3.21). The unknown source strength vector f_j may, therefore, be obtained from digital computer solution of the matrix equation using a straightforward elimination procedure. Equation (3.20) may be written

$$[f] = 2[\boldsymbol{\alpha} - \mathbf{I}]^{-1}[v_n] \tag{3:22}$$

where I denotes the unit matrix. Once $[\boldsymbol{\alpha} - \mathbf{I}]^{-1}$ is evaluated the vector f corresponding to several different v_n vectors may be computed with very little additional CPU time. The bulk of the CPU time is required for the evaluation of the α matrix and the inversion (or solution) of $[\boldsymbol{\alpha} + \mathbf{I}]$.

Using the same numerical scheme as outlined above, Equation (3.10) may be expressed as the sum

$$\phi_i = \beta_{ij} f_j, \qquad i, j = 1, 2, \ldots, N \tag{3.23}$$

in which

$$\beta_{ij} = \frac{1}{4\pi} \iint\limits_{\Delta S_j} G(x_i, y_i, z_i; \xi, \eta, \zeta)\, \mathrm{d}S \tag{3.24}$$

Equation (3.23) gives the value of ϕ at the ith node point on S. However, the i subscript may be dropped and ϕ evaluated at any point (x, y, z) in the fluid region if β is defined accordingly.

The fluid velocity in a given direction as defined by Equation (3.2) is equal to the spatial derivative of the velocity potential in that direction. The derivatives of ϕ are obtained by use of Equation (3.10) as

$$\begin{Bmatrix} \phi_x \\ \phi_y \\ \phi_z \end{Bmatrix} = \begin{Bmatrix} \alpha_{x_j} \\ \alpha_{y_j} \\ \alpha_{z_j} \end{Bmatrix} f_j \tag{3.25}$$

where the repeated index denotes summation and where

$$\begin{Bmatrix} \alpha_{x_j} \\ \alpha_{y_j} \\ \alpha_{z_j} \end{Bmatrix} = \frac{1}{4\pi} \iint\limits_{\Delta S_j} \begin{Bmatrix} \dfrac{\partial G}{\partial x}(x, y, z; \xi, \eta, \zeta) \\[2mm] \dfrac{\partial G}{\partial y}(x, y, z; \xi, \eta, \zeta) \\[2mm] \dfrac{\partial G}{\partial z}(x, y, z; \xi, \eta, \zeta) \end{Bmatrix} \mathrm{d}S \tag{3.26}$$

Finally, it is generally of primary interest to determine the hydrodynamic pressure on the immersed surface as well as the resulting forces and moments. The pressure is given in terms of the velocity potential as

$$P = \rho\omega \, \text{Re} \, [i\phi \, e^{-i\omega t}] - \tfrac{1}{2}\rho \, \text{Re} \, [(\phi_x^2 + \phi_y^2 + \phi_z^2) \, e^{-i2\omega t}]$$

$$- \tfrac{1}{2}\rho[|\phi_x|^2 + |\phi_y|^2 + |\phi_z|^2] - \rho g y \tag{3.27}$$

in which the second and third terms result from the quadratic velocity term in Bernoulli's equation and are, therefore, of second-order in the amplitude of the motion. Compared to the first term these terms are small and normally neglected. However, the second of the two terms is independent of time and is significant in the case of wave interaction with floating bodies because it, in part, gives rise to the so-called steady-state force. The last term in Equation (3.27) represents simply the hydrostatic pressure and gives rise to a buoyant force and contributes to the drift force.

The forces and moments acting on the immersed surface are obtained by integration of the pressure distribution. Using the first term in Equation (3.27), the force and moment vector are obtained from the surface integrals

$$\mathbf{F} = -\rho\sigma \iint_S \text{Re} \, [i\phi \, e^{-i\omega t}]\mathbf{n} \, dS \tag{3.28}$$

$$\mathbf{M} = -\rho\sigma \iint_S \text{Re} \, [i\phi \, e^{-i\omega t}](\mathbf{r}' \times \mathbf{n}) \, dS \tag{3.29}$$

in which $\mathbf{r}'$ denotes the position vector extending from the point where the forces and moment are resolved to the local point on the surface.

3.3.2   Numerical evaluation of the Green's function

The numerical solution to the hydrodynamics problem posed involves first the determination of the source strength distribution function f_i. Once f_i is known the problem is considered 'solved' in that ϕ and/or its derivatives at any point in the fluid region may be determined through the summations indicated in Equations (3.23) and (3.25).

The evaluation of the α and β matrices represents an important part of the numerical procedure because a great deal of CPU time is consumed in the process. The Green's function, G and derivatives of G, $\partial G/\partial x$, $\partial G/\partial y$ and $\partial G/\partial z$, may be evaluated most rapidly through use of the series given by Equation (3.14). The series converges rapidly because of the $K_0(\mu_k r)$ term and, therefore, the time requirements are less than for evaluation of the integral form. As a CPU time-saving measure, the functions $\sin[\mu_k(y+h)]$ and

$\cos[\mu_k(y+h)]$ may be generated once for all for each of the different values of y at the node points on the immersed surface and stored. With this information pre-computed the series may be evaluated rather efficiently.

When $kr \to 0$ the Bessel function $K_0(\mu_k r) \to \infty$ and, therefore, the series form of the Green's function cannot be used. It is necessary in this case to resort to the integral form given in Equation (3.13) when kr is less than some limiting number and contend with the more time consuming numerical integration of the infinite integral.

The evaluation of the infinite integral in Equation (3.13) may be accomplished in two different ways, either directly or after first rearranging. The direct approach involves breaking the interval up into two parts, 0 to $2k$ and $2k$ to ∞. The infinite upper limit on the second interval is replaced by a large number more or less arbitrarily selected as the point where contribution to the integral becomes small. The integrand of the first integral in Equation (3.13) over the 0 to $2k$ interval is singular like $1/(\mu - k)$ when $\mu = k$ and, therefore, the integral can be evaluated although a special scheme must be used to carry out the numerical integration. Monacella[17] has devised a method based on the '3/8 rule' for integration over the 0 to $2k$ interval which works well for computer evaluations.

An alternate method of evaluating the integral over the interval 0 to $2k$ is to write the integral in the form

$$\int_0^{2k} \frac{F(\mu)\,\mathrm{d}\mu}{\mu\,\tanh(\mu h)-\nu} = \int_0^{2k} \frac{F(\mu)-F(k)}{\mu\,\tanh(\mu h)-\nu}\,\mathrm{d}\mu + F(k)\int_0^{2k} \frac{1}{\mu\,\tanh(\mu h)-\nu}\,\mathrm{d}\mu$$

The second integral is further broken down into three intervals, 0 to $k-\varepsilon$, $k+\varepsilon$ and $k+\varepsilon$ to $2k$ where ε denotes a small number. All of the integrals are carried out by numerical integration with the exception of the interval $k-\varepsilon$ to $k+\varepsilon$ where the singularity occurs. In this interval the integrand is expanded in powers of $\mu - k$ as

$$\frac{1}{\mu\,\tanh(\mu h)-\nu} = \frac{a_{-1}}{\mu-k} + a_0 + a_1(\mu-k) + \cdots$$

and each term is integrated analytically giving

$$-\frac{\mathrm{sech}^2(kh)(1-kh\,\tanh(kh))}{[\tanh(kh)+kh\,\mathrm{sech}^2(kh)]^2}(2\varepsilon) + 0(\varepsilon^3)$$

For computing purposes $\varepsilon = 0\cdot1\,k$ respresents a reasonable value.

As an alternate procedure for evaluating the integral form of G, Faltinsen and Michelsen[18] have rearranged Equation (3.13) as follows:

$$G = \frac{1}{R} + \frac{1}{R'} + \frac{1}{[(x-\xi)^2 + (y-\eta)^2 + (z+\zeta)^2]^{1/2}}$$

$$+ 2\text{P.V.} \int_0^{u_1} \frac{(\mu+\nu)\,e^{-\mu h}\cosh\left[\mu(\eta+h)\right]\cosh\left[\mu(y+h)\right]J_0(\mu r)\,\mathrm{d}\mu}{\mu\sinh(\mu h) - \nu\cosh(\mu h)}$$

$$- \int_0^{u_1} e^{(y+\eta)\mu} J_0(\mu r)\,\mathrm{d}\mu$$

$$+ \frac{2\nu}{\pi} \int_0^{\pi} e^{\nu(y+\eta)+i\nu r\cos\theta} E_1\left[-(y+\eta+ir\cos\theta)(u_1-\nu)\right]\,\mathrm{d}\theta$$

$$+ i2\pi \frac{(k^2-\nu^2)\cosh\left[k(y+h)\right]\cosh\left[k(\eta+h)\right]}{k^2 h - \nu^2 h + \nu} J_0(kr) \qquad (3.30)$$

where E_1 denotes the exponential integral as defined by Abramowitz and Stegun[19] and in which the third term represents an image source with respect to the free surface. The value of u_1 must satisfy the following conditions:

(i) $u_1 \geq 2k$ because the principal value integral is evaluated by the special integration technique using the 0 to $2k$ interval,

(ii) $u_1 \geq 4 \cdot 5/h$ so that the assumed approximations,

$$\left.\begin{array}{c} \cosh(\mu h) \\ \sinh(\mu h) \end{array}\right\} \sim \tfrac{1}{2}e^{\mu h} \quad \text{for } \mu \geq u_1$$

are valid.

The form of G given by Equation (3.30) contains no infinite limits of integration and the time required for numerical evaluation tends to be less than that required for Equation (3.13), particularly in the case of small value of k (large period).

3.3.3 Evaluation of α and β

The definitions of α_{ij} and β_{ij} given by Equations (3.20) and (3.23) indicate that $\partial G/\partial n$ and G are to be integrated over the jth panel. It is convenient for discussion purposes to use the form $G = 1/R + G^*$ so that

$$\alpha_{ij} = \frac{1}{2\pi} \iint_{\Delta S_j} \frac{\partial}{\partial n}\left[1/R(x_i, y_i z_i; \xi, \eta, \zeta)\right]\mathrm{d}S$$

$$+ \frac{1}{2\pi} \iint_{\Delta S_j} \frac{\partial G^*}{\partial n}(x_i, y_i, z_i; \xi, \eta, \zeta)\,\mathrm{d}S \qquad (3.31)$$

and

$$\beta_{ij} = \frac{1}{4\pi} \iint\limits_{\Delta S_j} \frac{1}{R(x_i, y_i, z_i; \xi, \eta, \zeta)}\, dS + \frac{1}{4\pi} \iint\limits_{\Delta S_j} G^*(x_i, y_i, z_i; \xi, \eta, \zeta)\, dS$$

$$(3.32)$$

The integrands of the second terms in Equations (3.31) and (3.32), $\partial G^*/\partial n$ and G^*, are regular throughout the fluid domain and oscillate approximately with wave length L as defined by Equation (3.9). In practice L is generally large, (i.e.), at least comparable to the dimensions of the immersed surface, so G^* and $\partial G^*/\partial n$ vary slowly over S and, therefore, are nearly constant over a panel. Thus, a valid and convenient approximation to the integrals in the second terms of Equations (3.31) and (3.32) is to evaluate the integrands at the centroid (node point) of the panel and simply multiply by ΔS_j.

The integrands in the first integrals in Equations (3.31) and (3.32) are not gradually varying when the point i is near the panel over which the integration is to be carried out and are, in fact, singular as $R \to 0$. Thus, when point i is very near the jth panel or when $i = j$ the integrations must be evaluated properly; it is not adequate to make a simple approximation as in the case of the second term in Equations (3.31) and (3.32).

The first integral in (3.32) written in terms of local co-ordinates $\bar{x}$, $\bar{y}$, $\bar{z}$ and $\bar{\xi}$, $\bar{\eta}$ where the $\bar{x}$, $\bar{\xi}$ and $\bar{y}$, $\bar{\eta}$ axes lie in the plane of the panel as indicated in Figure 3.5, is

$$\iint\limits_{\Delta S} \frac{dS}{R} = \iint\limits_{\Delta S} \frac{d\bar{\xi}\, d\bar{\eta}}{[(\bar{x} - \bar{\xi})^2 + (\bar{y} - \bar{\eta})^2 + \bar{z}^2]^{1/2}}$$

$$(3.33)$$

Following a procedure similar to Hess and Smith[15] for the velocity components, Faltinsen and Michelsen[18] integrated this expression with respect to $\bar{\eta}$ arriving at

$$\iint\limits_{\Delta S} \frac{1}{R}\, dS = -\int_{\bar{\xi}_1}^{\bar{\xi}_2} d\bar{\xi} \ln\left\{ \bar{y} - \bar{\eta}_{12} + [(\bar{y} - \bar{\eta}_{12})^2 + (\bar{x} - \bar{\xi})^2 + \bar{z}^2]^{1/2} \right\}$$

$$- \int_{\bar{\xi}_2}^{\bar{\xi}_3} d\bar{\xi} \ln\left\{ \bar{y} - \bar{\eta}_{23} + [(\bar{y} - \bar{\eta}_{23})^2 + (\bar{x} - \bar{\xi})^2 + \bar{z}^2]^{1/2} \right\}$$

$$- \int_{\bar{\xi}_3}^{\bar{\xi}_4} d\bar{\xi} \ln\left\{ \bar{y} - \bar{\eta}_{34} + [(\bar{y} - \bar{\eta}_{34})^2 + (\bar{x} - \bar{\xi})^2 + \bar{z}^2]^{1/2} \right\}$$

$$- \int_{\bar{\xi}_4}^{\bar{\xi}_1} d\bar{\xi} \ln\left\{ \bar{y} - \bar{\eta}_{41} + [(\bar{y} - \bar{\eta}_{41})^2 + (\bar{x} - \bar{\xi})^2 + \bar{z}^2]^{1/2} \right\} \qquad (3.34)$$

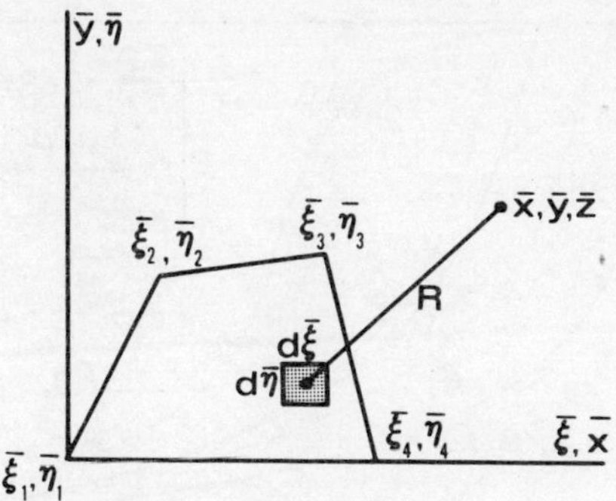

Figure 3.5 Integration over a panel

where

$$\bar{\eta}_{ij} = \bar{\eta}_i + \frac{\bar{\eta}_j - \bar{\eta}_i}{\bar{\xi}_j - \bar{\xi}_i}(\bar{\xi} - \bar{\xi}_i)$$

The integrals in Equation (3.34) may be directly evaluated through numerical integration in most cases. However, the integrand of the first integral is singular when $\bar{z} = 0$, $\bar{\xi} = \bar{x}$ and $\bar{y} - \eta_{12} < 0$ and in this case the first integral may be replaced by

$$\int_{\bar{\xi}_1}^{\bar{\xi}_2} \ln\left[(\bar{x} - \bar{\xi})^2 + \bar{z}^2\right] d\bar{\xi} - \int_{\bar{\xi}_1}^{\bar{\xi}_2} \ln\left[-(\bar{y} - \bar{\eta}_{12}) + \sqrt{((\bar{y} - \bar{\eta}_{12})^2 + (\bar{x} - \bar{\xi})^2 + \bar{z}^2)}\right] d\bar{\xi}$$

The first term of this alternate form can be integrated analytically and the second term can be integrated numerically since it contains no singularities. The difficulties with the integrands of the other integrals are handled in a similar manner.

In Figure 3.6 the integral indicated in Equation (3.33) and as evaluated through numerical integration of Equation (3.34) is compared with the approximation, $(1/R)\,\Delta S$, for panels of different aspect ratios. In this figure results for the particular case where $R_{\text{c.g.}}$ lies in the plane of the panel is studied as an example. The results suggest that for a panel of unit area it is necessary that the point (x, y, z) lie more than two units from the centre of the panel before the simple approximation becomes valid. Thus, for computing purposes the value of $R_{\text{c.g.}} = 2\sqrt{(\Delta S)}$ appears to represent a reasonable lower limit for the use of the simple approximation for panel aspect ratios up to at least 4·0. However, for more square shaped panels this limit can be reduced.

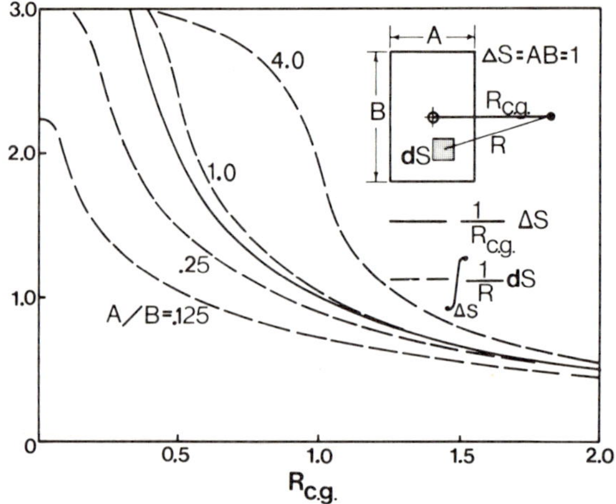

Figure 3.6   Comparison of the potential resulting from a distributed source of unit strength with that resulting from the same source concentrated at the centroid of the panel. (The distance between the centroid of the panel and the point in question is denoted by $R_{c.g.}$)

As a special case the integral of $1/R$ can be evaluated analytically when $R_{c.g} = 0$ provided the panel is rectangular. The result of this integration[21] is

$$\iint_{\Delta S} \frac{1}{R}\, dS = 2[\Delta S/b]^{1/2}\left\{\ln\,[b+(b^2+1)^{1/2}]+b\,\ln\left[\frac{1+(b^2+1)^{1/2}}{b}\right]\right\}$$

$$(3.35)$$

where b denotes the aspect ratio of the panel.

The singular integral in Equation (3.30) represents the velocity induced by a uniformly distributed source over a panel in the direction of the normal vector at some point on the immersed surface. Because of the singular nature of the source the evaluation must be carried out analytically when the point in question is near the panel just as in the preceding discussion regarding integration of $1/R$. This evaluation is most easily accomplished by computing the components in some convenient co-ordinate system attached to the panel and then resolving these components in the x, y, z co-ordinates. In terms of the local co-ordinates of Figure 3.5, Hess and Smith[20] have carried out the integrals analytically with the results,

$$\iint_{\Delta S} \frac{\partial}{\partial \bar{x}}\left(\frac{1}{R}\right) \mathrm{d}S = \frac{\bar{\eta}_2 - \bar{\eta}_1}{\bar{d}_{12}} \ln\left[\frac{\bar{r}_1 + \bar{r}_2 - \bar{d}_{12}}{\bar{r}_1 + \bar{r}_2 + \bar{d}_{12}}\right] + \frac{\bar{\eta}_3 - \bar{\eta}_2}{\bar{d}_{23}}$$

$$\times \ln\left[\frac{\bar{r}_2 + \bar{r}_3 - \bar{d}_{23}}{\bar{r}_2 + \bar{r}_3 + \bar{d}_{23}}\right]$$

$$+ \frac{\bar{n}_4 - \bar{\eta}_3}{\bar{d}_{34}} \ln\left[\frac{\bar{r}_3 + \bar{r}_4 - \bar{d}_{34}}{\bar{r}_3 + \bar{r}_4 + \bar{d}_{34}}\right] + \frac{\bar{\eta}_1 - \bar{\eta}_4}{\bar{d}_{14}} \ln\left[\frac{\bar{r}_4 + \bar{r}_1 - \bar{d}_{41}}{\bar{r}_4 + \bar{r}_1 + \bar{d}_{41}}\right]$$

$$(3.36)$$

$$\iint_{\Delta S} \frac{\partial}{\partial y}\left(\frac{1}{R}\right) \mathrm{d}S = \frac{\bar{\xi}_1 - \bar{\xi}_2}{\bar{d}_{12}} \ln\left[\frac{\bar{r}_1 + \bar{r}_2 - \bar{d}_{12}}{\bar{r}_1 + \bar{r}_2 + \bar{d}_{12}}\right] + \frac{\bar{\xi}_2 - \bar{\xi}_3}{\bar{d}_{23}} \ln\left[\frac{\bar{r}_2 + \bar{r}_3 - \bar{d}_{23}}{r_2 + r_3 + \bar{d}_{23}}\right]$$

$$+ \frac{\bar{\xi}_3 - \bar{\xi}_4}{\bar{d}_{34}} \ln\left[\frac{\bar{r}_3 + \bar{r}_r - \bar{d}_{34}}{\bar{r}_3 + \bar{r}_4 + \bar{d}_{34}}\right] + \frac{\bar{\xi}_4 - \bar{\xi}_1}{\bar{d}_{41}} \ln\left[\frac{\bar{r}_4 + \bar{r}_1 - \bar{d}_{41}}{\bar{r}_4 + \bar{r}_1 + \bar{d}_{41}}\right]$$

$$(3.37)$$

$$\iint_{\Delta S} \frac{\partial}{\partial \bar{z}}\left(\frac{1}{R}\right) \mathrm{d}S = \tan^{-1}\left[(\bar{m}_{12}\bar{e}_1 - \bar{h}_1)/\bar{z}\bar{r}_1\right] - \tan^{-1}\left[(\bar{m}_{12}\bar{e}_2 - \bar{h}_2)/\bar{z}\bar{r}_2\right]$$

$$+ \tan^{-1}\left[(\bar{m}_{23}\bar{e}_2 - \bar{h}_2)/\bar{z}\bar{r}_2\right] - \tan^{-1}\left[(\bar{m}_{23}\bar{e}_3 - \bar{h}_3)/\bar{z}\bar{r}_3\right]$$

$$+ \tan^{-1}\left[(\bar{m}_{34}\bar{e}_3 - \bar{h}_3/\bar{z}\bar{r}_3\right] - \tan^{-1}\left[(\bar{m}_{34}\bar{e}_4 - \bar{h}_4)/\bar{z}\bar{r}_4\right]$$

$$+ \tan^{-1}\left[(\bar{m}_{41}\bar{e}_4 - \bar{h}_4)/\bar{z}\bar{r}_4\right] - \tan^{-1}\left[(\bar{m}_{41}\bar{e}_1 - \bar{h}_1)/\bar{z}\bar{r}_1\right]$$

$$(3.38)$$

where

$$\bar{d}_{12} = [(\bar{\xi}_2 - \bar{\xi}_1)^2 + (\bar{\eta}_2 - \bar{\eta}_1)^2]^{1/2}$$
$$\bar{d}_{23} = [(\bar{\xi}_3 - \bar{\xi}_2)^2 + (\bar{\eta}_3 - \bar{\eta}_2)^2]^{1/2}$$
$$\bar{d}_{34} = [(\bar{\xi}_4 - \bar{\xi}_3)^2 + (\bar{\eta}_4 - \bar{\eta}_3)^2]^{1/2}$$
$$\bar{d}_{41} = [(\bar{\xi}_1 - \bar{\xi}_4)^2 + (\bar{\eta}_1 - \bar{\eta}_4)^2]^{1/2}$$

$$(3.39)$$

in which

$$\bar{m}_{12} = \frac{\bar{\eta}_2 - \bar{\eta}_1}{\bar{\xi}_2 - \bar{\xi}_1} \qquad \bar{m}_{23} = \frac{\bar{\eta}_3 - \bar{\eta}_2}{\bar{\xi}_3 - \bar{\xi}_2}$$

$$\bar{m}_{34} = \frac{\bar{\eta}_4 - \bar{\eta}_3}{\bar{\xi}_4 - \bar{\xi}_3} \qquad \bar{m}_{41} = \frac{\bar{\eta}_1 - \bar{\eta}_4}{\bar{\xi}_1 - \bar{\xi}_4}$$

$$(3.40)$$

and

$$\bar{r}_k = [(\bar{x} - \bar{\xi}_k)^2 + (\bar{y} - \bar{\eta}_k)^2 + \bar{z}^2]^{1/2}, \qquad k = 1, 2, 3, 4$$

$$\bar{e}_k = \bar{z}^2 + (\bar{x} - \bar{\xi}_k)^2, \qquad k = 1, 2, 3, 4 \qquad\qquad (3.41)$$

$$\bar{h}_k = (\bar{y} - \bar{\eta}_k)(\bar{x} - \bar{\xi}_k), \qquad k = 1, 2, 3, 4$$

Evaluation of Equations (3.36) and (3.37) causes no trouble in practice. They become infinite on the edges of the quadrilaterals but in practice one can avoid evaluation there. The result of Equation (3.38) tends to zero as $\bar{z} \to 0$ if the point is outside the panel. If $\bar{z} \to 0$ with the point inside the panel the result is $\pm 2\pi$, the $(+)$ sign corresponding to approaching the limit through positive values of $\bar{z}$ and the $(-)$ sign corresponding to approaching the limit through negative values. Positive values of $\bar{z}$ correspond to the outside of the body so for purposes of computer evaluation the integral in Equation (3.38) is set to $+2\pi$ when $\bar{z}$ is less than some arbitrarily small number in order to eliminate the ambiguous situation when $\bar{z} = 0$. It may be noted that the 2π contribution is accounted for in the integral Equation (3.17) by virtue of the $f(x, y, z)$ term and, therefore, would not be included in the diagonal elements of the α-matrix.

The radial velocity produced by the uniformly distributed source in the plane of the panel computed by Equation (3.36) is compared with the results of the simple approximation $[(x - \xi)/R^3]\,\Delta S$, in Figure 3.7 for panels of different aspect ratio. Similarly to the results shown in Figure 3.6, these results which

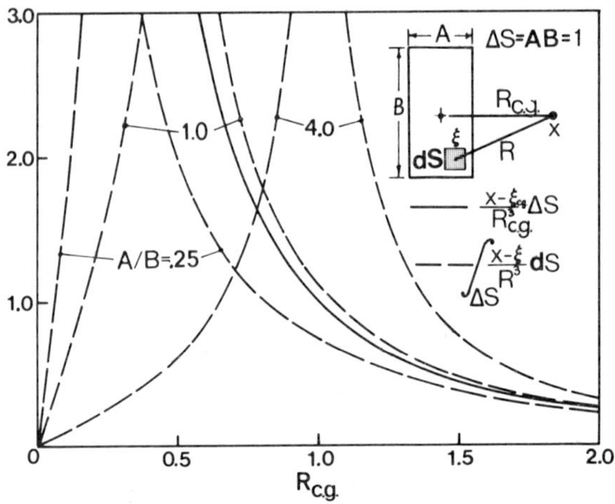

Figure 3.7 Comparison of the in-plane velocity induced by a distributed source of unit strength with that induced by the same source concentrated at the centre of the panel

are based on a panel of unit area indicate that the point where the induced velocities are to be computed must also be at least a distance of $2\sqrt{(\Delta S)}$ away from the centre of the panel before the simple approximation becomes valid for panel aspect ratios of up to 4·0.

3.4 WAVE LOADS ON FIXED STRUCTURES

3.4.1 General

The diffraction theory based on a Green's function solution procedure along with the Morison equation is well-suited to the calculation of wave loads on practical offshore platform designs. A rather versatile calculation procedure can be based on the use of diffraction theory for large displacement portions of the structures and the use of Morison's equation for relatively small diameter members.

For purposes of discussion consider the structure composed of a caisson with superstructure as shown in Figure 3.8. The influence of the superstructure on the caisson may be neglected and the wave interaction with the caisson alone considered first.

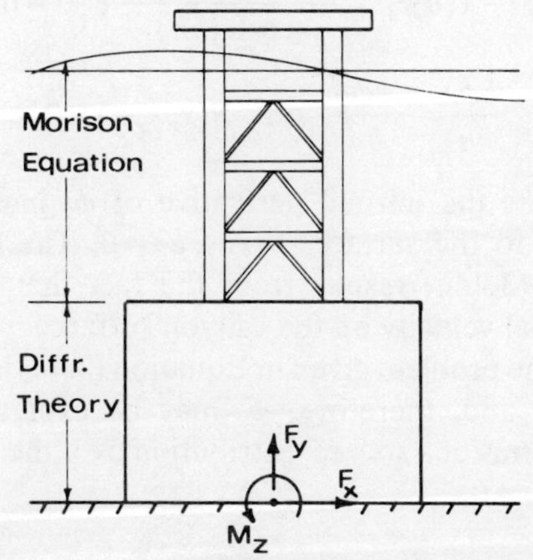

Figure 3.8 Zones of applicability of diffraction theory

The velocity potential associated with the wave/caisson interaction is written as the sum

$$\Phi = \Phi_0 + \Phi_s \tag{3.42}$$

where Φ_0 denotes the velocity potential corresponding to the incident wave and Φ_s represents the scattered wave potential which is a result of the presence of the caisson. The complex linear incident wave potential is well-known and is given by

$$\phi_0 = -i\frac{gH}{2\omega}\frac{\cosh\left[k(h+y)\right]}{\cosh(kh)}\,e^{i(kx\cos\gamma + kz\sin\gamma)} \tag{3.43}$$

where $\Phi_0 = \mathrm{Re}\left[\phi_0\,e^{-i\omega t}\right]$. The direction of propagation of the incident wave described by Equation (3.43) is defined by γ, the wave height is H and k denotes the wave number which is related to the frequency of the wave motion through Equation (3.9).

The boundary-value problem for the scattered wave potential takes the form,

$$\nabla^2\phi_s(x, y, z) = 0 \tag{3.44a}$$

$$\frac{\partial\phi_s}{\partial y}(x, -h, z) = 0 \tag{3.44b}$$

$$\frac{\partial\phi_s}{\partial y}(x, 0, z) - \frac{\omega^2}{g}\phi_s(x, 0, z) = 0 \tag{3.44c}$$

$$\phi_s(r, \theta, y) - \lambda(\theta)r_1^{-1/2}\frac{\cosh\left[k(y+h)\right]}{\cosh(kh)}\,e^{ikr_1} \to 0,\ r_1 \to \infty \tag{3.44d}$$

$$\frac{\partial\phi_s}{\partial n} = -\frac{\partial\phi_0}{\partial n} \quad \text{on} \quad S(x, y, z) = 0 \tag{3.44e}$$

where $\partial\phi_s/\partial n$ denotes the normal derivative of ϕ_s in the direction of the outward normal (**n**) to the surface, $S(x, y, z) = 0$. The kinematic boundary condition Equation (3.44e) results from the fact that $\partial\phi/\partial n = 0$ on $S = 0$ specifying zero normal velocity on the caisson surface.

The boundary-value problem given in Equation (3.44) is the same as the one discussed previously and, therefore, ϕ_s may be expressed in the form of Equation (3.10) in terms of a source distribution over the immersed surface of the caisson, $S(x, y, z) = 0$,

$$\phi_s = \frac{1}{4\pi}\iint_S f(\xi, \eta, \zeta)G(x, y, z; \xi, \eta, \zeta)\,\mathrm{d}S \tag{3.45}$$

The solution procedure is precisely as described in Section 3 with ϕ being replaced by ϕ_s and v_n being replaced by $-\partial\phi_0/\partial n$ in Equation (3.44e). The latter quantity is computed through differentiation of Equation (3.43).

The pressure on the immersed surface of the caisson is computed through Equation (3.27) replacing ϕ by $\phi_0 + \phi_s$ and disregarding the hydrostatic term since the buoyant force is accounted for separately. The first term in Equation

(3.37) is, of course, the most important but it is not difficult to include the effect of the quadratic terms. The derivatives of ϕ are given, for example, by $\phi_x = \phi_{0_x} + \phi_{s_x}$, etc. where ϕ_{s_x} is determined through Equation (3.25) and ϕ_{0_x} is obtained through straightforward differentiation of Equation (3.43). Although the inclusion of the velocity-squared terms is not mathematically consistent since in the linear theory second-order effects are neglected, in practice it is found that these terms account for the up-lift force being slightly larger than the downward force for the case of submerged caissons. The terms have no significant effect, however, on the horizontal force or overturning moment.

The forces and moments acting on the caisson may be computed through integration of the pressure distribution over the immersed surface according to Equations (3.28) and (3.29).

3.4.1 Wave loads on the superstructure

The wave loads on the superstructure may be computed by use of the Morison equation. However, the wave kinematics used in this calculation should reflect the effect of the caisson, an effect which tends to increase the velocity and acceleration and resulting superstructure loads due to the 'blockage' effect of the caisson.

The force vector acting on some differential length of member $d1$ may be expressed, in terms of the local fluid velocity and acceleration vector according to the Morison equation,

$$\mathbf{dF} = \rho(1+C_m)\frac{\pi D^2}{4}(\mathbf{a}.\mathbf{b}_a)\mathbf{b}_a\, d1 + \tfrac{1}{2}\rho C_d D(\mathbf{q}.\mathbf{b}_q)^2\mathbf{b}_q\, d1 \qquad (3.46)$$

in which D denotes the local diameter of the member, and $\mathbf{q}$ and $\mathbf{a}$ denote the velocity and acceleration vectors, respectively, of the fluid. In Equation (3.46) $\mathbf{b}_a$ represents a unit vector which gives direction to the inertia force and is defined by

$$|\mathbf{b}_a| = 1 \qquad (3.47a)$$

$$\mathbf{b}_a.\mathbf{c} = 0 \qquad (3.47b)$$

$$[d\mathbf{b}_a.\mathbf{a}\times\mathbf{c} = 0 \qquad (3.47c)$$

$$\mathbf{a}.\mathbf{b}_a > 0 \qquad (3.47d)$$

where $\mathbf{c}$ denotes a unit vector directed along the axis of the member; Equation (3.47c) insures that $\mathbf{b}_a$ will lie in the plane described by the acceleration vector and axis vector, $\mathbf{a}$ and $\mathbf{c}$, respectively, and Equation (3.7d) insures that $\mathbf{b}_a$ will be orientated in the direction of positive acceleration. The unit vector $\mathbf{b}_a$ is also defined by Equation (3.47) but with the acceleration being replaced by the velocity vector. An alternate but equivalent procedure to the use of the unit vectors $\mathbf{b}_a$ and $\mathbf{b}_q$ is to resolve the velocity and acceleration vector into local coordinates attached to the member, compute the drag and inertia force due to the cross flow components and then resolve the forces back to the global coordinates.

The unit vectors $\mathbf{b}_a$ and $\mathbf{b}_q$ as defined by Equations (3.47) are orientated in the direction of the cross-flow components of $\mathbf{a}$ and $\mathbf{q}$. Accordingly, $(\mathbf{a} \cdot \mathbf{b}_a)$ denotes the magnitude of the acceleration normal to the axis of the member and $(\mathbf{q} \cdot \mathbf{b}_q)$ denotes the component of the velocity in a direction perpendicular to the axis of the member. These cross-flow components are, therefore, used in the Morison equation, Equation (3.46), to compute the local forces.

3.4.2 Caisson/superstructure interaction

The superstructure is assumed to be composed of small diameter members and, therefore, assumed to have negligible effect on the caisson except possibly near the caisson superstructure-juncture. The caisson has a significant effect on the superstructure flow field, however; because of its blockage effect the velocity and acceleration of the flow over the top of the caisson is increased and loads acting on the superstructure are, therefore, also increased.

The velocity potential representing the combined effect of the incident wave and the scattered wave is given by Equation (3.42). The x-component of velocity, for example, is given by

$$q_x = \partial\Phi/\partial x = \partial\Phi_0/\partial x + \partial\Phi_s/\partial x = \mathrm{Re}\left[(\partial\phi_0/\partial x + \partial\phi_s/\partial x)\,e^{-i\omega t}\right] \quad (3.48)$$

where $\partial\phi_0/\partial x$ is obtained from the incident wave potential given by Equation (3.43). The contribution from the scattering potential, $\partial\phi_s/\partial x$, follows directly from (3.25) where the values of α_{x_j}, etc. are evaluated at the local value of x, y, z on the superstructure member. The addition of $\partial\phi_s/\partial x$ to $\partial\phi_0/\partial x$ accounts for the effect of the caisson on the incident wave motion.

For purposes of computer calculation it is convenient to identify node points and corresponding coordinates on the superstructure. In practice, a node point would always be placed at a joint and in cases where the members are very long, intermediate node points inserted. The diameter of members extending between the node points on the superstructure denoted, for example, by i and j, can then be represented by the D_{ij} element in a 'diameter matrix'. Since the effect of ϕ_s represents a secondary effect with respect to the superstructure loads, its evaluation at the superstructure node points only is adequate. A linear interpolation at intermediate points on the member may then be assumed.

Turning now to the effect of the caisson on the superstructure, it appears that if the superstructure legs which attach to the caisson are small, the error in the caisson loads caused by the covered area may be neglected. However if this is not the case it may be worthwhile to correct for this since it tends to affect the uplift force and moment. One simple procedure is to apply an axial force to the superstructure leg at the attachment point equal to the pressure at that location on the bare caisson times the area covered. This force cancels the force produced by the pressure times the area covered on the caisson.

A more accurate method of dealing with the caisson/superstructure juncture is to consider the lower part of the superstructure legs as part of the caisson and change over to application of the Morison equation some distance above the juncture. Figure 3.9 shows half of a tripod type platform which was treated in this manner; the structure is symmetrical so only half of the caisson is shown. The legs in this structure are so large that they actually represent an integral part of the caisson and cannot be simply treated independent of the caisson.

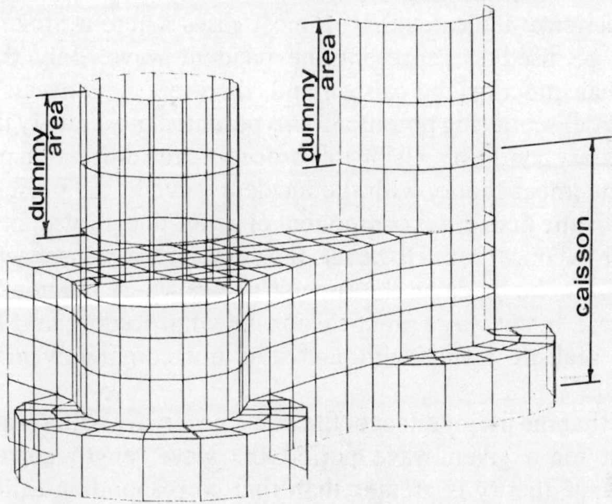

Figure 3.9　Subdivision of structure surface for source
calculation

As indicated in Figure 3.9 the legs are treated as part of the caisson and are terminated well above the top of the triangular base. However, the truncation of the leg will allow flow over the top of the stubs and, accordingly, the pressure in the region of the cut-off plane will be distorted. Thus, at least the top diameter of the shafts must be treated as 'dummy area'; sources are distributed over this area but in computing the caisson loads the pressure integration does not include this part of the caisson.

The loads on the portion of the shafts above the portion of the structure labeled 'caisson' are treated by use of the Morison equation. However, here again it is important not to introduce unnatural effects caused by the truncation of the shafts. The sources near the truncation level cause a rather high velocity flow over the top of the truncated shafts so to avoid this unnatural effect the contribution to the superstructure velocity and acceleration from the sources on the shafts are disregarded.

3.4.3 Non-linear effects

The procedures outlined in the preceding are based on linear theory to the extent that a linear incident wave and linear free surface boundary condition are assumed. There remains, therefore, the question of the importance of non-linear effects.

For a typical gravity platform the caisson is normally well-submerged so that the higher-order components in non-linear waves have little effect on the caisson. Experience indicates that in most cases where a Stoke's fifth-order wave[22] may be used to represent the incident wave, only the first-order component has effect on the caisson and, therefore, it is necessary to define only a first-order scattering potential. This potential must satisfy the linear free surface boundary condition. Using this procedure the diffraction problem is the same as in the linear theory with the incident wave in the present case being represented by the first-order component of a non-linear incident wave insofar as the caisson is concerned. However, the complete non-linear wave theory is used to compute the loads on the superstructure where the non-linear effects are the greatest. In this way a practical non-linear procedure may be developed which gives realistic results although it is not completely mathematically consistent.

It appears that the major effect of the non-linear wave with respect to caisson loads is that for a given wave period the wave length corresponding to non-linear wave theory is greater than that corresponding to linear theory. Since the horizontal force tends to decrease with the parameter $2\pi a/L$ above a given value, an increase in L results in a decrease in $2\pi a/L$ and corresponding increase in caisson loads. For a typical North Sea gravity platform this may be on the order of 5 per cent of the total horizontal load. Hogben and Standing[23] reached similar conclusions regarding non-linear effects in the case of North Sea gravity platforms.

3.4.4 Results of wave load calculations

It is appropriate to compare numerical results with the results of closed-form solutions for purposes of providing the validity and convergence of numerical procedures. In the case of wave force calculations only one strictly closed-form result from diffraction theory is known: that of MacCamy and Fuchs[10] for the vertical circular cylinder. Such comparisons for the circular cylinder and also for a semi-submerged sphere have been made by Garrison[21] so for the sake of brevity these will not be repeated here. Comparison with these results in general show excellent agreement.

However, for purposes of judging the rate of convergence of numerical results Table 3.1 is presented showing the horizontal force, vertical force and moment coefficients for a rectangular bottom-mounted caisson 100 metres by 100 metres in plan by 50 metres high placed in 100 metres of water. Numerical results presented for a total of 48, 108 and 192 panels show very rapid convergence, in general, and indicate that of the three components the moment is the most sensitive to subdivision size.

Table 3.1 Convergence of wave force results (caisson: 100 m × 100 m × 50 m high, depth = 100 m)

T(sec)	$F_x(\text{max})/\rho ga^3(H/2a)$			$F_y(\text{max})/\rho ga^3(H/2a)$			$M_z(\text{max})/\rho ga^4(H/2a)$		
	$N=48$	$N=108$	$N=192$	$N=48$	$N=108$	$N=192$	$N=48$	$N=108$	$N=192$
10·0	0·143	0·148	0·142	0·408	0·399	0·401	0·277	0·262	0·264
12·0	0·744	0·751	0·753	1·026	1·021	1·018	0·140	0·135	0·132
14·0	1·371	1·370	1·367	1·730	1·714	1·708	0·081	0·075	0·073
16·0	1·823	1·807	1·802	2·315	2·291	2·280	0·263	0·247	0·240
18·0	2·075	2·057	2·050	2·710	2·690	2·679	0·382	0·359	0·350
20·0	2·192	2·170	2·158	2·980	2·959	2·947	0·452	0·424	0·412
22·0	2·214	2·190	2·181	3·170	3·148	3·137	0·487	0·456	0·444
24·0	2·191	2·166	2·154	3·305	3·286	3·278	0·500	0·469	0·457

As indicated in Section 3.3, when the wave length becomes large relative to the dimensions of the caisson the free surface boundary condition reduces to the 'rigid wall' condition and the Green's function simplifies to the form given in Equation (3.16c). In order to give some insight into the range of applicability of such an asymptotic approximation, the horizontal force results based on the Green's function given in Equations (3.16c) and (3.13–3.14) are compared in Figure 3.10. Taking the results based on Equations (3.13–3.14) as a standard, Figure 3.10 shows that the results for the vertical cylinder of one radii height based on the asymptotic form of G for long waves approach the correct results, as expected, when $ka = 2\pi a/L \to 0$ and/or when the caisson is deeply submerged, i.e., when h/a becomes large. These results indicate that there are many instances in practice where the simple asymptotic form of the Green's function given by Equation (3.16c) would yield results of adequate accuracy. In fact, the proportions of the example caisson corresponding to Figure 3.10 are typical of current North Sea gravity platforms and assuming a typical water depth of 150·0 metres and wave period of 16·0 seconds gives the following dimensionless parameters: $ka = 0·8$ and $h/a = 3·0$. Reference to Figure 3.10

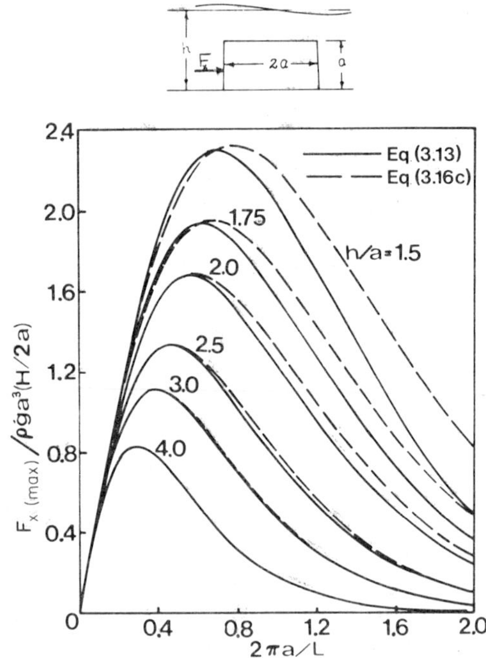

Figure 3.10 Comparison of wave force results based on the general and asymptotic form of the Green's function

indicates that the results based on the asymptotic and general form of the Green's function are in very close agreement at these conditions.

The use of the asymptotic Green's function [Equation (3.16c)] has the advantage that the CPU time requirements for both generation of the large arrays and inversion of the α-matrix are much less than for the general form. If out-of-core storage is used it is practical using this approximate procedure to make computations for highly complex configurations which require a large number of panels. The α-matrix is real and because of its very strong diagonal Equation (3.20) can be solved very efficiently by use of the Gauss-Siedel iteration technique.

Calculations have also been made for comparison with wave channel results reported by Iversen, Leivseth and Tørum[24] for two slightly different CON-DEEP structures which are defined in Figure 3.11. The two structures are alike except that one is constructed of 18·0 metre diameter cylinders and the other one of 20·0 metre diameter cylinders. The structures were symmetrical about one vertical plane so half the caisson and superstructure were input to the computer for numerical calculation as shown in Figure 3.12. In this instance the

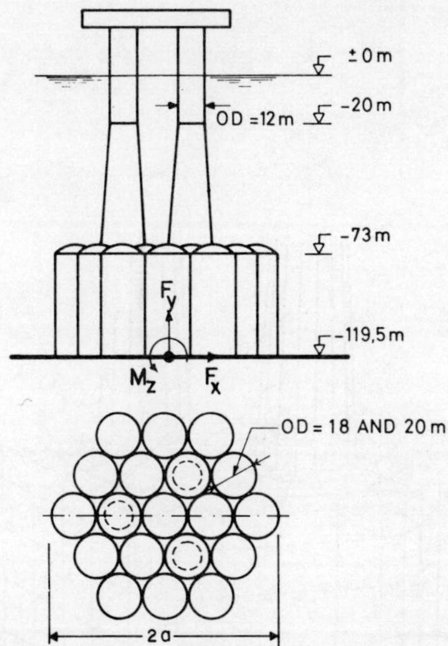

Figure 3.11 Definition of CONDEEP structures

lower part of the superstructure legs was treated as part of the caisson as previously described.

Calculations were compared with these same experimental results in an earlier paper,[25] but these calculations were somewhat approximate because the superstructure loads were based on the kinematics along the vertical centre line of the structure and a linear wave was used as the incident wave as opposed to the fifth-order wave used herein. The present calculations based on the discretized model of Figure 13.12 are presented in Figures 3.13–3.15 and show excellent agreement between the theory and experiment. Figure 3.13 shows the variation of the loads as a function of time over a complete cycle of the motion and, as indicated in the figure, the experimental and computed results are practically identical. In Figures 3.14 and 3.15 are shown the maximum values of the load components as a function of wave height and here again the agreement between theory and experiment is excellent. Hogben and Standing[23] have also presented comparisons of experimental results with diffraction theory which also show good agreement.

The calculations presented in Figures 3.13–3.15 were carried out using a Stoke's fifth-order incident wave. The pressure equation included contributions from the $\partial \phi / \partial t$ term from the incident wave plus that due to the scattering of the first-order component. In addition, the first-order component of the

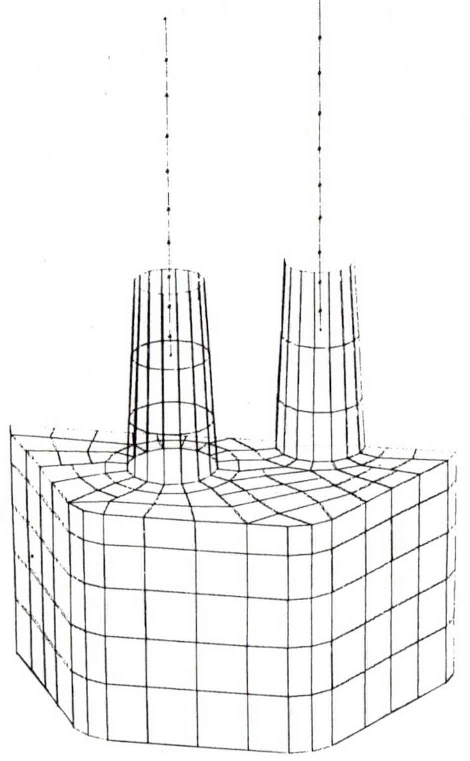

Figure **3.12** Mathematical model of CONDEEP platform

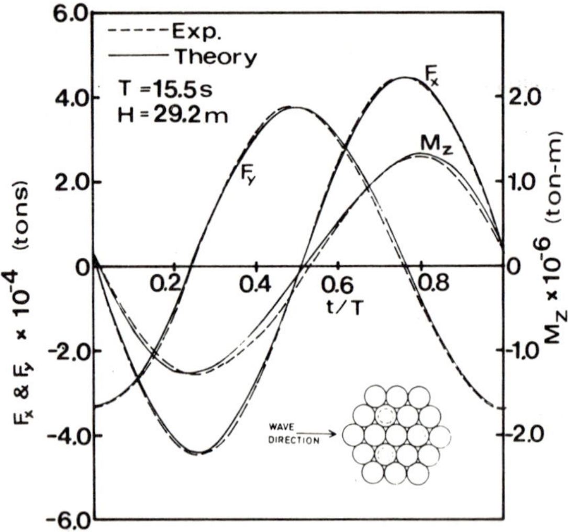

Figure **3.13** Wave loads on the 20 metre CONDEEP platform

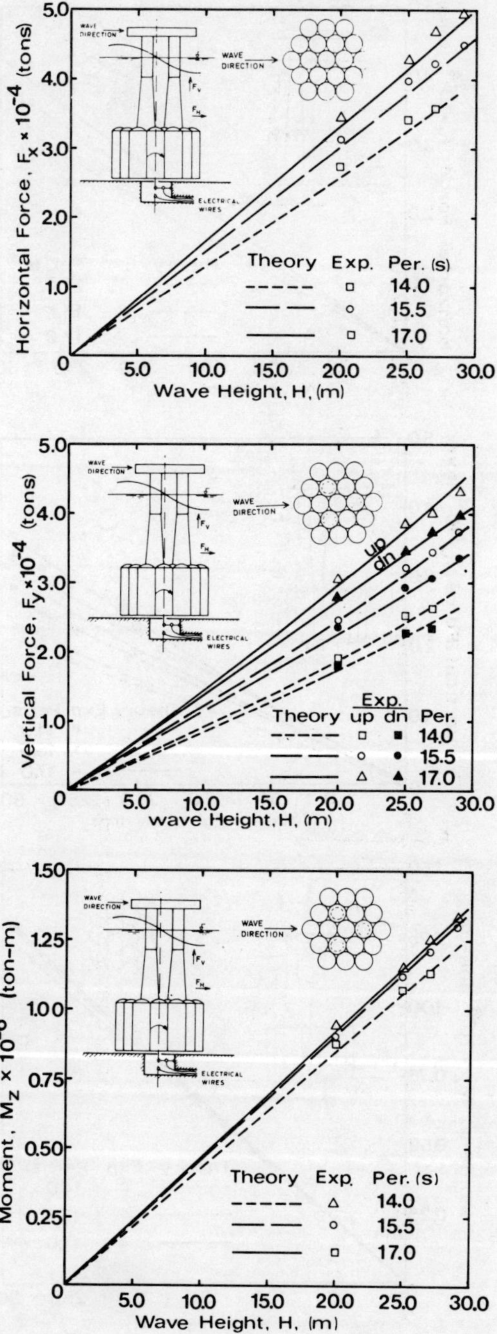

Figure 3.14 Wave loads on the 20 metre CONDEEP platform

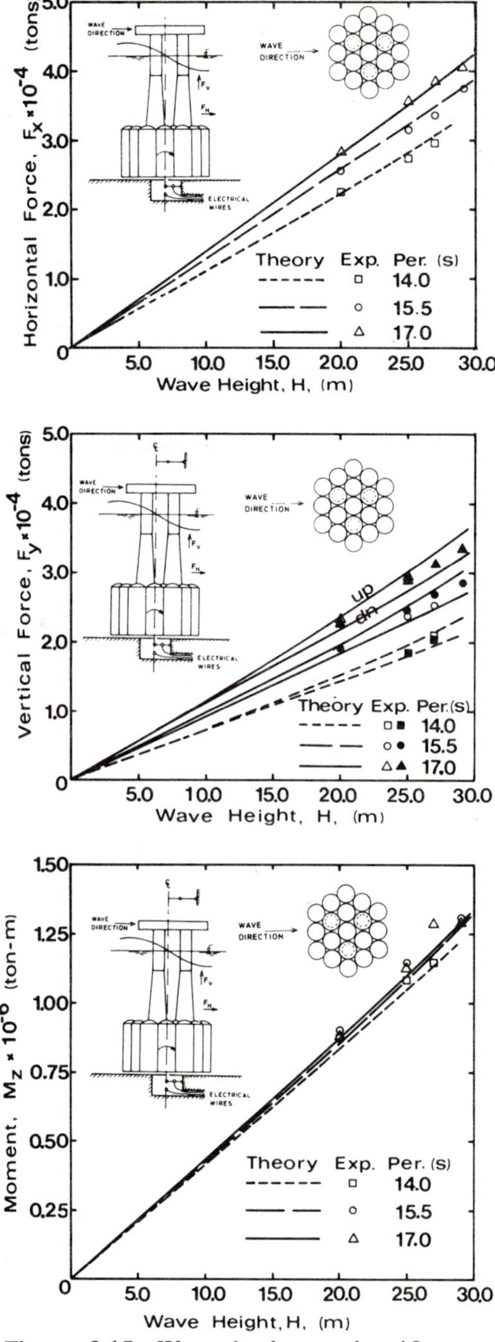

Figure 3.15 Wave loads on the 18 metre CONDEEP platform

incident and scattered wave velocities were included in the quadratic term. This accounts for the fact that the uplift force shown in Figures 3.13–3.15 is slightly greater than the downward force.

Comparison of Figures 3.11 and 3.12 indicates that the numerical distributed source model did not include the scalloped detail on the caisson but instead the caisson was represented by plane surfaces. The top surface was set at a mean elevation between the peaks and valleys and the sides were drawn tangent to the cylinders. This choice was made in order to properly model horizontal forces, but it did cause a small error in the vertical force and moment because the mathematical model had a larger plan form area than the physical model. In the calculations presented in Figures 3.13–3.15 corrections on the vertical force and moment for this area discrepancy were made by simply multiplying the computed caisson pressures times the area involved.

3.4.5 Spectral analysis

In wave force analysis the terms 'deterministic' and 'spectral' have come into common usage. Deterministic analysis refers to the direct representation of the force as a function of the incident wave motion whereas in a spectral analysis the fluid motion is represented by a spectrum and statistical properties of the loads are determined. [*vide* Chapter 2].

Linear systems have the fundamental property that if the random input is described by its energy spectrum, $S(\omega)$, then the spectrum of the response is characterized by a linear transfer function

$$S_R(\omega) = |R(\omega)|^2 S(\omega) \tag{3.49}$$

where $R(\omega)$ denotes the transfer function and $S_R(\omega)$ denotes the response spectrum. In the case of wave force analyses the random input is represented by the incident wave spectrum and the transfer function represents the ratio of any one of the components of wave load to the wave height as a function of frequency, $\omega = 2\pi/T$. Once the response spectrum for any one of the components of the wave loads is determined through Equation (3.49), certain statistical properties such as the mean, significant and maximum wave load associated with the computed response spectrum may be obtained from the variance or area under the spectrum,

$$m_0 = \int_0^\infty S_R(\omega)\, d\omega \tag{3.50}$$

Several theoretical wave spectra are in usage for computing loads on offshore structures. For the North Sea the JONSWAP spectrum, as described by Rye, Byrd and Tørum,[26] is the most common. This spectrum is very sharply peaked, having a great deal of its energy concentrated near the 16·0 second period.

The application of spectral analysis in wave force analysis, in general, is problematic because the basic assumption behind the use of Equation (3.49) is that the transfer function is linear. However, in the case of large displacement structures the loads tend to be highly linear as indicated in Figures 3.14 and 3.15, and therefore a straightforward application of spectral analysis should yield good results.

3.5 RESPONSE OF FLOATING BODIES

3.5.1 General formulation

The theoretical hydrodynamic analysis of the motion of floating bodies relies on the principle of superposition of linear solutions to problems of the type discussed in Section 3.3. The formal development begins with the assumption that the amplitude of the waves which excites the motion of a floating body is small, of order ε, and that the resulting induced motion of the body is also of order ε. The boundary-value problem is formulated with terms of $0(\varepsilon^2)$ and above disregarded in comparison to terms of order ε.

The result of this linearization procedure shows that it is possible to deal with the rather complex motion associated with wave interaction with a floating body which is at the same time responding in six degrees of freedom in a fairly simple way. Through linearization, the complex problem under consideration can be decomposed and treated as the sum of seven separate problems: the fluid motion produced by the (1) surge, (2) heave, (3) sway, (4) roll, (5) yaw and (6) pitch motion of the floating body, one degree of freedom at a time in otherwise still water and (7) the interaction of a regular wave with the restrained body.

Wehausen[27] has given an excellent review of the general problem emphasizing ship related hydrodynamics. This will not be repeated here; in the following the Green's function method of Section 3.3 will be applied to calculation of the motion of large three-dimensional floating caissons.

The small amplitude periodic motion of the floating body may be described by

$$X_j = \mathrm{Re}\,[X_j^0\,e^{-i\omega t}], \qquad j = 1, 2, \ldots, 6 \qquad (3.51\mathrm{a})$$

where X_j^0 $(j = 1, 2, 3)$ denotes the complex amplitude of the displacements in surge, heave and sway and

$$X_j^0 = a\Theta_j^0, \qquad j = 4, 5, 6 \qquad (3.51\mathrm{b})$$

where a represents the characteristic dimension of the body or floating caisson and θ_j^0 denotes the complex angular amplitudes of the motion in roll, yaw and pitch. The linear and angular displacement about the body axes $0(x', y', z')$ are noted in the definition sketch, Figure 3.16.

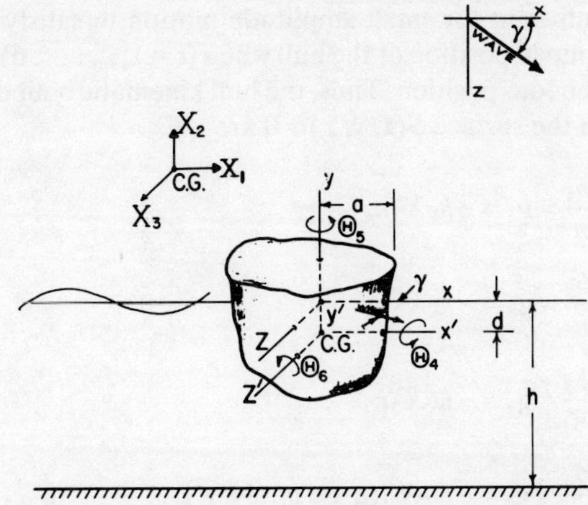

Figure 3.16   Definitions for floating body analysers

The complex potential for wave interaction with the floating body is given by

$$\phi = \sum_{j=0}^{7} \phi_j \tag{3.52}$$

where $\phi_0 =$ the complex potential associated with the incident wave [Equation (3.43)], ϕ_i ($i = 1, 2, \ldots, 6$) denote the complex potentials associated with the six degrees of freedom of the body and ϕ_7 denotes the complex potential associated with the scattering of the incident wave by the restrained body. The space dependent part of the potentials are related to the complete velocity potentials according to $\Phi_i(x, y, z, t) = \text{Re}\,[\phi_i(x, y, z)\,e^{-i\omega t}]$ for $i = 1, 2, \ldots, 6$.

Each of the complex potentials must satisfy the Laplacian,

$$\nabla^2 \phi_i = 0, \qquad i = 0, 1, \ldots, 7 \tag{3.53}$$

the kinematic boundary condition on the bottom,

$$\frac{\partial \phi_i}{\partial y}(x, -h, 0) = 0, \qquad i = 0, 1, 2, \ldots, 7 \tag{3.54}$$

and the linearized free surface boundary condition,

$$\frac{\partial \phi_i}{\partial y}(x, 0, z) - \frac{\omega^2}{g}\phi_i(x, 0, z) = 0, \; i = 0, 1, \ldots, 7 \tag{3.55}$$

On the surface of the body it is necessary that ϕ_i ($i = 1, 2, \ldots, 7$) satisfy the appropriate kinematic boundary condition. Linearization here again is used to show that it is sufficient for small amplitude motion to satisfy this boundary condition on the mean position of the hull when ($i = 1, 2, \ldots, 6$) rather than on the exact instantaneous position. Thus, the hull kinematic boundary conditions to be satisfied on the surface $S(x, y, z) = 0$ are

$$\frac{\partial \phi_1}{\partial n} = g_1 = -i\omega X_1^0 n_x \tag{3.56a}$$

$$\frac{\partial \phi_2}{\partial n} = g_2 = -i\omega X_2^0 n_y \tag{3.56b}$$

$$\frac{\partial \phi_3}{\partial n} = g_3 = -i\omega X_3^0 n_z \tag{3.56c}$$

$$\frac{\partial \phi_4}{\partial n} = g_4 = -i\omega \Theta_4^0 [(d + y)n_z + z n_y] \tag{3.56d}$$

$$\frac{\partial \phi_5}{\partial n} = g_5 = -i\omega \Theta_5^0 [z n_x - x n_z] \tag{3.56e}$$

$$\frac{\partial \phi_6}{\partial n} = g_6 = -\omega \Theta_6^0 [x n_y - (d + y)n_x]$$

$$\frac{\partial \phi_7}{\partial n} = g_7 = -\frac{\partial \phi_0}{\partial n} = \frac{ig\eta^0 k}{\omega \cosh (kh)} [n_y \sinh [k(h + y)]$$
$$+ i(n_x \cos \gamma + n_z \sin \gamma) \cosh [k(h + y)]]$$
$$\cdot e^{i(kx \cos \gamma + kz \sin \gamma)} \tag{3.56f}$$

where $\mathbf{n} = i n_x + j n_y + k n_z$ denotes the outward unit normal vector on $S(x, y, z) = 0$ and d denotes the depth of submergence of the origin of the body axes which is located at the centre of gravity of the body. Finally, the radiation ($i = 1, 2, \ldots, 6$) and scattering ($i = 7$) potentials must satisfy the radiation condition given in Equation (3.8).

The boundary-value problems defined by Equations (3.53–56) are of the type discussed in Section 3.3 and, therefore, the solution may be expressed in terms of the Green's function as in Equation (3.10),

$$\phi_k(x, y, z) = \frac{1}{4\pi} \iint_S f_k(\xi, \eta, \zeta) G(x, y, z; \xi, \eta, \zeta) \, dS \tag{3.57}$$

where $k = 1, 2, \ldots, 6$ denotes the six degrees of freedom and $k = 7$ denotes scattering. The Green's function is, of course, the same for all seven boundary-value problems but a different distribution function $f_k(\xi, \eta, \zeta)$ ($k = 1, 2, \ldots, 7$) is associated with each of the different problems.

The integral equations are also of the form given by Equation (3.17),

$$-f_k(x, y, z) + \frac{1}{2\pi} \iint_S f_k(\xi, \eta, \zeta) \frac{\partial G}{\partial n}(x, y, z; \xi, \eta, \zeta)\, dS = 2g_k(x, y, z),$$

$$k = 1, 2, \ldots, 7 \quad (3.58)$$

where g_k are given in Equations (3.56) and $\eta^0 = H/2$ denotes the amplitude of the incident wave.

The discretization and numerical procedure for the solution of the integral equation (3.58) has been discussed in detail in Section 3. Equation (3.20) is replaced by the equivalent form

$$-f_{k_i} + \alpha_{ij}f_{k_j} = 2g_{k_i} \qquad k = 1, 2, \ldots, 7 \qquad i, j = 1, 2, \ldots, N \qquad (3.59)$$

where N denotes the number of subdivisions on the immersed surface of the hull. After Equation (3.59) has been solved the value of the potential on the immersed surface may be obtained from Equation (3.23),

$$\phi_{k_i} = \beta_{ij}f_{k_j}, \qquad k = 1, 2, \ldots, 7 \qquad i, j = 1, 2, \ldots, N \qquad (3.60)$$

where β_{ij} is given by Equation (3.24).

The linearized form of Bernoulli's equation gives the dynamic pressure on the immersed surface in terms of ϕ_k according to

$$P_k = \rho\omega\, \mathrm{Re}\,[i\phi_k\, e^{-i\omega t}], \qquad k = 1, 2, \ldots, 6 \qquad (3.61)$$

for motion in the six degrees of freedom, and for wave interaction with the fixed hull

$$P_{07} = \rho\omega\, \mathrm{Re}\,[i(\phi_0 + \phi_7)\, e^{-i\omega t}] \qquad (3.62)$$

The forces and moments caused by the dynamic fluid pressure acting on the immersed surface of the body are determined from expressions similar to Equations (3.28) and (3.29),

$$F_{ij}(t) = -\iint_S P_j h_i\, dS, \qquad i, j = 1, 2, \ldots, 6 \qquad (3.63)$$

$$F_i(t) = -\iint_S P_{07} h_i\, dS, \qquad i = 1, 2, \ldots, 6 \qquad (3.64)$$

where F_i denotes the ith component of wave excitation force or moment and F_{ij}

denotes the ith components of force or moment arising from the jth component of body motion. The functions h_i are

$$h_1 = n_x, \qquad h_2 = n_y, \qquad h_3 = n_z, \qquad h_4 = (d+y)n_z - zn_y,$$

$$h_5 = zn_x - xn_z, \qquad h_6 = xn_y - (d+y)n_x$$

The excitation force and moment coefficients, respectively, may be represented by the complex coefficients

$$C_j = \frac{F_j(\max)}{\rho g a^3 \eta^0} e_j^{i\delta}, \qquad j = 1, 2, 3$$

$$C_j = \frac{F_j(\max)}{\rho g a^4 \eta^0} e_j^{i\delta}, \qquad j = 4, 5, 6$$

$$(3.65)$$

where δ_j denotes the phase shift of the force with respect to the crest of the incident wave. A positive value of δ_j denotes a lag.

In the case of the forces and moments due to the motion of the hull it is more common to describe the hydrodynamic force in terms of added means and damping coefficients

$$-M_{ij} - iN_{ij} = \frac{F_{ij}(\max)}{\rho \omega^2 a^3 X_j^0} e_{ij}^{i\delta} \qquad i = 1, 2, 3 \qquad j = 1, 2, \ldots, 6$$

$$(3.66)$$

$$-M_{ij} - iN_{ij} = \frac{F_{ij}(\max)}{\rho \omega^2 a^4 X_j^0} e_{ij}^{i\delta} \qquad i = 4, 5, 6 \qquad j = 1, 2, \ldots, 6$$

where M_{ij} and N_{ij} denote the added mass and damping tensors, respectively.

3.5.2 Equations of motion

The equations of motion linearized with respect to small angular displacements of the body described in Figure 3.16 may be written as follows:

$$F_1^T(t) = m\ddot{X}_1(t)$$

$$F_2^T(t) = m\ddot{X}_2(t)$$

$$F_3^T(t) = m\ddot{X}_3(t)$$

$$F_4^T(t) = I_{44}\ddot{\Theta}_4 - I_{45}\ddot{\Theta}_5 - I_{46}\ddot{\Theta}_6$$

$$F_5^T(t) = I_{55}\ddot{\Theta}_5 - I_{56}\ddot{\Theta}_6 - I_{54}\ddot{\Theta}_4$$

$$F_6^T(t) = I_{66}\ddot{\Theta}_6 I_{64}\ddot{\Theta}_4 I_{65}\ddot{\Theta}_5$$

$$(3.67)$$

where $F_i^T(t)$ $(i = 1, 2, 3)$ and $F_i^T(t)$ $(i = 4, 5, 6)$ denote the total external force and moment, respectively, acting on the body, m denotes the mass of the body and the moments of inertia are defined typically as

$$I_{45} = I_{54} = \int_m x'y' \, dm$$

$$I_{56} = I_{65} = \int_m y'z' \, dm$$

(3.68)

etc.

It is convenient to cast the equations of motion in dimensionless form by introducing the dimensionless parameters:

$$m_{11} = m_{22} = m_{33} = m/\rho a^3$$

$$m_{12} = m_{21} = m_{13} = m_{31} = m_{23} = m_{32} = 0$$

$$m_{44} = I_{44}/\rho a^5, \quad m_{55} = I_{55}/\rho a^5, \quad m_{66} = I_{66}/\rho a^5$$

$$m_{45} = m_{54} = -I_{45}/\rho a^5, \quad M_{46} = m_{64} = -I_{46}/\rho a^5$$

$$m_{56} = m_{65} = -I_{56}/\rho a^5$$

$$f_i^T(t) = F_i^T(t)/\rho g a^3, \qquad i = 1, 2, 3$$

$$f_i^T(t) = F_i^T(t)/\rho g a^4, \qquad i = 4, 5, 6$$

(3.69)

where a denotes the characteristic dimension of the hull and ρ denotes the density of the fluid. Using these definitions Equation (3.67) condenses to

$$f_i^T(t) = m_{ij}\ddot{X}_j/g, \qquad i, j = 1, 2, \ldots, 6$$

(3.70)

where $f_i^T(t)$ represents the total external force or moment coefficient which includes contributions from the surrounding fluid and any mooring lines which may be attached.

The fluid induced forces are of three types: (a) wave excitation forces and moments, (b) dynamic forces and moments caused by the motion of the body and (c) hydrostatic forces and moments resulting from displacements. Thus, we may write,

$$f_i^T(t) = c_i(t) + \sum_{j=1}^{6} [c_{ij}(t) + k_{ij}(t)]$$

(3.71)

where the excitation force or moment coefficient is $c_i(t) = F_i(t)/\rho g a^3$, $(i = 1, 2, 3)$ and $c_i(t) = F_i(t)/\rho g a^4$, $(i = 4, 5, 6)$, and the hydrodynamic force or moment in the i-direction associated with the motion of the body in the j-direction is

$$c_{ij}(t) = -M_{ij}\ddot{X}_j(t)/g - (\omega/g)N_{ij}\dot{X}_j(t)$$

(3.72)

The parameter $k_{ij}(t)$ represents the dimensionless force or moment associated with displacement of the body and can result from hydrostatic pressures as well as mooring line reactions provided these can be assumed to vary linearly. The dimensionless displacement dependent force or moment can be expressed in the form

$$k_{ij}(t) = (K'_{ij} + K_{ij})X_j(t)/a \qquad (3.73)$$

where k'_{ij} represents the mooring line spring constant tensor (force/$\rho g a^3$ when $i = 1, 2, 3$ and moment/$\rho g a^4$ when $i = 4, 5, 6$) and K_{ij} may be termed the hydrostatic restoration tensor. The non-zero values of K_{ij} are given, for example, by Garrison[21] as

$$K_{22} = -\frac{1}{a^2} \iint_S n_y \, \mathrm{d}S = A_w/a^2$$

$$K_{24} = K_{42} = \frac{1}{a^3} \iint_S z' n_y \, \mathrm{d}S$$

$$K_{44} = \frac{1}{a^4} \iint_S (y'z'n_z - z'^2 n_y) \, \mathrm{d}S$$

$$K_{46} = K_{64} = \frac{1}{a^4} \iint_S x'z' n_y \, \mathrm{d}S \qquad (3.74)$$

$$K_{26} = K_{62} = -\frac{1}{a^3} \iint_S x' n_y \, \mathrm{d}S$$

$$K_{66} = \frac{1}{a^4} \iint_S (x'y'n_x - x'^2 n_y) \, \mathrm{d}S$$

where A_w denotes the waterline area. Equation (3.74) represents a particular form of the standard hydrostatic restoration coefficients which are convenient for computer numerical evaluation as part of the numerical scheme discussed here.

Mooring line reactions depend, in general, on the mooring line configuration, weight, shape (catenary), hydrodynamic and elastic properties. However, it is often permissible to disregard dynamic effects and approximate the reactions by the linear relationship with the six components of body displacement as indicated by the generalized form given by Equation (3.73). K'_{ij} represents the dimensionless force (or moment) produced in the (negative) i-direction by a unit displacement of the body in the j-direction.

Using Equations (3.70–3.73) and (3.65) the complete equation of motion may be expressed in the compact indicial form

$$\sum_{j=1}^{6} [-\nu(m_{ij}+M_{ij})-i\nu N_{ij}+(K'_{ij}+K_{ij})]X_j^0/\eta_0 = |C_i|e_i^{i\delta}, \qquad i = 1, 2, \ldots, 6$$

$$(3.75)$$

where the frequency parameter $\nu = \omega^2 a/g$ and the complex amplitude of the response is given by $X_j^0 = |X_j^0| \, e^{i\psi}j$ in which ψ_j denotes the phase shift angle of the response. The surface elevation of the incidence wave at the origin is given by

$$\eta = \eta^0 \cos(\omega t)$$

and, therefore, all phase shifts are referenced to the wave crest with positive values indicating lag.

3.5.3 Numerical results

The results of calculations of hydrodynamic coefficients and dynamic response of several configurations is presented in the following as examples.

Figures 3.17–3.19 show the hydrodynamic coefficients needed to determine the motion of a free floating hemisphere in heave and surge. In Figure 3.17 the heaving force coefficient labelled Havelock's results which were presented for comparison with the present results were obtained indirectly from the damping coefficient presented by Havelock.[28] Haskind's[29] relations provide a relationship between the excitation force and damping coefficients and in the special case where the body is symmetrical with respect to the vertical axis Newman[30] derived the relationship between the heave force and damping coefficient. Havelock's results agree well with the present numerical results corresponding to large depth ($h/a = 10\cdot0$).

The dynamic response of the hemisphere is shown in Figure 3.20 as a function of the frequency parameter $\omega^2 a/g$. These results indicate that even though the effect of depth on the hydrodynamic coefficients was quite large, the resulting effect on the response was rather minor except in the case of the low frequency surge response.

Gravity structures generally have a large flat base and, therefore, to give some idea of the expected response for such structures a study was made for a short, circular cylinder (or circular dock).[21,36] The cylinder floats upright

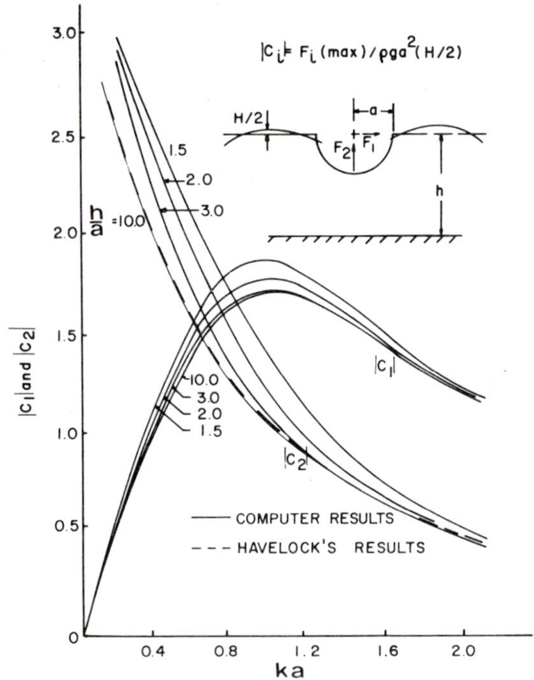

Figure 3.17   Surge and heave forces on a hemisphere

with its axis vertical and has a diameter which is four times its draft. For purposes of the calculations, its centre of gravity was taken at the mean water line level and the dimensionless moment of inertia in pitch was taken to be $m_{66} = 0.75$ which corresponds to a radius of gyration of $0.691a$.

Figure 3.21 shows some of the primary hydrodynamic coefficients which determine the response of the floating cylinder. These results show that the added moment of inertia in pitch, M_{66}, is almost independent of frequency and depth in the range of interest. The variation of M_{66} with depth varies from only approximately 0.225 to 0.255 as h/a goes from 10.0 to 1.0. The added mass coefficient in heave, M_{22}, on the other hand, shows a rather large effect of depth. Moreover, a very large increase in this parameter may be expected when the gap between the cylinder and bottom becomes small.

Figure 3.22 shows the horizontal excitation force coefficient for the vertical cylinder configuration. These results agree with those of Garrett[31] which were computed by a somewhat different method.

In Figures 3.23 and 3.24 the surge and heave response, respectively, are shown as a function of the frequency parameter and water depth ratio. These

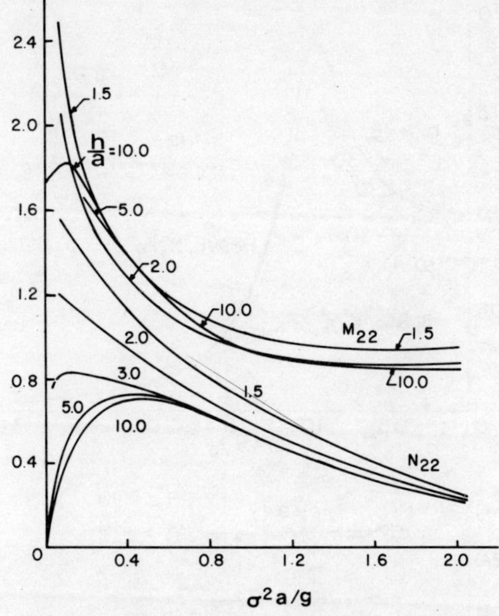

Figure 3.18　Added mass and damping in heave for a hemisphere

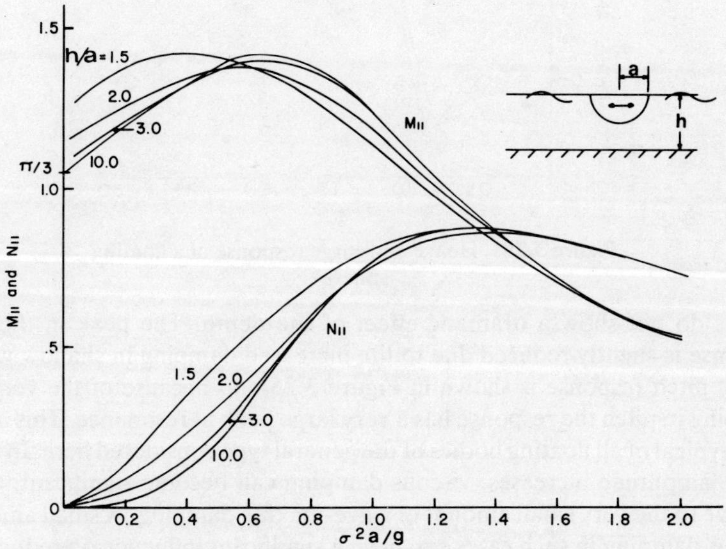

Figure 3.19　Added mass and damping in surge for a hemisphere

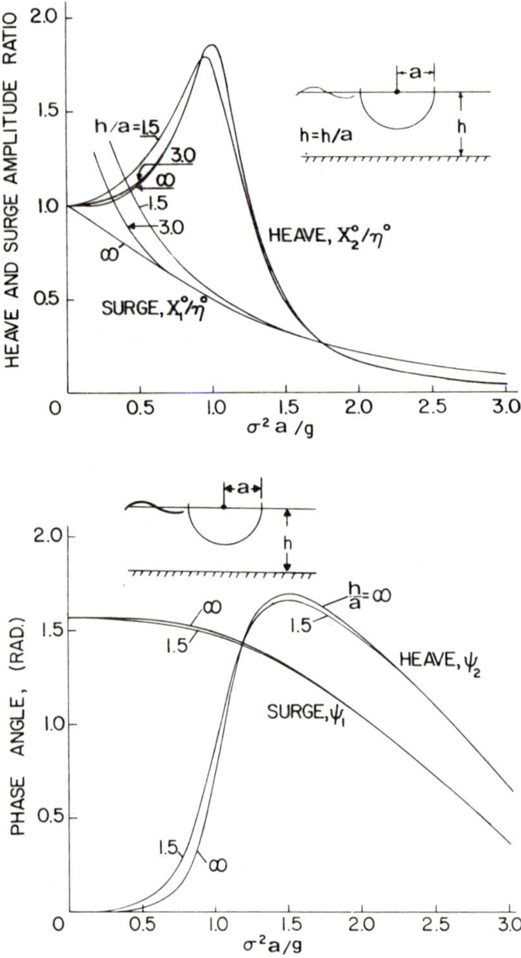

Figure 3.20   Heave and surge response of a floating sphere

results do not show a dramatic effect of the depth. The peak in the heave response is slightly reduced due to the increased damping in shallow water.

The pitch response is shown in Figure 3.25 and because of the very small damping in pitch the response has a very large peak at resonance. This appears to be typical of all floating bodies of the general type considered here. In reality, as the amplitude increases, viscous damping can become significant, at least relative to the very small amount of wave-making damping. A small amount of viscous damping in such cases can have a significant influence on reducing the large peak response.

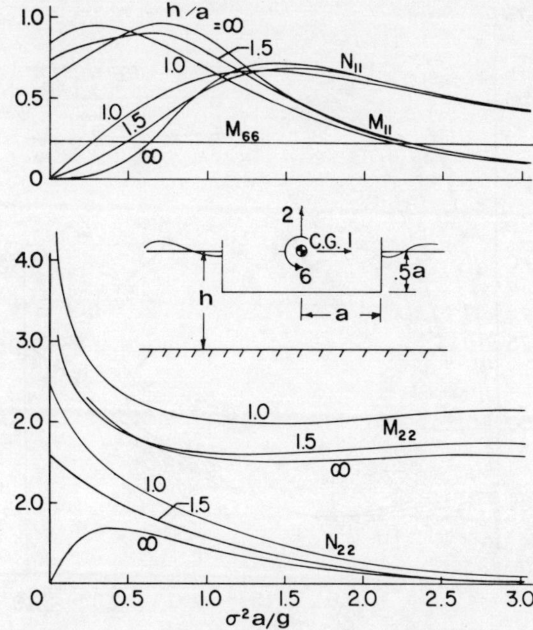

Figure 3.21   Added mass and damping coefficients
for a floating vertical circular cylinder

Experimental results obtained by wave channel model tests have been
published by Hoffman, *et al.*[32] for the case of a 'disc hull' buoy. The buoy was
circular in plan and otherwise as described by the definition sketch shown in
Figure 3.26. In this figure, the heave and pitch results obtained from tests is
compared with the present calculations and the agreement appears to be quite
good.

3.5.4   Steady state forces and moments

In addition to the periodic wave induced forces and moments, time-average
components also exist. These forces and moments are partially attributable to
the velocity-squared terms in Bernoulli's equation and partially attributable to
the surface elevation variation or run-up on the structure at the waterline when
the structure is surface piercing. The time-average components are of second-
order with respect to the waveheight but nonetheless may, under certain
circumstances, be of considerable significance, particularly during the sinking
operation of a large caisson which has little or no waterline area. The vertical
force is usually directed upwards and is greatest when the object is submerged

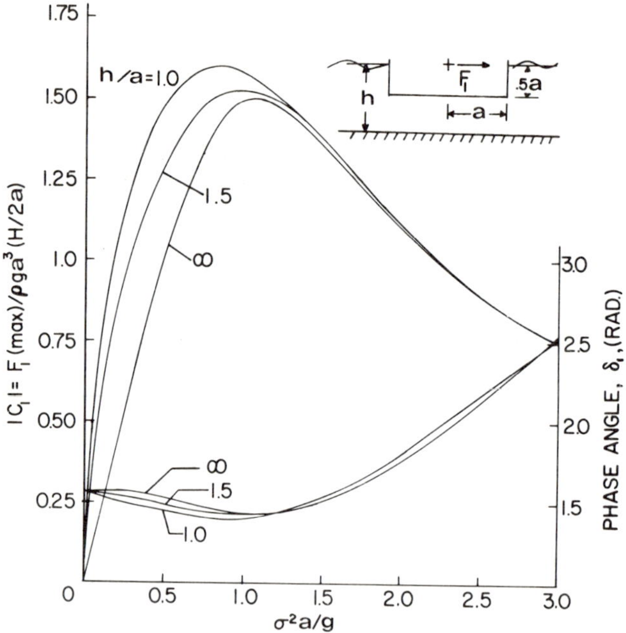

Figure 3.22   Horizontal force coefficient for a floating vertical circular cylinder

just beneath the free surface, decreasing rapidly with depth of submergence.
The horizontal component generally acts in the direction of wave propagation
and in ship hydrodynamics is referred to as the drift-force.

The steady state force may become a slowly varying force in irregular
waves[33] with a frequency near the resonance frequency of a body due to its
mooring lines since in such cases the resonance period is typically very large.
The large horizontal excursions of the body near resonance can result in failure
of the anchor system.

Faltinsen and Michelsen[18] have computed the steady-state horizontal drift
force and moment on floating bodies by application of the expressions
developed by Newman.[34] This procedure uses the far-field radiation potentials.
Garrison[35,36] has given all six components of the steady-state loads in terms of
the near-field potential integrated over the immersed surface and around the
waterline curve.

The total complex potential associated with the floating body is,

$$\phi^T = \phi_0 + \phi_7 + \sum_{j=1}^{6} e^{i\psi_j} \phi_j \tag{3.78}$$

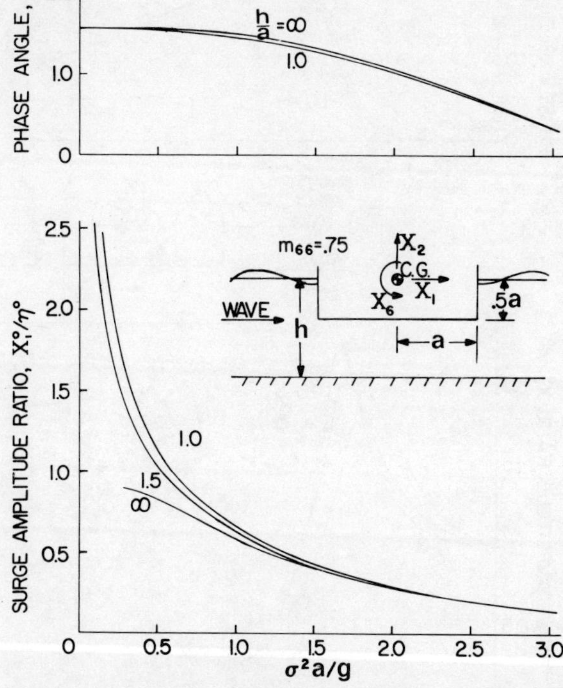

Figure 3.23 Surge response of a floating vertical circular cylinder

Introducing the dimensionless potential,

$$u^T = [-i\omega/gka\eta^0]\phi^T \tag{3.79}$$

the ith component of the steady-state load is given by,

$$C_i^{ss} = \left\{ \frac{k^2 a}{2} \int_{C_*} \{1/[2\sqrt{(1-n_y^2)}] - 1\}|u^T|^2 h_i \, dc \right.$$

$$\left. + \frac{k^2}{4\nu a} \iint_S [|au_x^T|^2 + |au_y^T|^2 + |au_z^T|^2 + (\nu/k)^2 - 1]h_i \, dS \right. \tag{3.80}$$

where C_i^{ss} = steady state force/$\rho g a^3 (H/2a)^2$ when $i = 1, 2, 3$ and where C_i^{ss} = steady state moment/$\rho g a^4 (H/2a)^2$ when $i = 4, 5, 6$.

The first term in Equation (3.80) represents a second-order term which arises from the integration of the hydrostatic pressure (which is of order

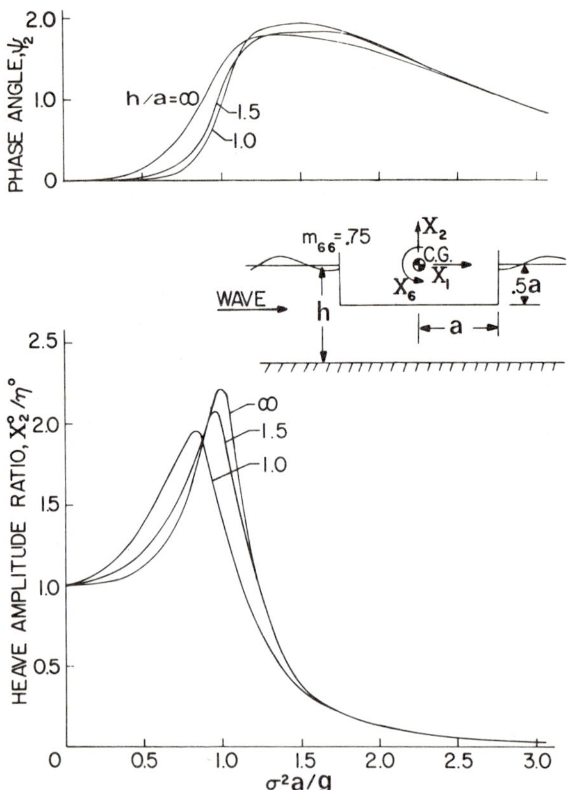

Figure 3.24   Heave response of a floating vertical circular cylinder

$\eta^0 = H/2$) over the wetted area above the MWL which is also of order η^0. Thus the resulting force is of order $(\eta^0)^2$ and C_* denotes the arc length around the waterline area. The second term represents the steady-state contribution from the velocity-squared terms, the integration being carried out over the immersed surface up to the MWL.

As an example of the steady state uplift force, Figure 3.27 shows results for a fixed and free floating sphere in deep water.[35]

Acknowledgements

The financial support of the National Science Foundation under Grant No. ENG73-04019 A01 and the Energy Research and Development Administration under the Ocean Thermal Energy Program, Interagency Agreement E(49-26)-1044, is gratefully acknowledged.

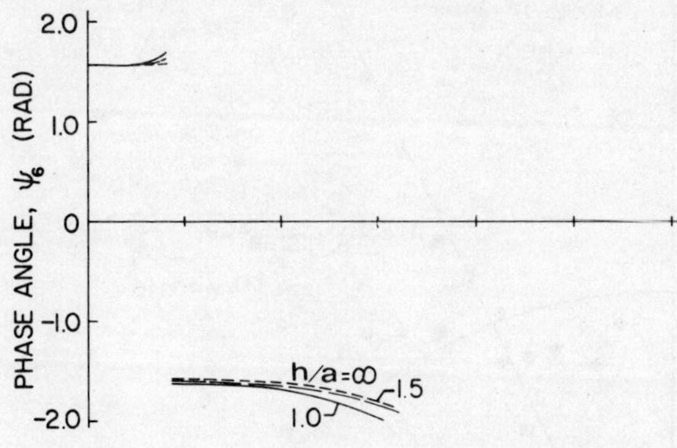

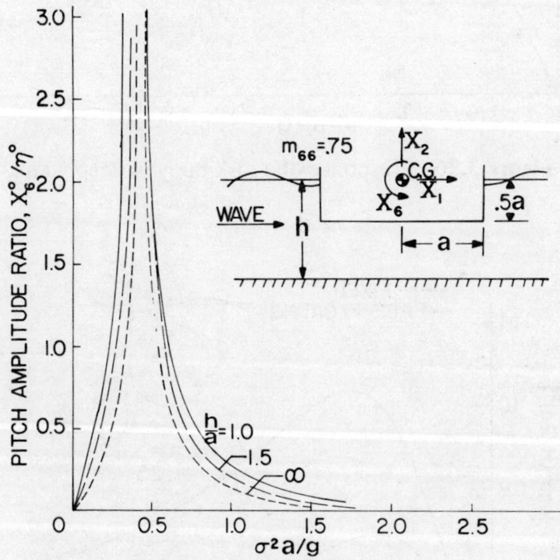

Figure 3.25   Pitch response of a floating vertical circular cylinder

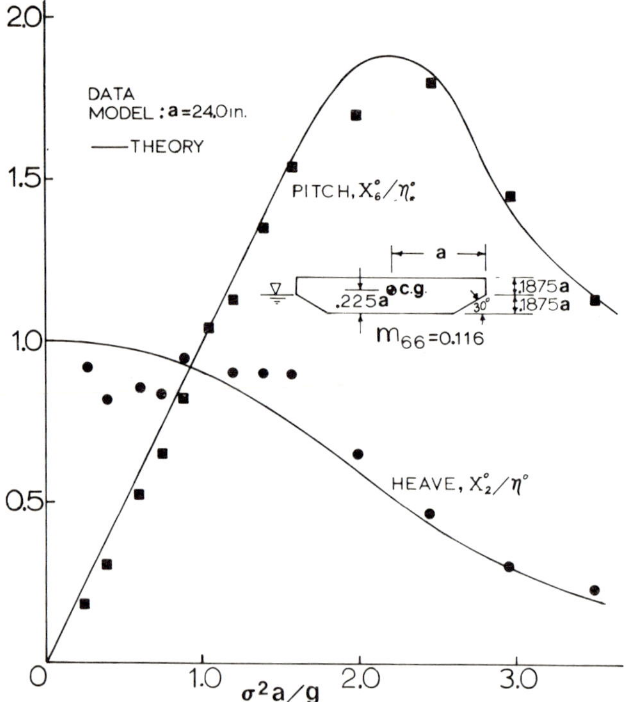

Figure 3.26   Response of a disk buoy in deep water

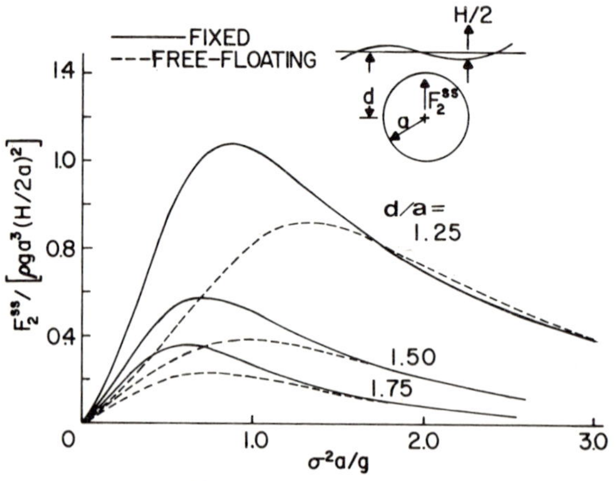

Figure 3.27   Steady-state uplift on a sphere in deep water

REFERENCES

1. Raman, H. and Venkalanarasaiah, P. (1976). 'Forces due to nonlinear waves on vertical cylinders', *Journal of the Waterways, Harbors and Coastal Engineering Division*, Proc. ASCE, Vol. 102, No. WW3.
2. Garrison, C. J. (1976). 'Consistent second-order diffraction theory', presented at the *Fifteenth Coastal Engineering Conference*, Honolulu, Hawaii.
3. Lee, C. M. (1968). 'The second-order theory of heaving cylinders in a free surface', *Journal of Ship Research*, 313–327.
4. Potash, R. L. (1970). 'Second-order theory of oscillatory cylinders', Dissertation, *University of California*, Berkeley, Ca. 157.
5. Apelt, C. J. and Macknight, A. (1976). 'Wave action on large offshore structures', *Proceedings Fifteenth Coastal Engineering Conference*, Honolulu, Hawaii.
6. Morison, J. R., O'Brien, M. P., Johnson, J. W., and Schaaf, S. A. (1950). 'The force exerted by surface waves on piles', *Petroleum Trans.*, **189**, TP2846, 149–154.
7. Garrison, C. J., Field, J. B., and May, M. D. (1976). 'Drag and inertia coefficients in oscillatory flow about cylinders', in Press, *Journal of Waterways, Port and Coastal and Ocean Division*, ASCE, **103**, WW2.
8. Van Dyke, M. (1964). *Perturbation Methods in Fluid Mechanics*, Academic Press.
9. May, M. D. (1975). 'Forces on an oscillating circular cylinder', *M.S. Thesis*, Naval Postgraduate School.
10. MacCamy, R. C. and Fuchs, R. A. (1954). 'Wave forces on a pile: A diffraction theory', Tech. Memo. 69, *U.S. Army Corps of Engineers Beach Erosion Board*, Washington, D.C.
11. Garrison, C. J., Rao, V. S., and Snider, R. H. (1970). 'Wave interaction with large submerged objects', *Proc. Offshore Technology Conference, Paper OTC1278*.
12. Lamb, H. (1932). Hydrodynamics, 6th Edition, *Cambridge University Press*.
13. Chakrabarti, S. K. and Naftzger, R. A. (1976). 'Wave interaction with a submerged open-bottom structure', Paper No. OTC 2534, *Offshore Technology Conference*, Houston, Texas.
14. Wehausen, J. V. and Laitone, E. V. (1960). 'Surface waves', *Encyclopedia of Physics*, Vol. 9, Springer-Verlag, Berlin, 446–778.
15. John, F. (1949). 'On the motion of floating bodies', *Comm. in pure and applied math.*, p. I.
16. Garrison, C. J. and Berklite, R. B. (1973). 'Impulsive hydrodynamics of submerged rigid bodies', *Journal of the Engineering Mechanics Division*, Proc. ASCE, Feb., 99–120.
17. Monacella, V. J. (1966). 'The disturbance due to a slender ship oscillating in waves in a fluid of finite depth', *Journal of Ship Research*, **10**, No. 4, 242–252.
18. Faltinsen, O. M. and Michelsen, F. C. (1974). 'Motion of large structures in waves at zero Froude number', *The Dynamics of Marine Vehicles and Structures in Waves*, Paper No. 11, The Institution of Mechanical Engineers.
19. Abramowitz, M. and Stegun, I. A. (1964). 'Handbook of mathematical functions', *National Bureau of Standards Mathematics Series*, 55, Washington, D.C.
20. Hess, J. L. and Smith, O. M. A. (1962). 'Calculation of non-lifting potential flow about arbitrary three-dimensional bodies', Rept. No. E.S. 40622 (Douglas Aircraft Division, Long Beach, CA), also *Journal of Ship Research*, 1964, **8**.
21. Garrison, C. J. (1974). 'Hydrodynamics of large objects in the sea, Part I—Hydrodynamic analysis', *Journal of Hydronautics*, **8**, 5–12.

22. Skjelbreia, L. and Hendricksen, J. (1960). 'Fifth-order gravity wave theory', *Proc. Coastal Engineering Conference*, 184–196.
23. Hogben, N. and Standing, R. G. (1975). 'Experience in computing wave loads on large bodies', Paper No. OTC 2189, *Offshore Technology Conference*, Houston, Texas.
24. Iversen, C., Leivseth, S., and Tørum, A. (1973). 'Condeep platform for Mobil North Sea Ltd. on the Beryl Field, model tests, Progress Rept. No. 1 on wave forces', *Vassdrags -Og Havenlaboratory, Ved Norges Tekniske Hogskole—Tilslullet Sintex*, Trondheim, Norway.
25. Garrison, C. J., Tørum, A., Iversen, C., Leivseth, S., and Ebbesmeyer, C. C. (1974). 'Wave forces on large volume structures—a comparison between theory and model tests', Paper OTC 2137, *Offshore Technology Conference*, Houston, Texas.
26. Rye, H., Byrd, R., and Tørum, A. (1974), 'Sharply peaked wave energy spectra in the north sea', Paper OTC 2107, *Offshore Technology Conference*, Houston, Texas.
27. Wehausen, J. V. (1971). 'The motion of floating bodies', *Annual Review of Fluid Mechanics*, **3**.
28. Havelock, T. (1955). 'Waves due to a floating sphere making periodic heaving oscillations', *Proc. of the Royal Society*, London, **231**, Sec. A, 1–7.
29. Haskind, J. D. (1957). 'The exciting forces and wetting of ships in waves' (in Russian), *Isvestra Akademic Nauk SSSR, Otdelenie Teknicheskikh Nauk*, No. 7, 65–78 (English translation: David Taylor Model Basin Translation, No. 30, March 1962), Washington, D.C.
30. Newman, J. N. (1962). 'The exciting forces on fixed bodies in waves', *Journal of Ship Research*, **6**, 10–17.
31. Garrett, C. J. R. (1971). 'Wave forces on a circular dock', *Journal of Fluid Mechanics*, **46**, Pt. 1, 129–139.
32. Hoffman, D., Geller, E. S., and Niederman, C. S. (1973). 'Mathematical simulation and model tests in the design of data buoys', *The Society of Naval Architects and Marine Engineers*, presented at the Annual Meeting of SNAME, New York, N.Y.
33. Hsu, F. H. and Blenkarn, K. A. (1970). 'Analysis of peak mooring force caused by slow drift oscillation in random seas', *Proc. Offshore Technology Conference, Paper No. OTC 1159*.
34. Newman, J. N. (1967). 'The drift force and moment on ships in waves', *Journal of Ship Research*.
35. Garrison, C. J. (1974). 'Dynamic response of floating bodies', *Proc. Offshore Technology Conference, Paper No. OTC 2067*.
36. Garrison, C. J. (1975). 'Hydrodynamics of large objects in the sea, Part II—Motion of Free-Floating Bodies', *Journal of Hydronautics*, **9**, 58–63.

Chapter 4

The Finite Element Method for Determining Fluid Loadings on Rigid Structures Two- and Three-Dimensional Formulations

O. C. Zienkiewicz,
P. Bettess and
D. W. Kelly

4.1 INTRODUCTION

With the great increase of interest in offshore structures in recent years the topic of fluid loading has received considerable attention and a number of reviews and surveys are now available.[1-5] The problem of wave loading on such structures is highly complex, but it is generally agreed that an important component is due to inertia forces and that these forces are often affected by wave diffraction. In the few computer programs currently available these forces are computed using full three-dimensional boundary integral approaches,[6-9] based on a Green's function for the wave equation in a homogeneous liquid of constant depth. Indeed the formulation and solution procedures in Chapter 3 of this book are based on such an approach. Here the wave diffraction and refraction problem will be placed in a more general context and the links with the methods of acoustics, and other disciplines will be shown. The special problems raised by unbounded domains are discussed, and four different ways of dealing with the exterior problem using finite elements will be shown.

The formulation using the finite element methodology has the advantage over pure boundary integral equation techniques in

(a) producing symmetric and banded (sparse) equation systems,
(b) being ideally suited for deriving the coupling with the structure for which, generally, a finite element discretization is used.

We shall first show the application of the finite element process in a two-dimensional approximation of the problem. It would have already been observed from the study of Chapter 3 that both the incident wave solution and the radiation boundary conditions are based on such two-dimensional approximations and for many structures it would appear advantageous to use this approximation irrespective of the actual solution techniques used.

In Part I of this Chapter we shall discuss therefore the techniques exclusively on this basis—and show the range of applicability of such approximations.

The treatment of the full three-dimensional solution follows precisely the same principles as that of the two-dimensional one. Indeed the basic formulation is simpler but clearly the computational effort is greatly increased. In Part II we discuss the details of such modelling and show how with relatively little expense full three-dimensional problems can be dealt with.

4.2   BASIC ASSUMPTIONS

The conventional approach to wave loading is to assume that the total force on an object in the waves is the sum of drag and inertia force components.[10] (Although other approaches have been proposed.[11]) This assumption, introduced by Morison, takes the drag term as a function of velocity and the inertia force as a function of acceleration, so that

$$F_{\text{total}} = F_{\text{drag}} + F_{\text{inertia}}$$

where

$$F_{\text{drag}} = C_D \rho \frac{u|u|}{2} A \qquad (4.1)$$

and

$$F_{\text{inertia}} = C_M \rho V \dot{u}$$

where

A is the projected area of the object,
V is the volume of the object,
ρ is the density of the liquid,
u is the velocity of the liquid,
$\dot{u}$ is the acceleration of the liquid,
C_D is a drag coefficient and
C_M is an inertia or mass coefficient.

It is generally accepted that the relative importance of the two forces depends upon the geometry of the object and the wave. Hogben and Standing[8] suggest that for a cylinder of diameter D, the drag force can be neglected if $D/H > 0.2$, where H is the wave height from crest to trough. The inertia force is then calculated on the assumption that the wave pattern is not affected by the presence of the object. This is the so-called Froude-Krylov force and this is given by the second term of Equation (4.1). However, where the size of the object is a considerable fraction of the wave length it will be found to modify the wave pattern, by diffraction (and variations in bathymetry may also cause refraction). In these cases the Froude–Krylov assumption is no longer adequate. Hogben and Standing[8] suggest that when $D/L > 0.2$, where L is the wavelength, then diffraction effects cannot be safely neglected. The dynamic force can still be calculated on the assumption that the flow is potential and inviscid. The further assumptions usually made are that the wave amplitude is small compared with the water depth, and bed friction is ignored. It is with such calculations that this Chapter is concerned.

The assumptions are at present subject to extensive discussion and investigation,[5] and are obviously an approximation. To quote Hogben and Standing[8] 'Reliability in application to very severe wave conditions will be limited because of linearization of the free surface conditions.' If a linearized theory is adopted fully transient loading can be dealt with, using Fourier techniques. The loading upon the structure can be calculated using periodic sinusoidal waves, for the complete range of frequencies, and these can be used to determine the loading for any known series of waves. Fourier spectra can be determined by records from typical sites or synthesized in some plausible way (vide Chapter 2).

The problem is thus to solve the potential equations with a forcing effect due to the incoming waves, and subject to a boundary condition on the structure boundary of zero normal velocity (or if the structure is allowed to move this is equated to the structure velocity).[12,13] The reflected waves are subject to the radiation condition.[12,14,15]

4.3   THE GOVERNING EQUATIONS

4.3.1   General

Equations of flow of an inviscid fluid are well known.[13,16,17]

In Chapter 3 these have been quoted without proof (Equations 3.2–3.7) but it is instructive to derive these briefly from a degenerate form of Navier-Stokes equations which *on neglecting of convective acceleration and viscosity terms* give

$$\frac{\partial u}{\partial t} = -\frac{1}{\rho}\frac{\partial p}{\partial x}$$

$$\frac{\partial v}{\partial t} = -\frac{1}{\rho}\frac{\partial p}{\partial y} \qquad (4.2)$$

$$\frac{\partial w}{\partial t} = -\frac{1}{\rho}\frac{\partial p}{\partial z}$$

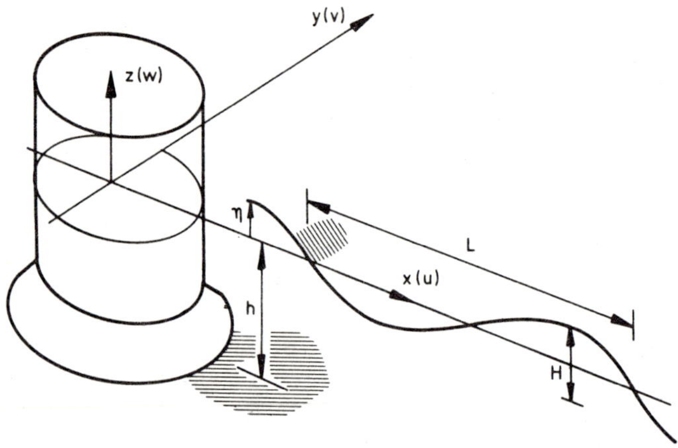

Figure 4.1   Problem geometry

where u, v and w are the velocity components, ρ is fluid density and p *is the excess of the pressure over hydrostatic.* Figure 4.1 shows the co-ordinate directions.

The continuity relation can be written as

$$\frac{\partial u}{\partial x}+\frac{\partial v}{\partial y}+\frac{\partial w}{\partial z}=\frac{1}{K}\frac{\partial p}{\partial t}\tag{4.3}$$

where K is the bulk modulus of the fluid.

Finally, on the boundaries of the region the following conditions have to be specified.

Solid boundaries

$$u_n=\bar{u}_n\tag{4.4a}$$

i.e. normal velocity component is prescribed by the motion of the face.

Free fluid surface; here the following condition is imposed

$$\frac{1}{g}\frac{\partial^2 p}{\partial t^2}+\frac{\partial p}{\partial z}=0\tag{4.4b}$$

(This is the linearized free surface condition derived as follows; assuming the fluid elevation above the mean surface as η, we can write

$$w=\frac{\partial \eta}{\partial t}$$

As

$$\frac{\partial w}{\partial t}=-\frac{1}{\rho}\frac{\partial p}{\partial z}\quad\text{and}\quad p=\rho g\eta$$

Equation (4.4b) follows by elimination of w and η).

Above have to be supplemented by specifying a radiation condition to be satisfied at infinity—we shall return to this later.

4.3.2 Pressure equations

Above equations can be simplified to a single pressure variable and yield on elimination of velocities from Equation (4.3) by Equation (4.2).

$$\nabla^2 p = -\frac{\rho}{K}\frac{\partial^2 p}{\partial t^2} \tag{4.5}$$

supplemented by boundary conditions

$$\frac{\partial p}{\partial n} = -\rho\frac{\partial \bar{u}_n}{\partial t} \quad \text{on solid boundary} \tag{4.6a}$$

and

$$\frac{1}{g}\frac{\partial^2 p}{\partial t^2} + \frac{\partial p}{\partial z} = 0 \quad \text{on the free surface} \tag{4.6b}$$

4.3.3 Velocity potential equation

Alternative form introduces a velocity potential Φ related to the pressure by

$$p = -\rho\frac{\partial \Phi}{\partial t} \tag{4.7}$$

Now the velocity components can be written simply as

$$u = \frac{\partial \Phi}{\partial x}\ \text{etc.}$$

and the governing equations become

$$\nabla^2 \Phi = -\frac{\rho}{K}\frac{\partial^2 \Phi}{\partial t^2} \tag{4.8}$$

with

$$\frac{\partial \Phi}{\partial n} = \bar{u}_n \quad \text{on solid boundary} \tag{4.9a}$$

and

$$\frac{1}{g}\frac{\partial^2 \Phi}{\partial t^2} + \frac{\partial \Phi}{\partial z} = 0 \quad \text{at free surface} \tag{4.9b}$$

In this Chapter we shall use the velocity potential formulation as this follows the precedent of Chapter 3 and in general is most widely adopted. In linearized situations which we are discussing here the relationship [Equation (4.7)] allows the transition from one to the other variable to be easily accomplished.

4.3.4 Radiation condition

The basic equations just outlined are characteristic of wave type solutions of the form

$$\Phi = f_1(x+ct)+f_2(x-ct) \qquad (4.10)$$

where c is the wave celerity. Two basic wave velocities exist; the first

$$c=\sqrt{K/\rho}$$

for the compression waves—and another velocity, which we shall characterize later, typical of surface disturbances.

At large distances r from a disturbance (such as presented by the structure) the wave are nearly plane and are only outgoing. We can thus write Equation (4.10) as

$$\Phi = f_2(r-ct) \qquad r \to \infty \qquad (4.11)$$

Noting that

$$\frac{\partial \Phi}{\partial t} = -cf_2' \quad \text{and} \quad \frac{\partial \Phi}{\partial r} = f_2' \qquad (4.12)$$

we observe that

$$\frac{\partial \Phi}{\partial t} + c\,\frac{\partial \Phi}{\partial r} = 0 \qquad (4.13)$$

This is the general radiation condition in the form given by Newton and Zienkiewicz.[12]

This condition for problems of complex response was first formulated by Sommerfield[14] and generalized by Rellich.[15] Implications of it are widely discussed in the literature.[3,16-22]

In the problems involving the computation of forces due to *incident waves* the radiation condition can only be applied to the scattered or reflected waves. If thus Φ_0 represents the potential due to the incoming wave the total potential can be written as

$$\Phi = \Phi_0 + \Phi_s \qquad (4.14)$$

where Φ_s is the scattered wave—and the radiation condition 4.13. is only applicable to Φ_s.

4.3.5 Periodic—complex formulation

In the study of general wave response it is necessary to determine a periodic form of response to an imput of angular frequency ω. Writing thus

$$\Phi = \phi\, e^{i\omega t}, \qquad \bar{u}_n = \hat{\bar{u}}_n\, e^{i\omega t} \tag{4.15}$$

where ϕ is a complex potential, the problem equations (4.8), (4.9) and (4.13) can be rewritten in a complex form as

$$\nabla^2 \phi = \frac{\rho}{K}\omega^2 \phi \tag{4.16}$$

with

$$\frac{\partial \phi}{\partial n} = \hat{\bar{u}}_n \quad \text{on solid boundary} \tag{4.17a}$$

$$-\frac{\omega^2}{g}\phi + \frac{\partial \phi}{\partial z} = 0 \quad \text{on free surface} \tag{4.17b}$$

and

$$\frac{\partial \phi}{\partial r} + \frac{i\omega}{c}\phi = 0 \quad \text{at infinity} \tag{4.17c}$$

(Sommerfield states the last requirement as

$$\lim_{r \to \infty} r^m \left(\frac{\partial \phi}{\partial r} - \frac{i\omega}{c}\phi \right) = 0 \tag{4.18}$$

where $m = (n-1)/2$, n being number of dimensions.) The pressure in its complex form is now related to ϕ as

$$p = -\rho i \omega \phi \tag{4.19}$$

In what follows we shall be concerned with solution of these equations but a simplification of incompressibility, i.e. $K = \infty$ will be taken, thus retrieving e.g. (3.1)–(3.7) of Chapter 3.

PART I: TWO-DIMENSIONAL IDEALIZATION

4.4 TWO-DIMENSIONAL EQUATIONS

When the depth of the sea bed h is constant (see Figure 4.1) then it is possible to integrate exactly the governing equations in the vertical direction z and thus reduce the problem to a two-dimensional one with x and y (plan form) being now the problem domain. Such integration can also be carried out introducing some approximations for a variable depth. We shall refer the reader to the original work by Berkhoff[23,24] for the derivation which, in the complex periodic form gives a degenerate form of Equations (4.16–4.19). Thus writing

$$\phi = \bar{\phi} Z(z)$$

and

$$p = \bar{p} Z(z)$$

where

$$Z = \frac{\cosh k(h+z)}{\cosh kh}$$

we obtain the governing equation referring to $\bar{\phi}$

$$\nabla^T(cc_g \nabla \bar{\phi}) + \frac{c_g}{c}\omega^2 \bar{\phi} = 0 \tag{4.21}$$

In the above we define the wave number as

$$k = \omega/c \tag{4.22a}$$

group velocity as

$$c_g = \frac{1}{2}\left(1 + \frac{2kh}{\sinh 2kh}\right)c \tag{4.22b}$$

and the wave velocity is computed from

$$c^2 = \frac{g}{k}\tanh(kh) \tag{4.22c}$$

When the wavelength $(L \equiv 2\pi/k)$ is large

$$c^2 \to gh$$

and the governing equation degenerates to the well known shallow water equation

$$\nabla^T(h\nabla\bar{\phi}) + \frac{\omega^2}{g}\bar{\phi} = 0$$

For a general case when the depth is constant the equation reduces to

$$\nabla^2\bar{\phi} + k^2\bar{\phi} = 0 \tag{4.23}$$

As the free surface condition is explicitly satisfied in the formulation given above the only boundary conditions that can be applied are (vide 4.17)

$$\frac{\partial\bar{\phi}}{\partial n} = \hat{\bar{u}}_n \quad \text{on solid boundaries} \tag{4.24a}$$

(here $\hat{\bar{u}}_n$ is the normal velocity at the surface) and

$$\frac{\partial \bar{\phi}}{\partial r} + \frac{i\omega}{c}\bar{\phi} \equiv \frac{\partial \bar{\phi}}{\partial r} + ik\bar{\phi} = 0 \quad \text{when } r \to \infty \qquad (4.24\text{b})$$

The problem specified by Equations (4.21) or (4.23) in homogeneous regions is that of the classical Helmholtz equation.

The two-dimensional equations have analytical solutions for the special cases of waves diffracted by a circular cylinder[25] which was apparently earlier derived in acoustics,[26] waves refracted and diffracted by a circular cylinder on top of a circular shoal,[27] and waves in a rectangular harbour in a straight infinite coastline.[28]

Numerical solutions to the Helmholtz equation for water waves in unbounded domains have been obtained using several techniques, applied to a range of problems. Hwang and Tuck[29] use a source distribution boundary integral method in solving oscillations with constant depth water. Olsen and Hwang[30] extend the technique to bays of variable depth, and Lee[31] uses a similar method. Shaw[19] discusses the method and shows that it can be linked to finite difference solutions in areas of varying depth, and discusses extensions, both to non-linear waves (like solitary waves) and also to fully transient problems which he deals with elsewhere.[32] Shaw solves again Homma's shoal problem,[27] and so do Vastano and Reid,[33] but they use a finite difference technique, and they apply the radiation boundary condition in a special finite difference form. A detailed discussion of the parabolic shoal problem is given by Jonsson.[34] The Green's function for the exterior wave propagation problem is simply a Hankel function of zero order. The Hankel functions form a complete set for the exterior problem (and satisfy the radiation boundary condition). Chen and Mei[17,35,36] use a set of Hankel functions and trigonometric functions as an exterior solution in a wave diffraction problem, and use a specially devised variational statement to link the exterior solution with finite element solutions in the interior domain. Zienkiewicz[37,38] generalizes these procedures which can be used to link finite element solutions to any kind of exterior solution, series, analytical, or boundary integral (both source distribution and Green's identity). An entirely different procedure is given by Zienkiewicz and Bettess,[13,39] where 'infinite elements' with shape functions extending over an infinite domain are used to solve wave diffraction and refraction problems.

The Helmholtz equation is also met in acoustics, where the constants in the governing equation depend upon the compressibility of the medium, usually air. It is also encountered in, among other applications, those of pressure waves in liquids,[12,40,41,42] and Love waves in elastic media[43,44] for which there are a number of analytical solutions. Of particular interest are those for pressure waves in unbounded media, generated from cavities of various shapes, as there is an almost complete analogy with the problem of diffraction round an object of the same shape.[45,46] Earthquake loading caused by movement of the bed rock beneath a structure is also very similar to 'loading' of the object by an incident wave train.[47,48] This has been studied using a finite element model coupled to either dampers[49] which model the energy absorbing effect of the exterior, or special finite elements which use Hankel functions.[50,51,52,53]

4.5 FINITE ELEMENT METHODS FOR WAVE DIFFRACTION AND REFRACTION. THE EXTERIOR REGION PROBLEMS

4.5.1 General

At the moment, as has been indicated in the previous section, there are four methods which can be used in conjunction with finite element methods to solve numerically diffraction and refraction problems so as to satisfy the infinity conditions. These are

1. boundary 'dampers,' based on the radiation boundary conditions applied at finite distance (viz., Equation (4.24b)).
2. exterior analytical or series solutions (boundary solutions)
3. exterior boundary integral formulation, using source distributions or Green's identity
4. 'infinite' elements.

The advantages of a finite element formulation of the equations in the inner domain are

1. Inhomogeneities (variations in depth, or even salinity or temperature) are easily dealt with.
2. Any geometry can be dealt with.
3. Extensions to non-linear formulations are straightforward.
4. Where a full dynamic analysis of both fluid and structure (and possibly the unbounded foundation) is needed, it is natural to model the structure and foundation using finite elements. When the fluid is also modelled at least partially using finite elements, it is simple to formulate the coupling equations, and indeed such coupling has been described by several authors. However, if the fluid is modelled using boundary integral techniques the formulation is less simple.
5. Boundary integral formulations are almost all subject to inaccuracies and difficulties at sharp corners, which are encountered in many offshore structures. No such difficulties attend finite element models.

The theory will be developed and examples will be given, with reference to the two-dimensional equations, the extension to three-dimensional problems being dealt with later.

As a finite element solution will be sought a variational form of the problem shall be used throughout. We shall deal only briefly with the finite element formulation *per se* as details of this are available in texts.

The general governing equation (4.21) can be cast as the variational problem of minimizing.

$$\Pi = \iint_{\Omega} \frac{1}{2}\left[cc_g(\nabla\bar{\phi})^2 - \frac{\omega^2 c_g}{c}\bar{\phi}^2 \right] dx\ dy - \int_{\Gamma} \bar{\phi}\left(\frac{\partial\bar{\phi}}{\partial n}\right)_p d\Gamma \qquad (4.25)$$

The surface integral is inserted to account for presence of boundary conditions on $(\partial\bar{\phi}/\partial n)_p$. Thus on the surface Γ, on which condition of Equation (4.24a) is imposed, the surface integral becomes

$$\int \bar{\phi}\hat{\bar{u}}\ d\Gamma$$

Similarly, if we wish to apply the condition Equation (4.24b) on equating the directions n and r a suitable term arises.

As we shall use special techniques for dealing with the radiation (exterior) region it is convenient to subdivide the total problem domains as shown in Figure 4.2.

Two main boundaries are distinguished. These are Γ_c which is the outer boundary of the finite element model of the wave field. Beyond Γ_c some special model of the waves is used. The possibilities have already been outlined. Γ_B marks the boundary outside which the wave field is divided into the incoming

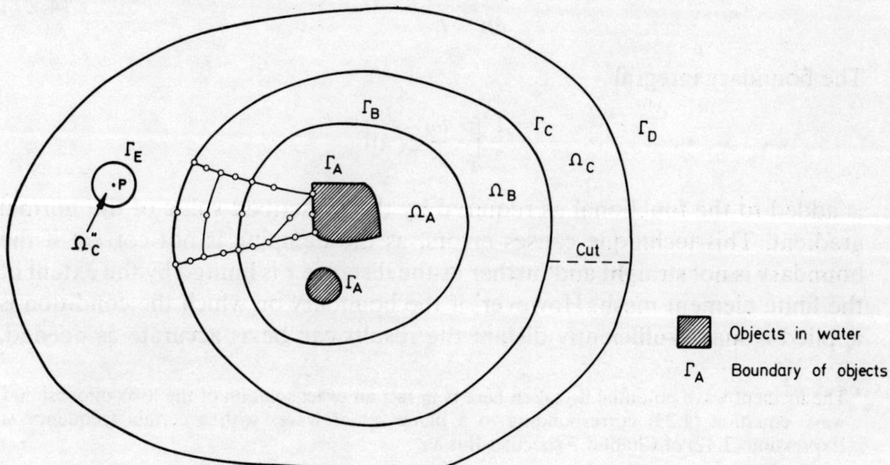

Figure 4.2   Plane of the wave problem domain

wave, which is known, with potential $\bar{\phi}_0$, and the reflected wave, $\bar{\phi}_s$.* Inside this boundary the wave field is modelled using $\bar{\phi}$, the velocity potential of the total wave. This boundary may, if desired, co-incide with the boundary of the objects situated in the water, Γ_A. Alternatively it may go out as far as Γ_c but no further. In the finite element model on the boundary Γ_B there must be a change of nodal variables from $\bar{\phi}$ to $\bar{\phi}_s$. This is easily achieved, by standard finite element techniques. In addition there is a line integral term arising from the incoming wave. When $\bar{\phi}_s + \bar{\phi}_0$ is substituted for $\bar{\phi}$, the functional of equation (4.25) becomes

$$\Pi = \iint\limits_{\Omega_B + \Omega_c} \frac{1}{2}\left[cc_g(\nabla\bar{\phi}_s)^2 - \frac{\omega^2 c_g}{c}\bar{\phi}_s^2 \right] dx\ dy$$

$$+ \int_{\Gamma_B} cc_g\left[\frac{\partial\bar{\phi}_0}{\partial x}\bar{\phi}_s\ dy - \frac{\partial\bar{\phi}_0}{\partial y}\bar{\phi}_s\ dx \right]$$

(4.26)

From now on in discussing the modelling of the outer domain we will deal entirely in terms of $\bar{\phi}_s$, and we will drop the subscript s for simplicity.

4.5.2 Boundary dampers

For the one-dimensional, or plane wave case, the radiation boundary condition is imposed on a finite boundary Γ_c

$$\frac{\partial\bar{\phi}}{\partial n} + \frac{i\omega}{c}\bar{\phi} = 0$$

(4.27)

The boundary integral

$$\frac{1}{2}\int_{\Gamma_c} \frac{i\omega}{c}\bar{\phi}^2\ d\Gamma$$

is added to the functional as required by the prescribed value of the normal gradient. This technique causes errors, as the damping is not correct if the boundary is not straight and further as the distance r is limited by the extent of the finite element mesh. However, if the boundary on which the condition is applied is made sufficiently distant the results can be as accurate as needed.

---

* The incident wave potential $\bar{\phi}_0$ taken here is in fact an exact solution of the 'two-dimensional' wave equation (4.23) corresponding to a plane set of waves with a certain frequency ω. Expression (3.42) of Chapter 3 specifies this as

$$\bar{\phi}_0 = -\frac{igH}{2\omega}\frac{\cosh k(h+z)}{\cosh kh}e^{i(kx\cos\gamma + ky\sin\gamma)}$$

Any adverse effect due to the boundary approximation can be estimated, by solving twice with the boundary at different radii. The method has the merits of conceptual and computational simplicity. It can also be used in fully transient rather than periodic problems.[20] It has the disadvantage that the domain which must be idealized using finite elements may be extensive. Such dampers, originally used by Zienkiewicz and Newton[12] for pressure wave problems using finite elements, have been used in many applications.[40,49]

4.5.3 Coupling with boundary solution

For the specific case of surface wave diffraction a suitable variational statement leading to coupling with an exterior series solution in Hankel functions was given by Chen and Mei.[17,35,36] Here however a more general formulation due to Zienkiewicz is given.[37,38] The domain of the problem is divided into two sections, inner and outer, denoted by $\Omega_A + \Omega_B$ and Ω_C in Figure 4.2. The outer domain is assumed to be of constant depth, so that the wave number, k, is constant, and Equation (4.23) is valid. In the outer domain, Ω_C, it is assumed that the boundary solution procedure is used. In the domain $\Omega_A + \Omega_B$ finite elements are used. The velocity potential in $\Omega_A + \Omega_B$ is approximated by a set of finite element shape functions, $\bar{N}_i$, and a discrete set of unknown parameters, b_i (in this case nodal values of the complex velocity potential $\bar{\phi}$).

$$\bar{\phi} \approx \hat{\phi}^b = \sum_{i=1}^{n} \bar{N}_i b_i = \bar{\mathbf{N}}\mathbf{b} \tag{4.28}$$

and within the finite element domain the potential energy can be approximated and desired minimizing equations written for the moment neglecting the interface Γ_c and the contribution of ϕ_0 as

$$\frac{\partial \Pi}{\partial \mathbf{b}_j} = \sum K_{ji} b_i + f_j \tag{4.29}$$

where

$$K_{ji} = \iint\limits_{\Omega_A + \Omega_B} \left\{ (\nabla \bar{N}_j)^T c c_g (\nabla \bar{N}_i) - \frac{\omega^2 c_g}{c} \bar{N}_j \bar{N}_i \right\} dx\ dy$$

$$f_j = \int_{\Gamma_A} \bar{N}_j \hat{\bar{u}}\ d\Gamma \tag{4.30}$$

In the outer domain Ω_c, ϕ is similarly approximated by a series of functions N_i which now satisfy exactly the governing equation (4.23) and the Sommerfeld radiation condition [Equation (4.24b)]. Thus in Ω_c

$$\bar{\phi} \approx \hat{\phi}^a = \sum N_i a_i = \mathbf{N}\mathbf{a} \tag{4.31}$$

where a_i are undetermined parameters.

The contribution of the outer domain to the total energy can now be evaluated as

$$\Pi_c = \frac{1}{2} \int_{\Gamma_c} \bar{\phi} \frac{\partial \bar{\phi}}{\partial n} \, d\Gamma \tag{4.32}$$

this result being obtained by application of Green's theorem to the domain in which the function $\bar{\phi}$ satisfies the governing equation.

Finally, as on the interface, the value of $\bar{\phi}$ as given by Equation (4.31) must be identical to that given by Equation (4.28), we must supplement the total functional by an additional term thus giving

$$\Pi_c^* = \frac{1}{2} \int_{\Gamma_b} \bar{\phi}^b \frac{\partial \bar{\phi}^a}{\partial n} \, d\Gamma - \int_{\Gamma_b} (\bar{\phi}^a - \bar{\phi}^b) \frac{\partial \bar{\phi}^a}{\partial n} \, d\Gamma \tag{4.33}$$

The total functional can now be written for the whole region and minimized with respect to parameters **a** and **b**. Details of this process are given elsewhere.[37] Defining the operator

$$\frac{\partial}{\partial n} \equiv P \tag{4.34}$$

we can write

$$\frac{\partial \Pi_c^*}{\partial b_j} = \sum \bar{K}_{ji}^T \mathbf{a}_i$$

$$\frac{\partial \Pi^*}{\partial a_j} = \sum \hat{K}_{ji} a_i + \bar{K}_{ji} b_i \tag{4.35}$$

where

$$\hat{K}_{ji} = \frac{1}{2} \int_{\Gamma_c} ((PN_j)^T N_i + N_j^T (PN_i)) \, d\Gamma$$

$$\bar{K}_{ji} = \int_{\Gamma_c} (PN_j)^T \bar{N}_j \, d\Gamma \tag{4.36}$$

The total system matrix can now be assembled and solved writing

$$\frac{\partial \Pi}{\partial \mathbf{a}} = 0 \qquad \frac{\partial \Pi}{\partial \mathbf{b}} = 0 \tag{4.37}$$

or

$$\begin{bmatrix} \mathbf{K} & \bar{\mathbf{K}}^T \\ \bar{\mathbf{K}} & \hat{\mathbf{K}} \end{bmatrix} \begin{Bmatrix} \mathbf{b} \\ \mathbf{a} \end{Bmatrix} + \begin{Bmatrix} \mathbf{f} \\ 0 \end{Bmatrix} = 0 \tag{4.38}$$

These being in a standard form of finite element equation.

In the wave problem various possibilities are open for the choice of the shape functions **N**. The series of Hankel and trigonometrical functions

$$\bar{\phi} = \sum_{j=0}^{m} H_j(kr)(\alpha_j \cos j\theta + \beta_j \sin j\theta)$$

satisfies Equation (4.23) term by term, and the radiation boundary condition, and so α_j and β_j can be used as the parameters **a**. The form of the matrix $\bar{\mathbf{K}}$ depends upon the shape functions $\bar{\mathbf{N}}$, in the adjacent finite elements, and in general numerical integration will be needed to evaluate the line integrals along Γ_c. For piecewise constant shape functions the matrices have been evaluated explicitly by Chen and Mei,[17] and are shown in Table 1.

Table 1   Matrices $\bar{\mathbf{K}}$ and $\mathbf{K}$ for exterior solution

$$\bar{\mathbf{K}}^T = \frac{-khL_c}{2} \begin{bmatrix} 2\,\acute{H}_0 \cdots \acute{H}_n\,(\cos n\theta_p + \cos n\theta_1), & \acute{H}_n\,(\sin n\theta_p + \sin n\theta_1) \cdots \\ 2\,\acute{H}_0 \cdots \acute{H}_n\,(\cos n\theta_1 + \cos n\theta_2), & \acute{H}_n\,(\sin n\theta_1 + \sin n\theta_2) \cdots \\ \vdots & \vdots \\ 2\,\acute{H}_0 \cdots \acute{H}_n\,(\cos n\,\theta_{p-1} + \cos n\,\theta_p), & \acute{H}_n\,(\sin n\theta_{p-1} + \sin n\theta_p) \cdots \end{bmatrix}$$

$$\hat{\mathbf{K}} = \pi r_c kh\{\mathrm{diag}\,|2H_0\acute{H}_0,\ H_1\acute{H}_1,\ H_1\acute{H}_1, \ldots H_m\acute{H}_m,\ H_m\acute{H}_m|\}$$

m is the number of terms in the Hankel function series
r_c is the radius of boundary Γ_c which must be circular
L_c is the distance between (equidistant) nodes on Γ_c
k is wave number
p is the number of nodes
H_n and $\acute{H}_n$ are Hankel functions and derivatives

4.5.4   Coupling with boundary integrals

As is shown by Zienkiewicz,[37,38] boundary integrals of the two types, source distributions and Green's identity, can be used as exterior solutions and coupled to finite elements exactly as was done in the previous section. However, a simpler linking is possible[38,54] where Green's identity is used as the basis of the boundary integral method, and this will now be shown. The starting point for boundary integral method is Green's second identity

$$\iint\limits_{\Omega} (\psi \nabla^2 \bar{\phi} - \bar{\phi} \nabla^2 \psi)\,\mathrm{d}x\,\mathrm{d}y = \int_{\Gamma} \left(\psi \frac{\partial \bar{\phi}}{\partial n} - \bar{\phi} \frac{\partial \psi}{\partial n} \right) \mathrm{d}\Gamma \tag{4.39}$$

which applies to any two functions $\bar{\phi}$ and ψ [subject to the restriction that if the domain is unbounded, ψ, $r(\partial\psi/\partial x)$, $r(\partial\psi/\partial y)$, $\phi r(\partial\phi/\partial x)$, $r(\partial\phi/\partial y)$ must be bounded in absolute value for large r.] The domain used is as shown in Figure 4.2. The boundary of Ω_C is now in three segments, Γ_c the interior boundary, Γ_D the exterior boundary (at infinity) with a cut joining it to Γ_c, and Γ_ε is a circular boundary of radius ε around a singularity point P (the pole of the Hankel function). Now let ψ be a solution of Equation (4.23). For this solution to be non-trivial ψ must have singularities somewhere in Ω_C (Stakgold,[55] Chapter 1, p. 33). The appropriate singular solution to Equation (4.23) is the Hankel function $H_0(kr)$, which satisfies it everywhere except at $r = 0$, which is the pole P. Now denote the domain Ω_C except for the area round the pole P, by Ω'_C, i.e.

$$\Omega'_C = \Omega_C - \Omega''_C$$

Now it is required that ϕ be a solution of Equation (4.23) in Ω'_C and ψ is known to be a solution of Equation (4.23), in Ω'_C so that

$$\iint_{\Omega'_c} (\psi \nabla^2 \bar\phi - \bar\phi \nabla^2 \psi)\, dx\, dy = \iint_{\Omega'_c} (-k^2 \psi \bar\phi + k^2 \psi \bar\phi)\, dx\, dy = 0 \qquad (4.40)$$

so that, writing ψ as H_0

$$\int_{\Gamma_c} \left(H_0 \frac{\partial \bar\phi}{\partial n} - \bar\phi \frac{\partial H_0}{\partial n} \right) d\Gamma + \int_{\Gamma_\varepsilon} \left(H_0 \frac{\partial \bar\phi}{\partial n} - \bar\phi \frac{\partial H_0}{\partial n} \right) d\Gamma = 0 \qquad (4.41)$$

from Equation (4.39). Since Γ_ε is circular the second integral can be written as

$$\int_0^{2\pi} \left(H_0(k\varepsilon) \frac{\partial \bar\phi}{\partial n} - \bar\phi \frac{\partial H_0(k\varepsilon)}{\partial n} \right) \varepsilon\, d\alpha \qquad (4.42)$$

where n is the outward normal to Ω''_C, so that

$$\frac{\partial}{\partial r} = -\frac{\partial}{\partial n}$$

After some manipulation,[56] the integral can be evaluated and taking the limiting case, as ε tends to zero, we obtain

$$\lim_{\varepsilon \to 0} \int_{\Gamma_\varepsilon} \left(H_0(kr) \frac{\partial \bar\phi}{\partial n} - \bar\phi \frac{\partial H_0(kr)}{\partial n} \right) d\Gamma = 4i\bar\phi(p) \qquad (4.43)$$

Equation (4.41) can now be written as[57]

$$\int_{\Gamma_c} \left(H_0(kr) \frac{\partial \bar\phi}{\partial n} - \bar\phi \frac{\partial H_0(kr)}{\partial n} \right) d\Gamma + 4i\bar\phi(p) = 0 \qquad (4.44)$$

where p is the pole, and r is the distance from it. If p is placed on the boundary, Γ_c, then Equation (4.44) must be modified. It becomes

$$\int_{\Gamma_c} \left(H_0(kr) \frac{\partial \bar\phi}{\partial n} - \bar\phi \frac{\partial H_0(kr)}{\partial n} \right) d\Gamma + 4i\bar\phi(p)(1 - \gamma/2\pi) = 0 \qquad (4.45)$$

where γ is the angle of rotation of the tangent at point p, being π for a smooth curve.

This completes the formulation of the boundary integral equations. We now discretize them for numerical solution.[54] Shape functions $\mathbf{N}$, $\mathbf{M}$ are introduced for both $\bar{\phi}$ and $\partial\bar{\phi}/\partial n$ so that

$$\bar{\phi} = \mathbf{N}\hat{\boldsymbol{\phi}} \qquad \frac{\partial\bar{\phi}}{\partial n} = \mathbf{M}'\frac{\partial\hat{\boldsymbol{\phi}}}{\partial\mathbf{n}} \tag{4.46}$$

where $\hat{\boldsymbol{\phi}}$ and $\partial\hat{\boldsymbol{\phi}}/\partial\mathbf{n}$ denote nodal values of these quantities. Equation (4.45) now becomes, in discretized form

$$\frac{i}{4}\int_{\Gamma_c}\left(\mathbf{N}\hat{\boldsymbol{\phi}}\frac{\partial H_0(kr)}{\partial n} - H_0(kr)\mathbf{M}\frac{\partial\hat{\boldsymbol{\phi}}}{\partial\mathbf{n}}\right)\,\mathrm{d}\Gamma + (1-\gamma/2\pi)\hat{\phi} = 0 \tag{4.47}$$

This equation can be recast in matrix form, after some integration, as

$$\mathbf{A}\hat{\boldsymbol{\phi}} = \mathbf{B}\frac{\partial\hat{\boldsymbol{\phi}}}{\partial\mathbf{n}} \tag{4.48}$$

Consider again the functional [Equation (4.32)]. This now becomes

$$\Pi_c = \frac{1}{2}\int_{\Gamma_c}\frac{\partial\hat{\boldsymbol{\phi}}^T}{\partial\mathbf{n}}\mathbf{M}^T\mathbf{N}\hat{\boldsymbol{\phi}}\,\mathrm{d}\Gamma \tag{4.49}$$

But the normal derivatives can be eliminated, using Equation (4.48), and identifying $\boldsymbol{\phi}$ with the variables $\mathbf{b}$ of the finite element region we can write

$$\Pi_c = \tfrac{1}{2}\mathbf{b}^T(\mathbf{B}^{-1}\mathbf{A})^T\int_{\Gamma_c}\mathbf{M}^T\mathbf{N}\,\mathrm{d}\Gamma\,\mathbf{b} \tag{4.50}$$

We set variations of this functional with respect to $\mathbf{b}$ equal to zero, to give

$$\frac{\partial\Pi_c}{\partial\mathbf{b}} = \frac{1}{2}\left\{\mathbf{B}^{-1}\mathbf{A}\int_{\Gamma_c}\mathbf{M}^T\mathbf{N}\,\mathrm{d}\Gamma + \left(\mathbf{B}^{-1}\mathbf{A}\int_{\Gamma_c}\mathbf{M}^T\mathbf{N}\,\mathrm{d}\Gamma\right)^T\right\}\mathbf{b} = \check{\mathbf{K}}\mathbf{b} \tag{4.51}$$

where $\check{\mathbf{K}}$ is 'stiffness' matrix for the region Ω_c. It can be assembled and solved like any other element matrix, and has the virtues of symmetry and continuity and is directly additive to the finite element region contribution. In the exterior elements used here quadratic shape functions were adopted for both $\mathbf{M}$ and $\mathbf{N}$. The integrations in Equations (4.47) and (4.51) were carried out numerically, using Gauss integration, with more integration points close to singularities.[56]

4.5.5 Infinite elements

These infinite elements are defined in a manner very similar to finite elements, the only difference being that their domain extends to infinity, and so exponential shape functions are used. Such shape functions, based on exponential decay were introduced by Bettess[58] for steady state problems. For periodic problems they have been used by Zienkiewicz, Bettess and Saini.[13,39,41,42] Elements extending to infinity, but based on a mapping method have been used by Gartling and Becker.[59,60]

The element is parametric, the parent shape being a rectangle which extends to infinity in the ξ direction, ξ and η being the reference co-ordinates. The element has 3 reference points, in each of the directions, giving a total of 9 such points. In the ξ direction the 3rd point is assumed to be at a large but not infinite distance. The element is shown in Figure 4.3.

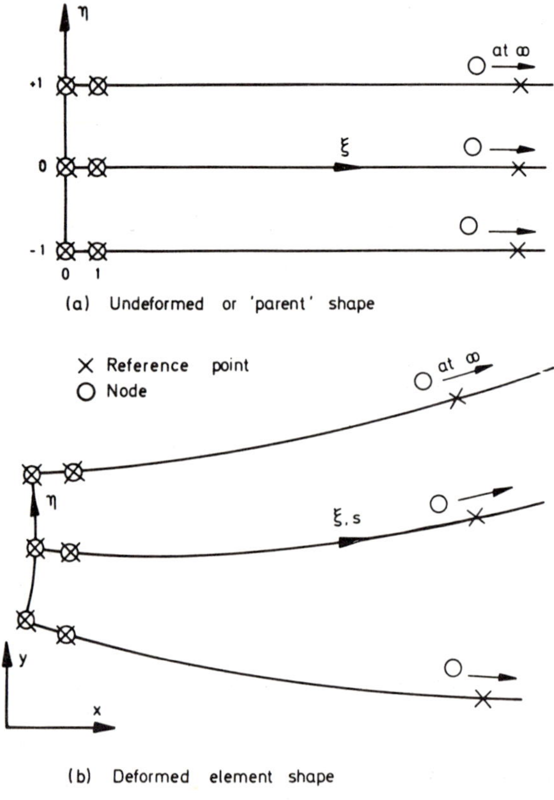

Figure 4.3   Infinite element geometry

The x, y co-ordinates of these reference points are used, as with finite elements, to set up a transformation from x, y to ξ, η, co-ordinate systems. In addition a new co-ordinate system is set up, the s direction, which is the same direction as the ξ co-ordinate, but distances along it are in the same measure as those in x, y co-ordinates.

In the η direction a conventional Lagrange polynomial is used. In the s direction a function of the form

$$p(s)\,\mathrm{e}^{-s/L}\,\mathrm{e}^{iks} \tag{4.52}$$

is used. Here $p(s)$ is a polynomial in s, L is a length which dictates the severity of the exponential decay, and k is the wave number. The first term allows a certain variation in the wave envelope, the second term makes the wave decay for large radii and the third term ensures a basically periodic shape. This shape function satisfies the radiation condition. The shape function and its nodal values are now complex. For the rth node the shape function can be written

$$N_r = \mathrm{e}^{(s_r - s)/L}\,\mathrm{e}^{iks} \prod_{\substack{q=1 \\ \neq r}}^{n-1} \left(\frac{s_q - s}{s_q - s_r} \right) \tag{4.53}$$

where the nth node is infinitely distant. The elements that are currently in use have 3 nodes in the s direction, the third node being placed at infinity, and indeed in this type of problem it is superfluous.

Numerical integration in the s direction has some novelties. At first Gauss Laguerre integration[61] was used, but this was fairly expensive, and later a special integration formula was devised.[39] In practice a six point integration rule in the s direction has been used most frequently.

4.5.6 Finite elements

Two types of finite element have been used in the interior of the model. These are the isoparametric 6 noded triangle and the isoparametric 8 noded quadrilateral. There is nothing particularly novel about the use of these elements. The element matrices obtained are entirely real.[62] Two additional elements are available however which are more unusual. The first is a linear element with 3 nodes. This again is isoparametric, and has a quadratic shape function. By applying this element along any boundary damping can be introduced, as was mentioned in describing the functional in an earlier section. This damping element is of very simple form and is obtained by considering the stationarity of a damping term

$$\frac{1}{2} \int_{\Gamma} \frac{\alpha i \omega}{c} \bar{\phi}^2 \, \mathrm{d}\Gamma \; . \tag{4.54}$$

along a boundary to which it is applied. If α is zero then the boundary is completely reflecting, if unity the boundary will absorb normally incident plane waves. This element has two purposes. Firstly it can be used to apply the radiation boundary condition, as described in Section 4.5.2. Secondly it can be used to model surfaces which do not reflect perfectly, for example rubble breakwaters. The exact values of α to choose in such cases are of course rather conjectural.

A second type of element is a development of the previous element, to the case where some kind of pervious boundary is present, as for example in storage tanks of the Doris design, which have an outer wall containing many holes. This effect is modelled by a simple approximation. It is assumed that the velocity through the previous boundary is directly proportional to the difference in wave elevations across it. The wave elevation η can be obtained from the velocity potential $\bar{\phi}$ because

$$\eta = -\frac{ig}{\omega}\bar{\phi}, \qquad \bar{\phi} = \frac{i\omega}{g}\eta$$

Now the velocity is $\partial\phi/\partial n$, so that

$$\frac{\partial\bar{\phi}}{\partial n} = \frac{\beta i\omega}{g}(\bar{\phi}_2 - \bar{\phi}_1) \tag{4.55}$$

where $\bar{\phi}_2$ and $\bar{\phi}_1$ are the two velocity potentials. This boundary can be modelled using a line element with 6 nodes, used to couple the two domains, the element matrix being obtained by stationarity of

$$\frac{1}{2}\int_\Gamma \frac{\beta i\omega}{g}(\bar{\phi}_2 - \bar{\phi}_1)^2 \, d\Gamma \tag{4.56}$$

The addition of such a term to the functional of Equation (4.25) leads to natural boundary conditions of the form of Equation (4.55) which is what is needed.

4.5.7　Wave forces

When the values of velocity potential have been determined at all nodes, using the finite elements, then the pressures can be found at any point, from Equations (4.19) and (4.20) as

$$p = -i\rho\omega\bar{\phi}(x, y)\frac{\cosh k(h+z)}{\cosh kh} \tag{4.57}$$

It is then possible to determine forces and moments acting on surfaces below the water by integrating over the area. The force $\mathbf{F}$ can be found multiplying the pressure by the unit normal vector to the surface at any point and integration over the surface noting that

$$d\mathbf{F} = \mathbf{n}p$$

and the momentum $\mathbf{M}$ from

$$d\mathbf{M} = \mathbf{r} \times \mathbf{n}p$$

where $\mathbf{r}$ is the vector from the point through which the axes about which the moments $\mathbf{M}$ are to be calculated pass to the point being considered.

4.6 RESULTS

The program WAVE[63] which contains all the features described in the earlier sections has been applied to an extremely wide range of wave problems,[13,39] and only some of the more interesting results will be given here.

The reader is invited to observe that for basically prismatic structures a very good approximation of the force calculation is obtained by the two-dimensional computations which are very cheap by comparison with the full three-dimensional treatment. In Sections 4.6.4 and 4.6.5 we shall show how this approximation fails when the structure is really three-dimensional and how this can be remedied by a partial three dimensionality.

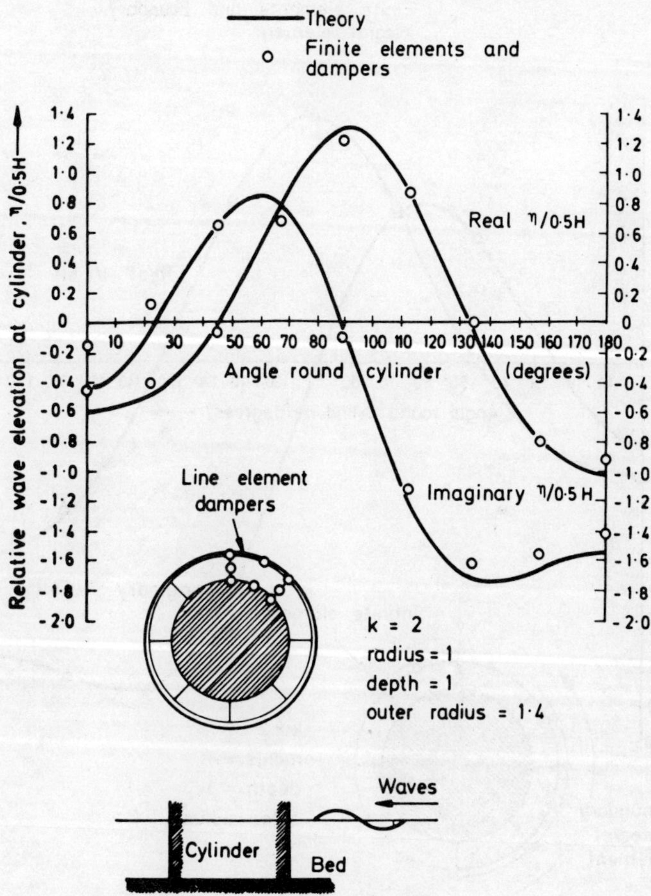

Figure 4.4 Waves diffracted by cylinder (analysis using dampers)

4.6.1 Waves incident upon cylinders

Figures 4.4–4.6 show a cylinder problem where waves of wave number 2 are incident upon a cylinder of radius 1. The relative amplitudes around the cylinder are determined using a mesh of interior finite elements, together with exterior solutions using dampers, infinite elements, boundary solutions, and boundary integrals. The results are all very accurate, with the exception of those obtained using dampers. The comparison is with the exact solution of MacCamy and Fuchs.[25]

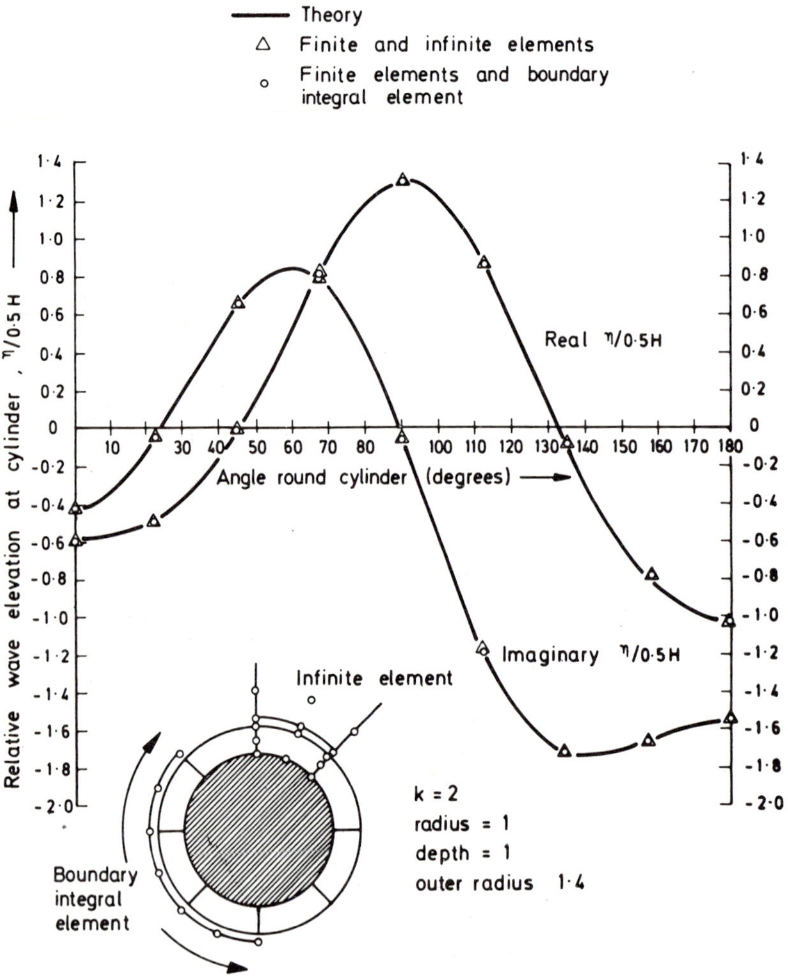

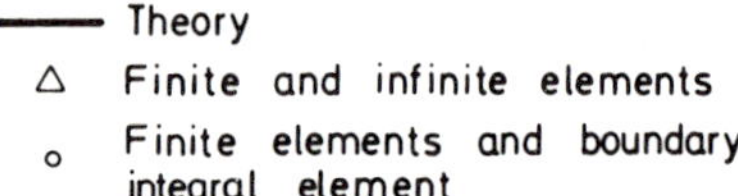

Figure 4.5 Waves diffracted by cylinder (analyses using both boundary integral elements and infinite elements)

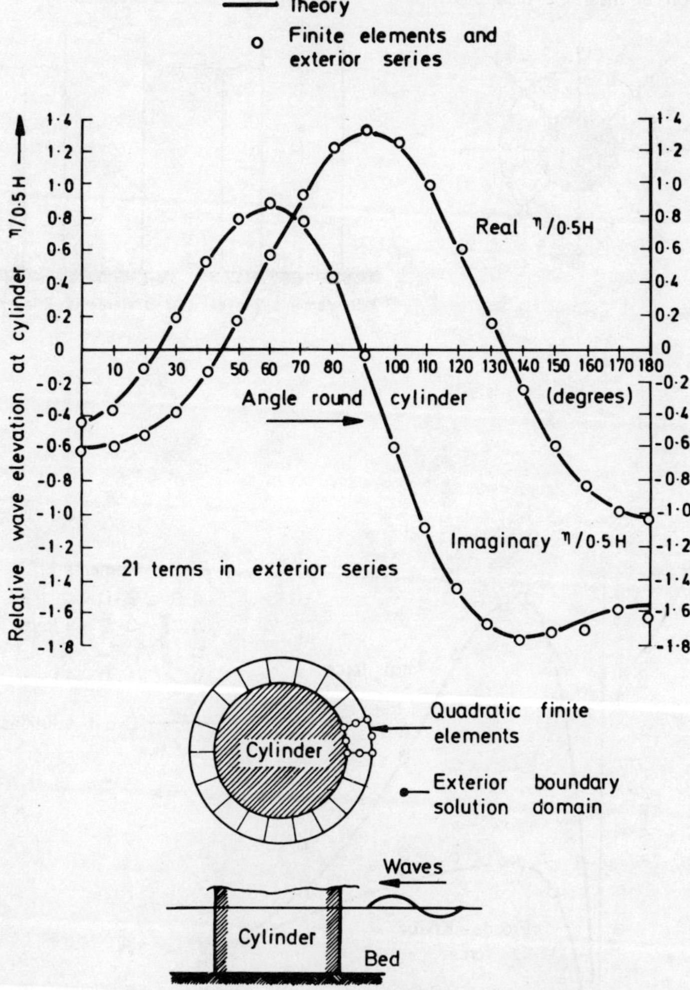

Figure 4.6 Waves diffracted by cylinder (analysis using exterior series solution)

Figures 4.7, 4.8 show the cylinder problem solved by Hogben and Standing[8] as a test case for their boundary integral program. This time the wave forces and the moments about the base of the cylinder are plotted in dimensionless form against wave number. For this case depth/radius $= 5\cdot0$. Very good agreement is obtained with the theoretical forces and moments. The Froude–Krylov forces are also evaluated to show the error involved in omitting the diffraction analysis.

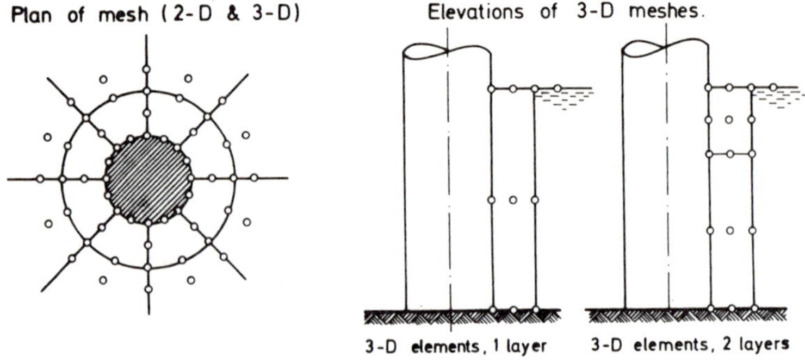

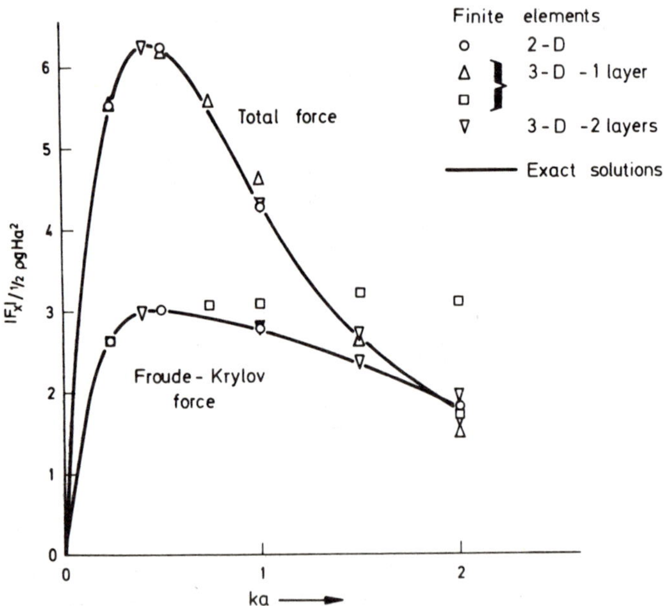

Figure 4.7 Horizontal forces on a surface piercing cylinder. k: Wavenumber; a: Radius of cylinder; H: Incident wave height; d: Height of cylinder; $d/a = 5$. Waves incident from x direction

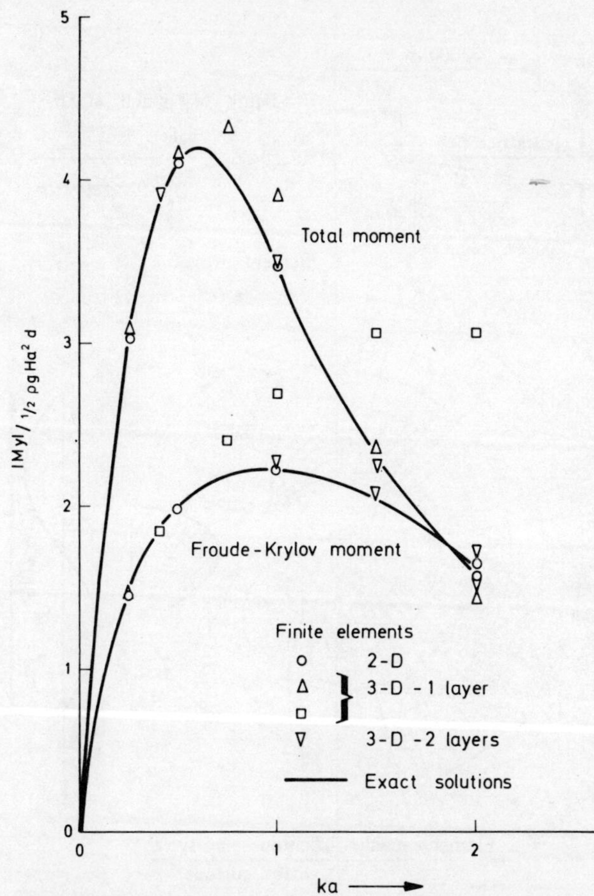

Figure 4.8 Moments on a surface piercing cylinder. k: Wavenumber; a: Radius of cylinder; H: Incident wave height; d: Height of cylinder; $d/a = 5$. Waves incident from x direction

4.6.2 Waves incident upon a parabolic shoal and island

This is the problem discussed by Homma,[27] Vastano and Reid,[33] and Jonsson *et al.*[34] It was solved here using various exterior solutions. Figure 4.9 shows the mesh used and the geometry of the problem. Figures 4.10–4.12 show the excellent agreement obtained between the analytical solution, and the solutions using interior finite elements and exterior boundary solutions, boundary integrals, and infinite elements.

 Numerical Methods in Offshore Engineering

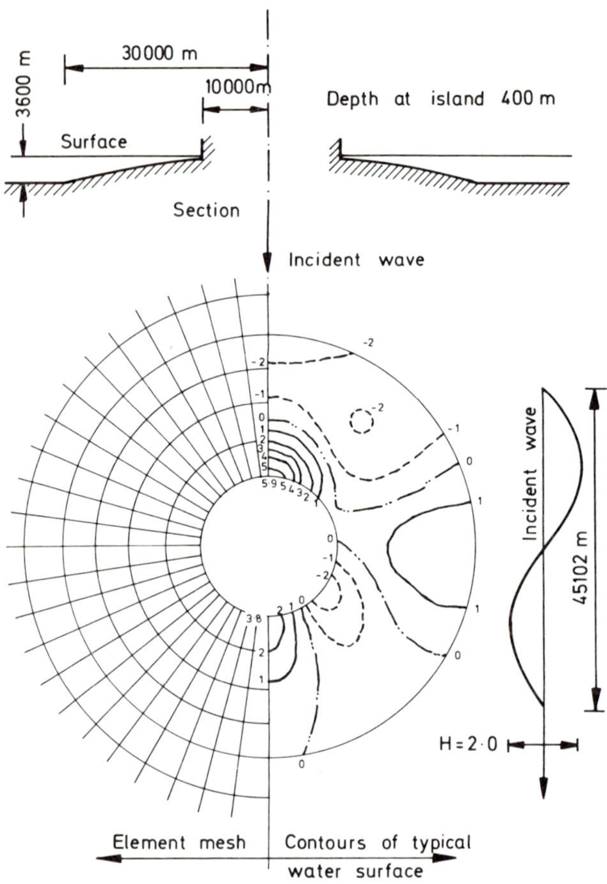

Figure 4.9 Waves incident upon a parabolic shoal

4.6.3 Waves incident upon a cylinder with pervious protecting wall

This problem is loosely based upon the Ekofisk storage tank. Figure 4.13 shows the geometry of the problem and Figure 4.14 shows a typical water surface for a 'permeability' constant $\beta/\beta_{\text{crit}}$ for the wall of 0·5. Wave forces on the protecting wall and the inner cylinder are shown in Figure 4.15 compared with the theoretical forces acting on impermeable cylinders of the inner and outer size for various values of $\beta/\beta_{\text{crit}}$. Figure 4.16 shows the corresponding moments.

4.6.4 Submerged cylinder problems

Problems which look approximately like submerged cylinders are of considerable interest in offshore structures, because they approximate closely storage

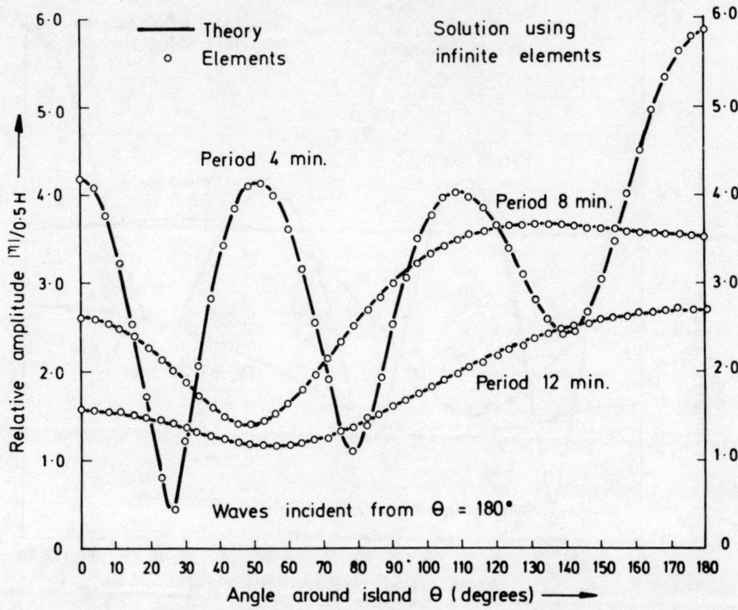

Figure 4.10 Relative amplitudes at island on parabolic shoal (analysis using infinite elements)

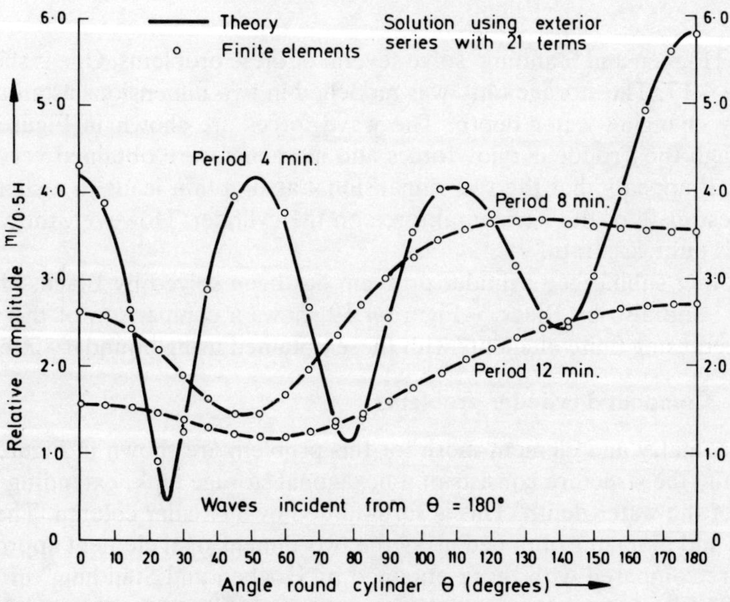

Figure 4.11 Relative amplitudes at island on parabolic shoal (analysis using exterior series solution element)

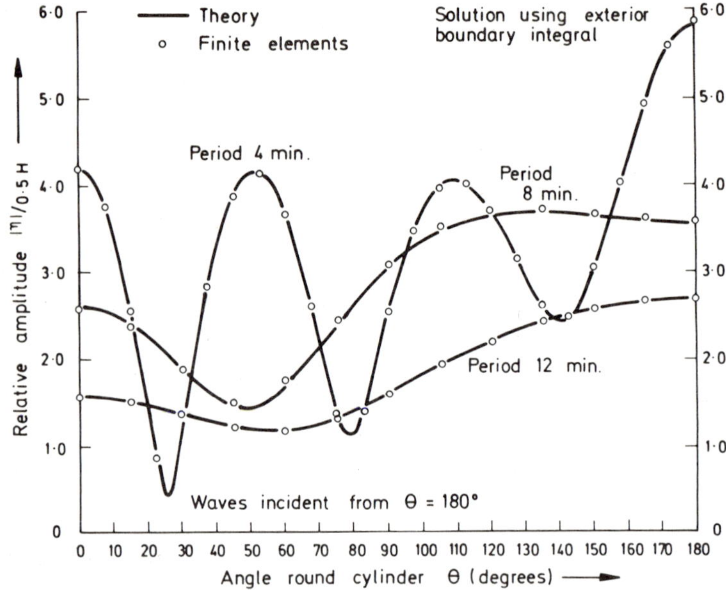

Figure 4.12   Relative amplitudes at island on parabolic shoal (analysis using boundary exterior boundary integral element)

tanks. Hogben and Standing[8] solve several of these problems. One is shown in Figure 4.17. The storage tank was modelled in two dimensions as an area of rapidly changing water depth. The wave forces are shown in Figure 4.18. Although the Froude–Krylov forces and moments were obtained very accurately, it appears that the two-dimensional assumption leads to a significant underestimate of the horizontal force on the cylinder. However, the vertical force is quite accurate.

Another submerged cylinder problem has been solved by Black, Mei and Bray,[64] and also by Black.[65] Figure 4.19 shows a comparison of the results obtained using finite elements with those obtained using boundary integrals.

4.6.5   Compound cylinder problems

The geometry and element mesh for this problem are shown in Figure 4.20. Basically the structure consists of a hexagonal storage tank, extending to one third of the water depth. This is surmounted by a circular column. The wave forces and moments obtained using the two-dimensional element approximation are compared with those obtained by Hogben and Standing,[8] in Figure 4.21. The horizontal force and the moments are still underestimated, but not too drastically.

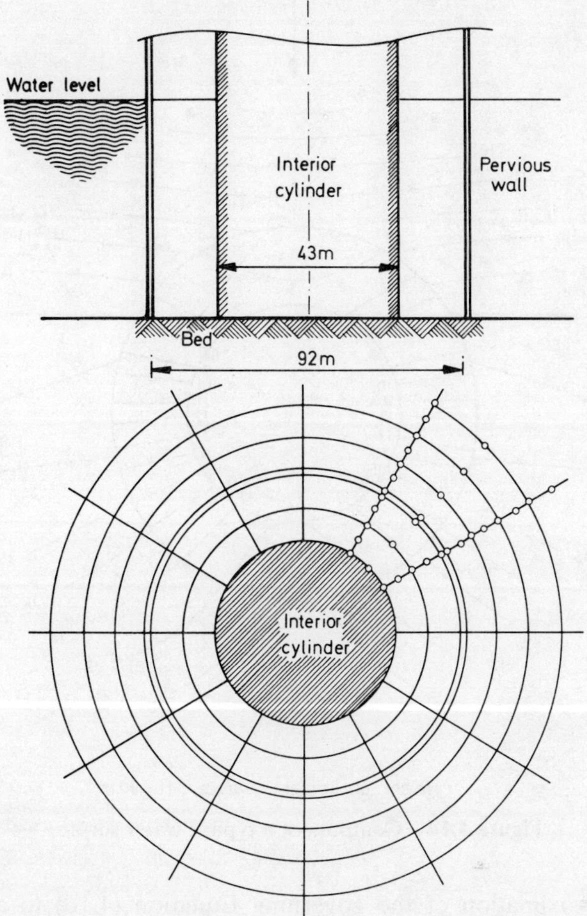

Figure 4.13   Geometry and mesh for Ekofisk-type structure

PART II: THREE-DIMENSIONAL SOLUTION

4.7   BASIC EQUATIONS AND FINITE ELEMENT DISCRETIZATION

Clearly the last two examples illustrate that for obstacles which are not simple prisms with vertical faces the two-dimensional idealization is at best a crude approximation. It is therefore obviously necessary to extend the general finite

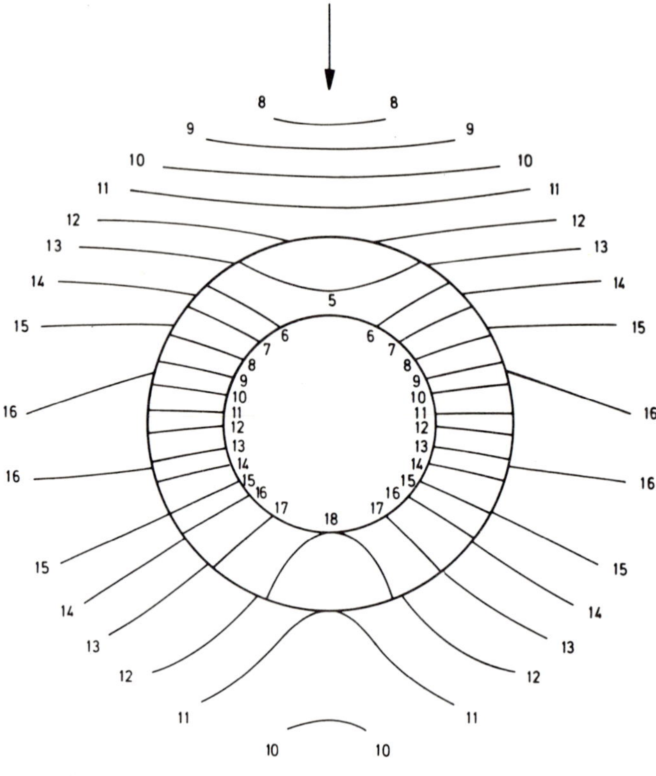

Figure 4.14   Contours of a typical water surface

element approximation of the governing Equation (4.16) to a full three-dimensional one.

Once again the basic variational principle corresponding to the Equation (4.16) is given by the expression Equation (4.26)—and the finite element approximation can be obtained following the process outlined before providing the shape functions $\bar{N}_i$ are understood to stand for a three-dimensional interpolation. We need now however to take now into account explicitly the free surface boundary condition Equation (4.17b) which in two-dimensional equations was implicitly satisfied.

Once again the energy derivatives are given by the standard relation (Π_a standing for the energy of the inner region)

$$\frac{\partial \Pi_a}{\partial \mathbf{b}} = \mathbf{Kb} + \mathbf{f} \qquad \phi = \sum \bar{N}_i b_i = \mathbf{Nb} \qquad b_i = \phi_i \qquad (4.58)$$

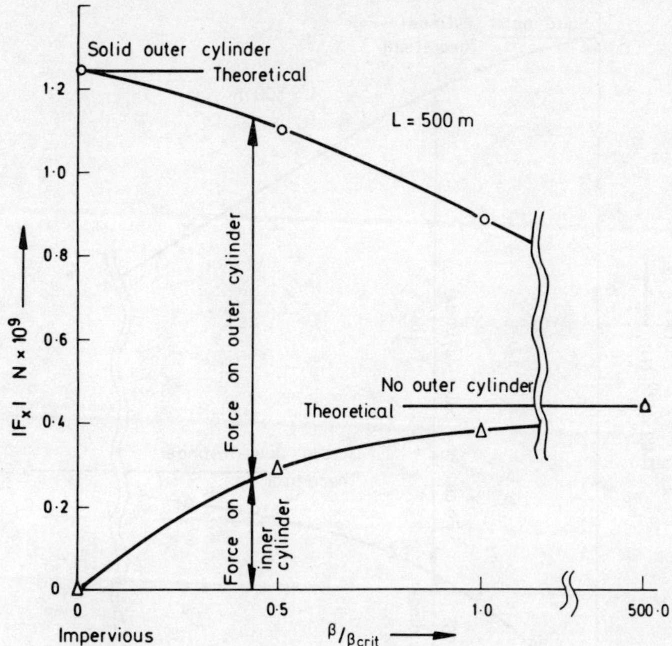

Figure 4.15 Forces on cylinder with protecting wall

where

$$K_{ij} = \int_{\Omega_A} \boldsymbol{\nabla}^T N_i \boldsymbol{\nabla}^T N_j \, dvol - \int_{\Omega_A} N_i \rho \frac{\omega^2}{K} N_j \, dvol - \int_{\Gamma_t} N_i \frac{\omega^2}{g} N_j \, d\Gamma$$

$$(4.59)$$

$$f_i = \int_{\Gamma_s} N_i \hat{\bar{u}}_n$$

In above Ω_A stands for the domain in which standard three-dimensional elements are used, Γ_t for its free surface and Γ_s for the solid boundary are used. In our computation use is made of the basic isoparametric quadratic 'bricks' shown in Figure 4.22 which indicates a typical problem domain. The second term derives from compressibility and is generally neglected in wave force calculation.

It is generally convenient to separate the effect of the incoming and reflected waves writing

$$\phi = \phi_0 + \phi_s \tag{4.60}$$

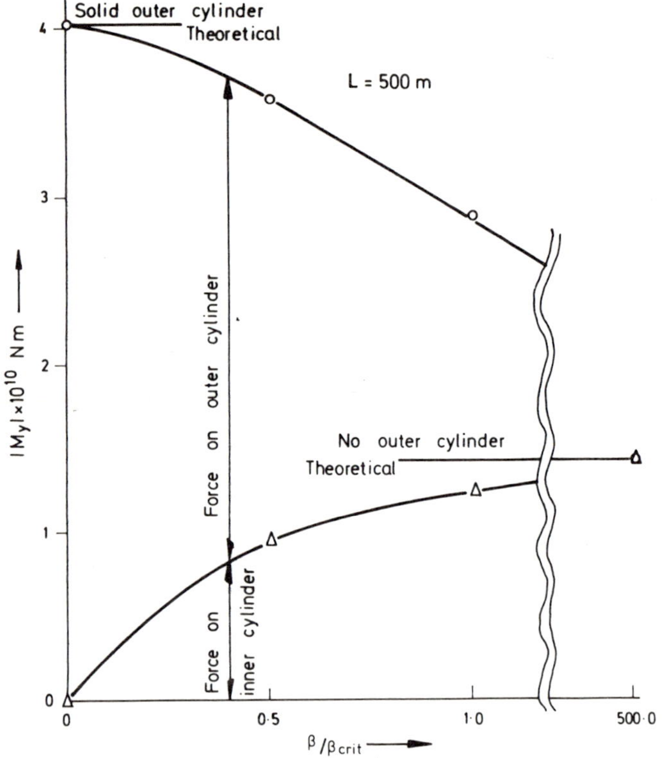

Figure 4.16   Moments on cylinder with protecting wall

and if the interpolation is deemed to refer to ϕ_s only then the only change occurring in the forcing expression on the solid boundary is

$$f_i = \int_{\Gamma_s} N_i \hat{\bar{u}}_n \, \mathrm{d}\Gamma - \int_{\Gamma_s} N_i \frac{\partial \phi_0}{\partial n} \, \mathrm{d}\Gamma \tag{4.61}$$

4.8   USE OF TWO-DIMENSIONAL APPROXIMATION IN THE EXTERIOR DOMAIN

If we assume that beyond a certain zone of influence of the obstacle the depth of water becomes constant—the full three-dimensional solution will tend to a two-dimensional one of the type discussed in Sections 4.4–4.5. In a full three-dimensional analysis it is convenient to take the finite element domain sufficiently far from the obstacle so that this two-dimensional approximation

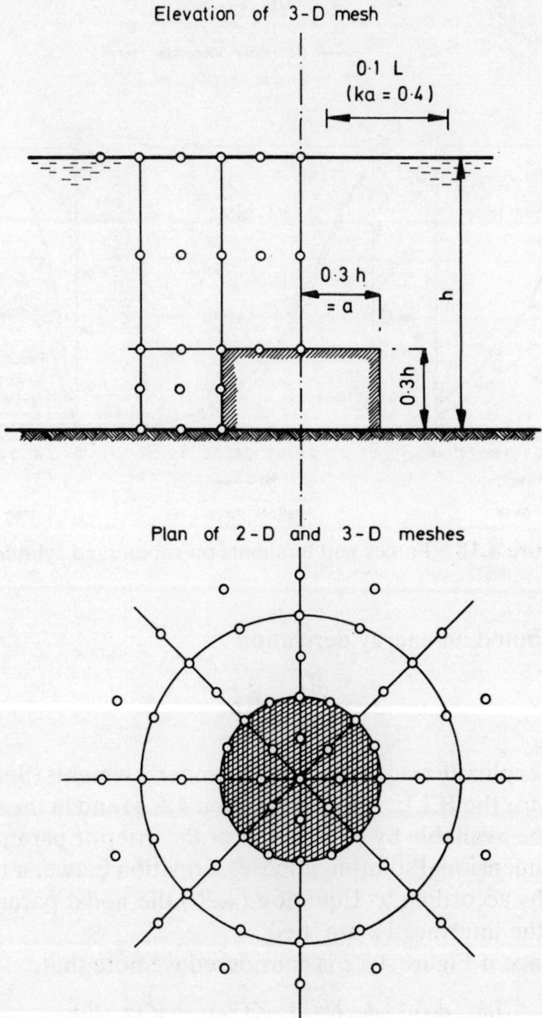

Figure 4.17 Geometry and meshes for submerged cylinder

can be made on the interface. We shall find in subsequent examples that this distance is quite moderate—and in Figure 4.22 showing a typical layout we would assume that two-dimensional conditions are applicable at Γ_c.

This allows the use of any of the two-dimensional exterior region forms derived in Section 4.5 where we found that parameters $\mathbf{b}$ defining $\bar{\phi}$ on the

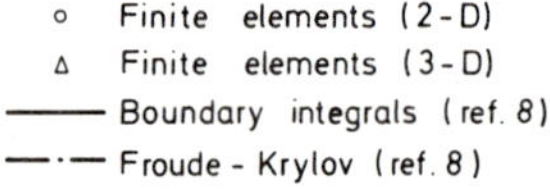

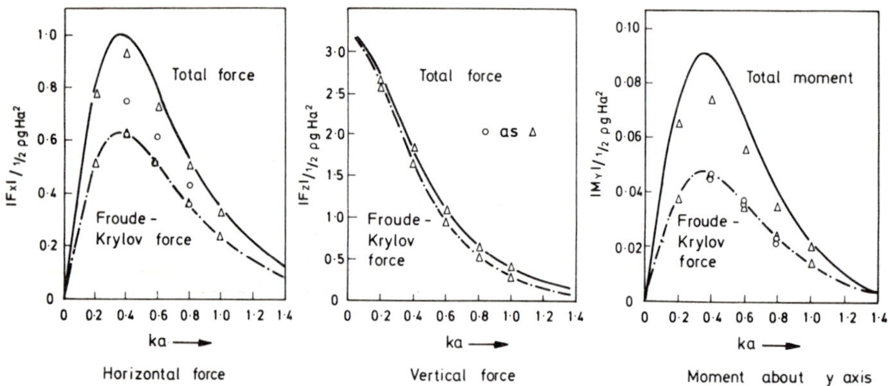

Figure 4.18   Forces and moments on submerged cylinder

interface contributed an energy derivation

$$\frac{\partial \Pi_c}{\partial \mathbf{b}} = \check{\mathbf{K}} \mathbf{b} \tag{4.62}$$

The matrix $\check{\mathbf{K}}$ is explicitly available for the 'damper' elements (Section 4.5.2) on the interface or for the B.I.E. solution (Section 4.5.4) and in the series solution method would be available by elimination of the interior parameters.

As the two-dimensional solution imposes a relation between the values of ϕ at various depths according to Equation (4.20) the nodal parameters on any vertical line of the interface Γ_c are 'tied'.

Thus for intance if Figure 4.22 is considered we note that

$$[\phi_1, \phi_2 \cdots \phi_m] \equiv [1, Z(z_2) \cdots Z(z_m)] \bar{\phi}_1 \tag{4.63}$$

and only the variable $\phi_1 = \bar{\phi}_1$ is independent.

It is of course possible to approach the solution in the exterior domain without imposing the two-dimensional restrictions. The use of 'boundary dampers' or simply of the radiation condition at a finite distance is simple in this context. The series or B.I.E. solution is however very much more complicated. The latter would follow the use of the singularity (source) functions used in Chapter 3 [Equations (3.9)] and reintroduce all the computational difficulties of such a formulation. In the examples we shall therefore use exclusively the two-dimensional exterior representation.

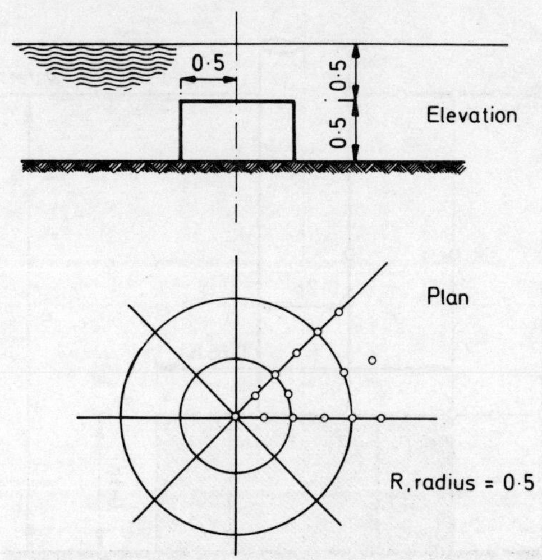

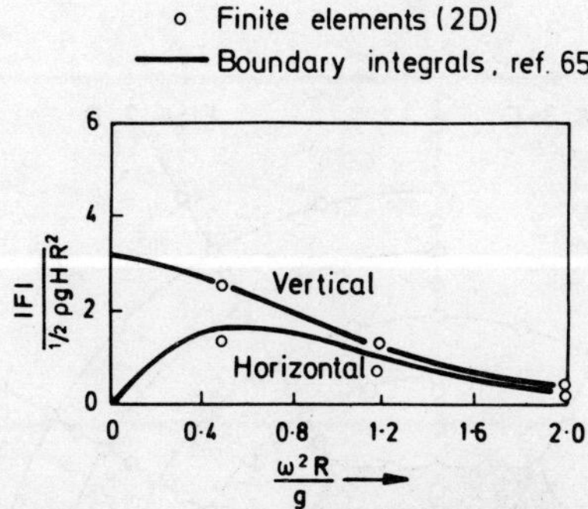

Figure 4.19 Wave forces on submerged cylinder

4.9 SYSTEM EQUATIONS

The equations for the compete system can be written noting that

$$\frac{\partial \Pi_a}{\partial \mathbf{b}} + \frac{\partial \Pi_c}{\partial \mathbf{b}} = 0 \tag{4.64}$$

i.e.

$$[\mathbf{K} + \check{\mathbf{K}}]\mathbf{b} + \mathbf{f} = 0 \tag{4.65}$$

Solution for parameters $\mathbf{b}$, which have been identified with the potential

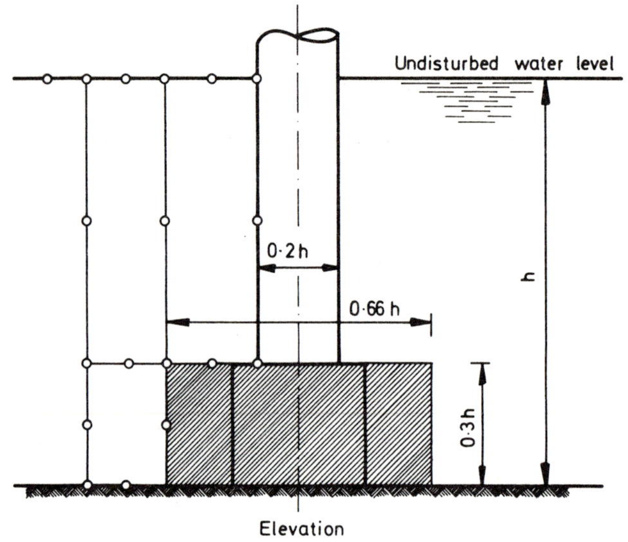

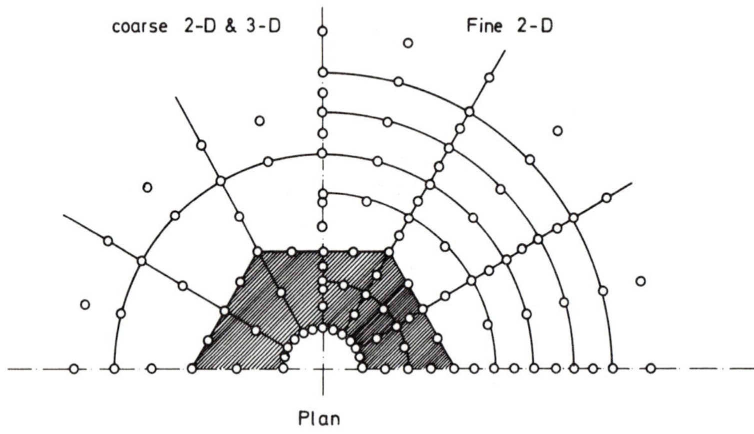

Figure 4.20   Geometry and meshes for compound column

functions at nodes of the finite element mesh are now available. Noting that **f** has components due to

(1) specified boundary motion
(2) incident wave forms

the solution of the system gives immediately both effects.

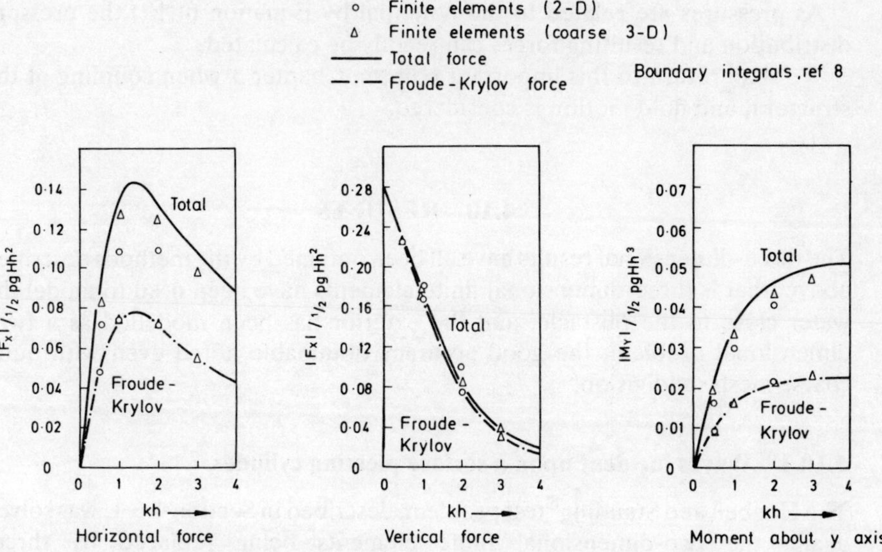

Figure 4.21 Wave forces on compound column

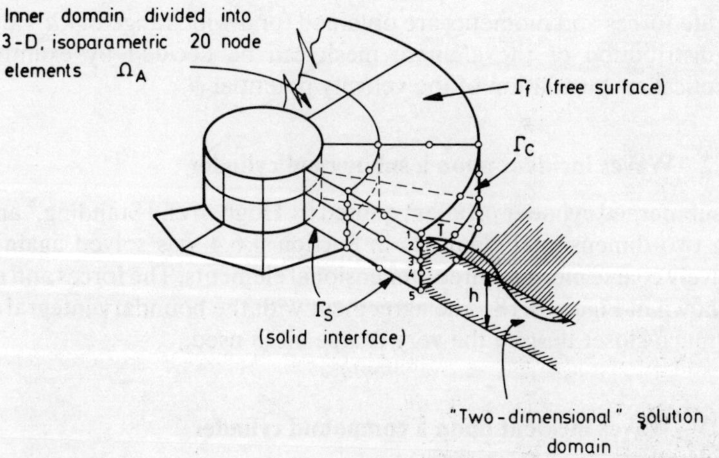

Figure 4.22 A typical three-dimensional discretization. On Γ_c nodes are tied to a two-dimensional infinite domain solution

$$[\phi_1, \phi_2 \ldots \phi_5] = [Z(z_1) \ldots Z(z_5)]\phi_1$$

As pressures are related to the potential by Equation (4.19) the pressure distribution and resulting forces can readily be calculated.

We shall return to this important point in Chapter 5 when coupling of the structural and fluid motion is considered.

4.10   RESULTS

The three-dimensional results have all been obtained by the methods described above, that is three-dimensional finite elements have been used to model the water close to the obstacle, and the exterior has been modelled as a two-dimensional problem, the good accuracy obtainable noted even with quite coarse mesh subdivision.

4.10.1   Waves incident upon a surface piercing cylinder

The Hogben and Standing[8] test problem, described in Section 4.6.1, was solved again, the two-dimensional finite elements being replaced by three-dimensional ones. The forces and moments obtained are shown in Figure 4.7 and 4.8, together wth the two-dimensional results. When only one layer of three-dimensional elements is used, the results agree well with exact theory[25] for $ka < 1$. For larger values of ka the parabolic variation of ϕ in the z direction does not model accurately the actual cosh variation and the results are not so accurate. However, if two layers of 3D elements are used as shown, accurate forces and moments are obtained for a wide range of ka values. The best distribution of the element mesh can be decided by examining the theoretical cosh variation of the velocity potential ϕ.

4.10.2   Waves incident upon a submerged cylinder

The submerged cylinder problem quoted by Hogben and Standing,[8] and solved using two-dimensional elements in Section 4.6.4 was solved again, using a relatively coarse mesh of three-dimensional elements. The forces and moments are shown in Figure 4.18. The agreement with the boundary integral results is now much closer despite the very coarse mesh used.

4.10.3   Waves incident upon a compound cylinder

The problem described in Section 4.6.5 was resolved, using a coarse mesh of 3D elements. Even with the coarse mesh a good agreement with the results of Hogben and Standing,[8] for forces and moments, was obtained. The results are shown in Figure 4.20.

4.11 CONCLUDING REMARKS

The procedures outlined in this Chapter generalize—and supplement—the three-dimensional source technique outlined in Chapter 3 for the computation of 'inertia' water loads or structures.

For relatively simple structures in a constant depth of water the choice of technique is purely a matter of computational economics and personal preference. Clearly if, however, a two-dimensional model is applicable the methods here outlined will be advantageous.

The formulation presented is such that it includes as a subset the pure boundary integral technique if the finite element region is omitted. This clearly would not be done if a large shoaling region were included where the equations cease to be homogeneous and the boundary integral method fails. Further if local damping barriers or indeed distributed viscous forces exist, the finite element approximation over a part of the region appears essential.

The full three-dimensional finite element modelling allows a very simple treatment for coupling of structural and fluid behaviour but clearly the computational savings are affected to some extent by the additional data needed to specify the mesh—the boundary domain technique reducing this to a surface specification. However, the few examples given show that a very coarse mesh suffices in most cases.

One aspect of the most general formulation given here needs comment. It appears that it may be possible with additional elaboration to include non-linear wave effects by modifying the free surface condition Equation (4.9b) in the zone close to the obstacle where their importance increases. Techniques suggested in another context by Kawahara[66] appear to indicate in which way such an extension is practicable.

With regard to the treatment of the exterior regions three general possibilities have been indicated all allowing for the energy dissipation by radiating waves. The reader may understandably be confused by the alternatives given in Section 4.5 all of which have proved workable. Some of these are simpler to programme and apply than others—the simple boundary damper being most efficient here but least accurate.

Many computational experiments are still needed to define the optimum technique but at the present moment all are applicable and available to the user.

REFERENCES

1. Wiegel, R. L. (1964). *Oceanographical Engineering*, Prentice Hall.
2. Myers, J. J., Holm, C. H., and MacAllister, R. F. (1969). *Handbook of Ocean and Underwater Engineering*, McGraw-Hill.
3. Hogben, N. 'Fluid loading on offshore structures, a state of art appraisal: wave loads', *Maritime Technology Monograph No. 1,* Royal Institute of Naval Architects.

4. Johnson, A. J. (1974). 'Design in relation to the environment', *Offshore Structures Conference at the Institution of Civil Engineers*, London, 7–8 October, 1974, 1–7.

5. British Ship Research Association Contract Report No. W. 278, August 1976. 'A critical evaluation of the data on wave force coefficients', *Department of Energy Report No. OT/R/7611*, with 2 Appendices.

6. Garrison, C. J. and Chow, P. Y. (1972). 'Wave forces on submerged bodies', *Journal of Waterways and Harbors Division Proc.*, ASCE, August 1972.

7. Boreel, L. J. (1974). 'Wave action on large off-shore structures', *Conference on Offshore Structures, Institution of Civil Engineers*, 7–8 October, 1974, 7–14.

8. Hogben, N. and Standing, R. G. (1974), 'Wave loads on large bodies', *Proc. International Symposium on Dynamics of Marine Vehicles and Structures in Waves*. Published by I. Mech. E., London.

9. Wehausen, J. V. and Laitone, E. V. (1960). 'Surface waves', *Encyclopedia of Physics*, Volume IX, Fluid Dynamics III. Springer Verlag.

10. Morison, J. R., O'Brien, M. P., Johnson, J. W., and Schaaf, S. A. (1950). 'The force exerted by surface waves on piles', *Petroleum Trans.*, **189**, TP 2846 (1950), 149–154.

11. Iverson, H. W. and Balent, R. (1951). 'A correlating modulus for fluid resistance in accelerated motion', *J. Applied Physics*, **22**, 3 (March 1951), 324–328.

12. Zienkiewicz, O. C. and Newton, R. E. (1969). 'Coupled vibrations of a structure submerged in a compressible fluid', *Proceedings of the Symposium on Finite Element Techniques*, held at the Institut fur Statik und Dynamik der Luft-und Baum-fahrtkonstruktionen, University of Stuttgart, Germany, June 10–12, 1969.

13. Zienkiewicz, O. C. and Bettess, P. (1975). 'Infinite elements in the study of fluid-structure interaction problems', *2nd International Symposium on Computing Methods in Applied Science and Engineering*, Versailles, France. 15–17 December, 1975.

14. Sommerfeld, A. (1949). *Partial Differential Equations in Physics*, Academic Press.

15. Rellich, F. (1943). 'Über das aymptotische Verhatten der Lösungen von $\Delta u + \lambda u = 0$ in Unedlichen Gebieten', *Jahresbericht der Deutschen Mathematiker Vereinigung*, **53**, 57–65.

16. Lamb, Sir. H. (1932). *Hydrodynamics*, 6th edition, Cambridge University Press.

17. Chen, H. S. and Mei, C. C. (1974). 'Oscillations and wave forces in an offshore harbour', Ralph M. Parsons, Laboratory for Water Resources and Hydrodynamics, *MIT, Report No. 190*, August 1974.

18. Newton, R. E. (1973). *Radiation Bounding Condition for Plane Strain*, private communication.

19. Shaw, R. P. (1975). 'Boundary integral equation methods applied to water waves', AMD, Vol. 11, Boundary Integral Equation Method: Computational Applications in Applied Mechanics, 1975 ASME. (Presented at 1975 *Applied Mechanics Conference, the Rensselear Polytechnic Institute*, Troy, New York, 23–25 June, 1975, edited T. A. Cruse and F. J. Rizzo).

20. Orlanski, I. (1976). 'A simple boundary condition for unbounded hyperbolic flows', *Journal of Computational Physics*, **21**, 251–269.

21. Stoker, J. J. (1957). *Water Waves*, Interscience, New York.

22. Courant, R. and Hilbert, D. (1962). *Methods of Mathematical Physics*, Vol. 2, Interscience, New York.

23. Berkhoff, J. C. W. (1972). 'Computation of combined refraction–diffraction', *13th International Conference on Coastal Engineering*, Vancouver, July 10–14, 1972.

24. Berkhoff, J. C. W. (1975). 'Linear wave propagation problems and the finite element method', pp. 251–280 *Finite Elements in Fluids Vol. 1*, edited by R. H. Gallagher *et al.*, Wiley, London.
25. MacCamy, R. C. and Fuchs, R. A. (1954). 'Wave forces on piles. A diffraction theory', *Beach Erosion Board Tech. Mem. No. 69*, 1954.
26. Lowan, A. N., Morse, P. M., Feshback, H., and Lax, M. (1946). *Scattering and Radiation from Circular Cylinders and Spheres C-67897, Tables of Amplitudes and Phase Angles*, U.S. Navy Dept., Office of Research and Inventions (July 1946).
27. Homma, S. (1950). 'On the behaviour of seismic sea waves around Circular Island', *The Geophysical Magazine*, published by the Central Meteorological Observatory, Tokyo, 1950, **XXI**, 199–208.
28. Ünluate, Ü. and Mei, C. C. (1975). 'Effects of entrance loss on harbor oscillations', *Journal of Waterways, Harbor and Coastal Engineering Division, ASCE*, WW2 May 1975, 161–180.
29. Hwang, L-S. and Tuck, E. O. (1970). 'On the oscillations of harbours of arbitrary shape', *J. Fluid Mech.*, **42**, part 3, 447–464.
30. Olsen, K. and Hwang, L. S. (1971). 'Oscillations in a bay of arbitrary shape and variable depth', *Journal of Geophysical Research*, **176**, 5048–5064.
31. Lee, J. J. (1971). 'Wave induced oscillations in harbours of arbitrary geometry', *J. Fluid. Mech.*, **145**, 375–394.
32. Shaw, R. P. (1975). 'Transient scattering by a circular cylinder', *Journal of Sound and Vibration*, **42**, part 3, 295–304.
33. Vastano, A. C. and Reid, R. O. (1967). 'Tsunami response for islands: verification of a numerical procedure', *Journal of Marine Research*, **25**, 129–139.
34. Jonsson, I. G., Skovgaard, O., and Brink-Kjaer, O. (1976). 'Diffraction and refraction calculations for waves attacking an island', *Journal of Marine Research*.
35. Chen, H. S. and Mei, C. C. (1974). 'Oscillations and wave forces in a man-made harbor in the open sea', presented at the *10th Naval Hydrodynamics Symposium*, June 1974.
36. Chen, H. S. and Mei, C. C. (1975). 'Hybrid-element method for water waves', *Proceedings of the Modelling Techniques Conference (Modelling 1975)*, San Francisco, 3–5 September, **1**, 63–81.
37. Zienkiewicz, O. C. (1975). 'The finite element method and boundary solution procedures as general approximation methods for field problems', *World Congresss on Finite Element Methods in Structural Mechanics*, Bournemouth, 12–17 October, 1975.
38. Zienkiewicz, O. C., Kelly, D. W., and Bettess, P. (1977). 'The coupling of the finite element method and boundary solution procedures', *International Journal for Numerical Methods in Engineering*, **11**. No. 2, 355–375.
39. Zienkiewicz, O. C. and Bettess, P. (1977). 'Diffraction and refraction of surface waves using finite and infinite elements', *International Journal for Numerical Methods in Engineering*, **11**, 1271–1290.
40. Newton, R. E. (1975). 'Finite element analysis of two-dimensional added mass and damping', pp. 219–232 *Finite Elements in Fluids*, Volume 1, Wiley, 1975, Gallagher, R. H. *et al.* (eds.).
41. Saini, S. S., Zienkiewicz, O. C., and Bettess, P. *Coupled Hydrodynamic Response of Concrete Gravity Dams using Finite and Infinite Elements*, Department of Civil Engineering, University College of Wales, Swansea, Research Report C/R/271/76.

42. Saini, S. S., Zienkiewicz, O. C., and Bettess, P. (1976). *Modelling of Hydrodynamic Effects on Dams using Finite Elements*, Department of Civil Engineering, University College of Wales, Swansea, Research Report C/R/269/76.
43. Graff, Karl F. (1975). *Wave Motion in Elastic Solids*, Clarendon Press, Oxford.
44. Ewing, M., Jardetsky, W., and Press, F. (1957). *Elastic Waves in Layered Media*, McGraw-Hill.
45. Selberg, H. L. (1952). 'Transient compression waves from spherical and cylindrical cavities', *Ark Fys*, **5**, 97–108.
46. Hopkins, H. G. (1960). 'Dynamic expansion of spherical cavities in metals', in *Progress in Solids Mechanics*, Vol. 1, Chap. 3 (Ed. I. Sneddon and R. Hill), North Holland.
47. Lysmer, J. and Drake, L. A. (1971). 'The propagation of love wave across non horizontally layered structures', *Bulletin of the Seismological Society of America*, **61**, No. 5, 1233–1251. October 1971.
48. Lysmer, J. and Drake, L. A. 'A finite element method of seismology', Chapter 6 of *Methods in Computational Physics*, Volume 11: Seismology, edited by Alder, B., Fernbach, S., and Bolt, A. Academic Press, New York and London.
49. Lysmer, J. and Kuhlemeyer, R. L. (1969). 'Finite dyamic model for infinite media', *Journal of the Engineering Mechanics Division, Proceedings of the American Society of Civil Engineers*, August 1969, EM4, 859–877.
50. Lysmer, J. and Waas, G. (1972). 'Shear waves in plane infinite structures', Journal of the Engineering Mechanics Division, *Proceedings of the American Society of Civil Engineers*, February 1972, EM1, 85–105.
51. Liang, V. C. (1974). 'Dynamic response of structures in layered soils', MIT Department of Civil Engineering, *Research Report R74-10*, January 1974.
52. Bai, K. J. (1972). 'A variational method in potential flows with a free surface', *Report No. NA72-2*, College of Engineering, University of California, Berkeley, September 1972.
53. Kausel, E. (1974). 'Forced vibrations of circular foundations on layered media', *Research Report R74-11*, Department of Civil Engineering, Massachusetts Institute of Technology, Cambridge, Massachusetts.
54. Zienkiewicz, O. C., Kelly, D. W., and Bettess, P. (1977). 'Marriage à la mode—or the best of both worlds. Boundary integrals and finite element procedures', *Conf. on Innovative Methods of Numerical Computation, Versailles*, to be published.
55. Stakgold, I. (1968). *Boundary Value Problems of Mathematical Physics*, MacMillan.
56. Karaiosifidis, S. (1976). 'A comparison of solution methods for exterior surface wave problems', *M.Sc. thesis*, University College of Wales, Swansea.
57. Banaugh, R. P. and Goldsmith, W. (1963). 'Diffraction of steady acoustic waves by surfaces of arbitrary shape', *Journal of the Acoustical Society of America*, **35**, No. 10, 1590–1601.
58. Bettess, P. (1977). 'Infinite elements', *International Journal for Numerical Methods in Engineering*, **11**, No. 1, 53–64.
59. Gartling, D. and Becker, Eric B. (1974). 'Computationally efficient finite element analysis of viscous flow problems', in *Computational Methods in Non Linear Mechanics*, ed. J. T. Oden *et al.*, Texas Institute for Computational Mechanics, 1974. (Proceedings of the International Conference on Computational Methods in Non Linear Mechanics held at the University of Texas, Austin, Texas, 23–25th September, 1974).

60. Gartling, D. and Becker, E. B. (1976). 'Finite element analysis of viscous incompressible fluid flow', *Computer Methods in Applied Mechanics and Engineering*, Vol. 8, 1976, 51–60.

61. Rabinowitz, G. P. and Weiss, G. (1959). 'Tables of abscissas and weights for numerical evaluation of integrals of the form $\int_0^\infty x^n f(x)\, e^{-x}\, dx$, *Mathematical Tables and Other Aids to Computation*, Vol. 13, 285–294.

62. Zienkiewicz, O. C. (1977). *The Finite Element method in Engineering Science*, McGraw-Hill (third edition).

63. 'WAVE, A finite element program for solving the wave equation', *Computer Report No. 81*, Department of Civil Engineering, University College of Wales, Swansea.

64. Black, J. L., Mei, C. C., and Bray, M. C. G. (1971). 'Radiation and scattering of water waves by rigid bodies', *J. Fluid Mech.*, **46**, 151–164.

65. Black, J. L. (1975). 'Wave forces on vertical axisymmetric bodies', *J. Fluid Mech.*, **67**, 369–376.

66. Kawahara, M. 'Periodic Galerkin finite element method of unsteady periodic flow of viscous fluid', *International Journal for Numerical Methods in Engineering*, to be published.

Chapter 5

Dynamic Fluid—Structure Interaction. Numerical modelling of the Coupled Problem

O. C. Zienkiewicz and P. Bettess

5.1 INTRODUCTION

In the preceding Chapter we have been concerned with the problem of computing the fluid loads due to wave and current action on a structure assumed immobile. When the structure itself is set in motion of periodic or transient character additional interactive forces are caused by this motion, and in turn effect it to a greater or lesser extent. The cause of the structure motion in offshore situations is primarily due to wave forces, but other causes such as earthquake excitation, vibration caused by machinery forces are often encountered.

If the relationship between these forces and the fluid motion is *linear* then a simple superposition of effects will occur. It was already remarked (Ch. 3 and Ch. 4) that for bluff bodies such as represented by gravity platforms, etc., the non-linear effects of the drag force can be neglected, but that the refraction and wave scattering effects are here important. It is this class of problem that is the basic concern of this chapter—and we feel that it is of primary importance to the understanding of the interaction effects especially if the high frequency content of the motion is to be considered. Indeed, we shall not exclude *a priori* the effects of fluid compressibility which at very high frequencies can be important.

For very slender structures with large amplitudes of motion the dynamic forces must include the drag component even when interaction forces are considered. Equation (4.1) of the previous chapter defining such forces has now to be written in terms of relative structure—fluid velocity. If we denote by **r** the structure displacement measured in the same direction as the velocity at its surface and by $\dot{\mathbf{r}}$ and $\ddot{\mathbf{r}}$ its velocity and acceleration respectively then Equation (4.1) can be rewritten as

$$F = \tfrac{1}{2}C_D\rho A \,|\mathbf{u}-\dot{\mathbf{r}}|(\mathbf{u}-\dot{\mathbf{r}}) + C_M\rho V(\dot{\mathbf{u}}-\ddot{\mathbf{r}}) \tag{5.1}$$

For structural analysis purposes it is customary to linearize the first term by methods suggested by Malhotra and Penzien.[1] We shall not consider further in this chapter such approximations which are discussed in the context of frame jacket structures in pages 252 of Chapter 8 and 298 of Chapter 9. In the present discussion we shall specifically omit the drag term of Equation (5.1).

5.2 THE STRUCTURE AND THE FLUID

It is invariably convenient to consider the structure (and its foundation) in isolation from the fluid surrounding it—and to introduce the coupling through the surface pressures through which this occurs. Typically thus the situation is illustrated in Figure 5.1, where the structure shown in Figure 5.1a is considered in its own idealization as separate from the fluid which is itself idealized by finite elements in Figure 5.1b.

The standard application of the finite element (or indeed of any discretization process) to the structure results in a set of equations of the general form (for linear systems)

$$\mathbf{M}_s\ddot{\mathbf{r}} + \mathbf{C}_s\dot{\mathbf{r}} + \mathbf{K}_s\mathbf{r} + \mathbf{f} = 0 \qquad (5.2)$$

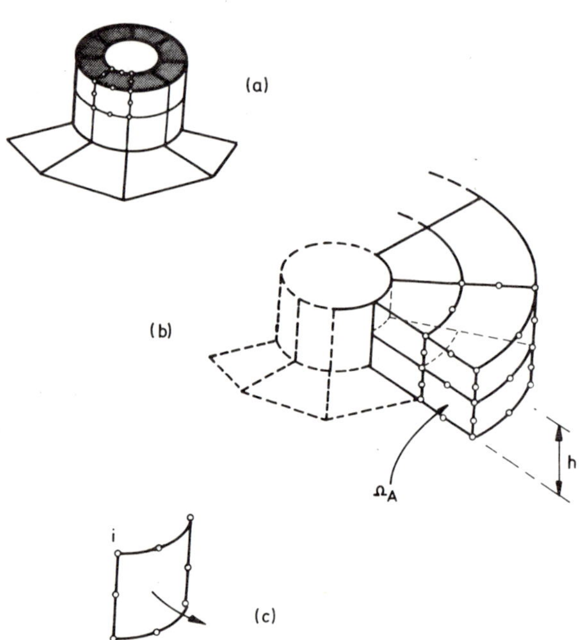

Figure 5.1 Structure–fluid interaction: (a) Structure discretized in suitable finite element mesh; (b) Fluid domain interacting with the structure, with its own finite element discretization; (c) Element of structure fluid interface.

Here $\mathbf{M}_s$, $\mathbf{C}_s$ and $\mathbf{K}_s$ stand respectively for the mass, damping and stiffness matrices of the structural system, and $\mathbf{r}$ is the nodal displacement vector. Formulation of such structural problems by the finite element method is discussed in full elsewhere,[2] and is referred to in more detail in this book in Chapters 6 and 7, so that any further particulars of these steps can be omitted. It is, however, important to note the nature of the force vector $\mathbf{f}$ which can be written as

$$\mathbf{f} = \mathbf{f}_0 + \mathbf{f}_p = \mathbf{f}_0 + \mathbf{L}\mathbf{p} \tag{5.3}$$

where $\mathbf{f}_0$ is the vector of independently specified static or dynamic forces, $\mathbf{p}$ is the vector of pressures on the structural interface and $\mathbf{L}$ is the appropriate coupling matrix. If $\bar{\mathbf{N}}_i$ is the shape function describing the structural displacements for a surface node i (Figure 5.1c), N_i is the interpolation function for pressures specified at the same nodes, and $\mathbf{n}$ is the unit normal vector to the surface, then we can write[2]

$$\mathbf{f}_i = \int_{\Gamma_s} \bar{\mathbf{N}}_i^T \mathbf{n} \mathbf{N} \mathbf{p} \, \mathrm{d}\Gamma \qquad p = \sum N_i p_i = \mathbf{N} \mathbf{p} \tag{5.4}$$

which specifies the element of the coupling matrix $\mathbf{L}$ as

$$L_{ij} = \int_{\Gamma_s} \bar{\mathbf{N}}_i^T \mathbf{n} N_j \, \mathrm{d}\Gamma \tag{5.5}$$

For periodic response computation we assume that all variables can be written in a complex form, i.e.

$$\mathbf{r} = \bar{\mathbf{r}} \, \mathrm{e}^{i\omega t} \qquad \mathbf{f}_0 = \bar{f}_0 \, \mathrm{e}^{i\omega t} \qquad \mathbf{p} = \bar{\mathbf{p}} \, \mathrm{e}^{i\omega t}$$

and the response equation becomes

$$|\mathbf{K} + i\omega \mathbf{C} - \omega^2 \mathbf{M}|\bar{\mathbf{r}} = \mathbf{L}\bar{\mathbf{p}} + \bar{\mathbf{f}}_0 \tag{5.6}$$

5.3 THE FLUID

We have already discussed in considerable detail in Chapter 4 the methods of computing the pressures (or complex potential) in a fluid subject to a wave force input and prescribed boundary motion. These procedures enable us immediately to compute the coupling forces.

To avoid unnecessary repetition we refer the reader for details of the fluid equations to Chapter 4, and here only will summarize the essential points for a full three-dimensional finite element model discussed in Section 4.7. For convenience we shall deal with periodic solutions only (i.e. in the frequency domain ω) and replace the potential by pressure, using the relationship Equation (4.18).

The governing equation of pressure distribution is (*vide* Chapter 4 e.g. 4.5, etc.) including compressibility effects

$$\nabla^2 \bar{p} - \frac{\omega^2 \rho}{K} \bar{p} = 0 \tag{5.7}$$

in the fluid domain Ω, together with the boundary condition

$$\frac{\partial \bar{p}}{\partial z} - \frac{\omega^2}{g} \bar{p} = 0 \tag{5.8a}$$

on the free surface, and

$$\frac{\partial \bar{p}}{\partial n} + i\rho\omega\hat{u}_n - i\rho\omega\frac{\partial \bar{\phi}_0}{\partial n} = \frac{\partial \bar{p}}{\partial n} - \rho\omega^2 \mathbf{n}^T \bar{\mathbf{N}}\mathbf{r} - i\rho\omega\frac{\partial \bar{\phi}_0}{\partial n} = 0 \tag{5.8b}$$

on solid surfaces, and

$$\frac{\partial \bar{p}}{\partial r} + \frac{i\omega}{c} \bar{p} = 0 \tag{5.8c}$$

at infinite distances (for the outgoing wave). In the above we have replaced the normal velocity of the solid surface comprising the structure, with the derivative of structural displacements $\bar{\mathbf{N}}\mathbf{r}$ thus providing the coupling, and have assumed that p stands for the scattered component of the pressure, the incoming wave if any being specified by ϕ_0 and contributing to the appropriate normal gradient on the boundary. $\mathbf{n}$ once again is the unit normal vector to the surface Γ_s.

In Equation (5.8c) the wave velocity c can be *either* that of the compressive wave or of the surface wave and so far no technique has been developed for filtering out both effects simultaneously. In what follows we shall assume the surface wave radiation to be of primary importance and neglect the outgoing compressive wave.

Presuming a finite element discretization in the fluid coupled to an exterior domain where one of the techniques described in Section 4.5 of the previous chapter is used, we can write a discretized form of the above equations, with

$$\bar{p} = \bar{\mathbf{N}}\bar{\mathbf{p}} = \sum \bar{N}_i \bar{p}_i \tag{5.9}$$

where $\bar{\mathbf{p}}$ stands for the vector of pressures at nodal points and $\bar{N}_i$ for appropriate shape functions, we can write (following the details of Section 4.7 and using standard finite element procedures)

$$(\mathbf{K}_f + \hat{\mathbf{K}}_f)\bar{\mathbf{p}} = \bar{\mathbf{g}} \tag{5.10}$$

where

$$K_{f_{ij}} = \int_{\Omega_A} (\nabla N_i)^T \nabla N_j \, d\Omega - \frac{\omega^2 \rho}{K} \int_{\Omega_A} N_i N_j \, d\Omega - \frac{\omega^2}{g} \int_{\Gamma_f} N_i N_j \, d\Gamma \tag{5.11a}$$

where Γ_f stands for the free surface which contributes to the third term while the second term of the 'stiffness' is due to compressibility of fluid and can often be neglected.

The terms of $\hat{\mathbf{K}}$ depend on the exterior discretization—and we shall not discuss them here except to note that $\hat{\mathbf{K}}$ will always contain complex terms indicating the radiation damping effect. Details of $\hat{\mathbf{K}}$ will be found in the previous chapter. Finally we can write

$$\mathbf{g} = \rho\omega^2 \int_{\Gamma_s} \mathbf{N}n^T\bar{\mathbf{N}}\,\mathrm{d}\Gamma\mathbf{r} + \mathbf{g}_0 = \rho\omega^2\mathbf{L}^T\mathbf{r} + \mathbf{g}_0 \tag{5.11b}$$

where $\mathbf{g}_0$ is the forcing vector due to the external oncoming wave.

$$\mathbf{g}_0 = i\rho\omega \int_{\Gamma_s} \mathbf{N}\frac{\partial\phi_0}{\partial n}\,\mathrm{d}\Gamma \tag{5.11c}$$

and matrix $\mathbf{L}$ is precisely the same as that occurring in the structural dynamic Equation (5.6).

The formal solution for the interface pressures can now be written from Equation (5.10), using the definitions of Equation (5.11) as

$$\bar{\mathbf{p}} = (\mathbf{K}_f + \hat{\mathbf{K}}_f)^{-1}(\mathbf{g}_0 + \rho\omega^2\mathbf{L}\mathbf{r}) \tag{5.12}$$

This 'inversion' need only be done for the $\mathbf{p}$ values relevant to the structural face Γ_s through which the structural equations can be coupled. We note from the previous chapter that often quite a coarse idealization can be used for pressure determination and, as the two interpolating functions $\bar{\mathbf{N}}$ and $\mathbf{N}$ can be quite different, use of this fact can simplify analysis considerably.

5.4 COUPLED STRUCTURAL DYNAMIC EQUATIONS

Inserting Equation (5.12) into the standard Equation (5.6) yields

$$[\mathbf{K} + i\omega\mathbf{C} - \omega^2\mathbf{M}]\bar{\mathbf{r}} = \rho\omega^2[\mathbf{L}(\mathbf{K}_f + \hat{\mathbf{K}}_f)^{-1}\mathbf{L}^T]\mathbf{r} + \mathbf{f}_0$$

$$+ \mathbf{L}(\mathbf{K}_f + \hat{\mathbf{K}}_f)^{-1}\mathbf{g}_0$$

$$\equiv \omega^2(\hat{\mathbf{M}} - i\hat{\mathbf{C}})\mathbf{r} + \mathbf{f}_f + \mathbf{f}_0 \tag{5.13}$$

where $\mathbf{f}_f$ is the wave force vector (discussed in detail in the previous chapter), $\hat{\mathbf{M}}$ is the 'additional mass matrix' and $\hat{\mathbf{C}}$ is the additional damping matrix which is given by the complex part of $\hat{\mathbf{K}}$ which only contributes to this term. The significance of these terms is clearer if we rewrite Equation (5.13) as

$$[\mathbf{K} + i\omega(\mathbf{C} + \hat{\mathbf{C}}) - \omega^2(\mathbf{M} + \hat{\mathbf{M}})]\mathbf{r} = \mathbf{f}_f + \mathbf{f}_0 \tag{5.14}$$

with

$$\hat{\mathbf{M}} = \rho \mathbf{L}(\mathbf{K}_f + \text{Re}\,\hat{K}_f)^{-1}\mathbf{L}^T \qquad (5.15a)$$

$$\hat{\mathbf{C}} = -\frac{\rho}{w}\mathbf{L}(\text{Im}\,\hat{K}_f)^{-1}\mathbf{L}^T \qquad (5.15b)$$

where Re and Im stand for real and imaginary parts. The nature of matrices $\mathbf{K}_f$ and $\hat{\mathbf{K}}_f$ is such that in general both $\hat{\mathbf{M}}$ and $\hat{\mathbf{C}}$ are frequency dependent.

5.5   A CRUDE SIMPLIFICATION

If compressibility is neglected one term of the $\mathbf{K}_f$ matrix disappears [Equation (5.11a)] reducing the frequency dependence of $\mathbf{M}$. If further it is assumed that surface wave effects can be neglected and the free surface condition [Equation (5.8a)] can be replaced by a simple condition,

$$\bar{p} = 0$$

then all frequency dependent terms vanish in the mass matrix and as no free surface waves can be generated the exterior matrix $\mathbf{K}$ becomes trivial, giving

$$\hat{\mathbf{K}} = 0$$

In this case the effect of the fluid is confined to giving a frequency independent added mass matrix

$$\hat{\mathbf{M}} = p\mathbf{L}\mathbf{K}_f\mathbf{L}^T$$

where $\mathbf{K}_f$ includes only the first term of the expression Equation (5.11a), this being simply a solution of a real Laplace equation in the fluid domain.

Such an approximation has been first suggested by Zienkiewicz *et al.*,[2,3] and used extensively in many studies.[4–11] Indeed, in this book when vibrations of slender pile-like structures are considered, the concept of 'added mass' is frequently used. Such an added mass is a direct consequence of a frequency independent mass matrix and represents simply the product

$$\hat{\mathbf{M}}\mathbf{r}$$

where $\mathbf{r}$ corresponds to a particular mode of vibration or more generally simply to an assumed rigid body displacement.

Clearly such approximations are useful but their validity must not be extended beyond certain limits. In the context of vibrations of fairly stiff submerged structures the importance of compressibility can be great, as shown by the work of Chopra[12–14] and others.[15,16]

In Figure 5.2[16] we show for instance the force caused by a rigid body displacement of a rigid wall in a compressible infinite fluid. The reader will

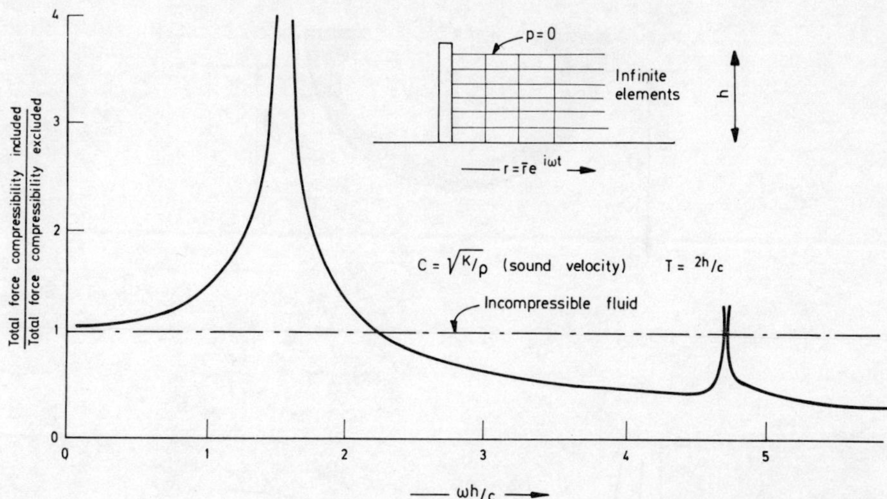

Figure 5.2 Force on a rigid wall due to presence of fluid effect of compressibility (Saini *et al.*[16])

note the resonance peaks occurring whenever the frequency approaches that of a multiple of h/c, the time for the compression wave to travel through the depth. Clearly the incompressible solution is not applicable in such problems over a wide range of frequencies.

Even if compressibility effects are negligible the neglect of radiation of surface waves can be of considerable importance in underestimation of damping and in incorrect added mass estimates. In Figure 5.3 a ship motion in a rigid displacement mode is studied.[17] Here the effect of surface waves and their radiation is included and again we see that the oversimplified model (for which damping would be identically zero and the mass matrix constant) introduces considerable errors.

5.6 FINAL REMARKS

This brief chapter indicates how quite generally the effect of incident forces on a structure vibrating in a fluid can be included. As in offshore structures compressibility effects are readily ignored in the frequency range of important wave and even earthquake forces—the computation of the coupling mass and damping matrices can be readily found as a by-product of the calculations involved in estimating the wave force effects.

Simplifying assumptions are possible but their advantage is questionable as they omit important damping and save little computational effort which for

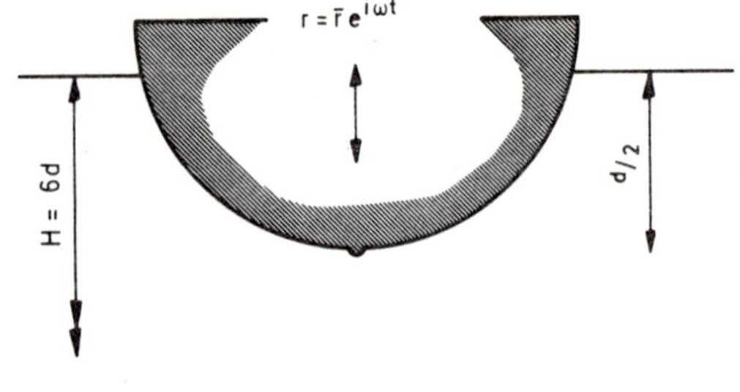

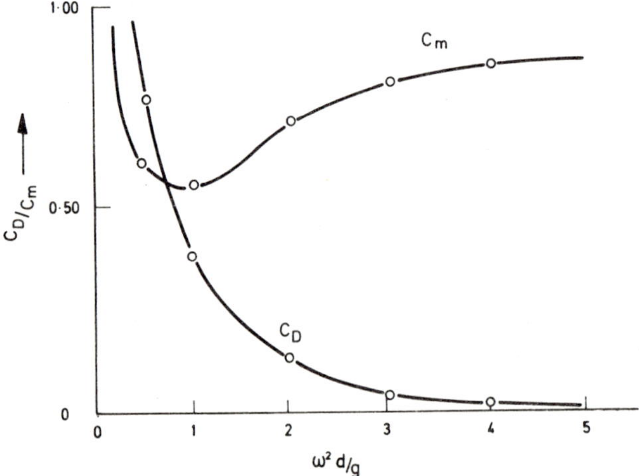

Figure 5.3   Added mass and damping coefficients for a rigid body vertical motion of a ship (Newton *et al.*[17])

most cases involves a complete frequency domain response calculation for the structure as a matter of course (see Chapter 7).

REFERENCES

1. Malhotra, A. K. and Penzien, J. (1970). 'Non deterministic analysis of offshore structures', *Proc. A.S.C.E.*, **96**, EM6, 985–1003.
2. Zienkiewicz, O. C. (1971). *The Finite Element Method in Engineering Science* (2nd Edn.), McGraw-Hill.

3. Zienkiewicz, O. C., *et al.* (1965). 'Natural frequencies of complex free or submerged structures by the finite element method', *Symposium on vibration in Civil Eng.*, Butterworths, London, 83–90.
4. Selby, A. and Severn, R. T. (1972). 'An experimental assessment of the added mass of some plates vibrating in water', *International Journal of Earthquake Engineering and Structural Dynamics*, **1**, 189–200.
5. Black, P. A. A., Cassels, A. C., Dungar, R., Gaukroger, D. R., and Severn, R. T. (1969). 'The seismic design of a double curvature arch dam', *Proc. I.C.E., London*, **43**, 217–298.
6. Shepherd, R. and McConnel, R. D. (1972). 'Seismic response predictions of a bridge pier', *Proc. ASCE*, **98**, EM3, 609–627.
7. Shepherd, R. (1973). 'The seismic response of elevated water tanks supported on cross-braced towers', *Proc. 5WCEE*, Rome, 640–649.
8. Silverman and Abramson. (1966). 'Lateral sloshing in moving containers', in the *Dynamic Behaviour of Liquids in Moving Containers*, ed. Abramson, NASA, SP-106.
9. Nath, J. H. and Harleman, D. R. F. (1967). 'The dynamic response of fixed offshore structures to periodic and random waves', *TR 102*, Hydrodynamics Laboratory, M.I.T.
10. Nath, B. (1971). 'Coupled hydrodynamic response of a gravity dam', *Proc. ICE, London*, **48**, 245–257.
11. Lindholm, U. S., *et al.* (1962). 'Breathing vibrations of a circular cylindrical shell with an internal liquid', *Journal of Aeronautical Science*, **29**, 1052–1059.
12. Chopra, A. K. (1970). 'Earthquake response of concrete gravity dams', *Proc. ASCE*, **96**, EM4, 443–454.
13. Chopra, A. K. and Chakrabarti, P. (1973). 'Dynamics of gravity dams—significance of compressibility of water and three dimensional effects', *International Journal of Earthquake Engineering and Structural Dynamics*, **2**, 103–104.
14. Chopra, A. K. and Liaw, C. Y. (1975). 'Earthquake resistant design of intake-outlet towers', *Proc. ASCE*, ST7, July 1975, 1349–1366.
15. Wylie, E. B. (1975). 'Seismic response of reservoir–dam systems', *Proc. ASCE*, **101**, HY3, 403–419.
16. Saini, S. S., Zienkiewicz, O. C., and Bettess, P. 'Coupled hydrodynamic response of concrete gravity dams using finite and infinite elements', to be published.
17. Newton, R. E. (1975). 'Finite element analysis of two dimensional added mass and damping', Ch. 11, 219–232, vol. I, *Finite Elements in Fluids*, ed. R. H. Gallagher *et al.*, Wiley.

Chapter 6

Numerical Methods for Dynamic Analysis

E. L. Wilson

6.1 INTRODUCTION

The value of the results of a dynamic analysis depends on the approximations involved in the establishment of mathematical models for the structure and foundation and in the selection of the various dynamic load conditions. In general the establishment of the models and the interpretation of the results are the most critical phases of a dynamic analysis. This assumes that the particular method of dynamic analysis used does not introduce additional errors in the solution of the model for the specified loads. It is important that the computer cost for any one analysis is not large in order that inexpensive re-analysis is possible and the results of the analysis can be used by the engineer to influence the basic design of the structure. Also an economical computer analysis will allow some of the basic assumptions used in selecting the model and loads to be varied and the sensitivity of the results evaluated.

The effectiveness of a numerical method for dynamic analysis depends primarily on two factors—the minimization of computer storage and the minimization of computer execution time. The purpose of this paper is to present a summary of various numerical methods which are used in the dynamic analysis of offshore structures and to comment with respect to their computer implementation and efficiency.

It should be noted that all structures, regardless of their simplicity, have an infinite number of degrees of freedom when subjected to dynamic loading. One of the main objectives of selecting a mathematical model is to reduce the infinite degree of freedom system to a model with a limited degree of freedom which will capture the significant physical behaviour of the system. Therefore a considerable insight into the expected dynamic behaviour of the system must be present if a realistic mathematical model is to be established. The model must be capable of representing both the significant wave propagation and structural vibration behaviour.

The force equilibrium of an offshore platform modelled as a system with a finite number of degrees of freedom may be written as

$$F_i + F_d + F_s = F \tag{6.1}$$

in which all forces are a function of time and defined as

F_i = The inertia force
F_d = The internal and external damping forces
F_s = The forces carried by the structural members
F = The external applied forces

Equation (6.1) holds for both linear and non-linear systems. However the appropriate numerical method for solution will depend on the degree of non-linearity which is present and if linearization is possible.

6.2   LINEAR DYNAMIC ANALYSIS

For linear systems with viscous damping Equation (6.1) can be written in the form

$$M\ddot{U}(t) + C\dot{U}(t) + KU(t) = F(t) \tag{6.2}$$

in which M is the mass matrix (lumped or consistent), C is the damping matrix (normally not given in the form) and K is the stiffness matrix for the system of structural elements. The time dependent vectors $U(t)$, $\dot{U}(t)$ and $\ddot{U}(t)$ are the displacements, velocities and accelerations respectively.

The time dependent external forces $F(t)$ may be due to wave or seismic forces. Normally the calculation of wave forces is a complicated procedure which depends on the shape of the structural elements. Other papers in these proceedings are referred to for the evaluation of wave forces. In the case of earthquake motion in three-dimensions the loading is of the form

$$F(t) = -M_x\ddot{U}_{xg} - M_y\ddot{U}_{yg} - M_z\ddot{U}_{zg} \tag{6.3}$$

where $\ddot{U}_{ig}$ is the ground acceleration in direction i and M_i is a column matrix which represents the sum of all columns in the mass matrix M associated with displacements in the i-direction.[1] This definition of the seismic loading is valid only if the vector is defined as the displacement relative to the displacement at the base of the structure.

In many cases wave and seismic loads in each direction are specified in terms of a spectrum of maximum response values. In the case of three-dimensional behaviour, however, great care must be taken in the application of the technique, since the basic input in the various directions must be statistically independent if the results are to be combined in a probabilistic manner.

Figure 6.1 indicates the possible solution techniques which can be employed for the dynamic analysis of linear systems. Four different solution approaches are possible. Two methods involve the evaluation of the undamped mode shapes and frequencies as the first step in the analysis—the mode superposition method and the spectrum analysis method. The frequency domain method

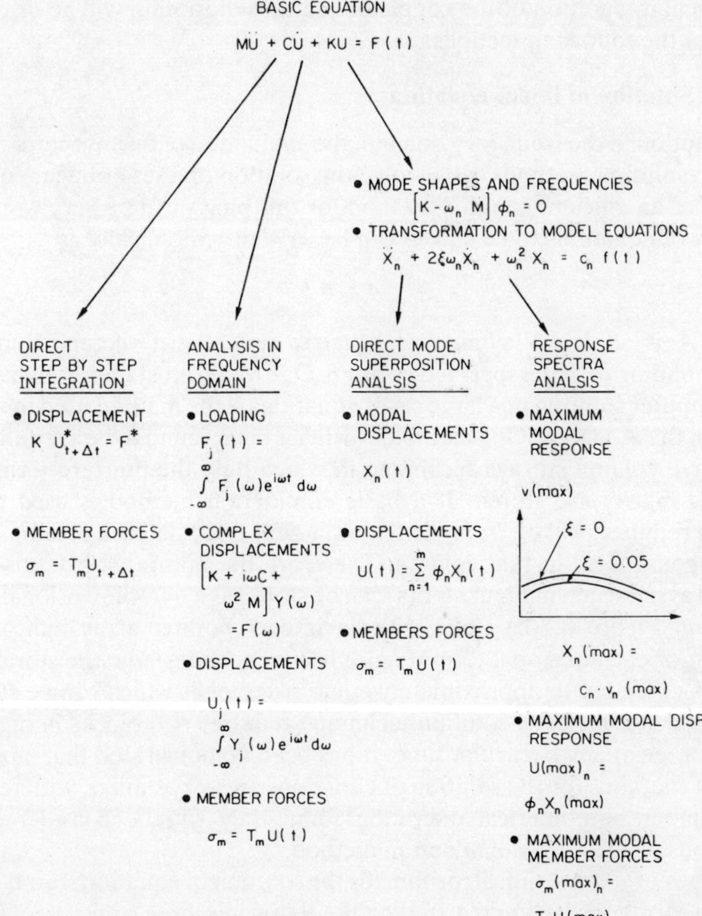

Figure 6.1 Methods for linear dynamic analysis

involves the expansion of the load in a Fourier Integral which reduces the dynamic analysis into a series of solutions of linear sets of complex equations. Another method which can be very efficient for certain systems is the direct step-by-step integration of the equations of motion.

Each phase of the possible solution methods suggested in Figure 6.1 involves a numerical method which must be formulated in effective form for computer implementation. Solution of linear equations, evaluation of eigenvalues and eigenvectors, numerical evaluation of integrals, transformation to the frequency domain, evaluation by the fast Fourier transform and the step-by-step numerical integration of the coupled equations of motion will be discussed in detail in the following sections.

6.2.1 Solution of linear equations

The solution in the frequency domain, the evaluation of eigenvectors and step-by-step solution methods can involve the solution of a set of linear equations; therefore an efficient solution method for this phase can be very worthwhile. The set of equations to be solved can be written symbolically as

$$AX = b \tag{6.4}$$

where A is an $N \times N$ symmetrical matrix and X is a vector of unknowns corresponding with the specified vector b. One of the most important aspects in the computer solution of a large set of equations is the method used to store the terms in the A matrix. One method which has been found to be very effective is the active column storage technique in which only the nonzero terms in the reduced matrix are stored. If a basic elimination method is used the only storage required will be from the first non-zero terms in each column down to the diagonal term in that column. Therefore the matrix with terms initially located as indicated in Figure 6.2(a) can be stored as a one-dimensional array as shown in Figure 6.2(b) along with an integer pointer array indicating the location of each diagonal term. Figure 6.2(c) indicates how the storage technique is extended to approximately equal size blocks which can be stored on low speed storage. The solution technique requires two blocks in high speed core storage at any particular time. It has been demonstrated that most of the popular methods for the solution of equations are very similar, with respect to the number of numerical operations, and they can be considered to be variations of the Gauss elimination method.[2]

The basic factorization algorithm for the solution of equation stored in active column form may be summarized by the following three steps:

(a) *Triangularization of A* $(j = 2 \ to \ N)$

$$A = LU \quad \text{or} \quad A = LDL^{T} \tag{6.5}$$

in which the jth column of upper triangular matrix U is evaluated from

$$U_{ij} = A_{ij} - \sum_{k=km}^{j-1} L_{ik}U_{kj} \qquad i = fj \ \text{to} \ j \tag{6.6}$$

and the jth row of lower triangular matrix is given by

$$L_{ji} = U_{ij}/D_{ii} \qquad i = fj \ \text{to} \ j-1 \tag{6.7}$$

where D is a diagonal matrix, fj is the first non-zero term in column j and fi is

$$AX = b \qquad A = LU = LDL^T$$

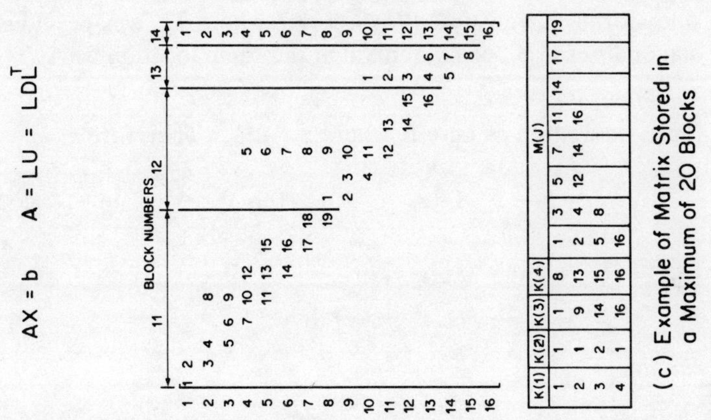

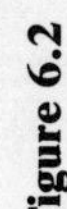

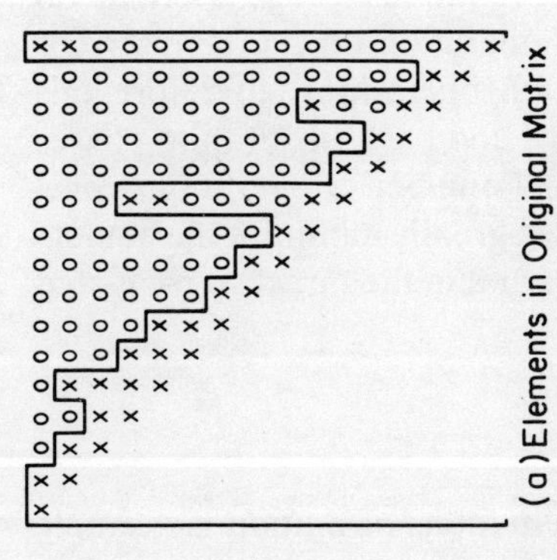

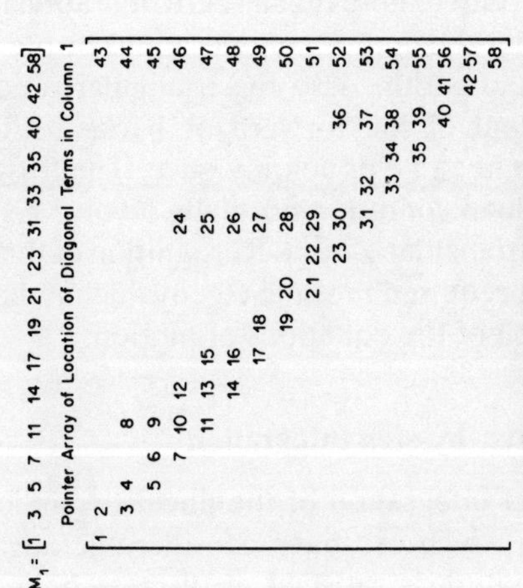

Figure 6.2

the first non-zero term in column i. The symbol km represents maximum of fj and fi.

The diagonal terms U_{ii} and D_{ii} are identical since L_{ii} is normalized as $1 \cdot 0$. It is most convenient to evaluate U_{ij} within a computer program column-wise with $i = fj$ to j. After each column is complete L_{ji} is evaluated row-wise with $i = fj$ to $j - 1$ and stored in transposed form as L_{ij}^* where U_{ij} was previously located. The diagonal term U_{ii} or D_{ii} remain at the same location as A_{ii}.

(*b*) *Forward reduction*
Equation (6.4) can be written as $Lz = b$, where $z = DL^T$. Therefore

$$z_i = b_i - \sum_{k=fi}^{i-1} L_{ki}^* z_k \qquad i = 1 \text{ to } N \qquad (6.8)$$

If y is defined as $y = D^{-1}z$ then $y = L^T x$; or

$$y_i = z_i / D_{ii} \qquad i = 1 \text{ to } N \qquad (6.9)$$

(*c*) *Backsubstitution*
From $y = L^T x$

$$x_i = y_i - \sum_{k=i+1}^{N} L_{ik}^* y_k \qquad i = N \text{ to } 1 \qquad (6.10)$$

It is important to note that all zero operations are skipped by this technique and that neither the number of operations or the the required storage is a function of the band width. Also the triangularization operation on the A matrix is independent of the forward or backsubstitution operations; therefore this operation need be done only once. The forward and backsubstitution phases for each load condition normally involve a small number of operations compared to triangularization. Recognition of this can greatly minimize the numerical effort required in some eigenvalue methods and in the direct step-by-step integration of the equations of motion.

6.2.2   Step-by-step integration

The direct integration of the linear dynamic equations of motion is a simple approach which can have considerable advantages for some problems. The basic equation is satisfied at discrete points in time, 0, Δt, $2\,\Delta t$, $3\,\Delta t, \ldots t$, $t + \Delta t, \ldots T$. The solution starts from a point in time where the displacements, velocities and accelerations are known. Based on an assumption on the behaviour of the system during the next small increment of time the displacements, velocities and accelerations at the next point in time can be evaluated. Many different methods have been developed for this purpose.[2–12] However only two techniques will be summarized in this paper.

(a) The central difference method

In the case of a diagonal mass and damping matrix and for systems where the shortest period is not too small the central difference method has proven to be most effective. At time t the equation to be satisfied is

$$M\ddot{U}_t + C\dot{U}_t + KU_t = F_t \tag{6.11}$$

The following standard finite difference relationships are used:

$$\ddot{U}_t = \frac{1}{\Delta t^2}(U_{t-\Delta t} - 2U_t + U_{t+\Delta t}) \tag{6.12}$$

$$\dot{U}_t = \frac{1}{2\,\Delta t}(U_{t+\Delta t} - U_{t-\Delta t}) \tag{6.13}$$

Equations (6.12) and (6.13) can be substituted into Equation (6.11) to form a set of linear equations of the form

$$M^* U_{t+\Delta t} = F^* \tag{6.14}$$

where

$$M^* = \frac{1}{\Delta t^2}M + \frac{1}{2\,\Delta t}C \tag{6.15}$$

$$F^* = F_t - KU_t + \frac{1}{\Delta t^2}M(2U_t - U_{t-\Delta t}) + \frac{1}{2\,\Delta t}CU_{t-\Delta t} \tag{6.16}$$

If M and C are diagonal one notes that the solution of Equation (6.14) is trivial; also computer storage will be minimized. Another very important technique which can be used to further minimize the number of numerical operations and computer storage requirements is not to form the complete stiffness K. The structural forces KU_t can be evaluated element by element, or

$$F_s = KU_t = \sum_m K_m U_t$$

in which K_m is the stiffness matrix for element m. If elements have identical stiffness matrices a further reduction in computer storage can be realized.

The solution $U_{t+\Delta t}$ is based on using the equilibrium at time t; therefore this approach is called an 'explicit integration method'. One of the most significant disadvantages of the central difference method is that it is only conditionally stable.[2] In order for the method to produce finite results the time step Δt must be less than T_n/π where T_n is the shortest period in the discrete model. For many structures this requires time steps so small that the method may be impractical.

(b) The Newmark–Wilson method

One of the most flexible step-by-step integration methods has been presented by Newmark.[8] This method is based on the following expressions for the velocity and displacement at the end of the time interval:

$$\dot{U}_{t+\Delta t} = \dot{U}_t + \Delta t\,(1-\delta)\ddot{U}_t + \Delta t\,\delta\ddot{U}_{t+\Delta t} \tag{6.17}$$

$$U_{t+\Delta t} = U_t + \Delta t\,\dot{U}_t + \Delta t^2\,(\tfrac{1}{2}-\alpha)\ddot{U}_t + \Delta t^2\,\alpha\ddot{U}_{t+\Delta t} \tag{6.18}$$

where α and δ are selected to produce the desired accuracy and stability. If $\delta = \tfrac{1}{2}$ and $\alpha = \tfrac{1}{6}$ the well known linear acceleration is produced, which is also a conditionally stable method. One of the most widely used methods is the constant-average-acceleration method ($\delta = \tfrac{1}{2}$ and $\alpha = \tfrac{1}{4}$) which is an unconditionally stable method without numerical damping.

This method is called an 'implicit integration method' since it satisfies the equilibrium equations of motion at time $t + \Delta t$, or

$$M\ddot{U}_{t+\Delta t} + C\dot{U}_{t+\Delta t} + KU_{t+\Delta t} = F_{t+\Delta t} \tag{6.19}$$

This equation can be solved by iteration; however Equations (6.17), (6.18) and (6.19) can be combined into a step-by-step algorithm which involves the solution of a set of equations at each time step of the form

$$K^* U_{t+\Delta t} = F^* \tag{6.20}$$

Since K^* is not a function of time it can be triangularized once at the beginning of the calculations. The computer solution time for this type of algorithm is basically proportional to the number of time steps required.

The Wilson θ method is a technique which can be used to modify the basic Newmark method in order to increase the stability limits and to add numerical damping.[13] The θ method was first applied to the linear acceleration method in order to improve stability and has been used to damp out high frequency oscillations which often develop in linear and non-linear step-by-step integration. The technique involves using the Newmark method to find the solution at $t + \theta\,\Delta t$, then, based on linear acceleration, calculating the results at $t + \Delta t$ for use as initial conditions for the next time step. The Newmark–Wilson algorithm is summarized in Table 6.1. With $\theta = 1$ the approach is the standard Newmark method. An unconditionally stable method with large damping in the higher modes is produced with $\delta = \tfrac{1}{2}$, $\alpha = \tfrac{1}{6}$ and $\theta = 1\cdot4$.[12]

6.2.3 Frequency domain approach

An alternative to the direct integration of the coupled linear equations of motion is to use a formal mathematical transformation to eliminate the time function from the equations before solution progresses.[1] The basic approach involves the expansion of the time-dependent loads in terms of a series of

Table 6.1   The Newmark–Wilson algorithm for linear step-by-step integration

A. INITIAL CALCULATIONS:
  1. Form stiffness matrix K, mass matrix M and damping matrix C.
  2. Initialize U_0, $\dot{U}_0$, and $\ddot{U}_0$.
  3. Specify algorithm parameters α, δ and θ

$$\delta \geq 0\cdot 50; \qquad \alpha \geq 0\cdot 25(0\cdot 5+\delta)^2; \qquad \theta \geq 1\cdot 0$$

  4. Calculate integration constants:

$$\tau = \theta\,\Delta t \qquad a_3 = \frac{1}{2\alpha}-1 \qquad a_7 = \Delta t\,\delta$$

$$a_0 = \frac{1}{\alpha\tau^2} \qquad a_4 = \frac{\delta}{\alpha}-1 \qquad a_8 = \Delta t^2(\tfrac{1}{2}-\alpha)$$

$$a_1 = \frac{\delta}{\alpha\tau} \qquad a_5 = \frac{r}{2}(\delta/\alpha-2) \quad a_9 = \alpha\,\Delta t^2$$

$$a_2 = \frac{1}{\alpha\tau} \qquad a_6 = \Delta t(1-\delta)$$

  5. Form effective stiffness matrix: $K^* = K + a_0 M + a_1 C$
  6. Triangularize K^*: $K^* = LDL^T$

B. FOR EACH TIME STEP:
  1. Calculate effective load vector at time $t+\tau$:

$$F^* = F_{t+\tau} + M(a_0 U_t + a_2 \dot{U}_t + a_3 \ddot{U}_t)$$
$$+ C(a_1 U_t + a_4 \dot{U}_t + a_5 \ddot{U}_t)$$

  2. Solve for displacements at time $t+\tau$:

$$LDL^T U_{t+\tau} = F^*$$

  3 Calculate accelerations, velocities and displacement at $t+\Delta t$:

$$\ddot{U}_{t+\tau} = a_0(U_{t+\tau} - U_t) - a_2 \dot{U}_t - a_3 \ddot{U}_t$$

$$\ddot{U}_{t+\Delta t} = \ddot{U}_t + \frac{1}{\theta}(\ddot{U}_{t+\tau} - \ddot{U}_t)$$

$$\dot{U}_{t+\Delta t} = \dot{U}_t + a_6 \ddot{U}_t + a_7 \ddot{U}_{t+\Delta t}$$
$$U_{t+\Delta t} = U_t + \Delta t \dot{U}_t + a_8 \ddot{U}_t + a_9 \ddot{U}_{t+\Delta t}$$

harmonic functions. One can use the standard Fourier Series in which the loads $F(t)$ are expanded in a series of the form

$$F(t) = \sum_{n=0}^{\infty} A_n \cos\frac{n\pi}{d}t + \sum_{n=0}^{\infty} B_n \sin\frac{n\pi}{d}t \qquad (6.21)$$

in which d is the duration of the loading. The Fourier Coefficients can be evaluated and exact solutions found for the harmonic functions $A_n \cos(n\pi/d)t$

and $B_n \sin (n\pi/d)t$. It is assumed that the loading can be approximated by a finite number of terms. Therefore the total solution in time is a summation of the exact solution for each harmonic function. For systems without damping this straightforward Fourier Series approach is numerically very effective since the response of an undamped structure to a harmonic sin or cos function loading is also sin or cos displacement solution.

An alternate method of eliminating the time variable from the dynamic equilibrium equations is to express the loads as an infinite integral, or

$$F(t) = \int_0^\infty (A(\omega)\cos \omega t + B(\omega)\sin \omega t)\, d\omega \tag{6.22}$$

The functions $A(\omega)$ and $B(\omega)$ in the Fourier integral are given by

$$A(\omega) = \frac{2}{\pi} \int_0^d F(t)\cos \omega t\, dt \tag{6.23}$$

$$B(\omega) = \frac{2}{\pi} \int_0^d F(t)\sin \omega t\, dt \tag{6.24}$$

Also the Fourier integral can be written in complex form as

$$F(t) = \int_{-\infty}^\infty \bar{F}(\omega)\, e^{i\omega t}\, d\omega \tag{6.25}$$

in which

$$\bar{F}(\omega) = \tfrac{1}{2}[A(\omega) - iB(\omega)] \tag{6.26}$$

$$\bar{F}(-\omega) = \tfrac{1}{2}[A(\omega) + iB(\omega)] \tag{6.27}$$

since

$$e^{i\omega t} + e^{-i\omega t} = 2\cos \omega t$$

$$e^{i\omega t} - e^{-i\omega t} = 2i\sin \omega t$$

The general equilibrium equations can now be written as

$$M\ddot{U} + C\dot{U} + KU = \int_{-\infty}^\infty \bar{F}(\omega)\, e^{i\omega t}\, d\omega \tag{6.28}$$

The solution is assumed to be of the form

$$U(t) = \int_{-\infty}^\infty Y(\omega)\, e^{i\omega t}\, d\omega \tag{6.29}$$

therefore

$$\dot{U}(t) = \int_{-\infty}^\infty i\omega Y(\omega)\, e^{i\omega t}\, d\omega \tag{6.30}$$

$$\ddot{U}(t) = \int_{-\infty}^{\infty} -\omega^2 Y(\omega)\, e^{i\omega t} \tag{6.31}$$

Hence the following complex set of equations must be solved for various values of ω:

$$(K + i\omega C - \omega^2 M)Y(\omega) = \bar{F}(\omega) \tag{6.32}$$

If $\bar{F}(\omega)$ requires a large number of points to define the complete function it is apparent that a large number of solutions of complex equations will be necessary. This large numerical effort can be minimized by solving for the basic eigenvalues of the system and transforming the equations to a smaller system expressed in model coordinants [Equation (6.67)]. The evaluation of the complex loads $\bar{F}(\omega)$ is not a major computational problem as compared to the multiple solution of a large system of complex equations. Furthermore the Fast Fourier transform algorithm can minimize both the evaluation of the Fourier transforms and the recovery of the displacements, Equation (6.29).

The major advantage of the Frequency Domain approach is in its application to substructure analysis. Structure-foundation or structure-fluid systems are considered by the development of the frequency-dependent matrices for the separate systems. Ritz techniques can also be used to effectively reduce the size of the system.

6.2.4 Numerical evaluation of mode shapes and frequencies

For large structural systems one of the most time consuming phases of a dynamic analysis may be the evaluation of eigenvalues and eigenvectors of the $N \times N$ matrix equation

$$MU + \bar{\omega}^2 KU = 0 \tag{6.33}$$

The undamped free vibration of the structural model has a solution form of

$$U(t) = \sum_{n=1}^{N} e^{i\bar{\omega}t} \phi_n$$

Therefore the resulting eigenvalue problem must be solved

$$(K - \bar{\omega}_n^2 M)\phi_n = 0 \tag{6.34}$$

(a) Static condensation

One technique which is often used to reduce the size of the system before the evaluation of eigenvalues is to eliminate the massless degrees of freedom from the system. For this case Equation (6.34) is rewritten as

$$\begin{bmatrix} K_{aa} & K_{ab} \\ K_{ba} & K_{bb} \end{bmatrix} \begin{bmatrix} \phi_a \\ \phi_b \end{bmatrix} = \bar{\omega}^2 \begin{bmatrix} 0 & 0 \\ 0 & M_b \end{bmatrix} \begin{bmatrix} \phi_a \\ \phi_b \end{bmatrix} \tag{6.35}$$

The first submatrix equation is

$$K_{aa}\phi_a + K_{ab}\phi_b = 0 \qquad (6.36)$$

Therefore the massless degrees of freedom are related to the degrees of freedom at the mass points by

$$\phi_a = T\phi_b \qquad (6.37)$$

where

$$T = -K_{aa}^{-1}K_{ab} \qquad (6.38)$$

The resulting eigenvalue problem is of the form

$$K_{bb}^{*}\phi_b = \bar{\omega}^2 M_b\phi_b \qquad (6.39)$$

in which

$$K_{bb}^{*} = K_{bb} + K_{ba}T \qquad (6.40)$$

Within a computer program however these submatrix operations are not necessary since it is more efficient to perform the 'static condensation' directly on the massless degrees of freedom, similar to the Gauss elimination procedure,[14] without requiring submatrix storage. One important disadvantage of the static condensation approach is that the matrix K_{bb}^{*} tends to fill as more massless degrees of freedom are eliminated. Therefore the reduction in size of the system may not be economical from a computational viewpoint. In addition if the mass is physically lumped at the time of creating the mathematical model additional errors may be introduced.

(b) Reduction of size of system by Ritz functions

A more general approach to the reduction of the size of the eigenvalue problem [Equation (6.34)] is the application of the Ritz method. This technique is not restricted to a particular mass distribution—a full mass matrix does not increase the computational effort significantly. However for systems with a limited number of masses the method can be identical to the static condensation method.[15]

The method can be very accurate if some physical insight into the behaviour of the structure is known. Static load patterns are selected and corresponding displacement vectors are calculated. Or

$$KR = P \qquad (6.41)$$

The true displacements of the system are approximated by a linear combination of the discrete Ritz vectors R. Or the true eigenvectors are approximated by

$$\phi = RX = R_1X_1 + R_2X_2 + R_3X_3 + \cdots R_LX_L \qquad (6.42)$$

where L is the number of load patterns and is smaller than the size of the system N. The load patterns can be very simple; however they must be linearly independent. In order to produce the lower frequencies the load pattern should activate the large masses and areas of maximum flexibility. If a single unit load is used as a load pattern the method mathematically lumps the consistent mass of the system at that degree of freedom.

The reduced eigenvalue problem is produced by the substitution of Equation (6.42) into Equation (6.34) and premultiplication by R^T. Or,

$$K^*X - MX\Omega = 0 \tag{6.43}$$

where

$$K^* = R^T KR = R^T P \tag{6.44}$$

$$M^* = R^T MR \tag{6.45}$$

$$\Omega = \text{diag}(\omega_i^2) \tag{6.46}$$

Both K^* and M^* are full matrices because the Ritz vectors are not orthogonal. Since all the eigenvalues and eigenvectors are required and the system is relatively small, less than 100, the Jacobi method is one of the most effective for this type of eigenvalue problem.[2]

One advantage of the Ritz method for the reduction of the number of degrees of freedom for a very large structure is that it involves only a solution of a set of linear equations which may have a large number of zero terms in the stiffness matrix, Equation (6.41). Therefore a large number of structural elements can be used to model the basic structural behaviour. The use of Ritz functions can be considered as a formal mathematical method of evaluating an approximate generalized mass matrix for the purpose of dynamic analysis.

Figure 6.3 illustrates the selection of static load patterns for a simple tower type structure. This structure has 6 unconstrained joints or 36 degrees of freedom. It is of interest to point out that for the lateral load pattern shown the resulting displacement is complex and involves movements in all directions in addition to torsional behaviour.

For a structure of this type one would expect the lateral behaviour to be expressed by the first six mode shapes. Because of the geometric arrangement of the members joints 1, 2 and 3 and joints 4, 5 and 6 will act as separate units with very little relative movement between joints. Therefore the six possible load patterns which will capture this fundamental behaviour are easily established and are shown in Figure 6.3a. Of course the first six vibrational mode shapes for this structure will be composed of a linear combination of the resulting displacement patterns. Figure 6.3b illustrates six other load patterns which could have been used to produce identical results.

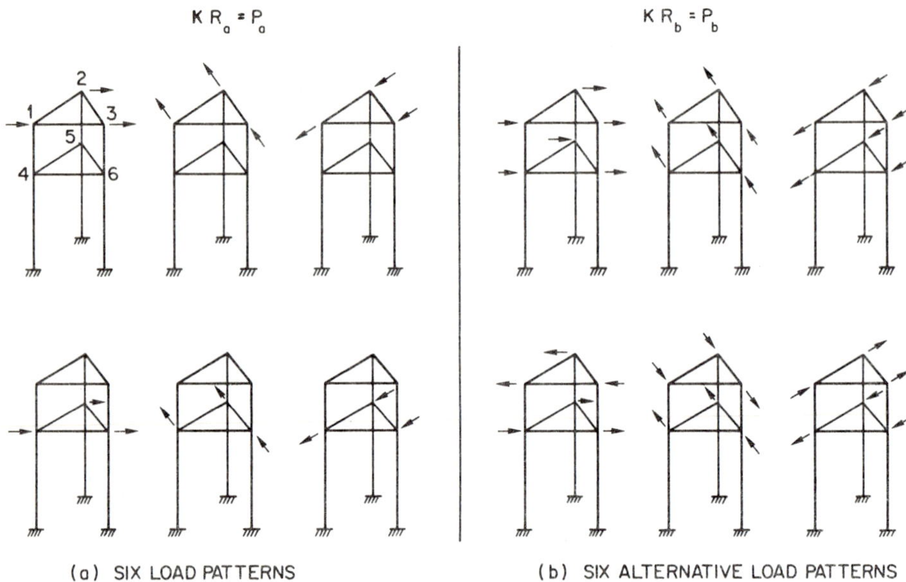

Figure 6.3   Example of selection of different load patterns for tower structure

The physical modes which have been neglected by this approach are the breathing modes between joints 1, 2 and 3 and joints 4, 5 and 6. Also the vertical vibrational modes have been omitted; but axial deformations are included in the six lateral modes.

(c) Subspace iteration

The subspace iteration method for the determination of eigenvalues and eigenvectors of very large structural systems is a significant extension of the Ritz reduction approach.[16,17,18] It is the only modern computer method for very large systems (over 5000 degrees of freedom) which will converge to the exact eigenvalues.

Table 6.2 summarizes the subspace iteration method. The computer implementation of the method and a FORTRAN listing is given in Reference 2. From Table 6.2 one notes that the initial calculations are identical to the Ritz method in which the first set of load patterns must be specified. After one Ritz solution is found and the first approximation of the first L eigenvalues of the system is calculated as

$$\phi^{(1)} = R_1 X_1 \tag{6.47}$$

an improved set of load patterns can be calculated from

$$P_2 = M \phi^{(1)}$$

Table 6.2   Summary of the subspace iteration algorithm for solution of large eigenvalue problem
$$K\phi_n = \omega_n^2 M\phi_n$$

A. INITIAL CALCULATION:
   1. Form stiffness matrix K and mass matrix M.
   2. Triangularize K:

$$K = LDL^T \qquad (N \times N)$$

   3. Specify initial load patterns:

$$P_1 = N \times L \text{ matrix} \quad \text{where} \quad L \ll N$$

B. FOR EACH ITERATION, $k = 1, 2, 3, \ldots$:
   1. Solve for Ritz vectors R_k:

$$LDL^T R_k = P_k \qquad (N \times L)$$

   2. Calculate generalized stiffness in subspace:

$$K^{(k)} = R_k^T K R_k = R_k^T P_k \qquad (L \times L)$$

   3. Calculate generalized mass in subspace:

$$M^{(k)} = R_k^T M R_k \qquad (L \times L)$$

   4. Solve eigenvalue problem in subspace:

$$K^{(k)} X_k = M^{(k)} X_k \Omega^{(k)} \qquad (L \times L)$$

   5. Calculate improved approximate eigenvectors:

$$\phi^{(k)} = R_k X_k$$

   6. Check for convergence:

$$\Omega^{(k)} \to \text{diag}(\omega_n^2) \quad \text{and} \quad \phi^{(k)} \to \phi \quad \text{as} \quad k \to \infty$$

      stop if converged—perform Sturm sequence check
   7. Calculate improved load patterns for next iteration:

$$P_{k+1} = M\phi^{(k)} \qquad (N \times L)$$

   8. Return to Step B-1.

It is assumed that P_2 is an improved estimation of the inertia forces associated with the basic mode shapes. From Table 6.2 it is clear that this iteration technique can be carried out to any desired degree of accuracy, assuming the method converges. The convergence of the method for various conditions is given in Reference 2.

Some additional comments associated with the advantages of the subspace iteration method with respect to numerical effort are:

(i) The total stiffness matrix for the system need be triangularized only once.

(ii) Since $K^{(k)}$ and $M^{(k)}$ tend to become diagonal as the iteration progresses the Jacobi method is very effective for this type of eigenvalue problem.

(iii) The size of the subspace L should be approximately 30 per cent more vectors more than the number of accurate eigenvalues required.

(iv) In order to insure participation of all modes one additional random vector should be added to the load pattern set before each iteration.

(v) Since the approach is basically a power method the lowest eigenvalues converge faster and are more accurate.

(vi) For most structures a high degree of accuracy is not required for the highest modes, because of their low participation in the dynamic response. Therefore the subspace iteration produces practical results with respect to required accuracy.

(d) Additional numerical techniques for eigenvalue problems
In a general computer program for the calculation of eigenvalues and eigenvectors several different numerical techniques may be useful in the solution strategy in order to improve the convergence and to minimize numerical effort.

(i) Inverse iteration
    If only one load pattern is used in the iteration procedure given in Table 6.2 the subspace iteration approach is the standard inverse iteration method and will converge to the lowest eigenvalue. For this case the eigenvalue problem in the subspace is trivial. Or

$$\lambda_1 = \omega_1^2 \simeq K^{(k)}/M^{(k)} \tag{6.49}$$

which is better known as the Rayleigh quotient.[6] After the first eigenpair λ_1 and ϕ_1 are determined inverse iteration can be used to calculate accurately additional eigenpairs if the techniques of shifting, determinant search, deflation and Sturm sequence checking are introduced.

(ii) Determinant search
    This numerical technique can be explained by considering the following equation:

$$[K - \lambda M]X = 0 \tag{6.50}$$

For any numerical value of λ_k the following triangularization is possible

$$[K - \lambda_k M] = L_k D_k L_k^T$$

The determinant of this matrix for λ_k is

$$\det(K - \lambda_k M) = D_{11} D_{22} D_{33} \cdots D_{NN}$$

where N is the order of the matrix.

    If the numerical value of the determinant is evaluated for a large number of different values of λ a function can be generated of the form shown in

Figure 6.4. This function $p(\lambda)$, which in this case is evaluated numerically, is a plot of the characteristic polynomial. $\lambda_1, \lambda_2, \lambda_3, \lambda_4, \ldots \lambda_N$, are the eigenvalues of the system and correspond to zero values of the det $(K - \lambda M)$.

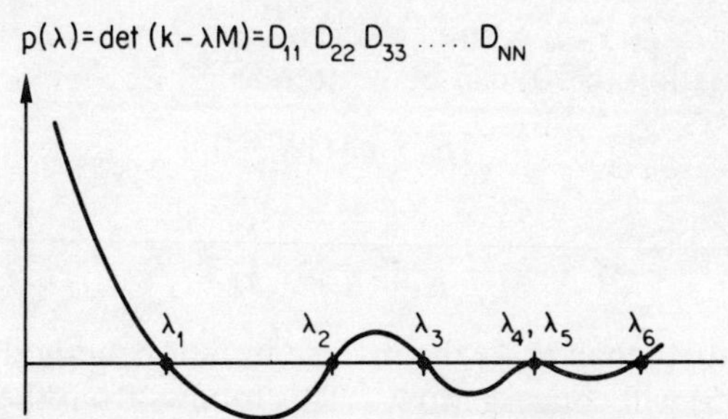

Figure 6.4　Characteristic polynomial $[K - \lambda M]$

(*iii*) *Deflation*

A numerical plot of a polynomial with the first root suppressed, or deflated, can be computed and would have the approximate shape as shown in Figure 6.5. The only reason for numerically evaluating $p_1(\mu_k)$ and $p_1(\mu_{k+1})$ is to obtain a better estimation for λ_s from the extrapolation equation

$$\lambda_s = \mu_k + \frac{\mu_k - \mu_{k+1}}{p_1(\mu_{k+1}) - p_1(\mu_k)} p_1(\mu_k) \tag{6.53}$$

It is this type of strategy which is used to obtain a value of λ_s which is close to a desired root. For this case two triangularizations are required in order to evaluate λ_s. Therefore this technique is effective for matrices with small band widths which can be triangularized with a minimum of numerical effort.[19]

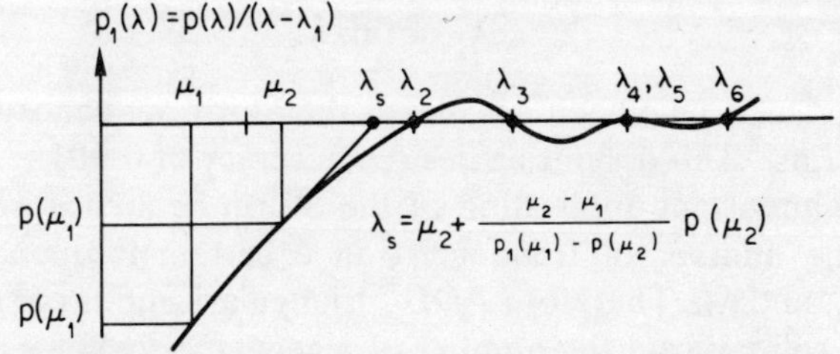

Figure 6.5　Deflated polynomial with λ_1 suppressed

(iv) Shifting

Inverse iteration converges to the numerically smallest eigenvalue. In order for the method to be used for other eigenvalues the following change of variable can be introduced

$$\lambda = \lambda_s + \rho \tag{6.54}$$

Therefore Equation (6.50) can be written as

$$[K_s - \rho M]X = 0 \tag{6.55}$$

where

$$K_s = K - \lambda_s M \tag{6.56}$$

If λ_s is closer to λ_2 than to λ_1 the inverse iteration method when applied to Equation (6.55) will converge to λ_2. This follows from the standard power method proof.[1,2]

(v) Sturm sequence check

One of the potentially serious problems which can develop in the numerical evaluation of frequencies and mode shapes in a practical dynamic analysis is if important frequencies are neglected. The Sturm Sequence Check is a method which allows the engineer to verify the results of an eigenvalue problem. One can prove the Sturm Sequence Theorem as presented here from a direct examination of the complete family of deflated polynomials, $p(\lambda)$, $p_1(\lambda)$, $p_2(\lambda)$, ... One notes that they are all derived from the basic sequence for a given value of λ, or

$$\det (K - \lambda_s M) = D_{11}D_{22}D_{33} \cdots D_{NN} \tag{6.57}$$

For a given value of λ the basic properties of this sequence of numbers are illustrated in Figure 6.6. It is apparent that the properties of this sequence of numbers can be used as a powerful technique in the numerical strategy of evaluating eigenvalues. If lowest eigenvalues are evaluated by any method the Sturm sequence can be calculated for

$$\lambda = 1 \cdot 001 \lambda_n \tag{6.58}$$

If all eigenvalues have been calculated the Sturm sequence should have n negative terms.[2] This assumes a desired accuracy of $0 \cdot 001$.

Another important application of the Sturm Sequence Technique is to evaluate the number of frequencies in a certain frequency range—say between λ_L and λ_H. Therefore LDL^T triangularizations of $[K - \lambda_H M]$ and $[K - \lambda_L M]$ will indicate the number of eigenvalues below each value. The difference will be the number of values within the range. In order to calculate the eigenpairs in the range one can shift into the range and use inverse iteration or subspace iteration to evaluate only the values of interest.

6.2.5 Transformation to uncoupled modal equations

The basic numerical properties of the undamped free vibration mode shapes are

$$M^* = \phi^T M \phi = I \tag{6.59}$$

$$K^* = \phi^T K \phi = \operatorname{diag}(\bar{\omega}_n^2) \tag{6.60}$$

In which the mode shapes have been normalized so the generalized mass is one. Or

$$\phi_n^T M \phi_n = 1 \tag{6.61}$$

The following transformation, change of variable, is introduced into the basic equilibrium equations.

$$U(t) = \sum_{n=1}^{M} \phi_n X_n(t) = \phi X(t) \tag{6.62}$$

Therefore the velocities and accelerations are

$$\dot{U}(t) = \phi \dot{X}(t) \tag{6.63}$$

$$\ddot{U}(t) = \phi \ddot{X}(t) \tag{6.64}$$

If Equations (6.62), (6.63) and (6.64) are substituted into Equation (6.2) and the resulting matrix equation premultiplied by ϕ^T we obtain

$$M^* \ddot{X}(t) + C^* \dot{X}(t) + K^* X(t) = P(t) \tag{6.65}$$

The matrices M^* and K^* are diagonal; however C^* is not diagonal unless an assumption is made on the basic form of viscous damping which exists in the structure. Since damping is normally small and is difficult to physically model and identify the following assumption is normally made:

$$C^* \simeq \operatorname{diag}(2\xi_n \bar{\omega}_n) \tag{6.66}$$

where ξ_n is the ratio of damping in mode n to the critical damping for the mode. With this assumption of uncoupled modal damping the typical modal equation can be written as

$$\ddot{X}_n(t) + 2\xi_n \bar{\omega}_n X_n(t) + \bar{\omega}_n^2 X_n(t) = p_n(t) = c_n f(t) \tag{6.67}$$

After the modal equations are evaluated the time dependent displacements are calculated from Equation (6.62).

The same technique can be used to uncouple Equation (6.32) in order to avoid the solution of a large number of complex linear equations. For this case the following transformation is introduced

$$Y(\omega) = \phi Z(\omega) \qquad (6.68)$$

The substitution of Equation (6.68) into Equation (6.32) yields a typical modal equation in the frequency domain.

$$[\bar{\omega}_n^2 + 2i\omega\bar{\omega}_n\xi_n - \omega^2]Z_n(\omega) = \phi_n\bar{F}(\omega) \qquad (6.69)$$

For structures which are formulated in the frequency domain and whose behaviour can be represented by a limited number of natural frequencies this is numerically the most efficient approach.

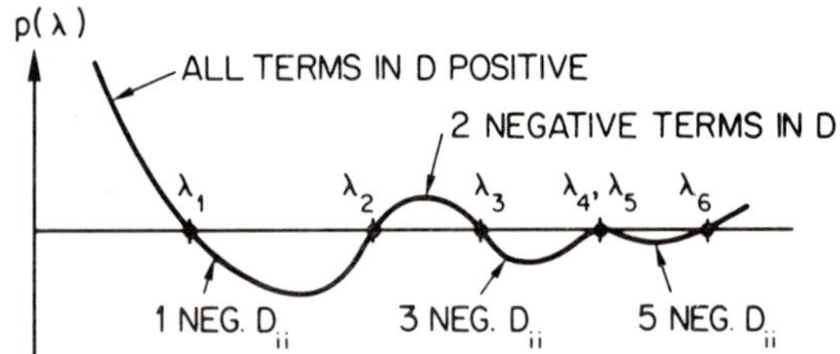

Figure 6.6   Sturm sequence properties of polynomial

6.2.5   Solution of modal equations

The solution to the single-degree of freedom modal equation given by Equation (6.67) can be accomplished by one of several methods. For certain loading which can be expressed as an analytic function exact mathematical solutions are possible.[1]

(a) Direct step-by-step solution
The most direct approach to the solution of this second order ordinary differential equation is to use a numerical finite difference method. The same techniques are possible which were used for the coupled equation. The step-by-step solution method given in Table 6.1 is numerically very efficient when applied to Equation (6.67).

(b) Duhamel Integral
In the case of arbitrary loading it is very common to express the solution in the form of the Duhamel Integral.[1] The Duhamel Integral is then numerically integrated. Since this numerical integration approach involves many numerical evaluations of trigonometric and exponential functions, which require series

expansions within a digital computer, the method cannot be considered a good numerical method. Also for high frequencies a very small integration interval is required for accuracy.

(c) Transformation to the frequency domain

The single degree of freedom modal equations can be transformed into the frequency domain. The Fourier integrals and transforms can be numerically evaluated by the Fast Fourier transform technique. This of course is identical to the approach suggested by Equation (6.69). This method introduces the same types of errors which are normally associated with the approximation of a function by a Fourier series or integrals.

(d) Piecewise exact method

Many types of loading can be represented by a series of straight lines between unequal time intervals. Most earthquake ground acceleration data is in this form. An exact mathematical solution is possible for a straight line loading subjected to displacement and velocity initial conditions. Therefore exact numerical values at any convenient time interval can be calculated. Table 6.3 summarizes the necessary· equations for this approach for the numerical evaluation of the modal equations. This may not be the most numerically efficient method; but it is definitely the most accurate. For most structures the numerical solution of the modal equation involves an insignificant amount of computer time compared to the computer time required for the other phases of the problem—formation of stiffness and mass matrices, solution of eigenvalue problem, calculation of displacement and member stresses. For these reasons the piecewise exact method should be used whenever possible.

(e) Response spectra analysis

For many structural problems the dynamic load is not given in terms of a time dependent function; but the load is specified as a response spectra. By definition a response spectra is a plot of maximum values of displacement response $v(\max)$ obtained from the solution of the following equation for various values of ω.

$$\ddot{v}(t) + 2\omega\xi\dot{v}(t) + \omega^2 v(t) = f(t) \tag{6.70}$$

A typical plot of the maximum, $v(\max)$, for specified ω and damping ratios ξ is shown in Figure 6.1.

The typical modal equation is of the form

$$\ddot{X}_n + 2\bar{\omega}_n\xi_n\dot{X}_n + \bar{\omega}_n^2 X_n = \phi_n F(t) = c_n f(t) \tag{6.71}$$

Therefore the maximum modal response can be calculated from

$$X_n(\max) = c_n v_n(\max) \tag{6.72}$$

where $v_n(\max)$ is the value obtained from Figure 6.1 for values of ω_n and ξ_n.

Table 6.3   Piecewise exact solution method

BASIC EQUATION:

$$\ddot{X} + 2\xi\omega\dot{X} + \omega^2 X = f(t) = a + bt$$

SPECIFIED LOAD $f(t)$

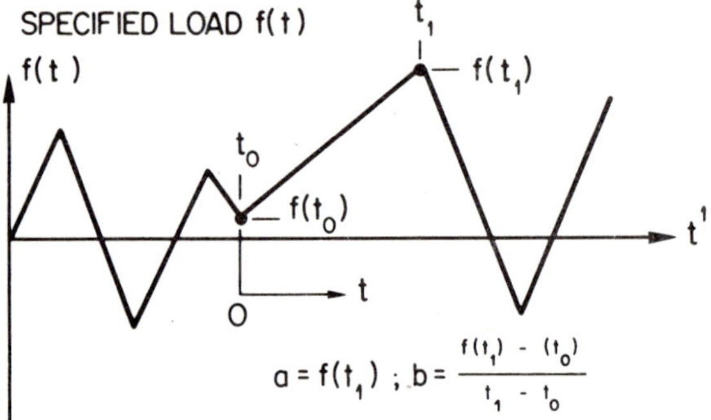

EXACT DISPLACEMENT SOLUTION:

$$X(t) = A_0 + A_1 t + A_2 e^{-\xi\omega t}\cos\omega^* t + A_3 e^{-\xi\omega t}\sin\omega^* t$$

where $A_0 = \dfrac{a}{\omega^2} - 2\dfrac{\xi b}{\omega^3}$

$A_1 = \dfrac{b}{\omega^2}$

$A_2 = X(t_0) - A_0$

$A_3 = \dfrac{1}{\omega^*}(\dot{X}(t_0) + \xi\omega A_2 - A_1)$

EXACT SOLUTION FOR VELOCITY:

$$\dot{X}(t) = A_1 + (\omega^* A_3 - \xi\omega A_2)e^{-\xi\omega t}\cos\omega^* t$$

$$- (\omega^* A_2 + \xi\omega A_3)e^{-\xi\omega t}\sin\omega^* t$$

After the maximum response in each modal equation is evaluated the maximum displacement in each modeshape for the complete structure is given by

$$U(\text{max})_n = \phi_n X_n(\text{max}) \tag{6.73}$$

For any particular degree of freedom a probabilistic value of displacement may be calculated from

$$u = \sqrt{(u_1^2 + u_2^2 + u_3^2 + \cdots u_m^2)} \tag{6.74}$$

Other methods exist for adding maximum modal response; however the square root of the sum of the squares of the maximum modal values is one of the most common.

In order to evaluate member stresses it is first necessary to evaluate the member stresses due to each of the maximum modal responses. Or

$$\sigma(\max) = TU(\max)_n \tag{6.75}$$

where T is a stress-displacement transformation matrix for the member. The probabilistic member stress is estimated by

$$\sigma = \sqrt{(\sigma_1^2 + \sigma_2^2 + \sigma_3^2 + \cdots \sigma_M^2)} \tag{6.76}$$

Note that the probabilistic member stress cannot be calculated from the probabilistic displacements.

6.3 NON-LINEAR ANALYSIS

For offshore fixed structures several different types of non-linear behaviour for both static and dynamic loads are possible. Non-linear behaviour implies that the displacements and stresses produced by the different load conditions cannot be directly added; or the basic principle of superposition does not hold. Because there are so many different types of non-linearities there is not one general method which can be applied to all problems.

Large structures for which their weight is significant may have a dead load stress distribution which is highly dependent on the method of construction and installation sequence. The correct theoretical method of evaluating these stresses is to perform a complete analysis at each stage of construction with the internal stresses from the analysis of the previous stage used as initial conditions for the new analysis. This analysis technique can be used for the evaluation of installation stresses also. For subsequent analysis of the structure in which additional non-linear stresses are developed due to static or dynamic loads it may be extremely important to start the analysis with an accurate estimation of the initial stress conditions.

Perhaps the most common type of non-linear behaviour is due to non-linear materials. Most structural design requires that the structural materials remain in the elastic range during the design loads. However foundation stresses in soil may be non-linear under low stress levels. Under dynamic loads soil non-linearities are often approximated by an effective damping factor to be used in a linear dynamic analysis. Also during earthquake or large sea conditions some non-linear material behaviour can be tolerated without the collapse of the structure.

One of the most unlikely types of non-linear behaviour to be expected in a practical structure is the existence of large strains which require an alternate definition of stress. This type of non-linearity exists in only rubber-like materials.

For tall structures or structures supported by cables large displacements may exist under design loading. For this case it is extremely important that the static or dynamic equilibrium equations are satisfied in the deformed geometry.

For fixed offshore structures where the velocities of the structure are comparable to the water particle velocities the wave forces are non-linear and may be expressed as

$$F(t) = F(U(t), \dot{U}(t)) \tag{6.77}$$

This type of non-linear behaviour can be considered by the linear step-by-step methods. Based on the previous increment the velocity of the structure can be predicted from

$$\dot{U}_{t+\Delta t} = \dot{U}_t + \Delta t\, \ddot{U}_t \tag{6.78}$$

and a good estimate of the non-linear drag forces can be estimated. Since the properties of the structures do not change the effective stiffness matrix need not be modified or triangularized at each time increment. Therefore the method given in Table 6.1 can be used for this form of non-linearity.

A general formulation for the non-linear analysis of a structural system can be developed if Equation (6.1) is rewritten at time

$$(F_t^i + \Delta F_t^i) + (F_t^d + \Delta F_t^d) + (F_t^s + \Delta F_t^s) = F_{t+\Delta t} \tag{6.79}$$

in which the force changes are given by

$$\Delta F_t^i = M_t\, \Delta \ddot{U}_t, \quad \Delta F_t^d = C_t\, \Delta \dot{U}_t, \quad \Delta F_t^s = K_t U_t \tag{6.80}$$

where M_t, C_t and K_t are the approximate mass, damping and stiffness matrices at time t. Therefore Equation (6.79) can be rewritten as

$$M_t\, \Delta \ddot{U} + C_t\, \Delta \dot{U}_t + K_t\, \Delta U_t = F^* \tag{6.81}$$

where

$$F^* = F_{t+\Delta t} - F_t^i - F_t^d - F_t^S \tag{6.82}$$

A direct step-by-step method, as presented for linear systems, can be used for the evaluation of $\Delta \bar{U}_t$, $\Delta \dot{U}_t$ and ΔU_t. Because the matrices M_t, C_t and K_t can only be approximated over the time interval, it is recommended that the forces F_t^i, F_t^d and F_t^s be recomputed at the end of the time increment. For example the structural forces F_t^s, which are consistent with U_t, should be evaluated as follows:

(a) From the displacement U_t calculate in member m the strain ε_m.

(b) From the specified non-linear stress strain behaviour calculate the stress τ_m.

(c) Using virtual work calculate the structural forces acting on element

$$F_m = \int A_m^T \tau_m \, dV_m$$

where A_m is the strain-displacement transformation matrix for member m.

(d) Calculate the total structural forces at time t from

$$F_t^s = \sum F_m$$

For structures with slight non-linearities this type of analysis may be accurate. However for very non-linear behaviour it may be necessary to iterate within a time step in order to minimize the accumulation of errors. For a more complete discussion of dynamic non-linear analysis methods the reader is referred to References 12, 20, 21, 22, 23, 24.

6.4 FINAL REMARKS

Several different numerical methods for the dynamic analysis of linear structural systems have been presented. Many of these methods have been incorporated into general purpose programs and have been successfully used in the solution of offshore structural systems.[24,25,26] General purpose programs for non-linear analysis have been developed based on the techniques presented.[22,24] However for non-linear analysis of complex offshore structures the development of special purpose computer programs is justifiable because of the unique nature of the structures and their loading.

REFERENCES

1. Clough, R. W. and Penzien, J. (1975). *Dynamics of Structures*, McGraw-Hill Book Company, New York, NY.
2. Bathe, K. J. and Wilson, E. L. (1976). *Numerical Methods in Finite Element Analysis*, Prentice-Hall, Inc., Englewood Cliffs, NJ.
3. Biggs, J. M. (1964). *Introduction to Structural Dynamics*, McGraw-Hill Book Company, New York, NY.
4. Hurty, W. C. and Rubinstein, M. F. (1964). *Dynamics of Structures*, Prentice-Hall, Inc., Englewood Cliffs, NJ.
5. Collatz, L. (1966). *The Numerical Treatment of Differential Equations*, Springer-Verlag, New York, NY, 116.
6. Crandall, S. H. (1956). *Engineering Analysis*, McGraw-Hill Book Company, New York, NY.
7. Forberg, C. E. (1969). *Introduction to Numerical Analysis*, Addison-Wesley Publishing Company, Inc., Reading, Massachusetts.
8. Newmark, N. M. (1959). 'A method of computation for structural dynamics', ASCE, *Journal of Engineering Mechanics Division*, **85**, 67–94.
9. Houbolt, J. C. (1959). 'A recurrence matrix solution for the dynamic response elastic aircraft', *Journal of Aeronautical Science*, **17**, 540–550.

10. Wilson, E. L. (1960). 'A computer program for the dynamic stress analysis of underground structures', *Report UC SESM 68-1*, Department of Civil Engineering, University of California, Berkeley.
11. Wilson, E. L. (1969). 'Elastic dynamic response of axisymmetric structures', *Report UC SESM 69-2*, Department of Civil Engineering, University of California, Berkeley.
12. Wilson, E. L., Farhoomand, I., and Bathe, K. J. (1973). 'Non-linear dynamic analysis of complex structures', *International Journal of Earthquake Engineering and Structural Dynamics*, **1**, 241–252.
13. Bathe, K. J. and Wilson, E. L. (1973). 'Stability and accuracy analysis of direct integration methods', *International Journal of Earthquake Engineering and Structural Dynamics*, **1**, 283–291.
14. Wilson, E. L. (1974). 'The static condensation algorithm', *International Journal of Numerical Methods in Engineering*, **8**, 199–203.
15. Shubinski, R. P., Wilson, E. L., and Selna, L. G. (1966). 'Dynamic response of deepwater structures', *Civil Engineering in the Oceans*, ASCE Conference, San Francisco.
16. Bathe, K. J. (1971). 'Solution Methods of Large Generalized Eigenvalue Problems in Structural Engineering', *Report UC SESM 71–20*, Civil Engineering Department, University of California, Berkeley.
17. Bathe, K. J. and Wilson, E. L. (1972). 'Large eigenvalue problems in dynamic analysis', ASCE, *Journal of Engineering Mechanics Division*, **98**, 1471–1485.
18. Bathe, K. J. and Wilson, E. L. (1973). 'Eigensolution of large structural systems with small bandwidth', ASCE, *Journal of Engineering Mechanics Division*, **99**, 467–479.
19. Bathe, K. J. and Wilson, E. L. (1973). 'Solution methods for eigenvalue problems in structural mechanics', *International Journal for Numerical Methods in Engineering*, **6**, 213–226.
20. Chopra, A. K. (1974). 'Earthquake analysis of complex structures', *Proceedings ASME Conference, Applied Mechanics in Earthquake Engineering*, November 1974, New York.
21. Felippa, C. A. (1976). 'Procedures for computer analysis of large non-linear structural systems', *International Symposium on Large Engineering Systems*, University of Manitoba, Winnipeg, Canada, August 9–12, 1976.
22. Bathe, K. J., Ozdemir, H., and Wilson, E. L. (1974). 'Static and dynamic geometric and material nonlinear analysis', *Report UC SESM 74-4*, Department of Civil Engineering, University of California.
23. Bathe, K. J. (1976). 'An assessment of current finite element analysis of nonlinear problems in solid mechanics', *Proceedings Symposium on the Numerical Solution of Partial Differential Equations*, May 1975, Academic Press, Inc.
24. Bathe, K. J. (1975). 'ADINA—A finite element program for automatic dynamic incremental nonlinear analysis', *Report 82448–1*, Acoustics and Vibration Laboratory, Department of Mechanical Engineering, Massachusetts Institute of Technology, Cambridge, Mass.
25. Bathe, K. J., Wilson, E. L., and Peterson, F. E. (1973). 'SAP IV—A structural analysis program for static and dynamic response of linear systems', *Report EERC 73-11*, College of Engineering, University of California, Berkeley, June 1973, revised April 1974.
26. *EAC/EASE 2 Dynamics—User Information Manual/Theoretical*, Control Data Corporation.

Chapter 7

Three-dimensional Dynamic Analysis of Fixed Offshore Platforms

J. Penzien and S. Tseng

7.1 INTRODUCTION

Fixed offshore structures are designed and constructed in regions of high seismicity and in areas where rough sea conditions often exist. Since these conditions develop high dynamic forces in the structure-foundation system it is important that they be predicted realistically for design purposes. In making such predictions the complete structure-foundation system must be modelled accurately, hydrodynamic forces must be considered properly and reliable numerical procedures must be used. It is the purpose of this paper to define an appropriate mathematical model and to develop effective and efficient numerical procedures for carrying out dynamic analyses. The basic procedure used is similar to that previously reported in the literature.[1–4]

7.2 MATHEMATICAL MODEL

A fixed offshore platform is typically either a steel framed structure as represented in Figure 7.1 or a reinforced concrete gravity structure as. represented in Figure 7.2. The steel framed structure is normally composed of tubular members joined together to form a three-dimensional truss system which supports the operation deck and it is usually supported on a pile foundation. The gravity structure normally consists of a number of tapered cylindrical legs which support the operation deck and are supported on a massive, cellular, reinforced concrete base of extremely high rigidity. The entire gravity structure rests directly on the sea bed and relies entirely on its weight to provide lateral load resistance.

For purposes of dynamic response analysis to seismic and/or wave and current excitations it is important that the entire platform structure and its supporting foundation, regardless of type, be properly modelled mathematically. Utilizing the substructuring concept, this modelling can be separated into

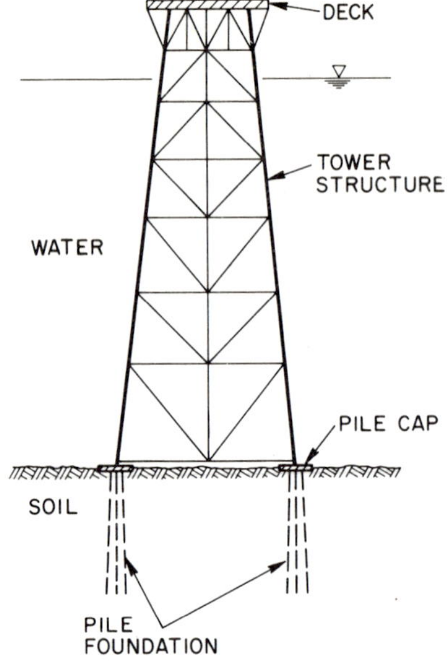

Figure 7.1   Framed tower

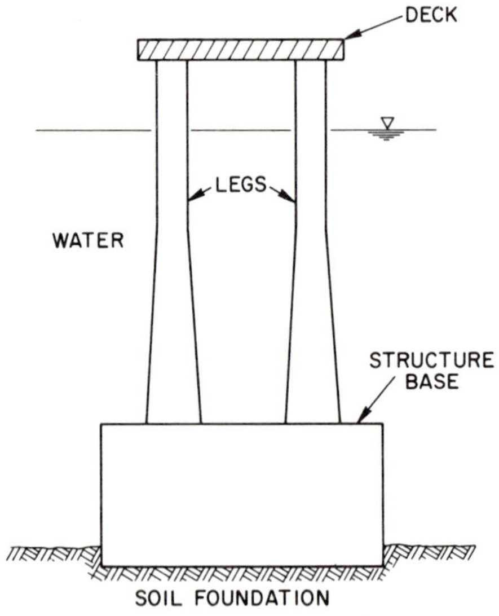

Figure 7.2   Gravity tower

a structure subsystem and a foundation subsystem. These two subsystems are then connected together at the structure-foundation interface boundary to form the complete structure-foundation model to be used in dynamic response analyses.

7.2.1 Structure subsystem

For purposes of dynamic response analysis, the tower structure is idealized by a discrete system consisting of a set of nodal points interconnected by linear elastic elements. All mass of the structure is lumped at the nodal points. For a general three-dimensional analysis, each nodal point has six degrees-of-freedom, three translational and three rotational. The total number of degrees-of-freedom n of the discrete structural model can be separated into a set of n_s degrees-of-freedom representing all n degrees-of-freedom exclusive of the n_b degrees-of-freedom associated with the nodal points at the structure-foundation interface boundary $(n = n_s + n_b)$; thus, the n component nodal displacement vector r can be partitioned into an n_s component vector and an n_b component vector as represented by

$$r = \left\{ \begin{matrix} r_s \\ r_b \end{matrix} \right\} \tag{7.1}$$

Examples of discrete models for steel framed and gravity structures are shown in Figures 7.3 and 7.4, respectively. The massive base of the gravity structure is

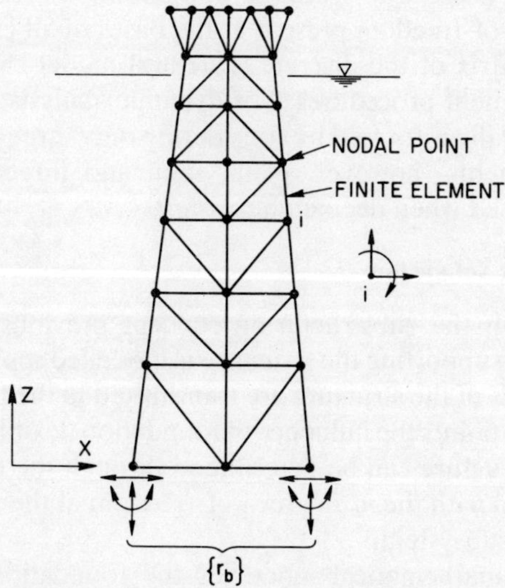

Figure 7.3 Discrete model of framed tower

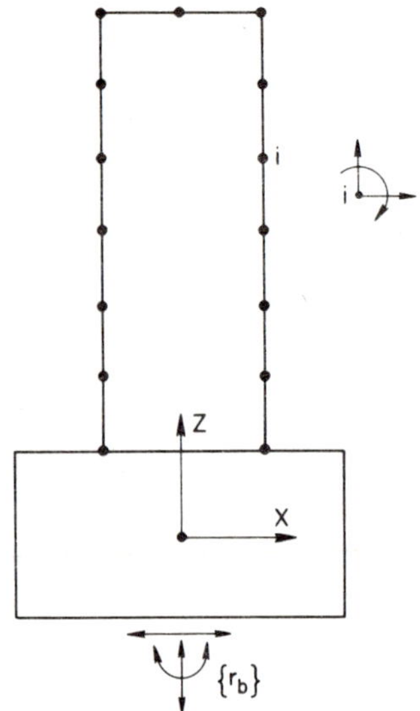

Figure 7.4 Discrete model of
gravity tower

modelled as a completely rigid body and its mass and mass moment of inertia
are lumped at its centre of gravity. The base degrees-of-freedom r_b for the
gravity structure consist of the six degrees-of-freedom at the centroid of its
base while the base degrees-of-freedom for the framed structure consists of the
combined degrees-of-freedom present at the bases of all legs.

The stiffness matrix of the discrete structural model can be generated by
standard finite element procedures. For dynamic analysis purposes the finite
element model usually is formed by considering only three-dimensional linear
elastic beam elements; however, plate, shell and three-dimensional solid
elements can be used when deemed necessary.

7.2.2 Foundation subsystem

In accordance with the substructuring concept previously introduced, the
foundation system supporting the structure is modelled separately as a subsystem. Since the loads of the structure are transmitted to the foundation through
the interface nodal points the influence of foundation flexibility on the dynamic
response of the structure can be determined through the force-displacement
relations associated with the n_b degrees-of-freedom at the interface boundary
of the foundation subsystem.

For a selected mathematical model of the foundation subsystem these
force-displacement relations (foundation impedance functions) can be derived

by performing a steady-state foundation substructure dynamic analysis under harmonic excitation at its interface boundary. This analysis produces a set of complex-valued, frequency dependent impedance functions forming an $n_b \times n_b$ impedance matrix which relates the set of steady-state harmonic forces f_b applied to the interface degrees-of-freedom at frequency ω with the resulting steady-state harmonic displacements r_b. This relationship is given by

$$F_b(\omega)\, e^{i\omega t} = [K^f_{bb}(\omega) + i\omega C^f_{bb}(\omega)] R_b(\omega)\, e^{i\omega t} \tag{7.2}$$

where $i = \sqrt{(-1)}$, $F_b(\omega)$ are the amplitudes of f_b at frequency ω, and $R_b(\omega)$ are the amplitudes of r_b at frequency ω. Matrices $K^f_{bb}(\omega)$ and $C^f_{bb}(\omega)$ represent the frequency dependent foundation stiffness and damping coefficients, respectively, which are associated with the interface boundary degrees-of-freedom. Since for gravity platforms these frequency dependent coefficients would be first calculated for translations and rotations of the rigid base structure about a point at the geometric centre of its bottom surface, they must be transformed to corresponding coefficients representing translations and rotations of the rigid base structure about its own centre of gravity. Depending upon the foundation model used, the damping coefficients represent energy dissipation due to both radiation into the far-field and due to heat generation within the foundation materials.

For the gravity structure, the foundation subsystem can be modelled conveniently using a uniform (or a layered) elastic (or viscoelastic) half-space as shown in Figure 7.5. The foundation impedance functions have been obtained

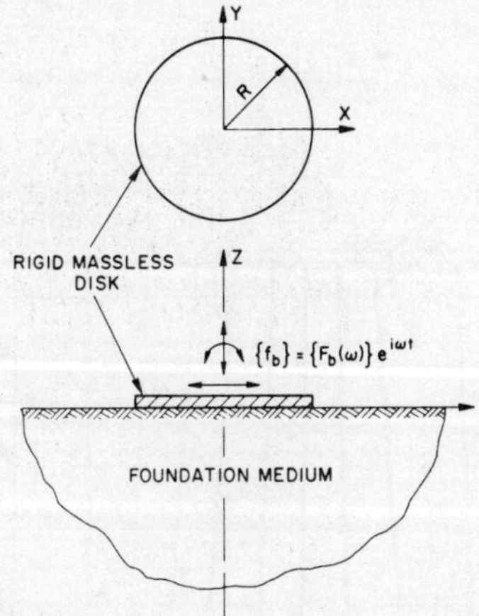

Figure 7.5 Translation and rocking of a rigid massless disc on the surface of a foundation medium

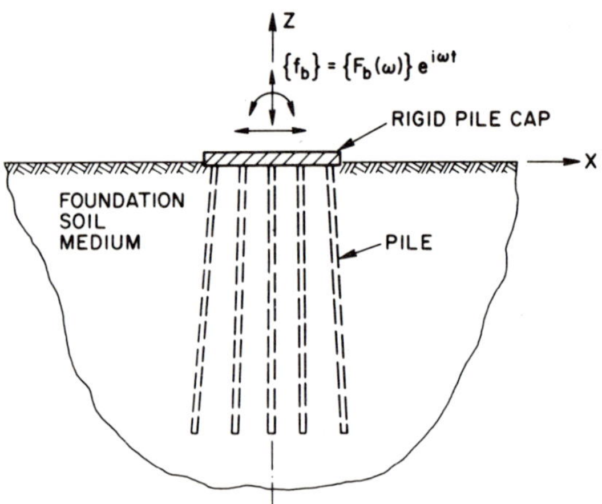

Figure 7.6 Translation and rocking of a rigid cap
on supporting pile group

for this model by Veletsos and Wei,[5] and Luco and Westman[6] assuming a
uniform elastic half-space and by Luco[7] assuming a layered viscoelastic half-
space. Figure 7.9 shows the impedance functions for a rigid massless circular
disc resting on the surface of a uniform elastic half-space as determined by
Veletsos and Wei while Figure 7.10 shows the corresponding functions for a
layered half-space as determined by Luco.

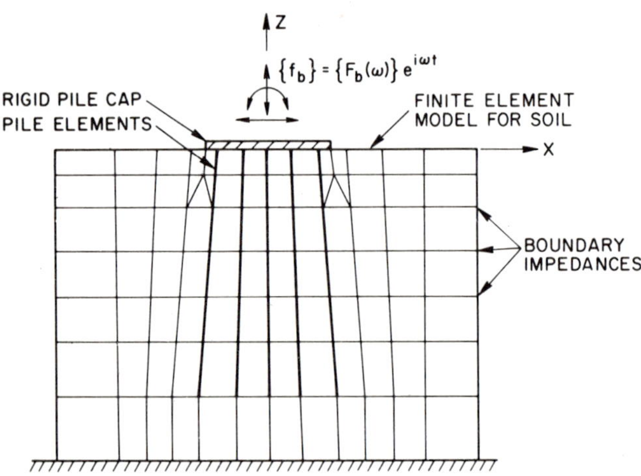

Figure 7.7 Finite element model of pile-soil foundation
system

Due to the presence of piles, realistic modelling of the foundation subsystem for framed structures is much more difficult than modelling of the corresponding system for gravity structures. Solutions yielding foundation impedance functions for pile cap motions as indicated in Figure 7.6 are available only for very simple cases. For example, Novak has generated these functions for a single pile in a uniform layer of soil founded on rigid rock.[8] Research is, of course, ongoing to develop these functions for more complex cases.

Finite element representations can also be used to obtain foundation impedance functions.[9] Figures 7.7 and 7.8 show example finite element representations for the foundations of framed and gravity structures, respectively. A finite element representation has the distinct advantage that it can effectively accommodate complex foundation geometries, such as those present in the case of a pile foundation, and can accommodate non-homogeneous soils. However, this representation does suffer from not being able to transmit waves above a certain 'cut-off' frequency, which is controlled by the individual finite element size, and from introducing artificial resonances at discrete frequencies which are a direct result of wave reflections from the artificial finite element mesh boundaries. The cut-off frequency can be raised by reducing the size of finite elements used and the resonances, which produce unrealistic peaks and valleys in the impedance functions, can be controlled by introducing 'wave transmitting elements' along the finite element mesh boundaries. These transmitting elements provide boundary impedances which approximate the dynamic characteristics of the far field outside the finite element mesh, thus, energy from the finite element mesh is permitted to radiate into the far field in a fairly

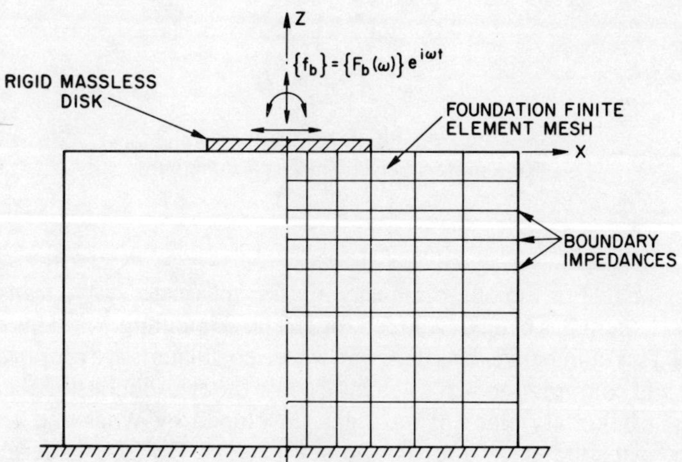

Figure 7.8 Finite element model of rigid massless disc on soil foundation

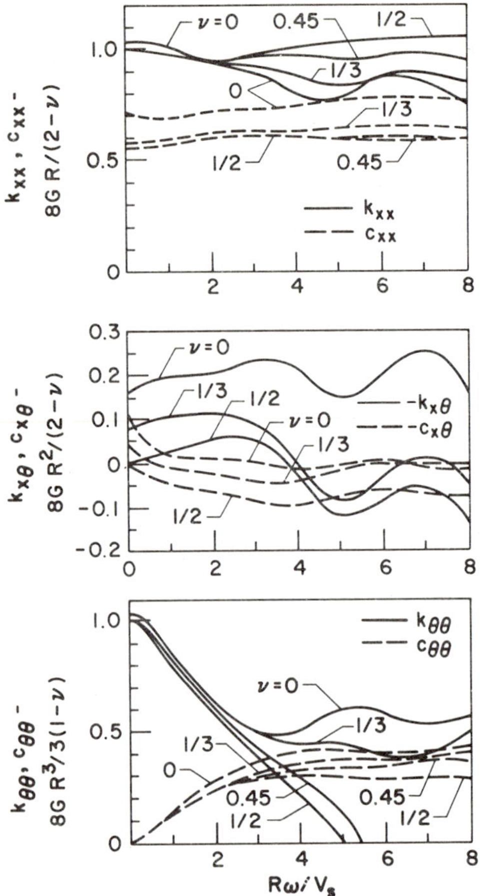

Figure 7.9 Foundation impedance functions
for a uniform elastic half-space medium

realistic manner. For the one-dimensional case, the 'exact' wave transmitting boundary element has been developed for an elastic medium.[10,11] This element consists of a set of linear viscous dampers whose coefficients are proportional to the shear and compression wave velocities. For the two-dimensional case, the transmitting boundary element has been developed by Waas and Lysmer[12] who later extended it to the axisymmetric case. For the general three-dimensional case, only approximate boundary elements have been developed to date.

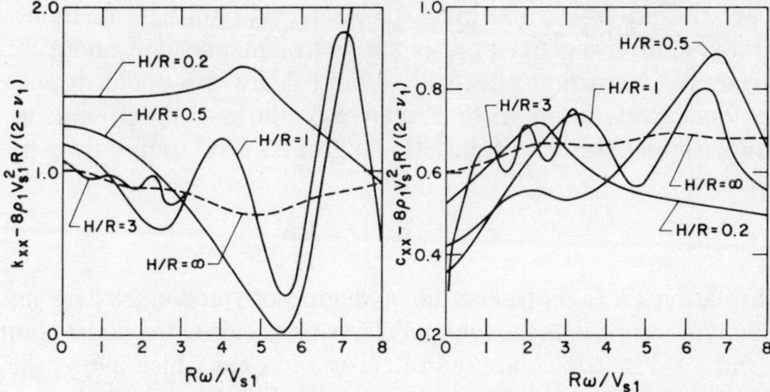

Figure 7.10(a)   Foundation impedance functions for a medium consisting of a layer of depth H resting on a uniform elastic half-space; horizontal vibrations ($V_{s1}/V_{s2} = 0\cdot8$; $\rho_{s1}/\rho_{s2} = 0\cdot85$; $\nu_1 = \nu_2 = 0\cdot25$)

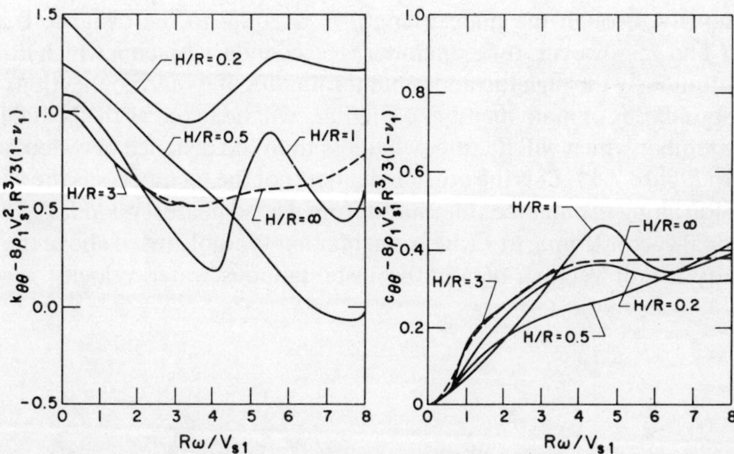

Figure 7.10(b)   Foundation impedance functions for a medium consisting of a layer of depth H resting on a uniform elastic half-space; rocking vibrations ($V_{s1}/V_{s2} = 0\cdot8$; $\rho_{s1}/\rho_{s2} = 0\cdot85$; $\nu_1 = \nu_2 = 0\cdot25$)

7.3   EQUATIONS OF MOTION

The equations of motion of a fixed offshore structure-foundation system subjected to simultaneous seismic and hydrodynamic (wave and current) loading can be written in the matrix form.[13]

$$\boldsymbol{M}\ddot{\boldsymbol{r}}_t + \boldsymbol{C}\dot{\boldsymbol{r}} + \boldsymbol{K}\boldsymbol{r} = \boldsymbol{f}_w \tag{7.3}$$

where $\boldsymbol{M}$, $\boldsymbol{C}$, and $\boldsymbol{K}$ are the mass, damping, and stiffness matrices of the structural system, respectively; f_w is the hydrodynamic load vector including fluid-structure interaction effects; $\boldsymbol{r}$, $\dot{\boldsymbol{r}}$, and $\ddot{\boldsymbol{r}}$ are the nodal displacement, velocity, and acceleration vectors relative to the moving ground; $\ddot{\boldsymbol{u}}_g$ is the three-dimensional vector of translational ground acceleration time-histories; and, where

$$\ddot{\boldsymbol{r}}_t = \ddot{\boldsymbol{r}} + \ddot{\boldsymbol{r}}_g = \ddot{\boldsymbol{r}} + \boldsymbol{B}\ddot{\boldsymbol{u}}_g \tag{7.4}$$

Since Equation (7.3) represents an n degree-of-freedom system and $\ddot{\boldsymbol{r}}_t$ in Equation (7.4) represents the total acceleration vector of the nodal points, $\boldsymbol{B}$ is simply the $n \times 3$ matrix composed of zeros and ones which allows the direct contribution of ground accelerations to the total nodal accelerations.

To formulate the hydrodynamic load vector f_w, consider first the hydrodynamic loading on a single, uniform, circular, cylindrical member i located between nodes I and J as shown in Figure 7.11. If nodes I and J lie along a principal member, such as a leg or a primary bracing where the member passes continuously through the nodes, length L is equal to the distance between nodes I and J; however, if the member is a secondary bracing which does not run continuously through the nodes but terminates at its end connections to the outer boundaries of main members, length L will be taken as the actual length of the member which will be somewhat less than the distance between nodes I and J. In Figure 7.11, D is the outside diameter of the member, s is the variable dimension along the member measured from its end nearest node I, $\hat{\boldsymbol{s}}$ is the unit directional vector from I to J (the triangular hat symbol is used above the letter to signify a unit vector), $\dot{\boldsymbol{u}}(s)$ is the instantaneous water velocity vector at

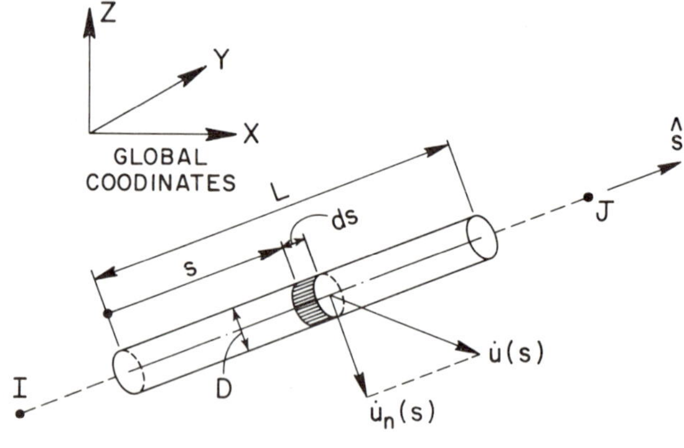

Figure 7.11   Water particle velocities along member i

location s, and $\dot{\mathbf{u}}_n(s)$ is the normal component of $\dot{\mathbf{u}}(s)$ lying in the plane of vectors $\dot{\mathbf{u}}(s)$ and $\hat{\mathbf{s}}$. Not shown in Figure 7.11, is the instantaneous water acceleration vector $\ddot{\mathbf{u}}(s)$ and its normal component $\ddot{\mathbf{u}}_n(s)$ which, of course, have directions different from vectors $\dot{\mathbf{u}}(s)$ and $\dot{\mathbf{u}}_n(s)$. Member i is moving in the water flow field with absolute velocity and acceleration as represented by vectors $\dot{\mathbf{r}}_t(s)$ and $\ddot{\mathbf{r}}_t(s)$, respectively.

Generalizing the one-dimensional form of the Morison equation,[14] the three-dimensional form of the hydrodynamic force per unit length along member i at location s is given by the vectorial relation.[15,16]

$$\mathbf{f}_w(s) = \rho A \ddot{\mathbf{u}}_n(s) + \rho A (K_M - 1)[\ddot{\mathbf{u}}_n(s) - \ddot{\mathbf{r}}_{tn}(s)] + \rho D K_D |\dot{\mathbf{u}}_n(s)$$
$$- \dot{\mathbf{r}}_{tn}(s)|(\dot{\mathbf{u}}_n(s) - \dot{\mathbf{r}}_{tn}(s)) \quad (7.5)$$

where ρ is the mass density of the surrounding water; A is the cross-sectional area of member i $(\pi D^2/4)$; K_M is the inertia coefficient; K_D is the drag coefficient; and, $\dot{\mathbf{r}}_{tn}(s)$ and $\ddot{\mathbf{r}}_{tn}(s)$ are the normal components of vectors $\dot{\mathbf{r}}_t(s)$ and $\ddot{\mathbf{r}}_t(s)$, respectively. Note that the symbol '$||$' denotes modulus of the vector contained therein.

It can easily be shown that

$$\dot{\mathbf{u}}_n(s) = \mathbf{N}\dot{\mathbf{u}}(s); \qquad \ddot{\mathbf{u}}_n(s) = \mathbf{N}\ddot{\mathbf{u}}(s)$$
$$\dot{\mathbf{r}}_{tn}(s) = \mathbf{N}\dot{\mathbf{r}}_t(s); \qquad \ddot{\mathbf{r}}_{tn}(s) = \mathbf{N}\ddot{\mathbf{r}}_t(s) \quad (7.6)$$

where

$$\mathbf{N} = \mathbf{I} - \hat{\mathbf{s}}\hat{\mathbf{s}}^T = \begin{bmatrix} (1 - s_1^2) & -s_1 s_2 & -s_1 s_3 \\ -s_1 s_2 & (1 - s_2^2) & -s_2 s_3 \\ -s_1 s_3 & -s_2 s_3 & (1 - s_3^2) \end{bmatrix} \quad (7.7)$$

and where s_1, s_2, and s_3 are the components of unit vector $\hat{\mathbf{s}}$ in the X, Y, and Z directions, respectively. Since

$$\dot{\mathbf{r}}_t(s) = \dot{\mathbf{r}}_g + \dot{\mathbf{r}}(s) \quad (7.8)$$

one can write

$$\dot{\mathbf{u}}_n(s) - \dot{\mathbf{r}}_{tn}(s) = \dot{\mathbf{u}}_n(s) - \dot{\mathbf{r}}_{gn} - \dot{\mathbf{r}}_n(s) \equiv \dot{\mathbf{q}}_n(s) - \dot{\mathbf{r}}_n(s) = \mathbf{N}[\dot{\mathbf{q}}(s) - \dot{\mathbf{r}}(s)] \quad (7.9)$$

Thus,

$$|\dot{\mathbf{u}}_n(s) - \dot{\mathbf{r}}_{tn}(s)|(\dot{\mathbf{u}}_n(s) - \dot{\mathbf{r}}_{tn}(s)) = |\dot{\mathbf{q}}_n(s) - \dot{\mathbf{r}}_n(s)|(\dot{\mathbf{q}}_n(s) - \dot{\mathbf{r}}_n(s)) \quad (7.10)$$

For the case when

$$\langle |\dot{\mathbf{q}}_n(s)| \rangle \gg \langle |\dot{\mathbf{r}}_n(s)| \rangle \quad (7.11)$$

(large triangular brackets indicate time average), i.e. when the drag term in

Equation (7.5) is the critical load term, it should be retained in its non-linear form.[1,17] This can be accomplished by substituting the approximate relation

$$\left|\dot{\mathbf{u}}_n(s) - \dot{\mathbf{r}}_{tn}(s)\right|(\dot{\mathbf{u}}_n(s) - \dot{\mathbf{r}}_{tn}(s)) \doteq \left|\dot{\mathbf{q}}_n(s)\right|\dot{\mathbf{q}}_n(s) - 2\left|\dot{\mathbf{q}}_n(s)\right|\dot{\mathbf{r}}_n(s) \tag{7.12}$$

thus, making use of Equations (7.6), Equation (7.5) becomes

$$\mathbf{f}_w(s) \doteq \rho A K_M \mathbf{N}\ddot{\mathbf{u}}(s) - \rho A (K_M - 1)\mathbf{N}\ddot{\mathbf{r}}_t(s)$$

$$+ \rho D K_D \left|\mathbf{N}\dot{\mathbf{q}}(s)\right|\mathbf{N}\dot{\mathbf{q}}(s) - 2\rho D K_D \left|\mathbf{N}\dot{\mathbf{q}}(s)\right|\mathbf{N}\dot{\mathbf{r}}(s) \tag{7.13}$$

For those conditions when Equation (7.11) is not satisfied, Equation (7.12) is no longer a good approximation. However, in that case the drag term in Equation (7.5) is no longer a dominating force so that one can still use the approximate relation given by Equation (7.13).

Using appropriate interpolation functions and standard finite element procedures, vector function $\mathbf{f}_w(s)$ along member i can be discretized into a 12-component force vector $\mathbf{f}_{wi}$ corresponding to the 6 nodal displacements at I and the 6 nodal displacements at J as shown in Figure 7.12. The standard lumping technique gives

$$f_{wi} \doteq \rho V K_M \mathbf{N}\ddot{\mathbf{u}} - \rho V(K_M - 1)\mathbf{N}\ddot{\mathbf{r}}_t$$

$$+ \rho A_D K_D \left|\mathbf{N}\dot{\mathbf{q}}\right|\mathbf{N}\dot{\mathbf{q}} - 2\rho A_D K_D \left|\mathbf{N}\dot{\mathbf{q}}\right|\mathbf{N}\dot{\mathbf{r}} \tag{7.14}$$

where $\ddot{\mathbf{u}}$, $\ddot{\mathbf{r}}_t$, $\dot{\mathbf{q}}$, and $\dot{\mathbf{r}}$ are the 12-component vectors in the member's nodal coordinates, V is the effective volume of the member ($V = AL = \pi D^2 L/4$), A_D is the effective drag area of the member ($A_D = DL$), and $\mathbf{N}$ is the 12×12 matrix given by

$$\mathbf{N} = \frac{1}{2}\begin{bmatrix} \mathbf{N} & \mathbf{0} & \mathbf{0} & \mathbf{0} \\ \mathbf{0} & \mathbf{0} & \mathbf{0} & \mathbf{0} \\ \mathbf{0} & \mathbf{0} & \mathbf{N} & \mathbf{0} \\ \mathbf{0} & \mathbf{0} & \mathbf{0} & \mathbf{0} \end{bmatrix} \tag{7.15}$$

Defining an effective volume matrix and an effective drag area matrix as given by

$$\mathbf{V} \equiv V\mathbf{N}; \qquad \mathbf{A} \equiv A_D\mathbf{N} \tag{7.16}$$

and approximating the last term in Equation (7.14) by

$$2\rho A_D K_D \left|\mathbf{N}\dot{\mathbf{q}}\right|\mathbf{N}\dot{\mathbf{r}} \doteq 2\rho A_D K_D \langle\left|\mathbf{N}\dot{\mathbf{q}}\right|\rangle\mathbf{N}\dot{\mathbf{r}} \tag{7.17}$$

Equation (7.14) can be written in the form

$$f_{wi} \doteq \rho K_M \mathbf{V}\ddot{\mathbf{u}} - \rho(K_M - 1)\mathbf{V}\ddot{\mathbf{r}}_t + \rho K_D \left|\mathbf{N}\dot{\mathbf{q}}\right|\mathbf{A}\dot{\mathbf{q}} - \mathbf{C_D}\dot{\mathbf{r}} \tag{7.18}$$

where

$$\mathbf{C_D} \equiv 2\rho K_D \langle\left|\mathbf{N}\dot{\mathbf{q}}\right|\rangle\mathbf{A} \tag{7.19}$$

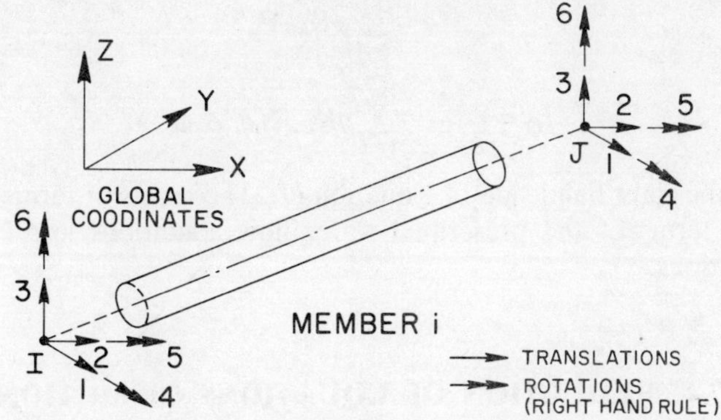

Figure 7.12 Nodal displacement components at nodal points I and J of member i

Equation (7.18) defines the 12-component nodal hydrodynamic force vector for one member, member i, in the complete system.

To obtain the total hydrodynamic force vector needed in Equation (7.3), all 12-component force vectors as defined by Equation (7.18) must be assembled into the n-component vector f_w. This process is indicated by the relation

$$f_w = \sum_{i=1}^{m} f_{wi} = \sum_{i=1}^{m} \{\rho K_{Mi} V_i \dot{u}_i - \rho (K_{Mi} - 1) V_i \ddot{r}_{ti}$$

$$+ \rho K_{Di} |N_i \dot{q}_i| A_i \dot{q}_i - C_{Di} \dot{r}_i \} \tag{7.20}$$

where each vector f_{wi} is as defined by Equation (7.18) but has been assembled into the n-component nodal system and where m is the total number of members in the structural system. For gravity platforms, this assemblage process should include the contribution from the structural base with appropriate values for K_M and K_D specially derived for the base [2]. The subscripts i have been added in Equation (7.20) to clarify that the terms refer to member i. Substituting Equation (7.20) into Equation (7.3) and rearranging terms give the equations of motion in the form

$$\bar{M}\ddot{r} + \bar{C}\dot{r} + Kr = M_w \ddot{u} - \bar{M}B\ddot{u}_g + f_D \tag{7.21}$$

where

$$\bar{M} = M + \sum_{i=1}^{m} \rho (K_{Mi} - 1) V_i \equiv M + M_a \tag{7.22}$$

$$\bar{C} = C + \sum_{i=1}^{m} C_{Di} \equiv C + \hat{C} \tag{7.23}$$

$$M_w \equiv \sum_{i=1}^{m} \rho K_{Mi} V_i \tag{7.24}$$

$$f_D \equiv \sum_i f_{Di} = \sum_{i=1}^{m} \rho K_{Di} |N_i \dot{q}_i| A_i \dot{q}_i \tag{7.25}$$

Note that the right hand side of Equation (7.21) consists of terms which are defined in terms of the prescribed water flow conditions and earthquake ground motions.

7.4　REDUCTION OF EQUATIONS OF MOTION

Corresponding to the same partitioning shown in Equation (7.1) which separates the vector r into the nodal displacements associated with the fixed base structure and the nodal displacements associated with the structure-foundation interface nodal displacements, matrices $\bar{M}$, $\bar{C}$, and K can be partitioned as given by

$$\bar{M} = \begin{bmatrix} \bar{M}_{ss} & \bar{M}_{sb} \\ \bar{M}_{sb}^{T} & \bar{M}_{bb} \end{bmatrix} \tag{7.26}$$

$$\bar{C} = C + \hat{C} = \left[\begin{array}{c|c} C_{ss} & C_{sb} \\ \hline C_{sb}^{T} & C_{bb} + C_{bb}^{f}(\omega) \end{array} \right] + \begin{bmatrix} \hat{C}_{ss} & \hat{C}_{sb} \\ \hat{C}_{sb}^{T} & \hat{C}_{bb} \end{bmatrix} \tag{7.27}$$

$$K = \left[\begin{array}{c|c} K_{ss} & K_{sb} \\ \hline K_{sb}^{T} & K_{bb} + K_{bb}^{f}(\omega) \end{array} \right] \tag{7.28}$$

In solving the coupled equations of motion, Equation (7.21), it is advantageous to use the lower normal modes and corresponding frequencies of the fixed base structure, i.e. the lower mode shapes and frequencies derived from the eigenvalue equation

$$|K_{ss} - \omega^2 \bar{M}_{ss}| = 0 \tag{7.29}$$

thus, one obtains the squared natural frequencies ω_i^2 $(i = 1, 2, \ldots, h)$ and the corresponding $n_s \times h$ mode shape matrix ϕ where h is the reduced number of fixed base modes required for engineering accuracy in the final solution. The h mode shapes in matrix ϕ should be normalized so that

$$\phi^T \bar{M}_{ss} \phi = I$$

where I is an $h \times h$ identity matrix. This normalization leads to

$$\phi^T K_{ss} \phi = \omega_i^2 \equiv \Omega \tag{7.30}$$

where matrix Ω is an $h \times h$ diagonal matrix containing the squared frequencies

$\omega_1^2, \omega_2^2, \ldots, \omega_h^2$. Further, it is advantageous to assume the damping matrix C_{ss} of the fixed base structure leads to uncoupled damped modes, i.e.

$$\boldsymbol{\phi}^T C_{ss} \boldsymbol{\phi} = \boldsymbol{\beta} \tag{7.31}$$

where matrix $\boldsymbol{\beta}$ is an $h \times h$ diagonal matrix containing the generalized damping coefficients $2\xi_1\omega_1, 2\xi_2\omega_2, \ldots, 2\xi_h\omega_h$. Quantities $\xi_1, \xi_2, \ldots, \xi_h$ are the damping ratios of the corresponding fixed base modes which are normally prescribed.

In making use of the h normal modes of the fixed-base structure, vector r_s can be separated into two sets of n_s components: (1) a set corresponding to a fixed base structure which are expressed in terms of the fixed-base structure's mode shapes and corresponding normal mode co-ordinates, and (2) a set corresponding to the pseudo-static displacements produced by the structure-foundation interface degrees of freedom. This separation is represented mathematically by

$$r = \left\{ \begin{matrix} r_s \\ r_b \end{matrix} \right\} = \begin{bmatrix} \boldsymbol{\phi} & \boldsymbol{\phi}_b \\ \boldsymbol{0} & \boldsymbol{I} \end{bmatrix} \left\{ \begin{matrix} x_s \\ x_b \end{matrix} \right\} \equiv Qx \tag{7.32}$$

where x_s are the h fixed-base normal mode co-ordinates, x_b is identical to r_b, and $\boldsymbol{\phi}_b$ is an $n_s \times n_b$ matrix containing the displacement influence coefficients for the n_s co-ordinates within the structure caused by unit displacements in the n_b interface co-ordinates. For towers having multiple supports, e.g. the pile supported framed tower represented in Figures 7.1 and 7.3, matrix $\boldsymbol{\phi}_b$ is generated using the relation[18]

$$\boldsymbol{\phi}_b = -K_{ss}^{-1} K_{sb} \tag{7.33}$$

For the single-base gravity tower represented in Figures 7.2 and 7.4, matrix $\boldsymbol{\phi}_b$ simply consists of the rigid body mode shapes produced by unit displacements in the base degrees-of-freedom.

Using the transformation defined by Equation (7.32), the n equations of motion given by Equation (7.21) can be written in the following reduced form which contains only $(h + n_b)$ equations:

$$\tilde{M}\ddot{x} + \tilde{C}\dot{x} + \tilde{K}x = \tilde{p} \tag{7.34}$$

where

$$\tilde{M} = Q^T \bar{M} Q = \left[\begin{array}{c|c} I & \boldsymbol{\phi}^T \bar{M}_{ss} \boldsymbol{\phi}_b + \boldsymbol{\phi}^T \bar{M}_{sb} \\ \hline \boldsymbol{\phi}_b^T \bar{M}_{ss} \boldsymbol{\phi} + M_{sb}^T \boldsymbol{\phi} & \boldsymbol{\phi}_b^T \bar{M}_{ss} \boldsymbol{\phi}_b + \boldsymbol{\phi}_b^T \bar{M}_{sb} + \bar{M}_{sb}^T \boldsymbol{\phi}_b + \bar{M}_{bb} \end{array} \right] \tag{7.35}$$

$$\tilde{C} = Q^T \bar{C} Q = Q^T C Q + Q^T \hat{C} Q \tag{7.36}$$

$$Q^T C Q = \left[\begin{array}{c|c} \boldsymbol{\beta} & \boldsymbol{\phi}^T C_{ss}\boldsymbol{\phi}_b + \boldsymbol{\phi}^T C_{sb} \\ \hline \boldsymbol{\phi}_b^T C_{ss}\boldsymbol{\phi} + C_{sb}^T\boldsymbol{\phi} & \boldsymbol{\phi}_b^T C_{ss}\boldsymbol{\phi}_b + \boldsymbol{\phi}_b^T C_{sb} + C_{sb}^T\boldsymbol{\phi}_b + C_{bb} + C_{bb}^f(\omega) \end{array} \right]$$

$$(7.37)$$

$$\tilde{K} = Q^T K Q = \left[\begin{array}{c|c} \boldsymbol{\Omega} & \boldsymbol{0} \\ \hline \boldsymbol{0} & K_{sb}^T\boldsymbol{\phi}_b + K_{bb} + K_{bb}^f(\omega) \end{array} \right] \tag{7.38}$$

and

$$\tilde{P} = Q^T M_w \ddot{u} - Q^T \bar{M} B \ddot{u}_g + Q^T f_D \tag{7.39}$$

All terms in Equation (7.34) can be evaluated directly by the methods previously given except for the terms involving damping matrices C_{ss}, C_{sb}, and C_{bb}. To evaluate these terms, it is reasonable to assume that the structural damping can be adequately characterized by its fixed-base damping C_{ss}, and that matrices C_{sb} and C_{bb} can be generated using the relations[1]

$$C_{sb} = -C_{ss}\boldsymbol{\phi}_b$$
$$C_{bb} = \boldsymbol{\phi}_b^T C_{ss}\boldsymbol{\phi}_b \tag{7.40}$$

Defining a new matrix

$$\boldsymbol{\Gamma} = \boldsymbol{\phi}^T \bar{M}_{ss}\boldsymbol{\phi}_b \tag{7.41}$$

it can be shown that

$$\boldsymbol{\phi}\boldsymbol{\Gamma} = \boldsymbol{\phi}\boldsymbol{\phi}^T \bar{M}_{ss}\boldsymbol{\phi}_b = \boldsymbol{\phi}_b \tag{7.42}$$

Therefore,

$$K_{sb}^T\boldsymbol{\phi}_b = -\boldsymbol{\phi}_b^T K_{ss}\boldsymbol{\phi}_b = -\boldsymbol{\Gamma}^T\boldsymbol{\Omega}\boldsymbol{\Gamma} \tag{7.43}$$

Substituting Equations (7.40), (7.41) and (7.43) into Equations (7.35)–(7.38) gives

$$\tilde{M} = \left[\begin{array}{c|c} I & \boldsymbol{\Gamma} + \boldsymbol{\phi}^T \bar{M}_{sb} \\ \hline \boldsymbol{\Gamma}^T + \bar{M}_{sb}^T\boldsymbol{\phi} & \boldsymbol{\Gamma}^T\boldsymbol{\Gamma} + \boldsymbol{\Gamma}^T\boldsymbol{\phi}\bar{M}_{sb} + M_{sb}^T\boldsymbol{\phi}\boldsymbol{\Gamma} + \bar{M}_{bb} \end{array} \right] \tag{7.44}$$

$$\tilde{C} = \left[\begin{array}{c|c} \boldsymbol{\beta} & \boldsymbol{0} \\ \hline \boldsymbol{0} & C_{bb}^f(\omega) \end{array} \right] + Q^T \hat{C} Q \tag{7.45}$$

$$\tilde{K} = \left[\begin{array}{c|c} \boldsymbol{\Omega} & \boldsymbol{0} \\ \hline \boldsymbol{0} & K_{bb} - \boldsymbol{\Gamma}^T\boldsymbol{\Omega}\boldsymbol{\Gamma} + K_{bb}^f(\omega) \end{array} \right] \tag{7.46}$$

7.5   SOLUTION OF REDUCED EQUATIONS OF MOTION

Since the reduced equations of motion given by Equation (7.34) contain the frequency dependent foundation impedance functions, i.e. matrices $K_{bb}^f(\omega)$

and $\tilde{\boldsymbol{C}}_{bb}^f(\omega)$, their solution must be carried out in the frequency domain. Only when these functions are approximated by constants, namely $\boldsymbol{K}_{bb}^f$ and $\boldsymbol{C}_{bb}^f$, can the solution be carried out in the time domain. Fortunately the frequency domain solution can be obtained very accurately and efficiently using the Fast Fourier Transform (FFT) technique[19] in which case the frequency dependent impedance functions cause no difficulty.

Taking the Fourier transform of Equation (7.34), the resulting equation can be written in the discrete form

$$(\tilde{\boldsymbol{K}} - \omega_j^2 \tilde{\boldsymbol{M}} + i\omega_j \tilde{\boldsymbol{C}})\boldsymbol{X}(\omega_j) = \tilde{\boldsymbol{P}}(\omega_j) \tag{7.47}$$

where

$$\omega_j = j\,\Delta\omega \qquad j = 0, 1, 2, \ldots, s/2 \tag{7.48}$$

$$\Delta\omega = \frac{2\pi}{s\,\Delta t}; \qquad s = 2^M; \ M = \text{integer} \tag{7.49}$$

$$\tilde{\boldsymbol{P}}(\omega_j) = \frac{1}{s} \sum_{k=0}^{s-1} \tilde{\boldsymbol{p}}(t_k) \exp(-i\omega_j t_k) \tag{7.50}$$

$$t_k = k\,\Delta t \qquad k = 0, 1, 2, \ldots, s-1 \tag{7.51}$$

Equation (7.47) can be easily solved by standard methods for each discrete frequency $\omega_j(j = 0, 1, 2, \ldots, s/2)$ giving the corresponding set of $(h + n_b)$ complex quantities contained in $\boldsymbol{X}(\omega_j)$. It should be noted that $\boldsymbol{X}(-\omega_j)$ is the complex conjugate of $\boldsymbol{X}(\omega_j)$, i.e.

$$\boldsymbol{X}(-\omega_j) = \boldsymbol{X}^*(\omega_j) \tag{7.52}$$

The time domain solution for $\boldsymbol{x}(t)$ is obtained by taking the inverse Fourier transform of $\boldsymbol{X}(\omega)$ which in discrete form is given by

$$\boldsymbol{x}(t_k) = \sum_{j=-s/2}^{s/2} \boldsymbol{X}(\omega_j) \exp(i\omega_j t_k) \tag{7.53}$$

Following the transformation defined by Equation (7.32), one obtains

$$\boldsymbol{r}(t_k) = \boldsymbol{Q}\boldsymbol{x}(t_k) \tag{7.54}$$

Should any other response quantity, say $z(t)$, , be required where

$$z(t) = \boldsymbol{a}^T \boldsymbol{r}(t) \tag{7.55}$$

in which $\boldsymbol{a}$ is the appropriate $n \times 1$ transformation vector, it can be obtained in discrete form through the relation

$$z(t_k) = \boldsymbol{a}^T \boldsymbol{Q}\boldsymbol{x}(t_k) \tag{7.56}$$

for $k = 0, 1, 2, \ldots, s-1$.

In the above procedure, the time interval Δt must be taken sufficiently small so that the time functions involved are accurately defined and the duration $s\,\Delta t$ must be taken sufficiently large so that convergence of response statistics is adequately reached; thus, it is necessary that integer M in the equation $s = 2^M$ be fairly large, say 10, 11, or 12 corresponding to $s = 1024$, 2048, and 4096.

7.6 EVALUATION OF STATIONARY STOCHASTIC RESPONSE

If the excitation of the structure-foundation system is prescribed as a stationary random process, a sample input excitation from this process can be generated and the corresponding response can be obtained deterministically using the above described frequency domain procedure. Assuming the excitation process is ergodic, one can characterize the random process of any response quantity, say $z(t)$, by (1) generating a single sample response time history $z(t_k)$ using Equation (7.56), (2) determining the dynamic fluctuations of this response about its means value through the relation

$$\bar{z}(t_k) = z(t_k) - \frac{1}{s}\sum_{k=0}^{s-1} z(t_k) \tag{7.57}$$

and (3) establishing the power spectral density function for $\bar{z}(t)$ in discrete form using

$$\begin{aligned} S_{\bar{z}\bar{z}}(\omega_j) &= |\bar{Z}(\omega_j)|^2 \\ S_{\bar{z}\bar{z}}(-\omega_j) &= S_{\bar{z}\bar{z}}(\omega_j) \end{aligned} \qquad j = 0, 1, 2, \ldots, s/2 \tag{7.58}$$

where

$$\bar{Z}(\omega_j) = \boldsymbol{a}^T \boldsymbol{Q} \boldsymbol{X}(\omega_j) - \boldsymbol{a}^T \boldsymbol{Q} \boldsymbol{X}(0) \tag{7.59}$$

The variance of random variable $\bar{z}(t)$ can now be obtained using the summation

$$\sigma_{\bar{z}\bar{z}}^2 = \sum_{j=-s/2}^{s/2} S_{\bar{z}\bar{z}}(\omega_j) \tag{7.60}$$

Also of major importance is the maximum (or extreme value) of $\bar{z}(t)$ over the duration $s\,\Delta t$. Adopting the notation

$$\bar{z}_m \equiv \bar{z}(t)_{\max} \tag{7.61}$$

the mean value of maximum response can be approximated using the relation[13,18]

$$E[\bar{z}_m] \doteq \sigma_{\bar{z}\bar{z}}\left(\sqrt{2 \ln \nu T} + \frac{0\cdot 5772}{\sqrt{2 \ln \nu T}}\right) \tag{7.62}$$

where

$$\nu \doteq \frac{1}{2\pi} \frac{\sum\limits_{j=-s/2}^{s/2} \omega_j^2 S_{\bar{z}\bar{z}}(\omega_j)}{\sum\limits_{j=-s/2}^{s/2} S_{\bar{z}\bar{z}}(\omega_j)} \tag{7.63}$$

and where T is the duration of the process. Normalizing the maximum response as defined by

$$\eta \equiv \frac{\bar{z}_m}{\sigma_{\bar{z}\bar{z}}} \tag{7.64}$$

the probability distribution (cumulative density) function for η can be approximated using the relation

$$P(\eta) \doteq \exp\left[-\nu T \exp\left(\frac{\eta^2}{2}\right)\right] \tag{7.65}$$

The probability density function for η can now be found by differentiation, i.e.

$$p(\eta) = \frac{dP(\eta)}{d\eta} \tag{7.66}$$

7.7 FATIGUE CONSIDERATIONS

During the life of a fixed platform, random wave action subjects the structure to an extremely large number of oscillations. Fatigue therefore is an important consideration.

If one adopts Miner's linear accumulative damage hypothesis,[20] accumulative damage (AD) for a narrow band response as normally produced by wave action can be expressed in the approximate form[13,18]

$$AD \doteq \nu \cdot T \int_0^\infty \frac{p(\bar{z}^*)}{N(\bar{z}^*)} \, d\bar{z}^* \tag{7.67}$$

where ν is given by Equation (7.63), T is the duration of the stationary response process $\bar{z}(t)$, $\bar{z}^*$ represents the individual single amplitudes of response (stress or force) present in process $\bar{z}(t)$, $p(\bar{z}^*)$ is the probability density function for random variable $\bar{z}^*$, and $N(\bar{z}^*)$ is the relationship expressing number of cycles to failure (N) versus amplitude of response ($\bar{z}^*$). For a narrow band process, $p(\bar{z}^*)$ can be approximated by the Rayleigh distribution

$$p(\bar{z}^*) = \frac{\bar{z}^*}{\sigma_{\bar{z}\bar{z}}^2} \exp\left(-\frac{\bar{z}^{*2}}{2\sigma_{\bar{z}\bar{z}}^2}\right) \tag{7.68}$$

Assuming a fatigue $\bar{z}^*$-N relationship which plots as a straight line on log–log paper, i.e.

$$N(\bar{z}^*) = \left(\frac{\bar{z}_1^*}{\bar{z}^*}\right)^b N_1 \tag{7.69}$$

where N_1 and $\bar{z}_1^*$ represent a convenient point on the $\bar{z}^*$-N curve and b is an even integer, Equation (7.67) becomes after substitution of Equations (7.68) and (7.69)

$$AD = \frac{\nu T}{N_1}\left(\frac{\sigma_{\bar{z}\bar{z}}}{\bar{z}_1^*}\right)^b 2^{b/2}\left(\frac{b}{2}\right)! \tag{7.70}$$

To estimate total accumulative damage over the full life of the structure (T_L), one must establish the joint probability density function $p(U, \theta)$ for the wind velocity U, which characterizes the wave height spectrum (or directional spectrum), and its horizontal directional angle θ. Having established this function, it follows that the total accumulative damage during the life of the structure is given by[13]

$$AD(\text{total}) = \frac{2^{b/2}\left(\dfrac{b}{2}\right)! T_L}{N_1(\bar{z}_1^*)^b} \int_{U=0}^{U_m} \int_{\theta=0}^{2\pi} \nu(U, \theta)\sigma_{\bar{z}\bar{z}}^b(U, \theta)p(U, \theta)\, \mathrm{d}\theta\, \mathrm{d}U \tag{7.71}$$

where U_m represents the maximum characteristic wind velocity expected during the life of the structure. Failure is predicted to occur when

$$AD(\text{total}) \geq 1 \tag{7.72}$$

7.8 EARTHQUAKE AND WAVE EXCITATIONS

For a seismic analysis the three-dimensional ground accelerations $\ddot{\mathbf{u}}_g$ must be fully prescribed as input to the structure-foundation system previously described. The simplest approach to defining these motions would be to assume a set of three-dimensional ground motions recorded during a past earthquake as representative of the future site ground motions to be expected. These motions would most likely be normalized to an appropriate intensity level. This simple approach can, of course, be questioned as two sets of accelerograms, even for the same site location, often have quite dissimilar characteristics. Another approach to defining the components of ground motion, would be to generate synthetic accelerograms derived from prescribed smooth response spectrum curves which have been obtained by a statistical analysis of response spectrum curves for many past recorded sets of accelerograms.[21,22,23] A third approach would be to generate sets of ground motions from an appropriate stochastic model.[18,24,25] This approach has the distinct advantage that a complete ensemble (or family) of possible motions can be generated which allows a direct evaluation of maximum structural response in probabilistic terms. Regardless of which approach is used, the prescribed earthquake motions should realistically represent the expected future ground motions at the particular site under consideration.

For a wave analysis, the water particle accelerations $\ddot{u}$ and velocities $\dot{u}$ must be prescribed at all nodal points of the structure within the fluid. One approach to defining these motions would be to start with fully prescribed wave surface condition and to calculate all required accelerations and velocities using whatever wave theory is judged to be appropriate. For deep water, Airy's linear wave theory is often adequate; however for shallower water cases it is recommended that a higher order wave theory be used.[26] Stoke's 5th order theory is commonly used for this purpose.

If the wave surface condition is prescribed by a wave height spectrum (or directional spectrum), a sampled sea surface condition can be generated by discretizing the spectrum at closely spaced frequencies and summing the individual harmonic surface waves which result after assigning a randomly sample (taken from a uniform probability density distribution between 0 and 2π) phase angle to each.[27] One then proceeds to calculate water particle accelerations and velocities in the appropriate manner described above.

If a steady sea current is present, its velocity at each nodal point in the flow can be added vectorially to the corresponding wave produced velocity so that its effect on structural response is included in the dynamic response analysis. As shown in the literature, current velocity can have a significant influence in certain cases.[28]

7.9 CONCLUDING STATEMENT

The methods of dynamic analysis presented herein make use of recently developed computational techniques and foundation modelling leading to effective and efficient procedures for obtaining seismic and wave induced structural response. They are similar but more general than the methods previously reported by the authors,[1,2,13,27,29] particularly with respect to the improved foundation modelling, the retention of drag in a non-linear form, and the three-dimensional form of solution. The same solution techniques were used by the authors for seismic response studies of concrete gravity platforms and were found to be extremely effective and efficient.[2,30]

Acknowledgement

The first author wishes to acknowledge with appreciation that much of his earlier work on the dynamic response of fixed offshore platforms was carried out under a series of grants from the Esso Education Foundation and the Standard Oil Company of California.

REFERENCES

1. Penzien, J. (1976). 'Structural dynamics of fixed offshore structures', *Proceedings, BOSS' 76*, International Conference on the Behaviour of *Offshore Structures*, Vol. 1, Trondheim, Norway, August 2–5, 1976.
2. Penzien, J. and Tseng, W. S. (1976). 'Seismic analysis of gravity platforms including soil-structure interaction effects', *Paper No. 2674*, Proceedings Offshore Technology Conference, Houston, Texas, May 1976.
3. Tseng, W. S. (1974). 'Soil-structure interaction analysis using fast fourier transform method', *Technical Report*, Bechtel Power Corporation, San Francisco, California, November 1974.
4. Chopra, A. K. and Gutierrez, J. A. (1974). 'Earthquake response analysis of multiple story buildings including foundation interaction', *International Journal of Earthquake Engineering and Structural Dynamics*, **3**, 65–77.
5. Veletsos, A. S. and Wei, Y. T. (1971). 'Lateral and rocking vibration of a footing', *Journal of the Soil Mechanics and Foundation Division*, ASCE, **97**, No. SM9, 1227–1249, September 1971.
6. Luco, J. E. and Westmann, R. A. (1971). 'Dynamic response of circular footings', *Journal of the Engineering Mechanics Division*, ASCE, **97**, No. EM5, 1381–1395, October 1971.
7. Luco, J. E. (1979). 'Impedance functions for a rigid foundation on a layered medium', *Nuclear Engineering and Design*, **31**, No. 2, 204–217.
8. Novak, M. (1974). 'Dynamic stiffness and damping of piles', *Canadian Geotechnical Journal*, **11**, 574–598.
9. Kausel, E. and Roesset, J. M. (1975). 'Dynamic effects of circular foundation', *Journal of the Engineering Mechanics Division*, ASCE, **101**, No. EM6, 771–785, December 1975.
10. Tsai, N. C. and Housner, G. W. (1970). 'Calculation of surface motions of a layered half-space', *Bulletin of the Seismological Society of America*, **60**, No. 5, 1625–1651, October 1970.
11. Lysmer, J. and Kuhlemeyer, R. L. (1969). 'Finite element model for infinite media', *Journal of the Engineering Mechanics Division*, ASCE, **95**, No. EM4, 859–877, August 1969.
12. Waas, G. and Lysmer, J. (1972). 'Vibrations of footings embedded in layered media', *Proceedings WES Symposium on Application of the Finite Element Method in Geotechnical Engineering*, U.S. Army Engineer Waterway. Experimental Station, Vicksburg, Mississippi, May 1972.
13. Penzien, J., Kaul, M. K., and Berge, B. (1972). 'Stochastic response of offshore towers to random sea waves and strong motion earthquakes', *International Journal of Computers and Structures*, **2**, 733–756.

14. Morison, J. R., O'Brien, M. P., Johnson, J. W., and Schaaf, S. A. (1950). 'The force exerted by surface waves on piles', *Petroleum Transactions*, **189**, TP2846, 149–154.
15. Borgman, L. E. (1958). 'Computation of the ocean-wave forces on inclined cylinders', *Journal of Geophysical Research, Transaction, AGU*, **39**, No. 5, 885–888, October 1958.
16. Chakrabarti, S. K., Tam, W. A., and Wolbert, A. L. (1976). 'Total forces on a submerged randomly oriented tube due to waves', *Paper No. 2495*, Offshore Technology Conference, Houston, Texas.
17. Taylor, E. R. (1976). 'Discussion of paper by J. Penzien: structural dynamics of fixed offshore structures', *Proceedings BOSS' 76*, International Conference on the Behaviour of Offshore Structures, Vol. 1, Trondheim, Norway, August 2–5, 1976.
18. Clough, R. W. and Penzien, J. (1975). *Dynamics of Structures*, McGraw-Hill.
19. Cooley, J. W. and Tukey, J. W. (1965). 'An algorithm for the machine calculation of complex fourier series', *Mathematics of Computations*, Vol. 19, 297–301, April 1965.
20. Miner, M. A. (1945). 'Cumulative damage in fatigue', *Journal of Applied Mechanics*, **12(1)**, A, 159–164.
21. Tsai, N. C. (1972). 'Spectrum compatible motions for design purposes', *Journal of the Engineering Mechanics Division*, ASCE, **98**, No. EM2, Paper 8807.
22. U.S. Atomic Energy Commission (1973). 'Design response spectra for seismic design of nuclear power plants', *Regulatory Guide 1.60*, Rev. 1, Directorate of Regulatory Standards, 1973.
23. Bechtel Power Corporation (1974). 'Topical report, seismic analyses of structures and equipment for nuclear power plants', *BC-TOP-4-A*, Rev. 3, San Francisco, CA.
24. Penzien, J. and Watabe, M. (1975). 'Characteristics of 3-dimensional earthquake ground motions', *International Journal of Earthquake Engineering and Structural Dynamics*, **3**, 365–373.
25. Kubo, T. and Penzien, J. (1976). 'Time and frequency domain analysis of three-dimensional ground motions, San Fernando earthquake', *Earthquake Engineering Research Center Report No. EERC 76-6*, University of California, Berkeley, CA., March 1976.
26. Dean, R. G. (1974). 'Evaluation and development of water wave theories for engineering applications', *Special Report No. 1*, Vols. 1 and 2, U.S. Army Corps of Engineers, Coastal Engineering Research Center, November 1974.
27. Malhotra, A. K. and Penzien, J. (1970). 'Nondeterministic analysis of offshore structures', *Journal of the Engineering Mechanics Division*, ASCE, EM6, 985–1003, December 1970.
28. Wu, S. C. and Tung, C. C. (1975). 'Random response of offshore structures to wave and current forces', *A University of North Carolina Sea Grant Program Publication*, *UNC-SG-75-22*, September 1975.
29. Penzien, J. (1975). 'Seismic analysis of platform structure-foundation systems', Paper No. 2352, *Proceedings Offshore Technology Conference*, Houston, Texas, May 1975.
30. Watt, B. J., Boaz, I. B., and Dowrick, D. J. (1976). 'Response of concrete gravity platforms to earthquake excitations', *Paper No. 2673*, Proceedings Offshore Technology Conference, Houston, Texas, May 1976.

Chapter 8

Dynamic Response of Framed and Gravity Structures to Waves

R. Sigbjörnsson, K. Bell and I. Holand

8.1. INTRODUCTION

This paper discusses dynamic response analysis of offshore structures with special reference to structures having a gravity type foundation and being of such dimensions that inertia wave forces dominate the overall loading, although drag forces are not negligible. The examples included refer to concrete gravity platforms of the CONDEEP type, of which three have been placed in the North Sea already, while the concreting of number 4 and 5 is completed, and number 6 is in its design phase.

A detailed mathematical modelling of the interaction problem, including structure, soil and sea, and accounting accurately for hydrodynamic and constitutive laws, is at present not practicable. The reasons are both lack of knowledge regarding the physical laws and lack of reliable computational procedures for problems of this complexity. The paper is largely a 'state of the art' report, describing a modelling that reflects present knowledge of physical properties, computational capabilities and economical restraints.

The prediction of structural safety and serviceability is the objective of the analysis. Hence the structural response is of primary concern. Nevertheless a substantial part of the paper is concerned with problems related to the surrounding media, sea and soil. The causality is evident: The modelling of the structure presents modest difficulties in comparison with an adequate description of sea and soil.

Even though the designer must be responsible for the structural analysis on which the design shall be based, modern society requires a certain level of safety, in particular in relation to risks for human lives and serious pollution. In Norway such safety requirements are laid down in governmental regulations, a draft of which[42] has been published for reviewing and criticism. In addition to establishing a minimum safety level, such regulations also form a framework regarding minimum requirements for static and dynamic analysis.

8.2　DESCRIPTION OF SEA STATE

8.2.1　Short-term model

The most common stochastic model used to describe the short-term state of the sea surface, in connection with offshore structures, is a zero mean ergodic Gaussian process, a thorough discussion of which may be found in References 6 and 26. The major advantage of this model, from a computational point of view, is its simplicity. It is completely defined by the correlation function

$$R_\eta(\chi, \tau) = E[\eta(\mathbf{x}, t)\eta(\mathbf{x}+\chi, t+\tau)] \tag{8.1}$$

Here η is the wave amplitude measured at point $\mathbf{x}$ and time t and at point $\mathbf{x}+\chi$ and time $t+\tau$.

In structural analysis it is more appropriate to use spectral densities than correlation functions. The three-dimensional wave spectral density is defined as the following Fourier transform of Equation (8.1)

$$S_\eta(\kappa, \omega) = (2\pi)^{-3} \int_\tau \int_\chi R_\eta(\chi, \tau) \exp\{-i(\kappa \cdot \chi + \omega\tau)\} \, d\chi \, d\tau \tag{8.2}$$

where ω is the circular frequency and κ is the two-dimensional wave number vector. Hence, the inverse Fourier transform yields again the correlation function.

If the sea elevation is measured only at a single point, i.e., $\chi = \mathbf{0}$, the one-dimensional spectral density is obtained

$$S_\eta(\omega) = (2\pi)^{-1} \int_{-\infty}^{\infty} R_\eta(\tau) \exp\{-i\omega\tau\} \, d\tau \tag{8.3}$$

Several analytical expressions have been proposed for this spectrum. Some of the most common spectra, for fully-developed seas, used in design of North Sea structures are plotted in Figure 8.1. For fixed, offshore platforms special attention should be given to the high frequency spectral contents in the waves. The reason is that for heavy sea states it is the high frequency tail of the spectrum that primarily determines the magnitude of the resonance contribution to the response. Therefore, double peaked spectra, commonly observed, should not be overlooked in this context.[39,46]

The one-dimensional spectral density does not furnish any information about the spatial distribution or directional distribution of waves. Such information may be obtained from the two-dimensional wave number spectral density, which may be derived from Equation (8.2). However, it is more common to express this property in terms of the following cross-spectral density, which can be derived from the wave number spectrum if it is assumed

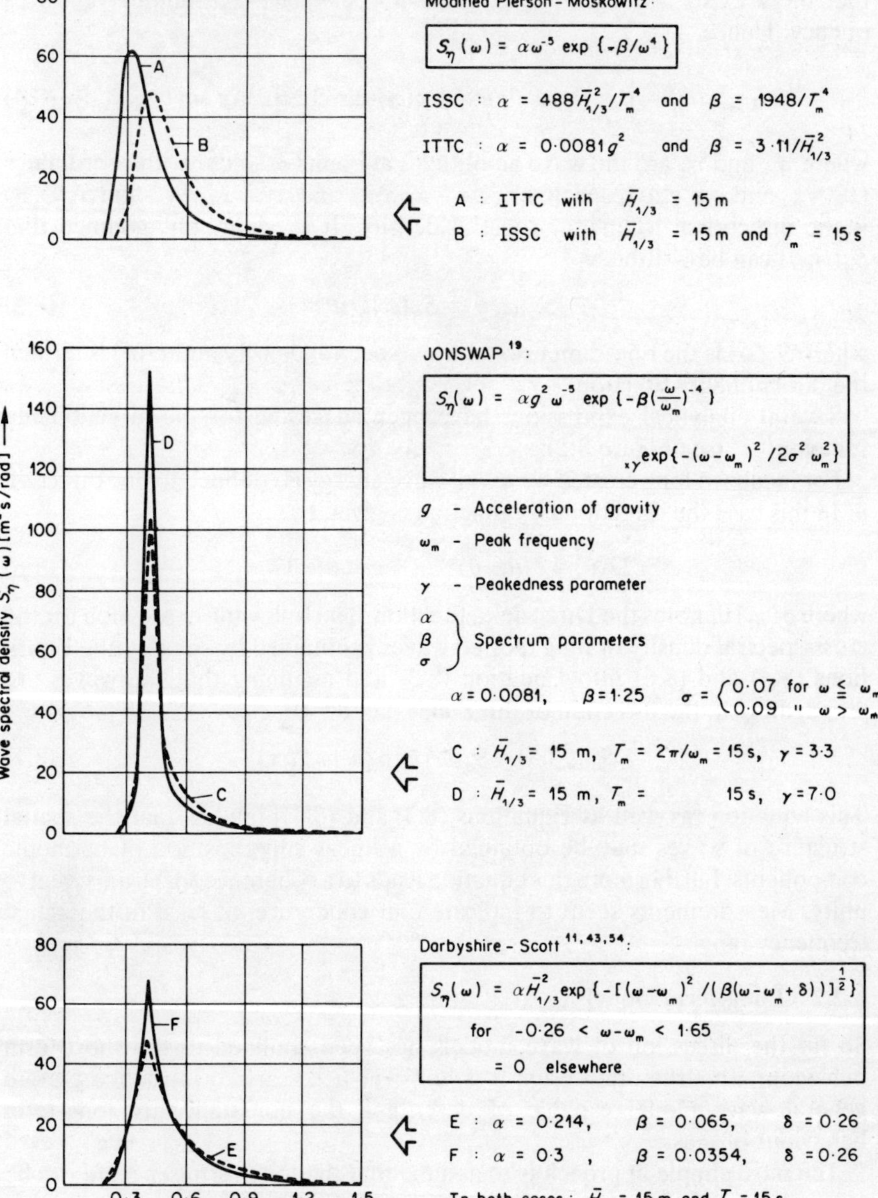

Figure 8.1   One-sided wave spectra

that there exists a one-to-one relationship between wave number and frequency. Hence

$$S_{\eta_m \eta_n}(\omega) = \int_{-\pi}^{\pi} S_\eta(\omega, \theta) \exp\left\{i\kappa(\omega)(\Delta x \cos\theta + \Delta y \sin\theta)\right\} d\theta \qquad (8.4)$$

where η_m and η_n are the wave amplitudes at points m and n with coordinates (x_m, y_m) and (x_n, y_n), respectively, $\Delta x = x_m - x_n$ and $\Delta y = y_m - y_n$, and $S_\eta(\omega, \theta)$ is the directional frequency spectral density. It is commonly assumed that $S_\eta(\omega, \theta)$ can be written as

$$S_\eta(\omega, \theta) = S_\eta(\omega)D(\theta) \qquad (8.5)$$

where $S_\eta(\omega)$ is the one-dimensional wave spectral density and $D(\theta)$ is termed the directionality function.

Several analytical expressions have been suggested for the directionality function[12,46] (see Figure 8.2).

For idealized long crested waves all wave energy is confined to one direction $\bar{\theta}$. In this case the directionality function is given by

$$D(\theta) = \delta(\bar{\theta} - \theta) \qquad -\pi < \bar{\theta} - \theta < \pi \qquad (8.6)$$

where $\delta(\ )$ denotes the Dirac delta function. The following expression for the cross-spectral density of long crested waves is obtained by substituting Equations (8.4) and (8.6) into Equation (8.3) and assuming that the waves are propagating in the direction of the x-axis (i.e. $\bar{\theta} = 0$),

$$S_{\eta_m \eta_n}(\omega) = S_\eta(\omega) \exp\left\{i\kappa(\omega)\,\Delta x\right\} \qquad (8.7)$$

This equation [as well as Equations (8.3) and (8.4)] implies that the spatial structure of waves may be obtained by a linear superposition of harmonic components. Furthermore this equation leads to a coherence spectrum equal to unity. Measurements seem to indicate unit coherence, at least in the actual frequency range.[36]

8.2.2 Long-term model

So far the discussion of wave models has been confined to the short-term behaviour. In structural design it is however necessary to introduce a more general wave model capable of describing the non-stationary long-term behaviour of seas.

The most simple approach is to assume that the long-term sea state can be approximated by series of stationary sea states. Furthermore it is assumed that each stationary sea state can be treated as a zero mean ergodic Gaussian process. This approach may be justified when the time scale governing the long-term fluctuations is much larger than the time scale characteristic for the

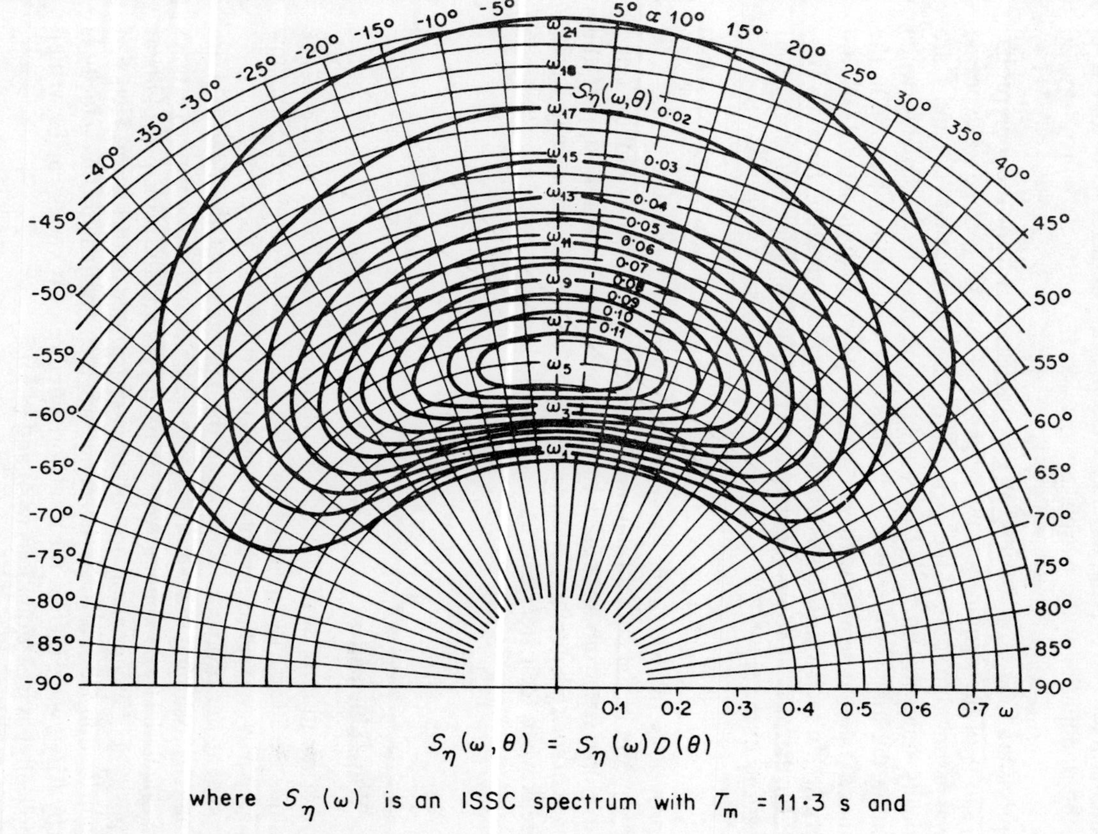

Figure 8.2 Directional wave spectrum in deep water[12]

short-term fluctuations. In open sea, where tidal effects are small, the shortest primary long-term fluctuation is due to macrometeorological high and low pressures. For instance the dominating period of high and low pressure movements in the North Atlantic and the North Sea is of the order of four days, while the dominating wave period in most actual cases is less than, say, 20 seconds.

Appropriate long-term models can be established by using the joint probability density of the significant wave height $\bar{H}_{1/3}$ and the average wave period $\bar{T}$. It was assumed in some of the earlier long-term models that $\bar{H}_{1/3}$ and $\bar{T}$ are statistically independent.[38,32] This simplification may be acceptable when estimating the long-term response of ships and semisubmersibles.[40] On the other hand it should be used with care in the response analysis of fixed offshore platforms.[45] In such a case it is essential to know the correlation between $\bar{H}_{1/3}$ and $\bar{T}$. Furthermore it seems reasonable to introduce a lower boundary for $\bar{T}$. Expressions of the type[23,45]

$$\bar{T} \geq c\bar{H}_{1/3}^{1/2} \tag{8.8}$$

have proved useful, where the parameter c is of the order 2 to 3.

The concept of the equilibrium range spectrum proposed by Phillips[41] and the envelope spectrum proposed by Bretschneider[7] seem to support the existence of such a boundary. Intuitively it seems reasonable since single waves will break when their steepness exceeds certain limits.

8.3 WAVE LOADING

8.3.1 Wave kinematics

The determination of wave forces on structures is a difficult problem. The difficulties are due partly to lack of knowledge about the kinematics of the flow field and partly to insufficient knowledge concerning viscous forces.

At present the problems concerning the kinematics of the flow field are overcome by assuming that velocities and accelerations of the water particles can be calculated by combining Airy's wave theory and the principles of linear superposition of independent and arbitrary distributed disturbances. Hence, if the contribution from a single harmonic wave component to the velocity is given by Airy's wave theory, the total horizontal velocity vector can be written (symbolically) as [see Equations (8.4) and (8.5)]

$$\dot{\mathbf{u}}(\mathbf{x}, z) = \begin{bmatrix} \dot{u}_x \\ \dot{u}_y \end{bmatrix} = \int_{-\infty}^{\infty} \int_{-\pi}^{\pi} \omega \frac{\cosh\{\kappa z\}}{\sinh\{\kappa h\}} \begin{bmatrix} \cos\theta \\ \sin\theta \end{bmatrix}$$

$$\times \cos\{\mathbf{\kappa} \cdot \mathbf{x} - \omega t + \zeta(\omega, \theta)\} \sqrt{(2 S_n(\omega) D(\theta)\, d\omega\, d\theta)} \tag{8.9}$$

where the z-coordinate is measured from the sea bottom, h is the water depth and $\zeta(\omega, \theta)$ represents a random independent phase angle identically and uniformly distributed between $-\pi$ and π.

From this equation the cross-spectral density matrix of the water particle velocity can be derived. This gives

$$\mathbf{S}_{\dot{u}_m \dot{u}_n}(\omega) = S_\eta(\omega)\omega^2 \frac{\cosh\{\kappa z_m\}\cosh\{\kappa z_n\}}{\sinh^2\{\kappa h\}}$$

$$\times \int_{-\pi}^{\pi} D(\theta) \begin{bmatrix} \cos^2\theta & \sin\theta\cos\theta \\ \sin\theta\cos\theta & \sin^2\theta \end{bmatrix} \exp\{i\kappa(\Delta x \cos\theta$$

$$+ \Delta y \sin\theta)\}\, d\theta \tag{8.10}$$

where $\Delta x = x_m - x_n$ and $\Delta y = y_m - y_n$.

The cross-spectral density of water particle acceleration can be obtained using properties of derived processes. This gives

$$\mathbf{S}_{\ddot{u}_m \ddot{u}_n}(\omega) = \omega^2 \mathbf{S}_{\dot{u}_m \dot{u}_n}(\omega) \tag{8.11}$$

Similarly, the cross-spectral density of velocity and acceleration is given as

$$\mathbf{S}_{\dot{u}_m \ddot{u}_n}(\omega) = -\mathbf{S}_{\ddot{u}_m \dot{u}_n}(\omega) = i\omega\, \mathbf{S}_{\dot{u}_m \dot{u}_n}(\omega) \tag{8.12}$$

It is worth pointing out that the velocities and accelerations of the water particles form a bivariate zero mean ergodic Gaussian process as a result of the assumptions made above.

This description of the kinematics refers to the undisturbed wave field. Nevertheless it is commonly used to predict velocities and accelerations used in the evaluation of wave forces. However it is known that the presence of the structure will impose disturbances in the flow field. These disturbances may result in loss of coherency in the kinematics. For instance, a slender structural member subjected to oscillatory flow may be immersed in its own wake, which can be quite turbulent.

Another point of controversy is the influence of the wavy sea surface on the kinematics, an influence which is commonly not accounted for. However Tung[48] has shown recently that these effects are significant.

8.3.2 Morison's equation

When the wave kinematics have been determined it is most common to evaluate the longitudinal wave forces by Morison's equation.[34] The modified Morison equation for a vertical slender cylinder is given by

$$q_t = \tfrac{1}{2}\rho C_D D |\dot{u} - \dot{r}|(\dot{u} - \dot{r}) + \frac{\pi}{4}\rho C_M' D^2 \ddot{u} + \frac{\pi}{4}\rho C_M'' D^2 (\ddot{u} - \ddot{r}) \tag{8.13}$$

where

q_t = total hydrodynamic force per unit length
ρ = density of water
D = cylinder diameter
C_D = drag force coefficient
C_M' = Froude–Kryloff force coefficient
C_M'' = inertia force coefficient
$\dot{u}$ = water particle velocity
$\ddot{u}$ = water particle acceleration
$\dot{r}$ = velocity of structure
$\ddot{r}$ = acceleration of structure

It is convenient to rewrite Equation (8.13) as

$$q_t = \tfrac{1}{2}\rho C_D D |\dot{u} - \dot{r}|(\dot{u} - \dot{r}) + \frac{\pi}{4}\rho C_M D^2 \ddot{u} + m\ddot{r} \tag{8.14a}$$

where $C_M = C_M' + C_M''$ is commonly taken equal to $2\cdot0$ for circular cylinders and m is the added mass per unit length given as

$$m = \frac{\pi}{4}\rho C_M'' D^2 \tag{8.14b}$$

C_M'' is commonly taken equal to $1\cdot0$ for circular cylinders. The drag coefficient C_D is of the order $0\cdot7$ for smooth cylinders at high Reynolds' numbers. It may however be as high as $1\cdot7$ for rough cylinders.[35]

Morison's equation is valid as long as the presence of a structure does not influence the flow field significantly, which is the case when the structural dimensions are small compared to the wave length.

It is seen that the wave force q is a non-Gaussian process due to the drag term, making the stochastic analysis more complicated. Borgman[4,5,6] has given the following expression for the cross-spectral density of wave forces taking $\dot{r} = \ddot{r} = 0$

$$S_{q_m q_n}(\omega) = k_D(m)k_D(n)\left[\frac{8}{\pi}\sigma_{\dot{u}_m}\sigma_{\dot{u}_n}S_{\dot{u}_m\dot{u}_n}(\omega) + \frac{4}{3\pi}(\sigma_{\dot{u}_m}\sigma_{\dot{u}_n})^{-1}\right.$$

$$\left. \times\{S_{\dot{u}_m\dot{u}_n}(\omega)\}^{*3} + \cdots\right]$$

$$+[\sigma_{\dot{u}_m}k_D(m)k_M(n) - \sigma_{\dot{u}_n}(n)k_D(n)k_M(m)]\sqrt{\left(\frac{8}{\pi}\right)}S_{\dot{u}_m\dot{u}_n}(\omega)$$

$$+k_M(m)k_M(n)S_{\ddot{u}_m\ddot{u}_n}(\omega) \tag{8.15}$$

Here

$$k_D(m) = \tfrac{1}{2}\rho C_D D(z_m)$$

$$k_M(m) = \frac{\pi}{4}\rho C_M D^2(z_m)$$

$$\sigma_{\dot{u}_m} = \text{standard deviation of water particle velocity at point } m$$

$$\{S_{\dot{u}_m\dot{u}_n}(\omega)\}^{*n} = \int_{-\infty}^{\infty} \{S_{\dot{u}_m\dot{u}_n}(\omega')\}^{*(n-1)} S_{\dot{u}_m\dot{u}_n}(\omega-\omega')\,d\omega'$$

It is worth noting that the equivalent linearization used by Foster,[15] Malhotra and Penzien[31] and others gives a spectral density identical to Equation (8.15) if the terms containing convolutions of the velocity spectra are neglected. However these terms may not always be negligible. The spectrum $\{S_{\dot{u}_m\dot{u}_n}(\omega)\}^{*3}$ has two peaks (for positive ω). The primary peak corresponds to the peak in $S_\eta(\omega)$, while the secondary peak corresponds to three times the peak frequency in $S_\eta(\omega)$. This secondary peak in $S_{q_m q_n}(\omega)$ is much smaller than the primary peak. However for fixed offshore structures in heavy seas the secondary peak may coincide with one of the natural frequencies of the structure, which in the case of narrow wave spectra might give a significant contribution to the response. This effect will nevertheless be small for most fixed platforms and may well be overshadowed by a secondary peak commonly present in the wave spectra.

In addition to the longitudinal force component acting on slender structures, a lateral (or lift) force component due to alternating vortex shedding may be present. It is difficult to treat these lateral forces and it appears that their behaviour is not sufficiently understood at present.[44,53]

8.3.3 Potential theory

The disturbances of the flow field imposed by structures with large dimensions make Morison's equation unsuitable for the calculation of wave forces. In such cases the forces may be evaluated by the linearized potential theory, assuming the viscous effects to be negligible.[13,16,20]

This approach assumes that a velocity potential Φ exists which satisfies the Laplace equation with appropriate linearized boundary conditions. Further it is assumed that Airy's wave theory is valid and that steady state conditions have been reached. The velocity potential can then be written as

$$\Phi = \Phi^{(w)} + \Phi^{(d)} + \sum_j \Phi_j^{(v)} \tag{8.16}$$

where

$$\Phi^{(w)} = \phi^{(w)} e^{-i\omega t} = \text{velocity potential of incident waves}$$

$$\Phi^{(d)} = \phi^{(d)} e^{-i\omega t} = \text{velocity potential of diffracted waves assuming restrained structure}$$

$$\Phi_j^{(v)} = \phi_j^{(v)} \dot{r}_j = \text{velocity potential of waves generated by the motion of the structure}$$

$$\dot{r}_j = |\dot{r}_j| e^{-i\omega t} = \text{velocity of the structure}$$

When this boundary value problem has been solved the wave pressure can be obtained from Bernoulli's equation neglecting second order terms. This yields the following harmonic wave force component in the x-direction

$$F_x(\omega) e^{-i\omega t} = Q_x(t) - m_{xj}(\omega)\ddot{r}_j - c_{xj}(\omega)\dot{r}_j \qquad (8.17a)$$

Here the excitation force is given as

$$Q_x(t) = i\omega\rho \int_A (\phi^{(w)} + \phi^{(d)}) n_x \, dA \, e^{-i\omega t} \qquad (8.17b)$$

where n_x denotes the x-component of the normal to the surface (positive outwards). The first term on the right hand side in Equation (8.17b) represents the excitation force due to an undisturbed wave field (the Froude–Kryloff force), while the second term gives the excitation force due to the distortion of the wave field imposed by the restrained structure (inertia forces). The second term on the right hand side in Equation (8.17a) is the hydrodynamic inertia forces due to the motion of the structure and the last term represents the hydrodynamic radiation damping. The added mass coefficient m_{xj} and the damping coefficient c_{xj} are given as

$$m_{xj} = \rho \int_A \text{Re} \, [\phi_j^{(v)}] n_x \, dA \qquad (8.17c)$$

$$c_{xj} = \omega\rho \int_A \text{Im} \, [\phi_j^{(v)}] n_x \, dA \qquad (8.17d)$$

By this approach it is possible to obtain the wave forces induced by a given harmonic component. Hence, by applying the principle of superposition, wave forces induced by irregular waves may be derived in terms of spectral densities. Wave forces obtained by this procedure will be zero mean ergodic Gaussian processes.

For a single vertical circular cylinder the total excitation force per unit length is given in Figure 8.3.[29] The hydrodynamic radiation damping is in most cases small compared to other damping forces and can therefore be neglected. The

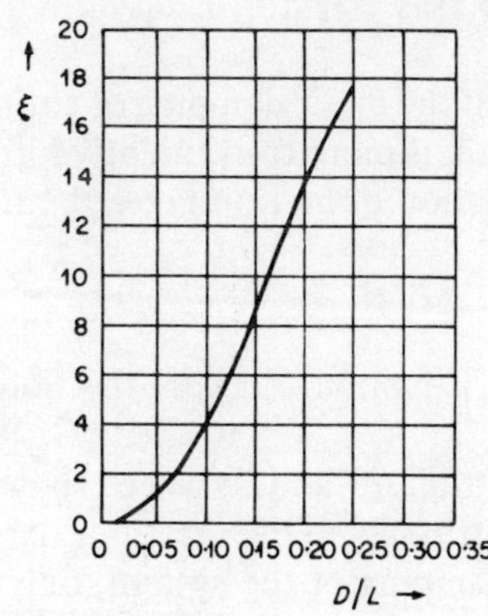
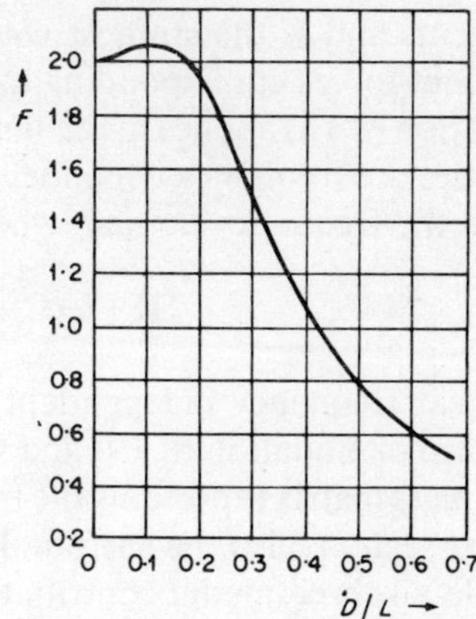

$$Q_x(t) = \int_{-\infty}^{\infty} A(\omega)\sqrt{2S_\eta(\omega)d\omega}\,\cos(\omega t - \xi + \zeta)$$

$$A(\omega) = \frac{\pi}{4}\rho F(\omega)\omega^2 \frac{\cosh(\kappa z)}{\sinh(\kappa h)} \qquad \text{where } \omega^2 = g\kappa\tanh(\kappa h)$$

$$S_{Q_x}(\omega) = A^2(\omega)S_\eta(\omega) \qquad\qquad \text{and } \kappa = \frac{2\pi}{L}$$

Figure 8.3   Inertia forces on a vertical circular cylinder

added mass can, disregarding three-dimensional effects (free surface), be taken equal to the mass of the water displaced by the cylinder. It is found that the effects of the free surface generally will reduce the added mass and thus tend to increase the natural frequency of the structure.

In cases where the solution of the boundary value problem is difficult model tests may be useful.[17] This may for instance be the case for groups of cylinders when proximity and mutual interaction effects are of significance.

8.4   STRUCTURAL MODEL

8.4.1   Mathematical statement of the problem

Limiting the discussion to a finite element type discretization and assuming viscous damping the problem is commonly stated as

$$\mathbf{M\ddot{r}}(t) + \mathbf{C\dot{r}}(t) + \mathbf{Kr}(t) = \mathbf{Q}(t) \tag{8.18}$$

in which $\mathbf{r}(t)$ is a vector of discrete, time dependent nodal point displacements and $\dot{\mathbf{r}}(t)$ and $\ddot{\mathbf{r}}(t)$ are the corresponding velocities and accelerations, respectively. $\mathbf{M}$ is the mass matrix of the system, $\mathbf{C}$ contains the viscous damping coefficients and **K** the stiffness coefficients. $\mathbf{Q}(t)$ is a vector of discrete, time dependent forces corresponding to $\mathbf{r}(t)$.

Equation (8.18) applies in the time domain. If the mass, damping or stiffness properties are frequency dependent the problem is more conveniently formulated in the frequency domain where the equation of motion takes the form

$$[\mathbf{K} + i\omega\,\mathbf{C} - \omega^2\mathbf{M}]\mathbf{r}(\omega) = \mathbf{Q}(\omega) \tag{8.19}$$

For linear frequency independent properties Equation (8.18) is the Fourier transform of Equation (8.19) and vice versa.

The mass matrix represents the mass of the structure, and, as will be shown in the next section, also the surrounding sea contributes to the mass matrix.

While all three media contributes to the damping of the system, only the structure and the soil contributes to the stiffness, and the wavy sea provides the forces.

8.4.2 Structural idealization

A typical gravity structure, consisting of a raft (caisson), one or more shafts (towers) and a deck, is shown in Figure 8.4 along with some idealizations of varying complexity. All models shown assume that the soil-structure interaction is simulated by a system of springs and dashpots (the latter are not shown in the figure), a representation which will be dealt with later in this section.

Figure 8.4b shows the simplest possible model. The entire structure is treated as a rigid block, and only two degrees of freedom, the rocking and sliding of the base, are considered. Clearly this model cannot provide any information about the structural response (in terms of stresses), but it may give an indication of the dynamic forces transmitted to the soil foundation. For a structure like the Ekofisk tank this may represent a sufficiently accurate dynamic model.[55]

A somewhat more realistic model is shown in Figure 8.4c. The structure is replaced by a simple cantilever beam of varying stiffness properties. The mass of the system is lumped at 5 nodal points. Each of these possesses three degrees of freedom, two translations and one rotation. While this model enables a more realistic representation of stiffness, mass and loading and thus may provide a reasonable prediction of the overall dynamic behaviour (period of natural vibration, forces between raft and soil), it is still too simple for the structural components.

It should be emphasized that the two simple models shown in Figure 8.4b and c can only provide reasonable results when combined with load models reflecting the true geometry of the structure.

In Figure 8.4d the platform is replaced by a space frame.[21] The raft, which is very stiff compared to the shafts, is modelled by a system of horizontal and

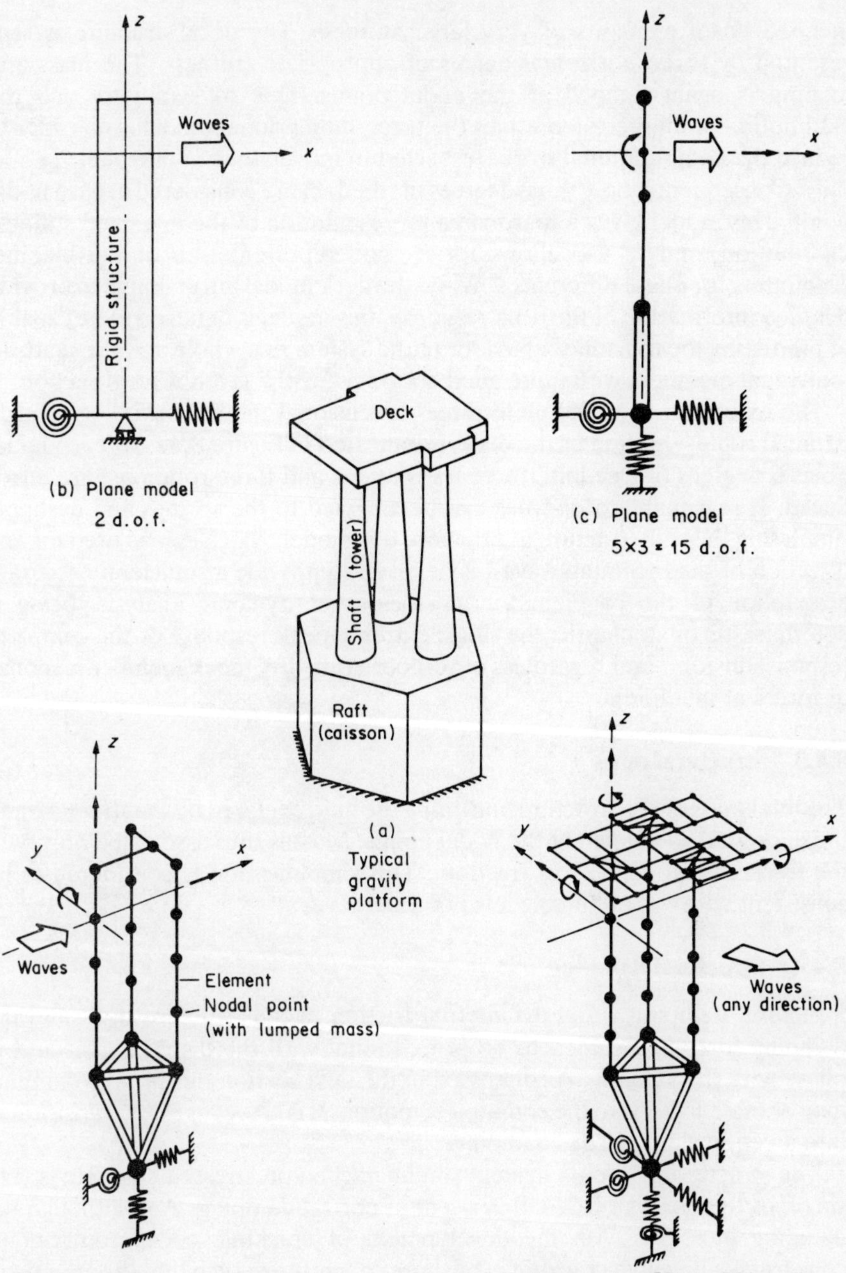

Figure 8.4   Structural idealizations

inclined beam elements of very large stiffness. The deck structure is represented by three horizontal beams of appropriate stiffness. The mass and loading is again lumped at the nodal points. Due to symmetry and the assumption of long-crested waves the three-dimensional model is, in order to reduce the computational work, restricted to movements in one plane ($y = 0$) only. Consequently only three degrees of freedom are considered at each nodal point. This model gives a reasonable representation of the mass and stiffness distribution, and it also allows for a good representation of the loading, accounting for phase differences. While the deck model is too simple to provide detailed information of the deck response, this model is believed to be capable of predicting the dynamic behaviour of the system as a whole and the shafts as individual members with quite good accuracy for the given wave direction.

The extension to a complete three-dimensional model, including a fairly detailed representation of the deck, is indicated in Figure 8.4e. At each nodal point 6 degrees of freedom (three translations and three rotations) are introduced. If reasonable properties can be assigned to the springs and dashpots simulating the soil-structure interaction, this model, which can be used for any direction of the incoming waves, is believed to provide a sufficiently accurate description of the problem for most cases of dynamic analysis, being it deterministic or stochastic, the single extreme peak response or the complete response history, and regardless of response quantity (deck member response or forces at mudline).

8.4.3 Structural mass

The total mass of the structure, including the fluid enclosed by the structure and the deck load, is lumped at the nodal points. Mass is thus associated only with the translational degrees of freedom. This simplification is considered to be consistent with the attainable level of accuracy.

8.4.4 Structural damping

Structural damping is due to internal friction in the structure itself and the damping force is assumed to be proportional with the displacement and in phase with the velocity. As discussed in the next section this type of damping may be introduced into the equation of motion. It is however often transformed into an equivalent viscous damping.

The structural damping in steel and uncracked prestressed concrete is very small, of the order of $0\cdot2$–$1\cdot0$ per cent of critical damping. And although the damping increases with the development of cracking, code requirements concerning allowable cracking in offshore concrete are such that the amount of structural damping must be assumed to be low, in the range of 1 per cent of critical damping.

If the structural damping is known in terms of modal damping ratios these may be introduced directly if a model technique is used. Alternatively this information may be transformed into an approximate damping matrix, for instance of the form

$$\mathbf{C}_{st} = a_0\mathbf{M}_{st} + a_1\mathbf{K}_{st} \tag{8.20}$$

where the coefficients a_0 and a_1 are expressed in terms of the modal frequencies and damping ratios.[10] Index st indicates that the matrices refer to the structure.

Conversely, once a damping matrix is known, the corresponding damping ratio λ_n of the nth normal mode $\boldsymbol{\psi}_n$ may be determined as

$$\lambda_n = \frac{c_n^*}{2m_n^*\Omega_n} \tag{8.21}$$

where

$$c_n^* = \boldsymbol{\psi}_n^T\mathbf{C}\boldsymbol{\psi}_n \quad \text{and} \quad m_n^* = \boldsymbol{\psi}_n^T\mathbf{M}\boldsymbol{\psi}_n \tag{8.22}$$

are the generalized damping and mass respectively and Ω_n is the circular frequency of the nth normal mode.

8.4.5 Structural stiffness

The structural stiffness is adequately expressed by an assemblage of beam elements. The assessment of the stiffness of a reinforced and usually prestressed concrete section is more controversial.

The Norwegian regulations[42] state that deterministic computational models shall aim at giving expected average values without introducing any increase or reduction in safety through the modelling, the uncertainty of the deterministic model being included in load and material coefficients. Thus expected average values are specified to be used for stiffness and mass in dynamic analyses.

This implies that the stiffness must be evaluated for the state of stress and strain in the specific case considered, the decisive distinction being between cracked and uncracked sections. In order to prevent corrosion of the reinforcement the amount of prestressing is usually determined so as to avoid significant cracking. Under such circumstances the full concrete section and the reinforcement should be included in the computation of stiffness. According to experiments made by Holand[23] there is no significant difference between the static modulus measured under short-term testing of concrete specimens and dynamic modulus measured on vibrating beams. Thus the static short-term modulus is recommended for uncracked conditions. For the Condeep platforms built at Stavanger this modulus has been found to be 30,000 MPa with only minor deviations and the stress–strain curve has been found to be remarkably linear up to half the cube strength.

8.4.6 Soil-structure interaction

The models in Figure 8.4 literally rest on the spring-dashpot system simulating the soil-structure interaction and they can only be justified if realistic properties can be assigned to these springs and dashpots.

Approximate values for the stiffness and damping coefficients may be obtained by considering the steady state response of a massless disc resting on the soil and subjected to harmonic excitation.

Assuming a rigid, circular disc supported by a homogeneous, linear elastic half-space, semi-analytical solutions are available for a wide range of the exciting frequency and for various values of Poissons's ratio, see for instance Veletsos and Wei,[49] and Veletsos and Verbic.[51]

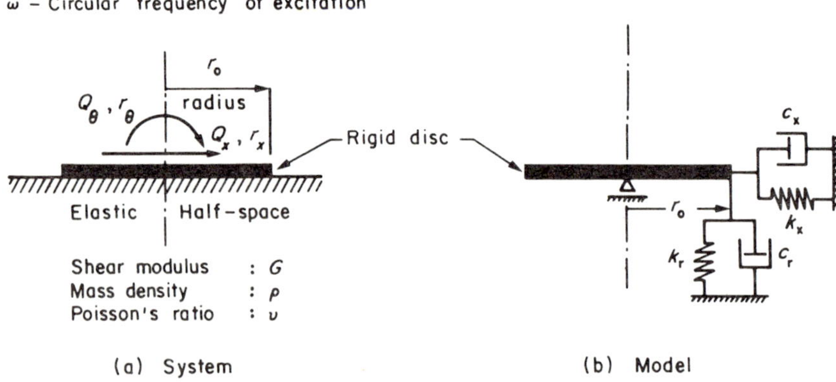

Figure 8.5 Elastic half-space approach

Considering the two most important motions of the disc, horizontal sliding (r_x) and rocking (r_θ), see Figure 8.5(a), these may, for a surface foundation (no embedment), be assumed to be uncoupled and the relation between the exciting forces and the corresponding displacements may be expressed as

$$\begin{bmatrix} Q_x \\ Q_\theta \end{bmatrix} = \begin{bmatrix} K_x & 0 \\ 0 & K_\theta \end{bmatrix} \begin{bmatrix} r_x \\ r_\theta \end{bmatrix} \tag{8.23}$$

K_x and K_θ are complex, frequency dependent impedance functions expressed in Reference 49 as

$$\left. \begin{aligned} K_x &= (k_1 + ia_0 c_1) K_{xs} \\ K_\theta &= (k_2 + ia_0 c_2) K_{\theta s} \end{aligned} \right\} \tag{8.24}$$

where

$$K_{xs} = \frac{8Gr_0}{2-\nu}$$

$$K_{\theta s} = \frac{8Gr_0^3}{3(1-\nu)}$$

$$(8.25)$$

are the 'static spring constants' and

$$a_0 = \frac{\omega r_0}{V_s} \qquad (V_s = \sqrt{(G/\rho)}) \tag{8.26}$$

is a dimensionless frequency parameter. V_s is the speed of propagation of shear waves in the half-space.

Equations (8.23) and (8.24) suggest that the displacements are out of phase with the corresponding force components. In other words the system is damped. This damping, which is referred to as geometrical damping, is caused by loss of energy through waves that originate at the structure-soil interface and that propagate away from the interface and are not reflected. The disc-foundation system may thus be modelled by a spring-dashpot system, as indicated in Figure 8.5(b), with the following coefficients

$$k_x = k_1 K_{xs}$$

$$k_r = \frac{1}{r_0^2} k_2 K_{\theta s} = \frac{2-\nu}{3(1-\nu)} k_2 K_{xs}$$

$$c_x = \frac{a_0 c_1}{\omega} K_{xs} = c_1 \frac{K_{xs} r_0}{V_s}$$

$$c_r = \frac{2-\nu}{3(1-\nu)} \frac{a_0 c_2}{\omega} K_{xs} = c_2 \frac{2-\nu}{3(1-\nu)} \frac{K_{xs} r_0}{V_s}$$

$$(8.27)$$

These coefficients along with similar coefficients for the other motions of the disc are introduced into (added to) the appropriate diagonal terms of $\mathbf{K}$ and $\mathbf{C}$.

The coefficients of Equations (8.24) are frequency dependent as indicated by Figure 8.6 for $\nu = 0$ [from (49)].

The linear elastic half-space analogy does not account for material (hysteretic) damping in the soil. This type of damping, which may be significant, particularly for the rocking motion, is introduced approximately by Veletsos and Verbic[50] who consider a linear visco-elastic half-space.

Kausel and Roësset[25] have used a finite element model to determine the equivalent spring and dashpot coefficients. Their model, which is axi-symmetric, may incorporate horizontal layers of different material properties

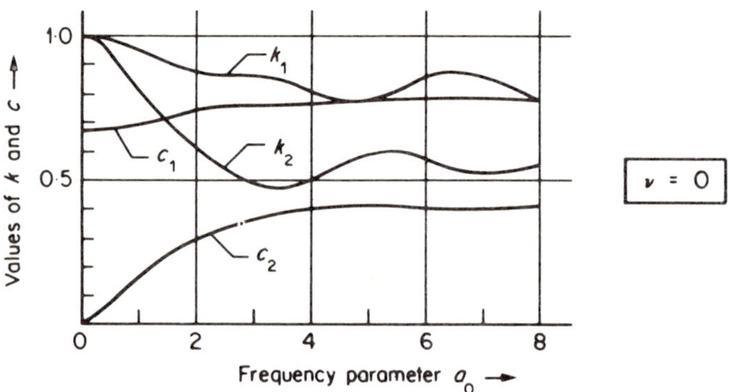

Figure 8.6 Stiffness and damping coefficients[49]

and it is limited in the vertical direction by a horizontal reflecting boundary (rigid-rock). The vertical boundaries consist of an energy absorbing layer ('semi-infinite elements') which enables waves to radiate into the far-field. Using a visco-elastic material model, material (hysteretic) damping may also be considered. The numerical formulation is by no means trivial and the demand on computer time is significant. Nevertheless this work suggests a more rational approach to the types of soil-structure problem encountered in practice.

Regardless of whether finite element or continuum theories are used there is the problem of choosing soil properties. Particularly the evaluation of the shear modulus is critical. To demonstrate the influence of this parameter on the dynamic behaviour, the frequency of the first natural mode of a typical gravity platform is shown in Figure 8.7 as a function of the shear modulus G. The results are obtained using the plane model in Figure 8.4d, with dimensions as shown in Appendix 8A and with spring stiffness determined by elastic half-space analogy.

So far a linear model has been assumed. Unfortunately soil is a strongly non-linear material, its properties depend upon the magnitude of the strain level, and hysteresis loops appear during cyclic loadings. It is not within the scope of this paper to deal with these problems, a paper by Whitman[52] is recommended for the interested reader, but generally speaking the effective shear modulus decreases while the internal damping increases with increasing strain amplitude.

8.4.7 Reduction of degrees of freedom

In case of a stochastic analysis and particularly in connection with long-term statistical models (fatigue analysis), the equation of motion may, as will be seen in the next section, have to be solved for many different sea states and also for different stiffness and damping properties. Even a fairly simple model like the one shown in Figure 8.4e may then prove rather costly and depending on the

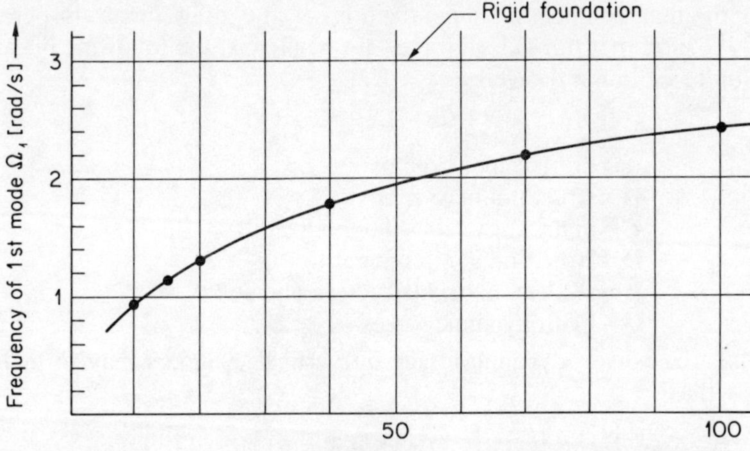

Figure 8.7   Lowest natural frequency of a typical gravity platform

method of solution it may be worthwhile to investigate the possibility of reducing the number of degrees of freedom.

If a lumped parameter model is used, both for mass and loading, all rotational degrees of freedom, except at the base point, may be eliminated by simple static condensation without introducing additional constraints on the model. Assuming that frequency dependent and non-linear properties are present only at the soil-structure interaction, this reduction to nearly half the number of degrees of freedom need be carried out only once. The disadvantage of this reduction is that the band form of the stiffness matrix is lost.

The in-plane stiffness of the deck is normally large and the error introduced by expressing all x- and y-displacements of the deck nodes in terms of three independent degrees of in-plane freedom is hardly serious.

Finally, by assuming a completely rigid raft, the degrees of freedom at the nodes on top of the raft may be expressed in terms of the base node degrees of freedom.

Applying all these reductions the 354 degrees of freedom of the model in Figure 8.4e may be reduced to 88. All matrices will in principle be full, but the gain is substantial, and this reduced model is a realistic proposition for even quite extensive parametric studies.

8.5 RESPONSE ANALYSIS

8.5.1 Formulation of basic equations

The equation governing the vibrations of an offshore platform may be obtained from Equation (8.18) [or Equation (8.19)] and Equations (8.13) and (8.17).

Assuming that a stochastic linearization of the drag forces is permissible[3,15,31,46] and that transversal forces are negligible, the following linearized equation of motion is derived

$$\tilde{\mathbf{M}}\ddot{\mathbf{r}} + \tilde{\mathbf{C}}\dot{\mathbf{r}} + i\,\tilde{\mathbf{D}}\mathbf{r} + \tilde{\mathbf{K}}\mathbf{r} = \tilde{\mathbf{Q}}(t) \tag{8.28a}$$

where

$$\begin{aligned}
\mathbf{r} &= \text{displacement vector} \\
\tilde{\mathbf{M}} &= \text{structural mass matrix} \\
\tilde{\mathbf{C}} &= \text{matrix of viscous damping (soil)} \\
\tilde{\mathbf{D}} &= \text{structural damping matrix} \\
\tilde{\mathbf{K}} &= \text{stiffness matrix (structure and soil)} \\
\tilde{\mathbf{Q}} &= \text{hydrodynamic forces}
\end{aligned}$$

The linearized hydrodynamic forces on vertical cylinders may formally be expressed as

$$
\begin{aligned}
\tilde{\mathbf{Q}}(t) = {}& \sum_j \mathbf{a}_j^T \left\{ \tfrac{1}{2}\rho C_D \sqrt{\left(\frac{8}{\pi}\right)} \int_0^l D(z)\mathbf{N}^T \{\sigma_{\dot{u}-\dot{r}}(z)\dot{u}(\mathbf{x},t)\}\,\mathrm{d}z \right\}_j \\
&+ \sum_j \mathbf{a}_j^T \left\{ \tfrac{1}{2}\rho C_D \sqrt{\left(\frac{8}{\pi}\right)} \int_0^l D(z)\mathbf{N}^T \{\sigma_{\dot{u}-\dot{r}}(z)N\}\,\mathrm{d}z \right\}_j \mathbf{a}_j\dot{\mathbf{r}} \\
&+ \sum_j \mathbf{a}_j^T \left\{ \frac{\pi}{4}\rho C_M \int_0^l D^2(z)\mathbf{N}^T \ddot{u}(\mathbf{x},t)\,\mathrm{d}z \right\}_j \\
&+ \sum_j \mathbf{a}_j^T \left\{ \frac{\pi}{4}\rho C_M'' \int_0^l D^2(z)\mathbf{N}^T\mathbf{N}\,\mathrm{d}z \right\}_j \mathbf{a}_j\ddot{\mathbf{r}} \\
&+ \int_\omega \sum_j \mathbf{a}_j^T \left\{ \rho \int_A i\omega(\phi^{(w)}(\omega)+\phi^{(d)}(\omega))\eta(\omega)\mathbf{N}^T\mathbf{n}\,\mathrm{d}A \right\}_j \exp\{i\omega t\}\,\mathrm{d}\omega \\
&+ \int_\omega \sum_j \mathbf{a}_j^T \left\{ \rho \int_A \mathbf{N}^T\,\mathrm{Re}\,[\phi^{(v)}(\omega)]\mathbf{n}^T\mathbf{N}\,\mathrm{d}A \right\}_j \mathbf{a}_j\ddot{\mathbf{r}}(\omega)\exp\{i\omega t\}\,\mathrm{d}\omega \\
&+ \int_\omega \sum_j \mathbf{a}_j^T \left\{ \rho \int_A \omega\mathbf{N}^T\,\mathrm{Im}\,[\phi^{(v)}(\omega)]\mathbf{n}^T\mathbf{N}\,\mathrm{d}A \right\}_j \mathbf{a}_j\dot{\mathbf{r}}(\omega)\exp\{i\omega t\}\,\mathrm{d}\omega
\end{aligned}
\tag{8.28b}
$$

where

$$\begin{aligned}
\mathbf{a} &= \text{connectivity matrix} \\
\mathbf{N} &= [N_1\,N_2\cdots] = \text{shape functions} \\
l &= \text{length of beam element} \\
A &= \text{area of element} \\
\sigma_{\dot{u}-\dot{r}} &= \text{standard deviation of the water particle} \\
&\quad\;\; \text{velocity relative to the structure} \\
\eta(\omega) &= \text{Fourier transform of wave amplitude} \\
\dot{\mathbf{r}}(\omega)\text{ and }\ddot{\mathbf{r}}(\omega) &= \text{Fourier transform of }\dot{\mathbf{r}}(t)\text{ and }\ddot{\mathbf{r}}(t)
\end{aligned}$$

The summation is carried out over all actual elements.

The first four terms in Equation (8.28b) yield the hydrodynamic loading on slender structural elements (beam elements) governed by Morison's equation, while the last three terms give the loading on large volume elements governed by the potential theory. It is suggested that for some elements the hydrodynamic loading may be described by the drag term in Morison's equation (i.e. the first two terms in Equation (8.28b)) and the potential theory (i.e. the last three terms in Equation (8.28b)).

It should be emphasized that the true geometry of the structure should be used when evaluating the hydrodynamic forces. For instance if the raft is modelled by a system of beam elements (see Figure 8.4), Equation (8.28b) should be applied to the raft itself with a liberal interpretation of the shape functions.

It is seen that the second and the last term in Equation (8.28b) are in phase with $\dot{\mathbf{r}}$, while the fourth and sixth term are in phase with $\ddot{\mathbf{r}}$. Therefore these terms may in principle be transferred to the left hand side of Equation (8.28a) giving contributions to the damping and mass matrices respectively. It is indicated in Equation (8.28b) that these contributions may depend on the frequency. It is also pointed out in Section 8.4 that the contribution from the soil to the system matrices is frequency dependent. Hence the equation of motion is more readily expressed in the frequency domain by Equation (8.19), introducing

$$
\begin{aligned}
\mathbf{M} &= \tilde{\mathbf{M}} + \sum_j \mathbf{a}_j^T \left\{ \frac{\pi}{4}\rho C_M'' \int_0^l D^2(z)\mathbf{N}^T\mathbf{N}\,dz \right\}_j \mathbf{a}_j \\
&\quad + \sum_j \mathbf{a}_j^T \left\{ \rho \int_A \mathbf{N}^T \operatorname{Re}\left[\phi^{(v)}(\omega)\right]\mathbf{n}^T\mathbf{N}\,dA \right\}_j \mathbf{a}_j \\
\mathbf{C} &= \tilde{\mathbf{C}} + \sum_j \mathbf{a}_j^T \left\{ \tfrac{1}{2}\rho C_D \sqrt{\left(\frac{8}{\pi}\right)} \int_0^l D(z)\mathbf{N}^T\{\sigma_{\dot{u}-\dot{r}}(z)\mathbf{N}\}\,dz \right\}_j \mathbf{a}_j \\
&\quad + \sum_j \mathbf{a}_j^T \left\{ \rho \int_A \omega \mathbf{N}^T \operatorname{Im}\left[\phi^{(v)}(\omega)\right]\mathbf{n}^T\mathbf{N}\,dA \right\}_j \mathbf{a}_j \\
\mathbf{K} &= \tilde{\mathbf{K}} + i\mathbf{D} \\
\mathbf{Q}(\omega) &= \sum_j \mathbf{a}_j^T \left\{ \tfrac{1}{2}\rho C_D \sqrt{\left(\frac{8}{\pi}\right)} \int_0^l D(z)\mathbf{N}^T\{\sigma_{\dot{u}-\dot{r}}(z)\dot{u}(\mathbf{x},\omega)\}\,dz \right\}_j \\
&\quad + \sum_j \mathbf{a}_j^T \left\{ \frac{\pi}{4}\rho C_M \int_0^l D^2(z)\mathbf{N}^T\ddot{u}(\mathbf{x},\omega)\,dz \right\}_j + \mathbf{P}(\omega)\eta(\omega) \\
\mathbf{P}(\omega) &= \sum_j \mathbf{a}_j^T \left\{ \rho \int_A i\omega(\phi^{(w)}(\omega)+\phi^{(d)}(\omega))\mathbf{N}^T\mathbf{n}\,dA \right\}_j
\end{aligned}
\tag{8.29}
$$

The solution of Equation (8.19) is given as

$$\mathbf{r}(\omega) = \mathbf{H}(\omega)\mathbf{Q}(\omega) \qquad (8.30a)$$

where

$$\mathbf{H}(\omega) = [\mathbf{K} - \omega^2\mathbf{M} + i\omega\mathbf{C}]^{-1} \qquad (8.30b)$$

is denoted the complex frequency response function.

The solution can also be expressed in the time domain in terms of the following convolution integral

$$\mathbf{r}(t) = \int_{-\infty}^{\infty} \mathbf{Q}(u)\mathbf{h}(t-u)\,\mathrm{d}u \qquad (8.31)$$

where $\mathbf{h}(t)$ is the impulse response function. It follows that $\mathbf{h}(t)$ and $\mathbf{H}(\omega)$ form a Fourier transform pair.

It is seen from Equation (8.29) that the equation of motion must be solved iteratively due to terms containing $\sigma_{\ddot{u}-\dot{r}}$. However the water particle velocity is usually much greater than the structural velocity, which ensures a fast convergence.

It can be shown, applying Equation (8.31), that the response of linear systems subjected to a zero mean ergodic Gaussian excitation is itself a zero mean ergodic Gaussian process. Therefore if the correlation matrix of the response can be estimated the multivariate distribution of the response is completely defined.

The correlation matrix can be derived from Equation (8.31) and expressed in terms of the impulse response function as

$$\mathbf{R_r}(t_2 - t_1) = \int_{-\infty}^{\infty} \int_{-\infty}^{\infty} \mathbf{h}(t_1 - \tau_1)\mathbf{R_Q}(\tau_2 - \tau_1)\mathbf{h}^T(t_2 - \tau_2)\,\mathrm{d}\tau_1\,\mathrm{d}\tau_2 \qquad (8.32)$$

where $\mathbf{R_Q}(\tau)$ is the correlation matrix of the load.

In the frequency domain the stochastic properties of the response are given by the cross-spectral density matrix $\mathbf{S_r}(\omega)$ defined as the Fourier transform of the correlation matrix. That is

$$\mathbf{S_r}(\omega) = \mathbf{H}(\omega)\mathbf{S_Q}(\omega)\mathbf{H}^{T*}(\omega) \qquad (8.33)$$

where the asterisk denotes the complex conjugate and $\mathbf{S_Q}(\omega)$ is the cross-spectral density of the load. $\mathbf{S_Q}(\omega)$ can be expressed in terms of the wave spectrum by applying Equation (8.29) combined with Equations (8.9) to (8.12). However the general expression is lengthy and is therefore omitted.

The covariance matrix is obtained as

$$\mathrm{cov}\,(\mathbf{rr}^T) = \int_{-\infty}^{\infty} \mathbf{S_r}(\omega)\,\mathrm{d}\omega \qquad (8.34)$$

which is a fundamental quantity in the subsequent statistical analysis.

8.5.2 Solution techniques

Most of the dynamic analyses of offshore platforms reported apply a modal analysis technique in one form or another. The simplest of these techniques uses the undamped normal modes* of the system as generalized coordinates. If the damping matrix satisfies certain orthogonality conditions[9,27] this will decouple the equations of motion.

Introduction of the damped normal modes† as generalized coordinates will however decouple the equations of motion[14,30] without imposing any restrictions on the damping matrix. It is however more complicated and laborious both to calculate and to apply the damped normal modes as compared to the undamped normal modes. The damped normal modes, for an underdamped system, are generally complex and the eigenvalue problem to be solved contains $2n \times 2n$ matrices if n degrees of freedom are involved. The damped normal modes have been used by Foster[15] in the analysis of fixed offshore towers.

The accuracy obtained by using the undamped normal modes may be increased if the coefficients in the diagonalized damping matrix are selected in an optimal manner, by minimizing the expected mean square error of the response. This technique is used by Malhotra and Penzien[31] among others.

The normal mode technique is generally only practically applicable if the system matrices in Equation (8.18) are constant. However Moan *et al.*[32,33] apply the undamped normal modes to a problem with moderate frequency dependent stiffness and damping matrices.

One of the alternatives to the modal technique is the direct frequency response method.[18,24] This method is especially attractive in stochastic analysis if the number of degrees of freedom is not too large.

The basic step of the direct frequency response method is the evaluation of the complex frequency response matrix $\mathbf{H}(\omega)$, i.e. the solution of the equation

$$[\mathbf{K} - \omega^2\mathbf{M} + i\omega\mathbf{C}]\mathbf{H}(\omega) = \mathbf{I} \tag{8.35}$$

where $\mathbf{I}$ is the unit matrix. A complete inversion is normally not necessary. Introducing a vector $\mathbf{e}_s$ containing only zero elements except for the sth element which is unity, it suffices to solve

$$[\mathbf{K} - \omega^2\mathbf{M} + i\omega\mathbf{C}]\mathbf{H}_s(\omega) = \mathbf{e}_s \tag{8.36}$$

---

* The undamped natural frequencies Ω and the undamped normal modes Ψ refer to the following eigenvalue problem

$$(\mathbf{K} - \Omega^2\mathbf{M})\Psi = 0$$

†The damped natural frequencies Ω and the damped normal modes Ψ refer to the following complex eigenvalue problem

$$(\mathbf{K} + i\Omega\mathbf{C} - \Omega^2\mathbf{M})\Psi = 0$$

for the degrees of freedom s that are of interest. $\mathbf{H}_s(\omega)$ is the sth column of $\mathbf{H}(\omega)$.

Equation (8.36) may be expressed in component form as[18]

$$\left[\begin{array}{c|c} \mathbf{K}-\omega^2\mathbf{M} & -\omega\mathbf{C} \\ \hline \omega\mathbf{C} & \mathbf{K}-\omega^2\mathbf{M} \end{array}\right]\left[\begin{array}{c} \mathrm{Re}\,[\mathbf{H}_s(\omega)] \\ \hline \mathrm{Im}\,[\mathbf{H}_s(\omega)] \end{array}\right] = \left[\begin{array}{c} \mathbf{e}_s \\ \hline \mathbf{0} \end{array}\right] \tag{8.37}$$

This system with $2n$ unknowns may be solved by a standard equation solver. It may easily be made symmetric and by rearranging the unknowns (mixing real and imaginary components) a more favourable bandwidth may be obtained. It is however found more convenient and efficient to solve Equation (8.36) directly using complex arithmetic.

It should be emphasized that frequency dependent system matrices cause no problems in the direct frequency response method.

Of other methods the step-by-step integration technique is worth mentioning. It is however not directly applicable in a stochastic analysis based on spectral representation or other statistical averages but must be combined with, for instance, Monte Carlo simulation.

In Table 8.1 an attempt is made to summarize some of the important properties and capabilities of the methods discussed.

8.5.3 Transfer functions

The wave loading on a typical gravity platform is often dominated by the inertia forces. Hence, the wave loading can be evaluated by potential theory employ-

Table 8.1   Comparison of methods commonly used in structural dynamics

Method \ Capability	General viscous damping	General structural damping	Frequency dependent system matrices	Non-linearities[a]	Stochastic analysis
Step-by-step integration	Yes	(No)	No	Yes	No[b]
Modal analysis			(No)	No[c]	Yes
—with undamped normal modes	No	No			
—with undamped normal modes + optimization	(Yes)				
—with damped normal modes	Yes	Yes			
Direct frequency response methods	Yes	Yes	Yes	No[d]	Yes

[a] Refers to non-linearities, which cannot be linearized.
[b] Yes, in connection with Monte Carlo simulation and statistical analysis of the response history.
[c] Yes, successive linearization.[37]
[d] Yes, in connection with the functional and perturbation method.

ing the last three terms in Equation (8.28b). The spectral density of the response can then be expressed as

$$\mathbf{S_r}(\omega) = \mathbf{H}(\omega)\mathbf{P}(\omega)\mathbf{P}^{T*}(\omega)\mathbf{H}^{T*}(\omega)S_\eta(\omega) \tag{8.38}$$

This equation can be simplified recognizing that the response induced by harmonic long-crested waves with unit amplitude is given by

$$\mathbf{r}_h(\omega) = \mathbf{H}(\omega)\mathbf{P}(\omega) \tag{8.39}$$

Equation (8.38) can then be rewritten as

$$\mathbf{S_r}(\omega) = \mathbf{r}_h(\omega)\mathbf{r}_h^{T*}(\omega)S_\eta(\omega) \tag{8.40}$$

The steady state response $\mathbf{r}_h$ to harmonic excitation with unit amplitude is sometimes termed the transfer functions of the system, including both mechanical and hydrodynamic effects. The modulus of some transfer functions derived from $\mathbf{r}_h$ for a typical gravity platform are shown in Figure 8.8 along with the corresponding quasi-static transfer functions. These results are obtained[1] using the model described in Appendix 8A with a soil shear modulus of

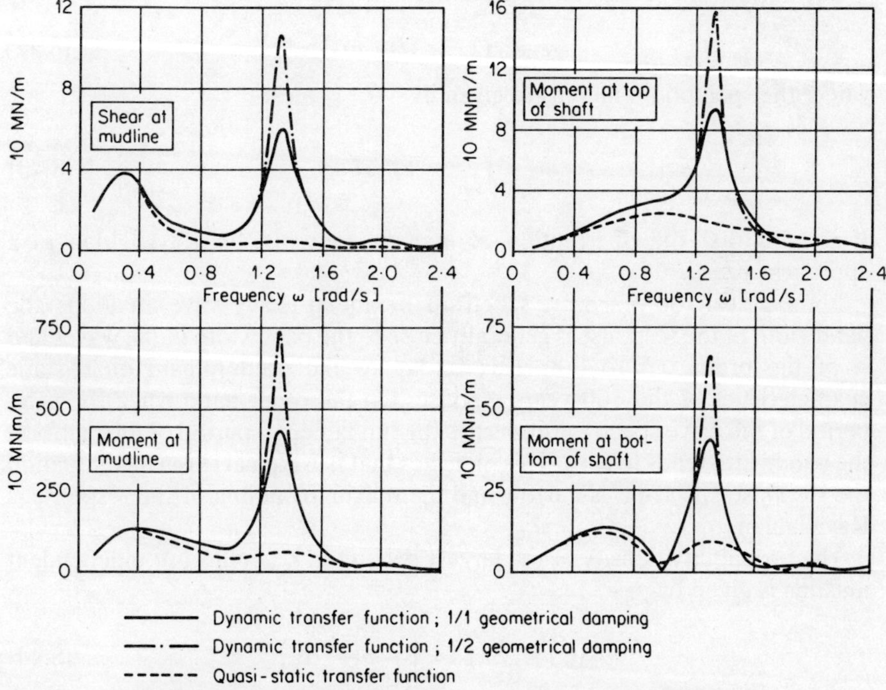

Figure 8.8   Transfer functions for a typical gravity platform

20 MPa. The effect of geometrical damping from the soil is indicated; internal damping in the soil is neglected and the combined structural and hydrodynamic damping is assumed to be 0·5 per cent of critical damping.

8.5.4 Short-term response

Once the response spectrum induced by a given short-term sea state has been obtained, it provides all information necessary to establish the distribution of the dynamic response for a linear (or linearized) system. Of special interest is the distribution of maxima (and minima) and the distribution of the largest maximum.

The probability density of maxima in a Gaussian process is given by [8,28]

$$p(\chi) = \frac{1}{\sqrt{(2\pi)}} \left[\varepsilon \exp\left\{-\tfrac{1}{2}\chi^2/\varepsilon^2\right\} \right.$$
$$\left. + \sqrt{(1-\varepsilon^2)}\chi \exp\left\{-\tfrac{1}{2}\chi^2\right\} \int_{-\infty}^{\chi\sqrt{(1-\varepsilon^2)}/\varepsilon} \exp\left\{-\tfrac{1}{2}t^2\right\} \mathrm{d}t \right] \quad (8.41)$$

Here $\chi = r_m/\sigma_r$ where r_m denotes the maxima of the response r, and where σ_r^2 is the variance of the response (i.e. the diagonal terms of the covariance matrix in Equation (8.34)).

The bandwidth parameter is defined as

$$\varepsilon = \sqrt{(1 - m_2^2/m_0 m_4)} \quad (8.42)$$

where the spectral moments are given by

$$m_k = \int_{-\infty}^{\infty} \omega^k S_r(\omega)\,\mathrm{d}\omega \quad (8.43)$$

It is seen that the distribution of maxima tends to be Rayleighian as ε approaches zero.

For a typical 100 year wave spectrum (or similar heavy wave situations) the bandwidth of the response is generally close to the bandwidth of the waves and is of the order 0·5 to 0·8. This is due to the predominant quasi-static characteristics of the 100-year response. On the other hand when the peak period of the wave spectra approaches the fundamental period of the platform the bandwidth tends to zero. This indicates that in a typical resonance situation the response spectrum is narrow and the maxima/minima of the response is Rayleighian.

The probability density of the largest maximum $r_{\max}$ out of N independent maxima is given by

$$p(\chi_m) = \frac{\mathrm{d}}{\mathrm{d}\chi_m}\left(1 - \int_{\chi_m}^{\infty} p(\chi)\,\mathrm{d}\chi\right)^N \quad (8.44)$$

where $\chi_m = r_{max}/\sigma_r$. It can be shown[8,10] that the expected largest maximum is given by

$$E[r_{max}] = \sigma_r g(\mu\,\Delta T) \tag{8.45}$$

where

$$g(\mu\,\Delta T) = \sqrt{(2\ln(\mu\,\Delta T))} + \frac{0.5772}{\sqrt{(2\ln(\mu\,\Delta T))}} \tag{8.46}$$

is termed the peak function,

$$\mu = \frac{1}{2\pi}\sqrt{\left(\frac{m_2}{m_0}\right)} \tag{8.47}$$

is the zero-upcrossing frequency and ΔT is the length of the time interval considered.

Table 8.2 shows the expected largest maxima of four response quantities of a typical platform (see Appendix 8A) induced by heavy sea states of various durations which approximately simulate the 'once in a 100 year' situation in the North Sea. The shear modulus of the soil is assumed to be 20 MPa and full geometrical damping is assumed, whereas internal (hysteretic) damping in soil is neglected. Combined structural and hydrodynamic damping is set to 0·5 per cent of critical damping

It is seen that the largest shear force at mudline is induced by the ITTC-spectrum which is the spectrum with the greatest energy in the low frequency range. This is plausible, recognizing that the transfer function for the shear force (see Figure 8.8) has a significant peak in the low frequency range, with a period of about 20 s, resulting chiefly from wave forces on the raft. The shear forces induced by the other spectra are more or less similar, mainly because they all have the same peak period, i.e. 15 s. This indicates that it may be questionable to use such a low period when estimating the design shear force at mudline.

It is seen in Figure 8.8 that the transfer function for the moment at top of shaft does not possess any peak in the low frequency range. The expected largest moment here is therefore sensitive to the high frequency tail of the heavy sea spectra. From Table 8.2 it is seen that the Darbyshire–Scott spectrum marked E, which is based on the parameters proposed by Scott[43] yields the most unfavourable moment.

The other two response quantities listed are apparently not affected to the same degree by the shape of the wave spectrum.

The response increases with increasing duration of the sea state. In the North Sea is seems reasonable to assume the duration of the '100 year sea state' of the order 6–10 hrs.[12]

Table 8.2   Expected peak response–extreme sea state

| Sea state | | Response quantity | | | |
Spectrum (see Figure 8.1)	Duration [hrs]	Shear at mudline $[10^2 \text{ MN}]$	Moment at mudline $[10^4 \text{ MNm}]$	Moment bottom shaft $[10^3 \text{ MNm}]$	Moment top shaft $[10^2 \text{ MNm}]$
(A)	3	4·47	1·85	1·51	2·29
ITTC	6	4·69	1·94	1·58	2·40
$\bar{H}_{1/3} = 15$ m	12	4·89	2·02	1·64	2·49
(B)	3	3·64	1·86	1·77	3·10
ISSC	6	3·81	1·94	1·85	3·24
$\bar{H}_{1/3} = 15$ m	12	3·97	2·02	1·93	3·37
$T_m = 15$ s					
(C)	3	3·86	1·88	1·72	2·80
JONSWAP	6	4·04	1·97	1·80	2·92
$\bar{H}_{1/3} = 15$ m	12	4·21	2·05	1·88	3·04
$T_m = 15$ s					
$\gamma = 3·3$					
(D)	3	4·01	1·90	1·70	2·60
JONSWAP	6	4·19	1·99	1·77	2·71
$\bar{H}_{1/3} = 15$ m	12	4·37	2·07	1·85	2·82
$T_m = 15$ s					
$\gamma = 7·0$					
(E)[a]	3	3·66	1·95	1·83	3·68
$\bar{H}_{1/3} = 15$ m	6	3·84	2·04	1·92	3·84
$T_m = 15$ s	12	4·01	2·12	2·00	3·99
(F)[a]	3	3·67	1·85	1·72	3·08
$\bar{H}_{1/3} = 15$ m	6	3·87	1·94	1·80	3·22
$T_m = 15$ s	12	4·04	2·02	1·88	3·35
Deterministic harmonic wave:					
$H = 30$ m		4·50	2·01	1·66	2·08
$T = 15$ s					

[a] Versions of the Darbyshire–Scott spectrum (see Figure 8.1).

Finally Table 8.2 indicates that a deterministic, harmonic or similar regular wave should be applied with care in the design of offshore structures.

The results presented here indicate that in order to be on the safe side the structure should be analysed for various possible wave spectra and designed for the most unfavourable response combination.[35] However such an approach can easily lead to an overconservative design if not applied critically. This may possibly be overcome by treating the shape of the wave spectrum as a stochastic variable. At present however more wave data are needed to provide the foundation of such an approach.

So far this discussion has been devoted to response induced by long crested waves. However the response induced by short crested waves shows roughly the same main trends, but is generally somewhat smaller. This reduction is particularly pronounced for small waves. For a heavy sea state the reduction is normally small as a result of the quasi-static characteristics of the response and long wave lengths compared to the actual structural dimensions.

In heavy sea the response at top of shafts, and of the superstructure in general, is more sensitive to the available amount of damping than the response at the lower part of the structure. However for small resonance waves this is not always the case.

8.5.5 Long-term response

The long-term structural response is obtained by applying the long-term wave model discussed in Section 8.2. Hence, the long-term distribution $F_r(\imath)$ of the response r can be expressed as

$$F_r(\imath) = \mathrm{Pr}\,[r < \imath]$$

$$= \int_{-\infty}^{\imath} \int_{\bar{\theta}} \int_{\bar{T}} \int_{\bar{H}_{1/3}} p_{\bar{H}_{1/3}\bar{T}\bar{\theta}}(\bar{H}_{1/3}, \bar{T}, \bar{\theta})p_{r|\bar{H}_{1/3}\bar{T}\bar{\theta}}(\imath|\bar{H}_{1/3}, \bar{T}, \bar{\theta})\, \mathrm{d}\bar{H}_{1/3}\, \mathrm{d}\bar{T}\, \mathrm{d}\bar{\theta}\, \mathrm{d}\imath$$

$$(8.48)$$

where $p_{\bar{H}_{1/3}\bar{T}\bar{\theta}}(\ \ )$ denotes the joint probability density of significant wave height $\bar{H}_{1/3}$ and zero crossing wave period $\bar{T}$ and mean wave directions $\bar{\theta}$, and $p_{r|\bar{H}_{1/3}\bar{T}\bar{\theta}}(\ \ )$ denotes the conditional probability density of the response induced by a given short-term situation characterized by $\bar{H}_{1/3}$, $\bar{T}$ and $\bar{\theta}$. It should be noted that $p_{r|\bar{H}_{1/3}\bar{T}\bar{\theta}}(\ \ )$ can be interpreted as the probability density of the parent or maxima or largest maximum response, and hence $F_r(\imath)$ as the corresponding long-term distribution.

It is commonly assumed that the waves are unidirectional and travelling in the most unfavourable direction.[33,47] However this may grossly overestimate the response. For North Sea conditions it seems more reasonable to assume $p_{\bar{H}_{1/3}\bar{T}\bar{\theta}}(\ \ )$ rotationally symmetric in $\bar{\theta}$, although this tends to underestimate the response.

In theory the long-term model provides the most logical approach to design. Especially, concerning fatigue problems, it seems to be the only rational approach. However at present it is not always possible to provide sufficient data to establish a reliable long-term model.

Figure 8.9 shows a long-term distribution of maxima of the response of the moment at top of shaft of the structure given in Appendix 8A. The long-term

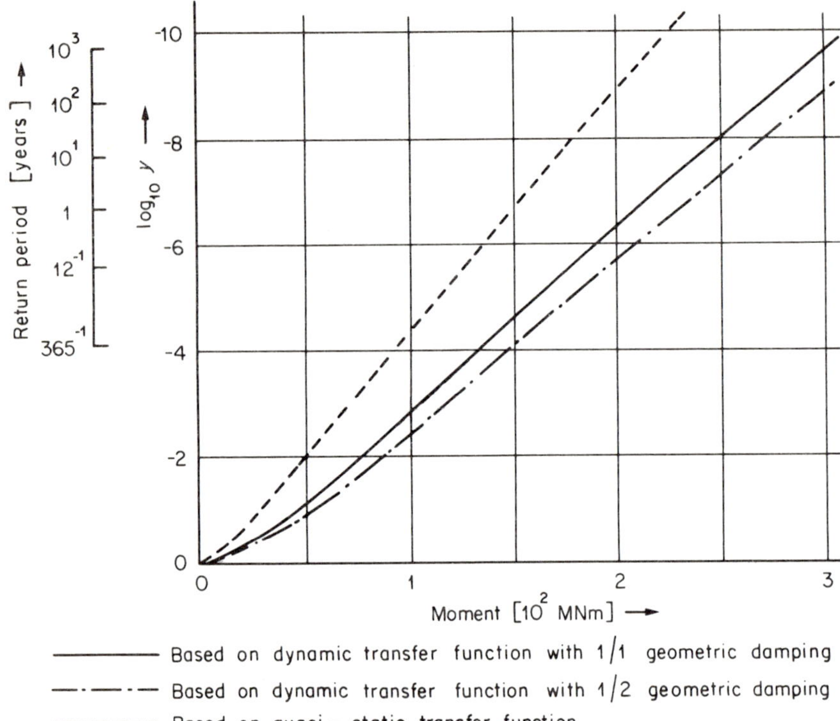

Figure 8.9 Long-term distribution of maxima of moment at top of shaft

distribution of the sea state applied is obtained from visual wave data, registered by the weather ship 'FAMITA',[40] combined with relations between visual and actual wave data given by Nordenström.[38] Further the parameter c in Equation (8.8) is taken equal to 2·4. More details concerning the data used can be found in Appendix 8B and References 33 and 45. All short-term sea states included in the analysis are assumed to be composed of unidirectional long crested waves described by a JONSWAP wave spectrum with a peakedness parameter equal to 3·3. The short-term distributions of the maxima of the response are taken to be Rayleighian.

The long-term distribution of maxima may be combined with statistics of extremes to predict the 'most probable' largest maxima out of N independent maxima, i.e.,

$$\Pr\left[r > r_N\right] = \frac{1}{N} \tag{8.49}$$

In Figure 8.9 the scale with return periods refers to this expression.

Figure 8.9 displays clearly the importance of accounting for the dynamic properties of the structure. It is worth noting that the '100 year' long-term moment is close to the moment induced by a '100 year' short-term situation (see Table 8.2).

8.6 CONCLUSION

It is demonstrated in the paper that methods accounting for the stochastic nature of the wave loading are available for linear dynamic response analysis of framed gravity structures. Wind and current loads that cannot be ignored in the design have been omitted here because of space limitations.

However numerous problems require a better solution before reliable predictions of the structural response can be made. This applies above all to the soil and the sea. The dynamic performance of the soil is at present not well understood, the result being that considerable uncertainties are connected with stiffness and damping in the soil. This refers both to the initial values of these parameters and to their changes during the cyclic loading over the considerable periods of time in question.

The choice of wave spectra for the one-dimensional description of the wavy sea surface is still open to discussion. This is even more the case regarding the spatial distribution of waves and the hydrodynamic transfer functions transforming the wave spectra into wave load spectra. Finally only limited knowledge is available concerning the long-term distribution of sea states, which is essential for the assessment of fatigue life.

In this situation a proper balance between computational accuracy and physical realities must be found. An ideal aim might be that the final accumulated error in the behaviour prediction should not be influenced significantly by errors introduced by the analytical work and that a uniform safety should be obtained in the various parts of the structure.

The final measure of analytical accuracy is obtained by relating observed prototype behaviour to analytical predictions. It is therefore of the greatest importance that prototype data, currently registered for instance in the North Sea, are made generally available for research and development.

Acknowledgement

The first author of this Chapter wants to express his gratitude to doctoral student E. K. Smith for suggestions and assistance.

APPENDIX 8A: EXAMPLE STRUCTURE

All numerical results presented in this paper are obtained using a structural model with the following characteristics (numerical results for this model are also presented in References 2,22,46):

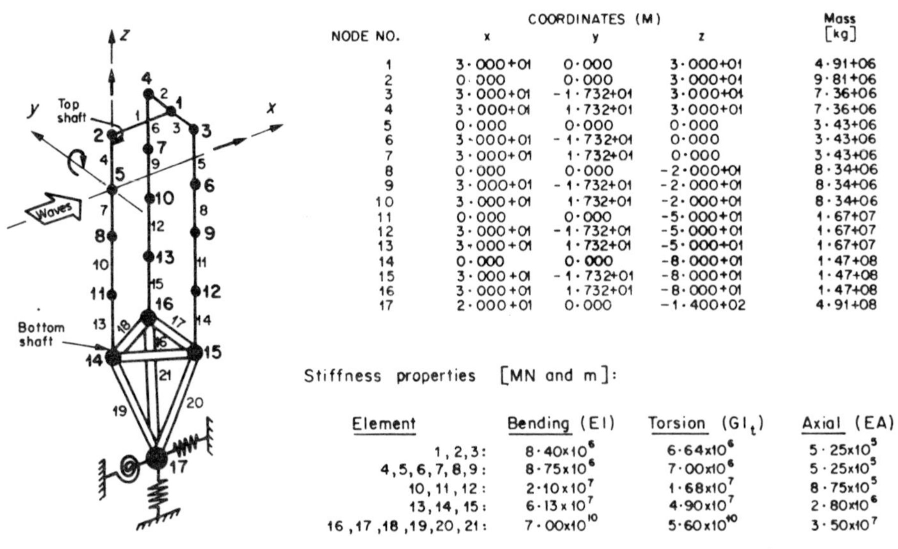

NODE NO.	COORDINATES (M) x	y	z	Mass [kg]
1	$3 \cdot 000 + 01$	$0 \cdot 000$	$3 \cdot 000 + 01$	$4 \cdot 91 + 06$
2	$0 \cdot 000$	$0 \cdot 000$	$3 \cdot 000 + 01$	$9 \cdot 81 + 06$
3	$3 \cdot 000 + 01$	$-1 \cdot 732 + 01$	$3 \cdot 000 + 01$	$7 \cdot 36 + 06$
4	$3 \cdot 000 + 01$	$1 \cdot 732 + 01$	$3 \cdot 000 + 01$	$7 \cdot 36 + 06$
5	$0 \cdot 000$	$0 \cdot 000$	$0 \cdot 000$	$3 \cdot 43 + 06$
6	$3 \cdot 000 + 01$	$-1 \cdot 732 + 01$	$0 \cdot 000$	$3 \cdot 43 + 06$
7	$3 \cdot 000 + 01$	$1 \cdot 732 + 01$	$0 \cdot 000$	$3 \cdot 43 + 06$
8	$0 \cdot 000$	$0 \cdot 000$	$-2 \cdot 000 + 01$	$8 \cdot 34 + 06$
9	$3 \cdot 000 + 01$	$-1 \cdot 732 + 01$	$-2 \cdot 000 + 01$	$8 \cdot 34 + 06$
10	$3 \cdot 000 + 01$	$1 \cdot 732 + 01$	$-2 \cdot 000 + 01$	$8 \cdot 34 + 06$
11	$0 \cdot 000$	$0 \cdot 000$	$-5 \cdot 000 + 01$	$1 \cdot 67 + 07$
12	$3 \cdot 000 + 01$	$-1 \cdot 732 + 01$	$-5 \cdot 000 + 01$	$1 \cdot 67 + 07$
13	$3 \cdot 000 + 01$	$1 \cdot 732 + 01$	$-5 \cdot 000 + 01$	$1 \cdot 67 + 07$
14	$0 \cdot 000$	$0 \cdot 000$	$-8 \cdot 000 + 01$	$1 \cdot 47 + 08$
15	$3 \cdot 000 + 01$	$-1 \cdot 732 + 01$	$-8 \cdot 000 + 01$	$1 \cdot 47 + 08$
16	$3 \cdot 000 + 01$	$1 \cdot 732 + 01$	$-8 \cdot 000 + 01$	$1 \cdot 47 + 08$
17	$2 \cdot 000 + 01$	$0 \cdot 000$	$-1 \cdot 400 + 02$	$4 \cdot 91 + 08$

Stiffness properties   [MN and m]:

Element	Bending (EI)	Torsion (GI_t)	Axial (EA)
1, 2, 3:	$8 \cdot 40 \times 10^{6}$	$6 \cdot 64 \times 10^{6}$	$5 \cdot 25 \times 10^{5}$
4, 5, 6, 7, 8, 9:	$8 \cdot 75 \times 10^{6}$	$7 \cdot 00 \times 10^{6}$	$5 \cdot 25 \times 10^{5}$
10, 11, 12:	$2 \cdot 10 \times 10^{7}$	$1 \cdot 68 \times 10^{7}$	$8 \cdot 75 \times 10^{5}$
13, 14, 15:	$6 \cdot 13 \times 10^{7}$	$4 \cdot 90 \times 10^{7}$	$2 \cdot 80 \times 10^{6}$
16, 17, 18, 19, 20, 21:	$7 \cdot 00 \times 10^{10}$	$5 \cdot 60 \times 10^{10}$	$3 \cdot 50 \times 10^{7}$

The raft has an equivalent radius of 45 m and a height of 60 m. The three shafts are identical with an outer diameter of 20 m at the top of the raft, varying linearly to 12 m at a point 20 m below mean water surface. Poisson's ratio for soil is put equal to 0·5 throughout.

The following assumptions should be observed:

The model is restricted to a plane motion (in the xz-plane).

The exciting wave forces on the shafts are determined by the strip theory using results from two-dimensional potential theory for a single cylindrical pile. Drag forces are neglected.

Wave forces on the raft are obtained by empirical formulas determined experimentally.[17]

Mutual interaction and interference effects are not accounted for.

Added mass is taken to be constant (equal to the mass of displaced fluid).

APPENDIX 8B: LONG-TERM DISTRIBUTION OF WAVES

The conditional Weibull distribution of visual wave height is given by

$$F_{H_v|T_v}(H|H_v) = 1 - \exp\{-((H-H_0)/(H_c-H_0))^v\}$$

where H_v denotes the visual wave height and T_v denotes the visual wave period. The wave data used in the present analysis are given below.

Weibull parameters: Visual wave data from the weather ship "FAMITA"
(57°30' N, 3° E)

$T_v[s]^a$	$H_0[m]$	$H_c[m]$	v	Frequency of occurrence
4·5	1·2	1·3	0·62	0·3764
6·5	1·2	1·83	1·04	0·3430
8·5	2·0	2·80	0·99	0·1989
10·5	1·5	3·30	1·32	0·0626
12·5	1·5	3·95	1·20	0·0164
14·5	0·0	4·50	3·15	0·0032

[a] Midpoint of visual period interval.

The connection between the visual and significant wave height is given by

$$\bar{H}_{1/3} = 1 \cdot 68 H_v^{0.75}$$

The relationship between the average wave period $\tilde{T}$ and the visual wave period is assumed to be given by a log-normal distribution with the following characteristics

$$E[\tilde{T}|T_v] = 2 \cdot 83 T_v^{0.44}$$

$$\mathrm{Var}\,[\tilde{T}|T_v] = 1 \cdot 0 \text{ second}^2$$

The correlation between $\bar{H}_{1/3}$ and $\tilde{T}$ is introduced according to Equation (8.8).

Further details concerning the long-term model may be found in References 32, 33, 38, and 45.

REFERENCES

1. Bell, K., Sigbjörnsson, R., and Smith, E. K. (1975). *CONVIB—A computer program for dynamic analysis of gravity type off-shore platforms (and arbitrary frame structures)*, SINTEF, Report STF71 A75042, Trondheim.
2. Bell, K., Hansteen, O. E., Larsen, P. K., and Smith, E. K. (1976). 'Analysis of a wave-structure-soil system, case study of a gravity platform', *The First International Conference on the Behaviour of Offshore Structures, BOSS-76*, Trondheim.
3. Berge, B. and Penzien, J. (1974). 'Three-dimensional stochastic response of offshore towers to wave forces', *Offshore Technology Conference, Houston*.

4. Borgman, L. E. (1967). 'Spectral analysis of ocean wave forces on piling', *J. Waterways and Harbors Div.*, ASCE, **93**, No. WW2.

5. Borgman, L. E. (1967). 'Random hydrodynamic forces on objects', *Ann. Math. Statistics*, **38**.

6. Borgman, L. E. (1972). 'Statistical models for ocean waves and wave forces', *Advances in Hydroscience*, Academic Press, Vol. 8.

7. Bretschneider, C. (1975). 'Envelope spectra', *Int. Conf. on Port and Ocean Eng. under Arctic Conditions*, Fairbanks (to be published).

8. Cartwright, D. E. and Longuet-Higgins, M. S. (1956). 'The statistical distribution of the maxima of a random function', *Proceedings of the Royal Society of London, Series A*, **237**.

9. Caughey, T. K. and O'Kelly, M. E. J. (1965). 'Classical normal modes in damped linear dynamic systems', *J. Applied Mechanics*, **32**.

10. Clough, R. W. and Penzien, J. (1975). *Dynamics of Structures*, McGraw-Hill.

11. Darbyshire, J. (1959). 'The spectra of coastal waves', *Deutsche Hydrographische Zeitschrift*, **12**, No. 1.

12. *Environmental conditions of the Norwegian Continental Shelf (special emphasis on engineering applications)*. (1976). Norwegian Petroleum Directorate, Stavanger.

13. Faltinsen, O. M. and Michelsen, F. C. (1974). 'Motions of large structures in waves at zero Froude number', *International Symposium on the Dynamics of Marine Vehicles and Structures in Waves*, London.

14. Foss, K. A. (1958). 'Co-cordinates which uncouple the equations of motion of damped linear dynamic systems', *J. Applied Mechanics*, **25**.

15. Foster, E. T. (1970). 'Model for nonlinear dynamics of offshore towers', *J. Engineering Mech. Div.*, ASCE, **96**, No. EM1.

16. Garrison, C. J. and Chow, P. Y. (1972). 'Wave forces on submerged bodies', *J. Waterways and Harbors Div.*, ASCE, **98**, No. WW2.

17. Hafskjold, P. S., Törum, A., and Eie, J. (1973). 'Submerged offshore concrete tanks', *The Eighth International Navigation Congress of PIANC*, Ottawa.

18. Handa, K. and Clarkson, B. L. (1971). 'Application of finite element method to the dynamic analysis of tall structures', *J. Sound and Vibration*, **18**, No. 3.

19. Hasselmann, K. *et al.* (1973). 'Measurements of wind-wave growth and swell decay during the Joint North Sea Wave Project', *Ergänzungsheft zur Deutschen Hydrographischen Zeitschrift*, Reihe A (8°), Nr. 12.

20. Hogben, N. and Standing, R. G. (1974). 'Wave loads on large bodies', *International Symposium on the Dynamics of Marine Vehicles and Structures in Waves*, London.

21. Holand, I. (1975). Discussion of 'Structural dynamics of offshore platforms' by R. E. Taylor, *Off-shore Structures*, Proc. of the conference held in London, Oct., 1974, the Institution of Civil Engineers.

22. Holand, I. (1976). *Svingning av marine konstruksjoner, stokastisk respons*, SINTEF Report STF71 A76001, Trondheim.

23. Holand, I. (1976). *Dynamic and static Young's moduli for prestressed concrete*, SINTEF Report STF71 A75037, Trondheim.

24. Hurty, W. C. and Rubinstein, M. F. (1964). *Dynamics of Structures*, Prentice-Hall.

25. Kausel, E. and Roësset, J. M. (1975). 'Dynamic stiffness of circular foundations', *J. Engineering Mech. Div.*, ASCE, **101**, No. EM6.

26. Kinsman, B. (1975). *Wind-waves*, Prentice-Hall.

27. Lin, Y. K. (1966). Discussion of 'Classical normal modes in damped linear dynamic system' by Caughey, T. K. and O'Kelly, M. E. J., *J. Applied Mechanics*, **33**.

28. Lin, Y. K. (1967). *Probabilistic Theory of Structural Dynamics*, McGraw-Hill.

29. MacCamy, R. C. and Fuchs, R. A. (1954). 'Wave forces on piles: A diffraction theory', *Beach Board Tech. Mem.*, No. 69.
30. Meirowitz, L. (1967). *Analytical Methods in Vibrations*, Macmillan.
31. Malhotra, A. K. and Penzien, J. (1970). 'Nondeterministic analysis of offshore structures', *J. Engineering Mech. Div.*, ASCE, **96**, No. EM6.
32. Moan, T., Haver, S., and Vinje, T. (1975). 'Stochastic dynamic response analysis of offshore platforms; with particular reference to gravity-type platforms', *Offshore Technology Conference*, Houston.
33. Moan, T., Syvertsen, K., and Haver, S. (1976). *Stochastic Dynamic Response Analysis of Gravity Platforms*, Division of Ship Structures, The University of Trondheim, Report SK/M 33.
34. Morison, J. R., Johnson, J. W., O'Brien, M. P., and Schaaf, S. A. (1950), 'The forces exerted by surface waves on piles', *J. Petroleum Technology*, **II**, No. 12, Dec. 1950.
35. Moshagen, H., Rye, H., Sigbjörnsson, R., Törum, A., and Houmb, O. G. (1976). *Environmental loads for fixed structures on the Norwegian continental shelf*, SINTEF Report STF71 F76018.
36. Neuman, G. and Pierson, W. J. (1966). *Principles of Physical Oceanography*, Prentice-Hall.
37. Nickell, R. E. (1976). 'Nonlinear dynamics by mode superposition', *Computer Methods in Applied Mechanics and Engineering*, **7**.
38. Nordenström, N. (1972). *Methods for predicting long-term distribution of wave loads and probability of failure for ships*, Det norske Veritas, Report 71-2-S.
39. Ochi, M. K. and Wang, S. (1976). 'Prediction of extreme wave-induced loads on ocean structures', *The First International Conference on the Behaviour of Off-shore Structures*, *BOSS-76*, Trondheim.
40. Pedersen, B., Egeland, O., and Langenfeldt, J. N. (1973). 'Calculation of long-term values for motions and structural response of mobile drilling rigs', *Offshore Technology Conference*, Houston.
41. Phillips, O. M. (1958). 'The equilibrium range in the spectrum of wind-generated waves', *J. Fluid Mechanics*, **4**, No. 4.
42. *Regulations for the structural design of fixed structures on the Norwegian Continental Shelves* (unofficial translation), Norwegian Petroleum Directorate, Stavanger, 1976.
43. Scott, J. R. (1965). 'A sea spectrum for model tests and long-term ship prediction', *J. Ship Research*, **9**, No. 3.
44. Sigbjörnsson, R. (1974). *Stochastic dynamics of offshore structures*, SINTEF, Trondheim.
45. Sigbjörnsson, R. and Smith, E. K. (1975). *Statfjord platform "A": Stochastic dynamics*, SINTEF Report STF71 F75027, Trondheim.
46. Sigbjörnsson, R. and Smith, E. K. *Wave-induced vibrations of gravity platforms, a linear approach*, SINTEF Report to appear.
47. Tickell, R. G., Burrows, R., and Holmes, P. (1976). 'Long-term wave loading on offshore structures', Proc. *Institution of Civil Engineers*, **61**, part 2.
48. Tung, C. C. (1975). 'Statistical properties of wave forces', *J. Engineering Mech. Div.*, ASCE, **101**, No. EM1.
49. Veletsos, A. S. and Wei, Y. T. (1971). 'Lateral and rocking vibrations of footings', *J. Soil Mech. and Foundations Div.*, ASCE, **97**, No. SM9.
50. Veletsos, A. S. and Verbic, B. (1973). 'Vibration of viscoelastic foundations', *Earthquake Engineering and Structural Dynamics*, **2**.

51. Veletsos, A. S. and Verbic, B. (1974). 'Basic response functions for elastic foundations', *J. Engineering Mech. Div.*, ASCE, **100**, No. EM2.
52. Whitman, R. V. (1976). 'Soil-platform interaction', *The First International Conference on the Behaviour of Off-shore Structures, BOSS-76*, Trondheim.
53. Wiegel, R. L. (1974). 'Ocean wave spectra, eddies, and structural response', *Flow-Induced Structural Vibrations*, ed. E. Naudascher, Springer-Verlag.
54. Wiegel, R. L. (1975). 'Design of offshore structures using wave spectra', *Oceanology International 75*, Brighton (to be published).
55. Öner, M. and Janbu, N. (1975). 'Dynamic soil-structure interaction in offshore storage tanks', *Proc. Int. Conf. on Soil Mechanics and Foundation Engineering*, Istambul (to be published).

Chapter 9

Spectral Fatigue Analysis for Offshore Structures

D. J. Cronin, P. S. Godfrey, P. M. Hook and T. A. Wyatt

9.1 INTRODUCTION

Offshore structures of all types are subject to cyclic loading which cause time varying stresses in the material. The ability of a material to withstand these stresses decreases with accumulating number of fluctuations and may lead to the development and propagation of cracks.

The fatigue damage sustained at a point in a structure will depend on the complete stress history during its lifetime. The calculation of this stress history and its effect on the material is an extremely complex task. The irregular nature of the sea, size of the structure, complexity of the welded details and possible dynamic effects all contribute to the problem.

Due to the difficulties and expense of inspection and repair of offshore structures it is extremely important to the operator that a satisfactory fatigue performance is achieved by his installations over their working life. For this reason many papers have been published in recent years relating to the fatigue analysis of offshore structures, however none of these has offered an entirely satisfactory solution. The paper presented here is intended to present and review the various techniques available, to highlight the areas of uncertainty which arise during the analysis and to suggest means of utilizing the best available practice.

9.2 GENERAL DESCRIPTION OF FATIGUE ANALYSIS

From theoretical and experimental investigations of cyclically loaded structures the following parameters have been found to be relevant to fatigue behaviour:

  (i) Stress cycle amplitude (stress range)
  (ii) Ratio of minimum applied stress
  (iii) Residual stress
  (iv) Number of stress cycles
  (v) Corrosive environment

(vi) Material defects

(vii) Nature of stress cycle history.

In full size welded structures the behaviour may be governed by the fabrication procedure used. In as welded structures residual stresses exist of magnitude up to the yield stress of the material. Thus under applied cyclic deformation the actual stress cycle to which the material adjacent to the weld is subjected varies from yield stress downwards, regardless of the nominal stress. For example, if the nominal stress cycle is $+S_1$ to $-S_2$, giving a total range of $S_1 + S_2$, the actual stress cycle will vary from S_y to $S_y - (S_1 + S_2)$.

It is therefore assumed that parameter (ii) which was included in the tabular presentation of BS 153: 1972 is not regarded as a significant factor for offshore structures and that the propagation of cracks may be expressed in terms of stress range and number of cycles only. Parameters (v), (vi) and (vii) are allowed for by using cycles to failure curves derived under appropriate laboratory conditions.

Since it is the repeated application of a stress range that causes fatigue damage in offshore structures, the extreme stresses associated with storm loadings are not critical. Instead it is the cumulative effect of medium height waves with high occurrence that do most damage. This is illustrated in Figure 9.1 which shows a typical damage density distribution with wave height (after Aruliah,[20] and Williams and Rinne[1]).

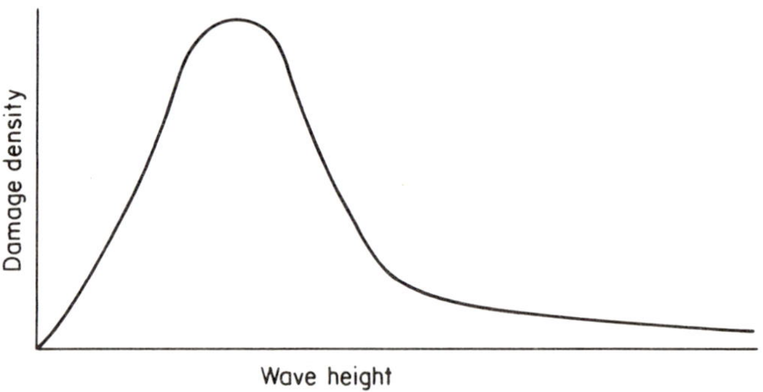

Figure 9.1

The use of the word cumulative implies that the analysis must cover the entire working life of the structure. This represents a formidable task in computation, and the determination of the physical parameters which govern the structural response to its environmental loadings also creates difficulties for the analyst, as will be seen later.

To cope with these problems various calculation procedures have been developed. However all the methods are essentially different ways of approaching (or sometimes avoiding) the same problems and consist of the same basic stages as shown on Figure 9.2.

Stage 1   Wave environmental model

Stage 2   Hydrodynamic loading model

Stage 3   Structural model

Stage 4   Joint stress model

Stage 5   Fatigue damage model

Figure 9.2

9.2.1   Wave environmental model

In reality the wave environment offshore is a random process dependent on wind speed, water depth, mudline characteristics and various other factors. This means that any attempt to model the sea must be simplified if the number of variables is to be kept to a reasonable level.

In practice the parameters used to define the state of the sea at any time are the wave heights and periods existing at that time. These are usually measured by means of a moored accelerometer buoy or pressure transducers on the sea bed. Figure 9.3 shows typical wave record taken from Draper.[2]

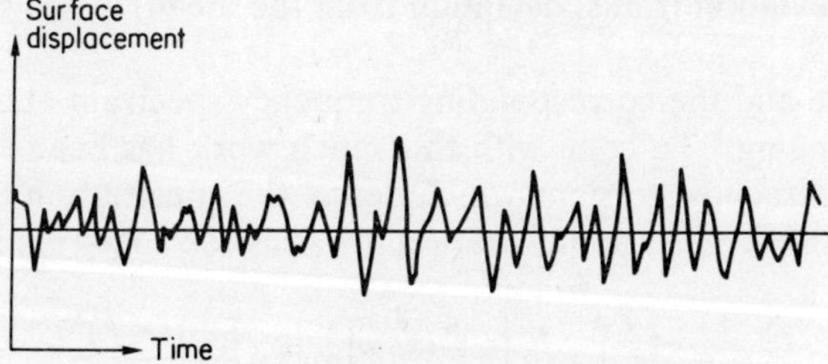

Figure 9.3

The simplest way to use this data in a fatigue analysis is simply to count the heights and associated zero crossing periods from the record and to produce a scatter diagram (Figure 9.4).

Each wave height/period combination can now be applied individually to the structural model and the effect summed to give the expected fatigue life of the structure. This will be referred to as the discrete wave method of analysis.

H \ T	4·5	5·0	5·5	6·0	6·5	7·0	7·5	8·0	8·5	9·0	9·5	10.0	10.5
2				3	2								
3	1	8	4	3	5	5	3						
4	1	12	10	20	12	12	9	2					
5		1	12	16	33	18	10	5					
6			5	13	22	19	14	3	5	1			
7			4	14	18	15	7	14	3	2			
8				6	20	18	16	9	7		1		
9				1	15	25	10	6	5	4			

Figure 9.4

The alternative is to recognize the random nature of Figure 9.3 and to model the process statistically. The time history of sea surface elevation over a period of a few hours (three to six hours is probably the optimum duration) has been shown to have a Gaussian Probability distribution,[3] and to be well represented as a stationary random process with a rather narrow banded power spectrum. The power spectral density function can be thought of as a measure of the contribution of the various frequency components to the overall wave record as quantified by the mean square value.

The most common practical measure of a sea state is the significant wave height H_s, which is generally found to coincide with visual estimates. This is analytically defined as the average height of the highest third of the waves. If the sea elevation is a narrow band Gaussian process then $H_s = 4\sigma_\eta$ where σ_η is the standard deviation (r.m.s. deviation from the mean) of sea surface elevation.[3]

The sea state and the corresponding frequency spectrum at a location are constantly changing.[4] To cope with this much work has been carried out to determine algebraic expressions which define the spectrum in terms of relatively few parameters. As an illustration the Jonswap spectrum is as follows:

$$\frac{fS^{\eta\eta}}{\sigma_\eta^2} = 5\left(\frac{\hat{f}}{f}\right)^4 \exp\left[\frac{-5}{4}\left(\frac{\hat{f}}{f}\right)^4\right] \gamma \exp\left[\frac{-(f-\hat{f})^2}{2B^2\hat{f}^2}\right] \tag{9.1}$$

where

$$S^{\eta\eta} = \text{spectral density of surface elevation}$$
$$\sigma_\eta = \text{standard deviation of surface elevation}$$
$$f = \text{frequency}$$
$$\hat{f} = \text{peak energy frequency} = 1/T_{\text{DOM}}$$
$$g = \text{gravitational constant}$$

$$H_s = \text{significant wave height}$$
$$\alpha = \text{constant}$$
$$\gamma = \text{constant}$$
$$B = \text{constant} = B_A \text{ for } f \le \hat{f}$$
$$= B_B \text{ for } f > \hat{f}$$

By appropriate choice of α, γ, B_A, B_B this spectrum has a high degree of generality; α relates the dominant period to H_s and the 'peakiness' of the spectrum is controlled by γ, B_A and B_B. Thus the model spectrum can be fitted to coincide closely with that measured.

Putting $\gamma = 1$ the expression transforms to the somewhat simpler Pierson–Moskowitz form as shown below. The spectrum has been evaluated for H_s of 1, 2 and 3m on Figure 9.5 for illustrative purposes.

$$\frac{fS^{\eta\eta}}{\sigma_\eta^2} = 5\left(\frac{\hat{f}}{f}\right)^4 \exp\left[\frac{-5}{4}\left(\frac{\hat{f}}{f}\right)^4\right] \qquad f = \frac{1}{2\pi}\left[\frac{\alpha g^2}{5\sigma_\eta^2}\right]^{1/4} \qquad (9.2)$$

where

$$\hat{f} = \text{dominant frequency}$$
$$\sigma_\eta = \frac{H_s}{4}$$
$$\alpha = \text{constant which relates } H_s \text{ to } \hat{f}$$

As can be seen from Figure 9.5 the 'effective' bandwidth (the bandwidth given by a rectangle height $\hat{S}^{\eta\eta}$ and with area equal to the area under the spectrum) is small, which supports the narrow band assumption.

The area under the spectrum is equal to the variance of the quantity described by the spectrum, the square root of which is the standard deviation of the quantity. In the spectral method, spectra of sea surface elevation are converted by multiplication by 'transfer functions' into spectra of stress; the variance or the standard deviation of stress in each sea state can then simply be obtained from the area under the stress response spectrum. The reciprocal of the frequency giving the peak spectral ordinate can be referred to as the dominant period, T_{DOM}. A measure of periodicity which can more directly be made in practise is the average zero crossing period (i.e. duration divided by the number of times the record crosses the mean value with positive slope), T_z. For the Pierson–Moskowitz spectrum:

$$T_{\text{DOM}} = 1{\cdot}4 \times T_z$$

Changes in the sea state may be presented as a scatter diagram of H_s and T_z similar to Figure 9.4. Ideally one should carry out the analysis for each combination of H_s and T_z. In practise the effect is relatively small and usually an average T_z is used for each H_s.

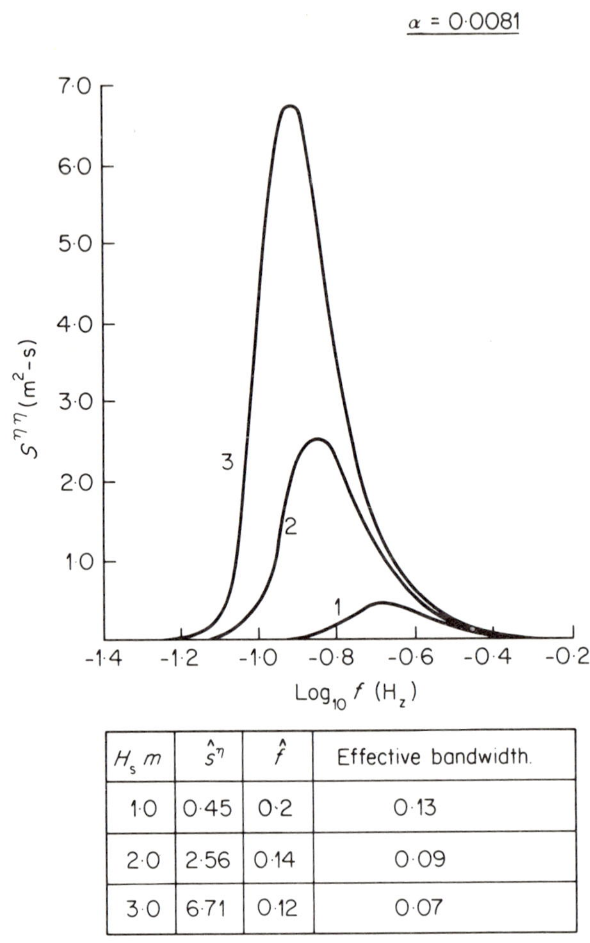

H_s m	$\hat{s}^{\eta}$	$\hat{f}$	Effective bandwidth.
1·0	0·45	0·2	0·13
2·0	2·56	0·14	0·09
3·0	6·71	0·12	0·07

Figure 9.5

The motions of individual particles within the wave are related to the surface displacement by an appropriate wave theory. The various wave theories are all solutions to the differential equations of fluid flow. The first order (Linear or Airy) solution gives a sinusoidal variation of sea surface and the particle motions are linearly related to wave height. Solutions such as the Stokes fifth order give a more realistic wave profile with greater curvature at the crest than the trough and crest displacement from mean sea level greater than the corresponding trough displacement. Considerable complications arise in spectral methods however, when non-linear wave forms are introduced.

9.2.2 Wave/Current loading model

Stress fluctuations may be caused by several different effects of the fluid motion around an offshore structure, for example:

(a) Wave loading
(b) Current loading
(c) Vortex shedding loads
(d) Wave slam loading
(e) Buoyancy loading

All of these can cause fatigue damage in structural members but it will be considered in this paper that topics (c), (d) and (e) constitute special cases of loading which although sometimes of governing importance are not directly relevant to the majority of well designed offshore structures. Current loading can be included in the analysis[5] but for simplicity will not be considered here.

The modelling of wave loading is usually carried out in two basic ways, the Morison Equation method and the Wave Diffraction method. The Morison Equation is applicable to members where the diameter is small compared to the wave length. Wave Diffraction is used when the diameter is large compared to the wave length and reflection of the wave becomes significant.

9.2.3 Structural model

A jacket type structure is generally modelled as a space frame of three-dimensional beam elements with every member of the structure modelled by one or more elements. This modelling is familiar to the engineer and is the basis of many commercial computer packages. For full analysis the dynamic effects will be included.

The dynamic equations of motion for such a system may be written as follows:

$$[M]\{\ddot{x}\}+[D]\{\dot{x}\}+[K]\{x\}=[P] \tag{9.3}$$

where

$\{x\}$ is the displacement vector
$[M]$ is the mass matrix
$[K]$ is the stiffness matrix
$[D]$ is the damping matrix
$[P]$ is the load vector which varies with time.

If a frequency of the loading is close to a structure natural frequency then the responses will be amplified due to resonant effects. Due to a large number of degrees of freedom in the 3D beam element model of a typical structure the solution of the dynamic equations given by Equation (9.3) is not carried out. Instead the static equations for the model are written as follows

$$[K]\{x\} = [P] \tag{9.4}$$

where $[P]$ is now the static loading corresponding to a particular position of the wave. If the structure is such that dynamic effects may be significant then it is allowed for in an approximate fashion using dynamic load factors.

However if the number of degrees of freedom of the structure are greatly reduced then the solution of the dynamic equations for 'lumped' system is practical. This procedure is described in more detail later.

9.2.4 Stress model

The results from the previous stages are curves of stress range (or standard deviation of stress range) against wave height for each wave direction and member to be analysed.

In jackets the analyst is primarily interested in the welded joints between the members as these are the weakest points as far as fatigue is concerned. Two modes of failure can occur in tubular joints, cracking at the toe of the weld joining the brace to the chord (brace fatigue) and cracking in the wall of the chord itself (punching shear fatigue).

When considering brace fatigue the geometry of the joint becomes important since stress concentrations will occur due to the non-uniform stiffness of the chord wall as 'seen' by the brace. Concentrations also occur at the toe of the weld due to the material discontinuities in this area.

Tests have been carried out on tubular joints to determine the stress concentrations occurring in a given joint geometry. It has been found from the tests that it is very difficult to predict the concentration factor using simple empirical equations and if an accurate value is required either full scale tests or a finite-element analysis should be carried out.

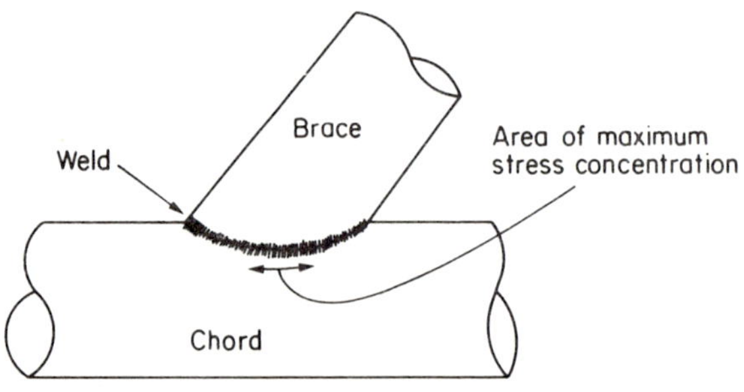

Figure 9.6

In practise values of stress concentration factors of between 1·5 and 3·0 are used in the analysis of brace fatigue in K and T joints, depending on the joint geometry.

In the calculation of punching shear stress, stress concentration is taken into account in the S/N curves. To convert the nominal stress in the brace to shear stress in the chord wall a simple equation based on the ratio of chord wall thickness to brace wall thickness and the intersection angle is used.[6,7]

Stress range vs. Number of Cycles to Failure (S/N) curves have been presented for numerous types of joints using laboratory and full scale tests. Various organizations (American Welding Society, Lloyds, Welding Institute etc.) have produced curves which often differ due to different philosophies of data handling and presentation. Care should be taken when using a curve that it represents what is required; Gurney and Maddox[8] have published curves on the basis of the mean values of test results, which imply a 50 per cent survival rate, and have also published 'mean minus two standard deviations' curves which imply a 95 per cent survival rate. Thus it must be decided by the designer which probability of survival is required before a curve is chosen.

In unwelded materials an 'endurance limit' has been found to exist. This is the stress range below which no crack initiation occurs. Once a crack is formed it will propagate at stress levels lower than the endurance limit. However there is a lower stress range at which no further propagation will occur, this is known as the 'stress cut off'.

In welded material microcracks already exist due to the welding process and therefore the entire damage history is one of crack propagation. The curves for welded details thus have a lower stress cut off than those for unwelded details. The environment also affects the cut off, differing values being used depending on whether the joint is in air or sea water.

9.2.5   Fatigue damage model

As mentioned above it has been found for welded steel structures that fatigue failure is mostly dependent on the stress ranges and number of cycles of each stress range applied.

In 1945 Miner[9] following work by Palmgren[10] developed the hypothesis that the fraction of 'fatigue damage' at a given time could be formulated as follows:

$$\frac{n_i}{N_i} = \text{fraction for damage } i$$

when

$$n_i = \text{number of cycles applied at stress range } S_i$$
$$N_i = \text{number of cycles to failure at stress range } S_i$$

Clearly if a constant single stress range S_i is applied for N_i cycles the fraction will be unity. The assumption is now made that if a number of different stress ranges S_i, $i = 1 \rightarrow r$ are applied, that at failure:

$$\sum_{i=1}^{i=r} \left(\frac{n_i}{N_i} \right) = 1 \cdot 0 \tag{9.5}$$

This relationship can now be used to sum the cumulative effects of the stress ranges applied to the structure.

In the discrete wave approach individual waves are applied to the structure and the associated stress ranges obtained for each member end. The curve of stress range versus wave height is then split into regimes of stress range ΔS_i (where $\Delta S_i = S_i - S_{i-1}$) represented by the average stress range S_{iav} (where $S_{iav} = (S_i + S_{i-1})/2$). From the wave exceedance data the occurrence of this stress range n_{iav} can be obtained as well as the number of cycles to failure N_{iav} from the S/N curve. The damage summation is then carried out for all the regimes of stress range as described above.

In the spectral approach slightly different processes are used but in essence the spectral and discrete methods of summation are identical. It can be said that 'stress cycles are merely transformed wave cycles.'[11] Where the transformation is linear, the stress peaks will conform to a probability distribution that can be derived theoretically from the spectral bandwidth[12]; in practise the bandwidth is sufficiently narrow to permit the fatigue damage summation to be based on the distribution of peaks occurring in a narrow band Gaussian process, namely the Rayleigh distribution (Figure 9.7) which is fully defined by the rms value of the response process σ_s. Where the transformation is non-linear (for example where drag forces are important, or higher-order wave theory has to be used) the probability of occurrence of very high response levels may differ greatly from the Rayleigh distribution. Reference can be made to the work of Holmes and Tickell,[13] but as the greatest contribution to fatigue damage is made by waves of moderate height it is common to retain the simple Rayleigh solution.

It will be seen from the specimen wave record (Figure 9.3) that the wave history, and thus the stress history, appears markedly different from a train of

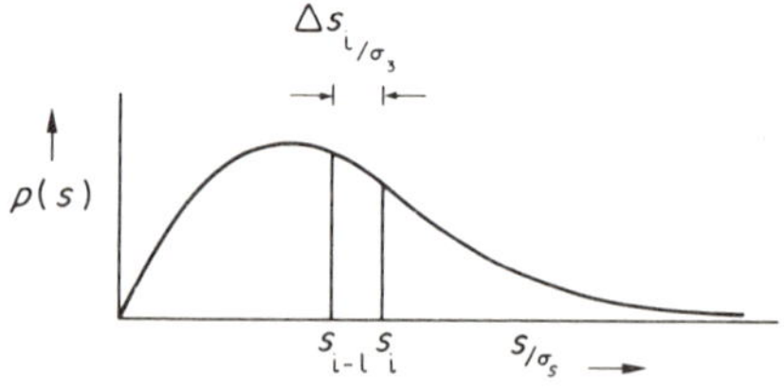

Figure 9.7

discrete waves, each crest being associated with an adjacent trough of corresponding magnitude. However methods have been established for resolution into range occurrences for application of Miner's rule, notably the 'rainflow' algorithm.[14] This has the effect on a long record of this type of coupling each peak with a corresponding trough elsewhere in the record, and the distribution of ranges is closely equal to the Rayleigh distribution for a process of rms value twice the rms value of the actual continuous stress process. Hence the stress history at a joint can be assumed to be a narrow band Gaussian Process whose peaks are Rayleigh distributed.

Knowing the standard deviation of stress range for a given sea state, a Rayleigh curve of probability density against stress range can be constructed (Figure 9.7).

The fraction occurrence within the sea state of the range of stress range represented by the average stress range S_{iav} is now given by

$$\int_{S_{i-1}/\sigma_s}^{S_i/\sigma_s} p(S)d\left(\frac{S}{\sigma_s}\right)$$

which can be obtained analytically.

From exceedence data for H_s the fraction occurrence of the sea state is obtained and used to calculate the total fraction occurrence of the range of stress range ΔS_i.

To relate this fraction occurrence to number of stress cycles it is necessary to calculate the effective response frequency for the sea state. This will depend on the form of the response spectrum and the method of calculation will be explained later.

The number of cycles applied in a year (say) from sea state j and at range of stress range ΔS_i is given by:

$$n_{ji} = \text{No. of seconds in a year} \times \text{Total fraction occurrence in a year} \times f_e.$$

Where f_e the effective frequency of vibration is defined later in Equation (9.39).

The damage caused in a year by range of stress range ΔS_i in a sea state j is thus given by:

$$\text{Fraction damage } ji = \frac{n_{jiav}}{N_{iav}}$$

This damage is then summed for each range of stress range and each sea state to produce the total fatigue damage.

9.3 METHODS OF ANALYSIS

We set out to review the different approaches to the fatigue analysis of offshore structures. The disadvantages of each method are outlined and a hybrid method combining their best features is suggested.

9.3.1 Discrete wave analysis

The discrete wave method of analysis makes no attempt to model the wave loading process as a narrow band random excitation but assumes that the process can be split into discrete waves each of which has specified periods associated with it.

This approach has been used widely in the past. It has advantages as far as detailed analysis of members and joints is concerned but it is difficult to include dynamic effects in a realistic way. Thus for the smaller jacket platforms and other dynamically insensitive structures results from this method can be regarded with some confidence.

No rational evaluation of the dynamic effect of the random sequence of differing waves on the structure is possible unless a time history response is carried out.

The wave theory used to estimate the water particle velocities and accelerations can be of any type, preferably the most accurate for the site (i.e. Stokes 5th order for deep water, Cnoidal for shallow water etc.). To calculate the forces on each member the particle motions are inserted in the familiar Morison equation:

$$P = \rho C^M \frac{\pi D^2}{4} \ddot{u} + \rho C^D \frac{D}{2} \dot{u}|\dot{u}| \tag{9.6}$$

where

$$
\begin{aligned}
P &= \text{force per unit length} \\
C^D &= \text{drag coefficient} \\
C^M &= \text{inertia coefficient} \\
\rho &= \text{mass density of water} \\
D &= \text{diameter of member} \\
\dot{u} &= \text{particle velocity} \\
\ddot{u} &= \text{particle acceleration}
\end{aligned}
$$

Thus having defined the structure it is possible to calculate the external forces on each member due to wave and/or current action. This is done using a computer program since the number of members in a jacket is usually of the order of several hundred. The stress range caused by a specified wave for each member end is then studied using an integrated suite of programs. For discrete fatigue analysis it is required that the stress range be defined for each member end in the structure for a given wave. Waves are applied from the different directions of attack so that for a given wave from a given direction the stress range at the ends of every member is defined. Logically a wave loading analysis should be carried out with the wave crest moving through the structure until the maximum and minimum stresses in the current member of interest are found, the process being repeated for each member in the structure. This is not practical for any structure with more than a few members so it is usual to assume that the position of wave crest that causes minimum or maximum total base shear (or overturning moment) also causes the minimum or maximum stresses in all the members in the structure. This assumption is good for most members in the structure but should be used with caution when members near the surface are being considered since peak local loadings on the member can be much greater than those indicated by the overall method.

At this stage an attempt is sometimes made to include dynamic effects by factoring the stresses by a 'dynamic load factor'. This is usually done by regarding the structure as a damped single degree of freedom system oscillating in a uniform wave train whose period is the period of the discrete wave applied.

The method of fatigue damage summation has been outlined above; In the discrete method it can be carried out by hand but it is usual to employ a computer program to carry out this somewhat tedious calculation. Programs have been written which interface automatically with the output from the structural analysis stage which means that an entire jacket can be analysed rapidly with very little manual effort, an important consideration for the designer or when checking for quality assurance.

9.4 SPECTRAL FATIGUE ANALYSIS

The essential feature of a spectral method is that it recognizes the probabilistic nature of real sea states. The sea elevation is not a truly random process since its statistical properties can be assumed to be constant with small time and is thus often referred to as stochastic. Using Fourier analysis techniques the wave record may be reduced to a superposition of a large number of sinusoidal components. This information may be presented as a frequency spectrum the ordinate of which at a particular frequency is the variance of the component sinusoid at that frequency. As the number of component frequencies is taken to infinity the plot becomes essentially a variance density and shows how the process is spread over the frequency range. The units of the ordinate will be (quantity)^2 per unit frequency. Since power is directly related to variance this curve is often referred to as the power spectral density function.

The response of a system to a stochastic input will also be stochastic and will in turn be described by its frequency spectrum. The area under the response spectrum is of course the response variance. If the variable can be assigned a known probability distribution, this provides a very powerful design tool as the probability of various levels of response being exceeded may be calculated.

As the process has been resolved into its sinusoidal components any known solutions for steady state sinusoidal input may be used in evaluation of the response. The spectra of input and response are related via the transfer function. This may be written:

$$S^{RR}(f) = J^2(f)S^{XX}(f) \tag{9.7}$$

If the system is linear then the transfer function is unique and is simply the amplitude squared of the response to unit amplitude input with variable frequency. For example the force and displacement spectra for a simple single degree of freedom spring mass system is:

$$S^{\dot{X}X}(f) = \left[\frac{H(f)}{K}\right]^2 S^{FF}(f) \tag{9.8}$$

where K is the spring stiffness and $H(f)$ is the familiar dynamic magnification factor.

The basis for the evaluation of fluid force on a jacket structure is usually the Morison equation. The force P on an element of structure of volume V and area A perpendicular to the direction of particle motion being given by:

$$P = \rho V C^M \ddot{u} + \rho A C^D \dot{u}|\dot{u}|/2 \tag{9.9}$$

where $\dot{u}$ is the fluid velocity and $\ddot{u}$ the fluid acceleration. The drag term is proportional to $|\dot{u}|^2$ but retains the sign of $\dot{u}$. This non-linear term causes considerable problems in applying the spectral method but first, to illustrate the step-by-step procedure of transferring from one spectrum to the next, we will consider the inertia force contribution only to the base moment of a uniform cantilever diameter D in water depth d. The sea elevation spectrum is plotted in diagram A (Figure 9.8). For linear wave theory the acceleration at the surface for each frequency is $(2\pi f)^2$ times the displacement amplitude so the transfer function from $S^{\eta\eta}(f)$ to $S_d^{\ddot{u}\ddot{u}}(f)$ is $(2\pi f)^4$. This acceleration spectrum is shown in diagram B. Allowing for the attenuation of acceleration with depth, the base moment is related to surface acceleration as follows:

$$BM = \left[\frac{\pi D^2}{4} C^M \int_0^d z \frac{\cosh kz}{\cosh kd} \, dz\right] \ddot{u}_s \tag{9.10}$$

where k is the wave number for a particular frequency. In deep water, for large k, $k = (2\pi f)^2/g$. The square of the term in brackets is the transfer function from surface acceleration to base moment. This function is plotted in diagram C. The final response curve is the product of B by C to give diagram D.

To show how the non-linear drag term is dealt with consider the equation of motion for a single degree of freedom system, this is a special case of the general equation for multi degree of freedom systems.

$$M\ddot{x} + D\dot{x} + Kx = \rho C^M V\ddot{u} + \rho C^D A\dot{u}|\dot{u}|/2 \tag{9.11}$$

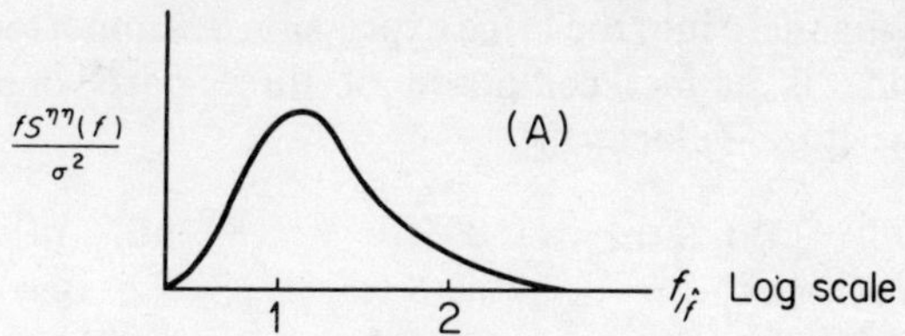

Non-dimensional wave spectrum

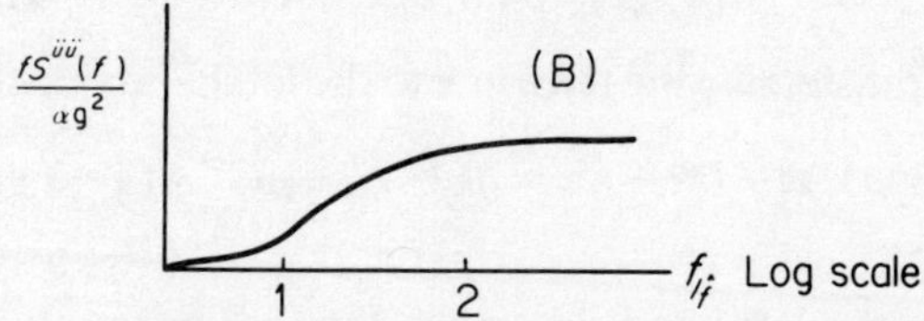

Spectrum of horizontal acceleration at surface

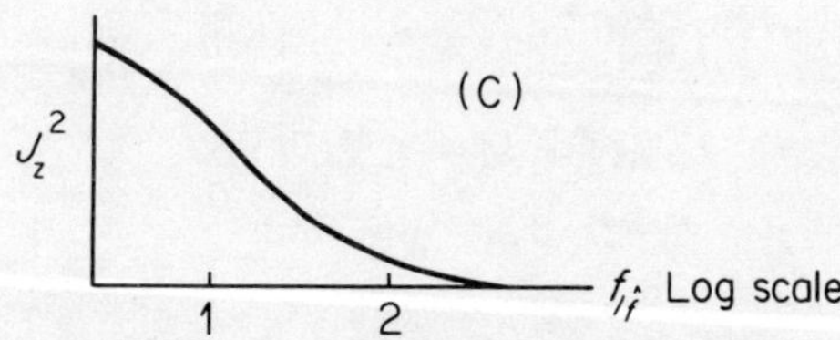

Transfer function for vertical variations

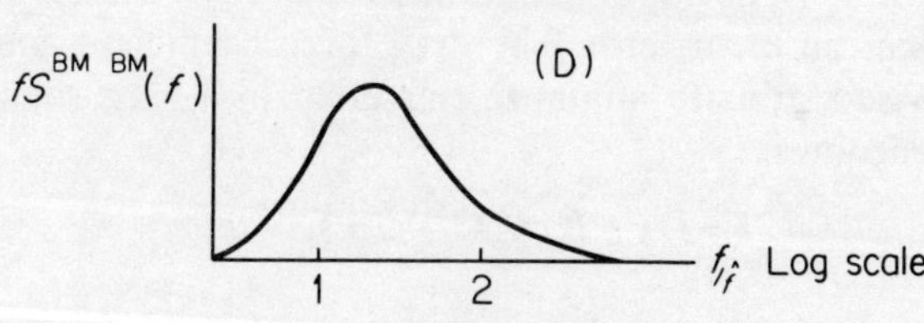

Spectrum of base movement

Figure 9.8

For structures which have considerable movement it is necessary to write Equation (9.11) in terms of relative motion r. It follows that:

$$r = u - x$$

$$\dot{r} = \dot{u} - \dot{x}$$

$$\ddot{r} = \ddot{u} - \ddot{x}$$

Before introducing these into the force expression it is important to note that the right hand side is in fact composed of three parts one of which is independent of structural velocity.

$$P = \rho(C^M - 1)V\ddot{u} \quad + \quad \rho V\ddot{u} \quad + \quad \rho C^D A \dot{u}|\dot{u}|/2 \qquad (9.12)$$

Added mass   Froude–Krylov   Drag
force   force   force

In terms of relative velocity this becomes:

$$P = \rho(C^M - 1)V(\ddot{u} - \ddot{x}) + \rho V\ddot{u} + \rho C^D A(\dot{u} - \dot{x})|\dot{u} - \dot{x}|/2 \qquad (9.13)$$

Re-arranging and transferring the term in $\ddot{x}$ to the left hand side this becomes:

$$(M + (C^M - 1)\rho V)\ddot{x} + D\dot{x} + Kx = \rho C^M V\ddot{u} + \rho C^D A(\dot{u} - \dot{x})|\dot{u} - \dot{x}|/2$$
$$(9.14)$$

The first term on the left hand side is the familiar 'mass plus added mass'. This may be written:

$$M^A\ddot{x} + D\dot{x} + Kx = C^{MV}\ddot{u} + C^{DA}\dot{r}|\dot{r}| \qquad (9.15)$$

where

$$M^A = M + (C^M - 1)\rho V$$

$$C^{MV} = \rho C^M V$$

$$C^{DA} = \rho C^D A/2$$

The drag term linearization is now carried out following the method given by Penzien and Malhotra.[15] The non-linear term is replaced by a linear term which of course introduces an error term. The drag force coefficient and damping term are then revised so as to minimize this error in a least squares sense. Equation (9.15) becomes:

$$M^A\ddot{x} + \mathbf{D}\dot{x} + Kx = C^{MV}\ddot{u} + \mathbf{C}^{DA}\dot{u} \qquad (9.16)$$

where

$$\mathbf{D} = D + C^{DA}\sqrt{\left(\frac{8}{\pi}\right)}\sigma_{\dot{r}}$$

$$\mathbf{C}^{DA} = \sqrt{\left(\frac{8}{\pi}\right)}\sigma_{\dot{r}}C^{DA}$$

The additional damping term is generally referred to as the hydrodynamic damping. Since the relative motion involves the structural motion any solution technique must involve an iterative process with the relative motion being taken equal to water particle motion on the first iteration. In practise this iteration is required only when the structure is very flexible and drag sensitive. Using this linearized equation of motion the drag force contribution in the simple example used earlier may be found and the total response is simply the sum of the variances.

9.4.1 Full spectral method

Consider the application of this method to a three-dimensional structure where the waves are assumed to be uni-directional and long crested. The development here is restricted to the horizontal translational degrees of freedom at each node. The individual member masses, areas and volumes are lumped at the member ends, the areas being projected in the directions of the degrees of freedom retained. For a wave approaching from a given angle the water particle motions are resolved in the directions of these degrees of freedom. Subject to the above restrictions the equations of motion of the structure may be written in matrix notation.

$$[M^A]\{\ddot{x}\} + [D]\{\dot{x}\} + [K]\{x\} = [\rho C^M V]\{\ddot{u}\} + [\rho C^D A/2]\{\dot{u}|\dot{u}|\} \tag{9.17}$$

The reduced stiffness matrix $[K]$ may be defined as the inverse of the flexibility matrix $[F]$. A typical term of this matrix is the response in degree of freedom j due to unit load in degree of freedom i.

By analogy with Equation (9.16) the linearized form of the matrix equation above is simply

$$[M^A]\{\ddot{x}\} + [\mathbf{D}]\{\dot{x}\} + [K]\{x\} = [C^{MV}]\{\ddot{u}\} + \mathbf{C}^{DA}]\{\dot{u}\} \tag{9.17}$$

where

$$\mathbf{D}_{ij} = D_{ij} \, i \neq j$$

$$\mathbf{D}_{jj} = D_{jj} + C_{jj}\sqrt{\left(\frac{8}{\pi}\right)}\sigma_{\dot{u}j}$$

$$\mathbf{C}_{jj}^{DA} = C_{jj}^{DA}\sqrt{\left(\frac{8}{\pi}\right)}\sigma_{\dot{u}j}$$

The matrix Equation (9.17) represents a set of n coupled differential equations. For a lightly damped system it is possible to uncouple these by using the system's eigenvector matrix as a coordinate transformation matrix. The free undamped equations of motion may be written:

$$[M^A]\{\ddot{x}\} + [K]\{x\} = 0 \tag{9.18}$$

This reduces to the classical eigenvalue form:

$$[M^A]^{-1}[K] - [\lambda][I] = 0 \tag{9.19}$$

The eigenvectors (modes) and eigenvalues (natural circular frequencies squared) may then be found. The linear transformation is written:

$$\{x\} = [\phi]\{q\} \tag{9.20}$$

Substituting into Equation (9.17) and pre-multiplying by $[\phi]^T$ gives

$$[\phi]^T[M][\phi]\{\ddot{q}\} + [\phi]^T[\mathbf{D}][\phi]\{\dot{q}\} + [\phi]^T[K][\phi]\{q\}$$
$$= [\phi]^T[C^{MV}]\{\ddot{u}\} + [\phi]^T[\mathbf{C}^{DA}]\{\dot{u}\} \tag{9.21}$$

Taking account of the orthogonality of the modes the mass and stiffness matrices diagonalize. The generalized equations are written:

$$[M^*]\{\ddot{q}\} + [D_C]\{\dot{q}\} + [K^*]\{q\} = \{P^*\} \tag{9.22}$$

$$[M^*] = [\phi]^T[M][\phi] \tag{9.23}$$

$$[D_C] = [\phi]^T[\mathbf{D}][\phi] = [\phi]^T\left([D] + \left[C^D\sqrt{\left(\frac{8}{\pi}\right)}\sigma_{\dot{u}}\right]\right)[\phi] \tag{9.24}$$

$$[K^*] = [\phi]^T[K][\phi] \tag{9.25}$$

$$\{P^*\} = [\phi]^T[C^{MV}]\{\ddot{u}\} + [\phi]^T[\mathbf{C}^{DA}]\{\dot{u}\} \tag{9.26}$$

The damping matrix will not diagonalize since it is independent of the modal matrix. This matrix is composed of two parts, the structural and the hydrodynamic damping, which are normally specified as a proportion of critical. The generalized structural damping in each mode is written:

$$D^*_{sjj} = 2\omega_j M^*_{jj}\zeta_j \tag{9.27}$$

The hydrodynamic matrix cannot be diagonalized so conveniently. It may be forced to be diagonal and the error minimized in a least squares sense as in the drag term linearization. This however introduces a further iteration. The first step of the iteration consists of using the diagonal elements of the coupled matrix as the elements of the uncoupled hydrodynamic damping matrix. This is considered sufficiently accurate for normal engineering practice and further information is given by Penzien and Malhotra.[15] Adopting the procedure outlined above the equations uncouple and each mode of vibration may be treated as a separate single degree of freedom system. A typical equation may be written:

$$M^*_{jj}\ddot{q}_j + D^*_{jj}\dot{q}_j + K^*_{jj}q_j = P^*_j \tag{9.28}$$

The response of this system to a unit amplitude sinusoidal input of frequency f is

$$\alpha^*(f) = \frac{1}{K^*_{jj} - (2\pi f)^2 M^*_{jj} + i(2\pi f)D^*_{jj}} \tag{9.29}$$

This is known as the complex receptance of the system. From the theory of random vibrations[16] it follows that:

$$[S^{qq}(f)] = [\bar{\alpha}^*(f)][S^{p^*p^*}(f)][\alpha^*(f)] \tag{9.30}$$

where $\bar{\alpha}$ denotes the complex conjugate. It is important to note that the complex receptance may be reduced to the normal dynamic magnifier form.[16] From the definition of spectral density the matrix of response variances is written:

$$[V^{qq}] = \int_0^\infty [S^{qq}] \, df \tag{9.31}$$

The diagonal elements of this matrix give the variance of response in each mode. The off-diagonals define the covariance of response between the modes. These terms arise because correlation between the input in the various modes gives rise to off-diagonal terms in the generalized force spectral density matrix.

The response matrix in generalized co-ordinates is then transferred back to the Cartesian system as follows:

$$[V^{xx}] = [\phi][V^{qq}][\phi]^T \tag{9.32}$$

Having found the structural displacements in statistical terms it is now possible to obtain corresponding information about the internal stresses.

In an undamped vibration the elastic forces corresponding to the deformation of the structure must be such as to provide the accelerations of the masses. The acceleration at any point i in mode j is simply

$$\phi_{ij}\frac{d^2 q_j}{dt^2}$$

which reduces to $\phi_{ij}\omega_j^2 q_j$ where ω_j is the circular frequency in mode j. Therefore the static 'inertia loading' which has the same effect as unit displacement in mode j is simply

$$\{F\}_j = [M^A]\{\phi\}_j\omega_j^2 \tag{9.33}$$

In a fatigue analysis the detailed member bending stresses are required and so the lumped mass plus added mass matrix $[M^A]$ in this equation is replaced by the actual mass and added mass distribution to produce a distributed lateral load case corresponding to $[P]$.

If the stresses are required at internal points then this inertia load vector is applied to the structure to obtain a set of influence coefficients B_{ij} relating stress at point i to displacement in mode j. This is done for each mode and the internal stress vector may be written:

$$\{S\} = [\beta]\{q\} \tag{9.34}$$

$$[V^{ss}] = [\beta][V^{qq}][\beta]^T \tag{9.35}$$

The generalized force spectral density matrix used in Equation (9.30) is stated here. The derivation of this is rather lengthy and has been put in Appendix A.

$$[S^{p^*p^*}] = [\phi]^T[C^{MV}][S^{\dot u \dot u}][C^{MV}][\phi] + [\phi]^T[\mathbf{C}^{DA}][S^{\ddot u \ddot u}][\mathbf{C}^{DA}][\phi] \quad (9.36)$$

Typical terms of the water velocity and acceleration spectral density matrices are given by

$$S_{ij}^{\dot u \dot u}(f) = (2\pi f)^2 \frac{\cosh kv_i \cosh kv_j}{\sinh^2 kd} \exp(i(kh_i - kh_j))S^{\eta\eta}(f) \quad (9.37)$$

$$S^{\ddot u \ddot u}(f) = (2\pi f)^2 S_{ij}^{\dot u \dot u}(f) \quad (9.38)$$

where v is the vertical co-ordinate of the node and h is the horizontal co-ordinate resolved in the direction of wave attack.

The procedure described up until now has been applicable to a 3D structure with the restrictions stated. However, the large number of nodes in real structures necessitates further simplification of the structural model.

9.4.2 Typical steel jacket

A slender jacket structure of simple form is shown in Figure 9.9. Such a structure is likely to have very stiff horizontal bracing at various levels, such that it is reasonable to assume that all points at a given level share the same displacements. Thus a simple dynamic model with a single horizontal degree of freedom in each of two selected orthogonal directions plus a rotation in plan, at each bracing level, may be constructed. As a further simplification the rotation

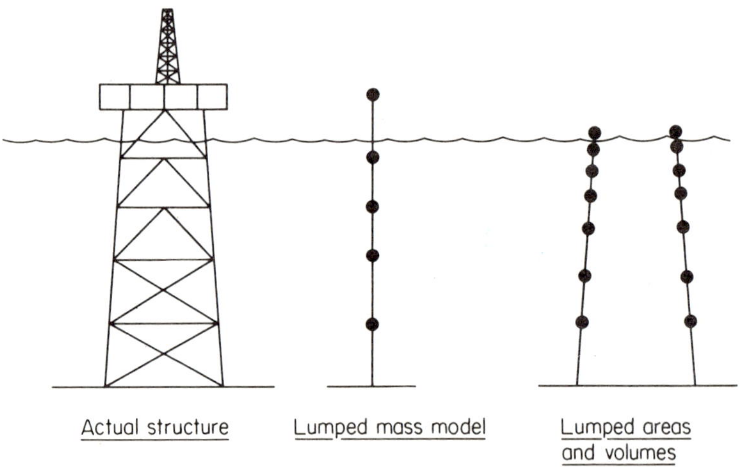

Figure 9.9

is usually neglected and the vibration in each of the two orthogonal planes is treated independently.

The structure mass is assigned to the nearest bracing level. To evaluate the hydrodynamic loading the projected areas parallel to the wave direction and the displaced volume may be lumped in two or more columns.

Because of the rapid attenuation of wave motion with depth, the modal generalized loadings are dominated by forces acting on the upper part of the structure, and it is generally convenient to use a finer sub-division of the structure for this purpose than for the basic dynamic analysis. This is examined in more detail in Appendix A. The direction of wave approach is restricted to the two principal directions of the platform.

It has been shown how the variance of stress may be obtained for a given sea state as defined by its spectrum. To carry out the fatigue damage summation it is necessary to associate this response level with a corresponding frequency of vibration.

An effective frequency of vibration[17] is defined in terms of the moments of the first mode response spectrum as follows

$$f_e = \sqrt{\frac{m_2}{m_0}} \tag{9.39}$$

$$m_2 = \int_0^\infty f^2 S(f)\, \mathrm{d}f \tag{9.40}$$

$$m_0 = \int_0^\infty S(f)\, \mathrm{d}f \tag{9.41}$$

Since for structures of this type the majority of the response is first mode this is a reasonable approximation. In a given sea state the stress range is taken to be Rayleigh distributed and damage summation carried out as described earlier.

The procedure outlined above constitutes the general spectral response analysis, whereby the variance of stress at any point in the structure can be computed for a given sea state. Although some adaptation of the modal analysis would still be required to permit treatment of cases where the wave direction does not coincide with one of the principal directions in plan for the natural modes. It will be appreciated from the broad discussion of fatigue analysis given earlier in the paper, however, that a full investigation of the many points in a real structure where fatigue may govern design will require detailed modelling of the structure. In many cases bending stresses produced in members by local fluid loading must be evaluated with reasonable accuracy and there is little physical justification for expecting the natural modes of the lumped structure to give sufficient modelling for this purpose.

When considering stresses (especially those in bracings) rather than displacements it is found that higher modes have greater effect. It may be important to consider the contribution of these to the low frequency components of the loading where the response is not greatly modified by the behaviour of the structure. The dynamic magnification is, however, generally only significant in the lowest one or two modes in each direction (and possibly in torsion) as the natural frequencies of the higher modes are such that the wave energy at these frequencies is small and rapidly attenuated with distance below the surface. Thus local effects will be basically quasi-static unless either the member is so slender that transverse deformation modes resulting primarily from bending of local members have natural periods approaching the fundamental value (which will not in general be associated with a satisfactory design for members where the local fluid loading is significant), or where impulsive or slamming effects produce high frequency load effects (which the above method based on the sea surface spectrum does not claim to consider).

As the fatigue analysis requires multiple repetition of any such spectral analysis to cover the range of sea conditions (both significant wave height and wave direction, and possibly differing spectral shape parameters) over the design life, it is generally concluded that to create a method suitable on grounds of cost and of computation capacity for application to practical structures in design or certification procedures, further modification is necessary. Some proposals to this end are described later. In special cases, however, the direct spectral procedure may be applicable. To assist understanding of the direct procedure prior to discussion of the possible approximations, consider an example.

9.4.3   Single uniform column

Not all offshore structures in which fatigue must be considered require analysis as complex space frames. It is instructive to consider the case of a uniform diameter free-standing cantilever. The response of such a structure will be dominated by the first mode. Consider the inertia component of the generalized force spectral density matrix and remove the first mode term.

$$S_{11}^{p^*p^*} = \{\phi\}_1^T [C^{MV}][S^{\ddot{u}\ddot{u}}][C^{MV}]\{\phi\}_1 \tag{9.42}$$

Since for a vertical cantilever there is no horizontal spacing the acceleration matrix term becomes:

$$S_{ij}^{\ddot{u}\ddot{u}} = \frac{(2\pi f)^4}{\sinh^2 kd} \cosh kv_i \cosh kv_j S^{\eta\eta}(f) \tag{9.43}$$

On expansion the matrix operations of Equation (9.42) are equivalent to:

$$\frac{(2\pi f)^4}{\sinh^2 kd}\left(\sum_{i=1}^{NLM}\phi_i\cosh kz_i C_i^{MV}\right)^2 S^{\eta\eta}(f)$$

Taking the number of lumped masses to infinity and substituting for C^{MV} gives:

$$S_{11}^{p^*p^*}=\left(\frac{(2\pi f)^2\pi D^2}{4\sinh kd}\right)^2\left(\int_0^d\phi(z)\cosh kz\,dz\right)^2 S^{\eta\eta}(f)\tag{9.44}$$

The first term of Equation (9.30) is simply:

$$S_{11}^{qq}=\bar{\alpha}_{11}^*S_{11}^{p^*p^*}\alpha_{11}^*\tag{9.45}$$

This equals:

$$S_{11}^{qq}=|\alpha_{11}^*|^2 S_{11}^{p^*p^*}$$

$$=\left|\frac{H_1^*}{K_{11}^*}\right|^2 S_{11}^{p^*p^*}$$

where:

$$H_1^*(f)=\frac{1}{\sqrt{\left(1-\left(\dfrac{f}{f_1}\right)^2\right)^2+\left(2\zeta_1^*\dfrac{f}{f_1}\right)^2}}\tag{9.46}$$

The derivation for the drag force follows an entirely analogous procedure. The proportion critical damping ζ_1^* in Equation (9.46) is the total of structural and hydrodynamic. In this case it may be written:

$$\zeta^*=\zeta_s^*+\sqrt{\frac{8}{\pi}\frac{\rho C^D D}{2\omega_1 M_{11}^*}}\int_0^d\sigma_{\dot{u}}(z)\phi^2(z)\,dz\tag{9.47}$$

The generalized mass and stiffness and modal equivalent loading are simply:

$$M_{11}^*=\int_0^d m(z)\phi^2(z)\,dz\tag{9.48}$$

$$K_{11}^*=\omega_1^2 M_{11}^*\tag{9.49}$$

$$F=\omega_1^2\int_0^d m(z)\phi(z)\,dz\tag{9.50}$$

9.4.4 Two uniform towers

The importance of retaining the horizontal spacing in the force calculation may be illustrated by considering two uniform vertical cantilevers separated by a distance Δ in the direction of wave propagation. Let J_1^2 and J_2^2 be the sea elevation—generalized first node force transfer functions for the two columns individually. Then the total generalized force is simply:

$$\{J_1^2 + 2J_1^2 J_2^2 \cos{(k\Delta)} + J_2^2\} S^{\eta\eta}(f)$$

for two equal towers this reduces to

$$J_1^2 \{2 + 2 \cos{(k\Delta)}\} S^{\eta\eta}(f)$$

The term inside the brackets introduces an oscillation into the response spectrum. This represents the cancelling effect obtained when the wave length is certain multiples of the leg spacing.

9.5 REDUCED SPECTRAL METHODS

It is helpful at this stage, to bring together the discussion that has been given on discrete wave methods in comparison with the basic spectral methods to indicate the relative advantages of each.

The discrete wave method facilitates the use of existing static wave programs, which can give a very detailed stress analysis at acceptable cost. This will include the bending of members by fluid forces applied between nodes of the framework. Various non-linearities (higher order wave forms, integration to instantaneous surface level, drag forces) can readily be included. However, a train of discrete waves is not even by visual inspection a very convincing model representation of actual wave records, and the association of wave periods with specific height waves is to some degree arbitrary. Strictly interpreted, the discrete wave theory provides no rational basis for estimation of dynamic magnification of response; the suggestion that the response to a discrete wave of specified period should be multiplied by the steady-state dynamic magnifier for continuous loading of that periodicity applied to the structure has clearly no validity, since it pays no attention to the actual frequency range of the excitation, and is virtually insensitive to damping, which are both highly significant in a rational analysis.

The spectral approach is based on a stochastic model of the wave motions that reproduces actual wave records very much more convincingly. Parameterization of wave spectra, initially in the form of sea state characterized by the significant height H_s related probabilistically to dominant period or average zero crossing frequency by a scatter diagram (Figure 9.4) or more recently in multi-dimensional form to include shape parameters of a more generalized spectral formulation such as the Jonswap spectra,[18] is now such that the excitation to be expected at relatively high frequencies, such as are associated with dynamic magnification, is well modelled.

The disadvantages are primarily questions of cost, time and the computer capacity required, although problems arising from non-linearities in the wave and wave loading processes have also been mentioned earlier. It will be appreciated that the economic constraints effect the application of the spectral method to fatigue analysis far more severely than they affect application to the conventional collapse criterion (analysis of maximum response), because of the repetition required to build up the total response history over a long period covering all sea states and wave directions. On the other hand the non-linearities are less serious, because of the larger influence of relatively modest seas rather than the maximum.

The conditions within which the number of degrees of freedom in the model used for spectral analysis may be reduced to a level at which the method is economically acceptable have been indicated. The method has, for example, been economically applied to fatigue studies of slender vertical cantilever structures acting predominantly as bending members. Where the fatigue problem centred on the total base moment. For routine application on a large scale to typical jacket structures, however, further simplification is clearly required. It is sensible to attempt to combine favourable features from the two approaches.

The fine detail of the basic static wave load analysis package used in the discrete wave method can be retained by using this package to estimate the transfer function linking the sea surface elevation spectrum to the spectrum of the stress response under investigation, treating this initially as a quasi-static process.[11] A separate approach is used to assess the mechanical admittance or dynamic contribution to the response.

This approach is summarized by Figure 9.10. A series of unit amplitude waves of differing frequency are applied to the full 3D static model of the structure. Each wave is passed through the structure to find the positions which maximize the response in positive and negative directions. The difference of these two quantities defines the range of variation for each frequency. Ideally the maximum and minimum should be found for each response of interest, but (as commonly in the discrete method itself) to reduce computer time in general all responses are assumed proportional to overall base shear or moment. The ordinate of the sea elevation—stress range transfer function for a given location at a particular frequency is simply the square of the stress range caused by a unit wave at that frequency. In this way the static transfer function for a particular point and given angle of attack is built up. The result is a curve of the type shown as curve A in Figure 9.10. It will be noted that this does not invoke modal decomposition of the response. However, on the assumption that the first mode dominates the response at all frequencies within the range of significant wave loading, an approximation to the dynamic amplification can be obtained by applying the mechanical admittance appropriate to the first mode

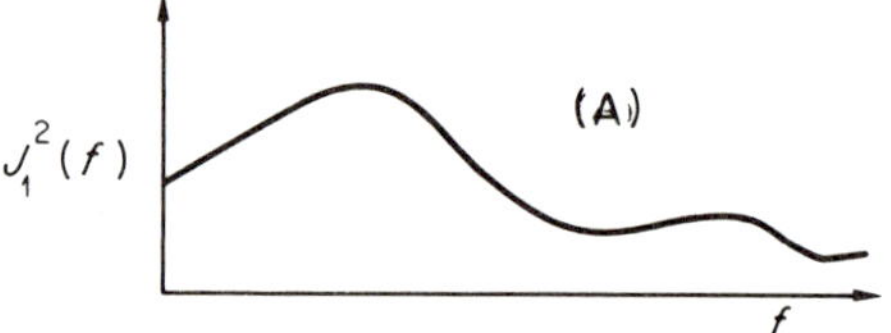

Quasi static sea elevation / Stress transfer function

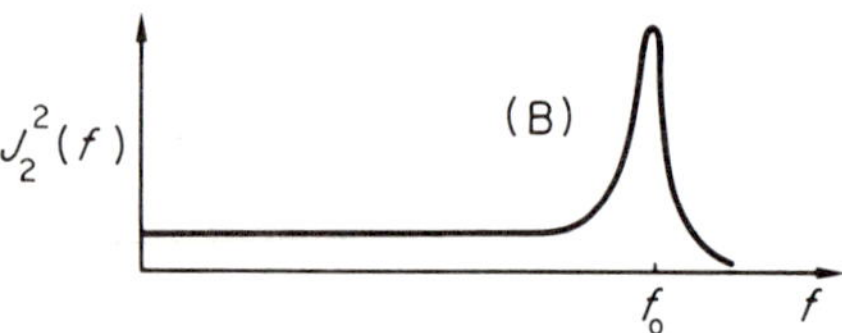

First mode mechanical admittance

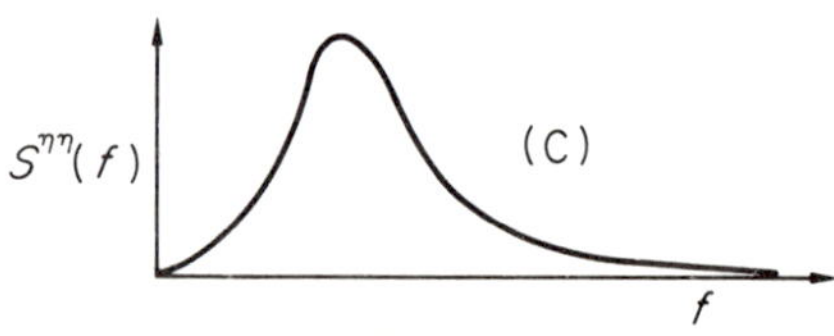

Sea elevation spectrum

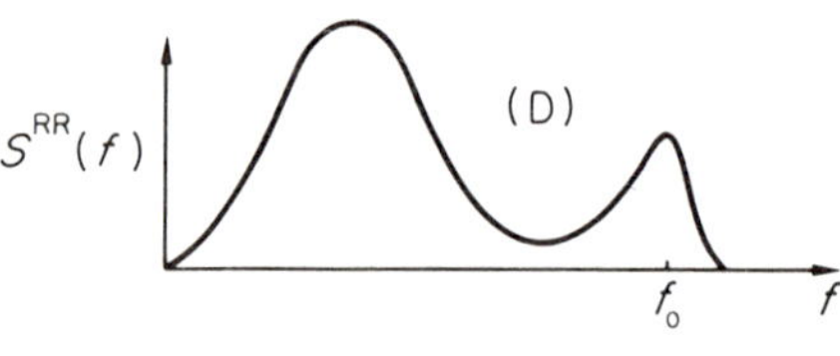

Response spectrum

Figure 9.10

as shown as curve B. The response to a particular sea spectrum (for example curve C) is then simply the product of ordinates A, B, and C, giving curve D and the response variance is the area under this curve.

The concept of a simple transfer function as defined can only be applied strictly to a linear system. There are three basic non-linearities inherent in the above procedure.

(i) The wave theory used

(ii) Most wave programs integrate to local sea elevation whereas a strictly linear approach would require integration to m.s.l. only.

(iii) The non-linear drag term in the Morison equation.

In deep water linear wave theory is acceptable. The effect of number (ii) should be small as most damage is caused by the lower sea states. Number (iii) is more difficult to deal with and will cause serious error for drag sensitive structures. Linearization in the optimum sense for prediction of response variance, as described for the full spectral method, would require sufficient re-programming to eliminate the advantage of the procedure. A method has been proposed,[11] whereby an effective transfer function is evaluated as the square of the quotient of the response range calculated for the highest possible wave at each given frequency divided by the corresponding wave height. As the non-linearities generally lead to a response that increases more rapidly than linearly with wave height, this gives a conservative (safe) estimate of the fatigue damage rate. For this purpose the maximum possible wave height at any frequency is taken as the breaking wave height $H = g/14\pi f^2$, or as the maximum wave height predicted to occur at that location during the design life if that is less than the breaking height. It is difficult to estimate the degree of conservatism thus introduced.

The dynamic magnifier as predicted by this method presumes the stress response in question to be equally sensitive to the inertial loading (including added mass) of the first mode as to the actual distribution of wave pressures. In the case of stresses arising due to bending of members between nodes near the sea surface, in the case of deck connections and of certain bracing connections this may not be a good assumption, and should be used with caution on dynamically sensitive structures.

The oscillatory form of the quasi-static response transfer function at high frequencies is a result of the spatial separation of the main legs in relation to the wavelength; this effect may well be significant in the vicinity of the natural frequency and thus for dynamic response, but is not very significant in respect to the total quasi-static response. The resonant response in overall cantilever modes higher than the fundamental is seldom significant although it may be necessary to consider the lowest mode in each orthogonal direction on plan, and/or the lowest torsion mode. The form of the response spectrum shown (curve D) is thus typical, notably the clear distinction between quasi-static and dynamic contributions to the variance of the response. This is, of course, dependant on the frequency separation between modes and it may be necessary to compute further modes to justify the approach. This opens the possibility of an alternative approach for dynamically sensitive structures, whereby the quasi-static part of the response is estimated by the detailed static programmes and the resonant contributions are assessed separately by the spectral method. An appropriately simplified model can be used for the latter since it is only the overall modal generalized loading which is required, not detailed stressing. The actual stresses from dynamic response should be obtained by multiplying the modal displacement by the stress influence factor calculated from the 'inertia loading' by the detailed static program (see above).

The dynamic contribution will, of course, be characterized by its variance. For a lightly damped system this may be evaluated from a knowledge of the spectral ordinate at the resonant frequency alone, since if the generalized force spectrum does not vary greatly within the range of major dynamic magnification the response displacement variance is

$$V^{qq} = \frac{\pi}{4\zeta K^{*2}} f_0 S^{p^* p^*}(f_0) \tag{9.51}$$

Since the stress range (S) in dynamic response is twice the amplitude, the stress range variance is

$$V^{ss} = 4\beta_s^2 V^{qq} \tag{9.52}$$

The need for repetitive evaluation of spectral ordinates at narrow intervals of frequency to permit numerical integration over the resonance peak is thus avoided. The response variances in further modes can be evaluated in the same way, and added to the total, since the correlation of modal resonance effects will be negligible unless the frequencies are effectively equal.

The static contribution could be evaluated by the process illustrated by Figure 9.10. Repetitive calculation to cover the frequency range prior to numerical integration can again be eliminated, by noting that the quasi-static part of the response spectrum is shaped mainly by the sea surface spectrum, and the corresponding variance is dominated by a narrow frequency band. The response variance can thus be estimated by multiplying the sea surface elevation variance (which follows directly from the significant wave height H_s which defines the sea state $V^{mm} = \frac{1}{16} H_s^2$) by a transfer factor based on the detailed static analysis applied to a 'representative' discrete wave of period equal to the dominant period for that sea state. Where non-linearities are likely to be significant, the representative wave height can be selected to minimize the probable error in damage estimation; this is discussed in Appendix B. If the representative wave height is taken as $H_R = 1 \cdot 25 H_s$, and this causes static response range S_R, the static contribution to the variance is taken as

$$V^{ss} = \tfrac{1}{25} S_R^2$$

As this procedure fits a correct analysis to the wave heights causing most damage in each sea state, it is a less severe approximation than the reduced method described earlier where a single fit covers the whole possible range of wave heights.

The total response variance is the sum of static and dynamic contributions, and the fatigue damage is then computed on the assumption that the stress ranges follow the Rayleigh distribution defined by the total variance. The total variance may be obtained by direct summation of the estimates made as described above. Alternatively some of the shortcomings of the simplification of the real structure in constructing the dynamic model may be reduced by the following indirect approach. The static analysis is repeated using the simplified model, giving an estimate of the static contribution to variance V^{ss} (static, simple model). The dynamic contribution as calculated for the same model is compared to this, and a pro-rata dynamic effect is then applied to the detailed model calculation, giving the estimate of total variance as

$$V^{ss} = \left\{ V^{ss}(\text{static, full model}) \right\} \times \left\{ 1 + \frac{V^{ss}(\text{dynamic, simple model})}{V^{ss}(\text{static, simple model})} \right\}$$

$$(9.53)$$

The choice of approach will depend on judgement of the likely errors caused by the simplified modelling, for example relative importance of local effects in comparison with effects of the overall complexity in the real structure. The designer must also exercise judgement as to the degree to which all stages must be repeated for stress responses at different points in the structure or whether some stages be short cut—for example by the assumption that the effective dynamic magnifier is the same for the whole of a definable class of stress responses. Assumptions of this type are common in respect of ultimate load design, but it has been suggested at various points in this paper that greater caution must be exercised in considering fatigue because the various critical stresses are less directly related to the overall cantilever action of the structure. In some of the methods the designer also has the option of integrating the damage expected in identifiable sea states prior to summing to the total damage, as opposed to summation of numbers of wave or stress cycles in each range prior to consideration of damage.

None of the methods described are backed by the length of experience that the designer of conventional terrestrial structures expects as the basis for such decisions. It is clear that many lessons will be learnt each time a structure is subjected to examination by these methods, and it is to be hoped that these lessons will rapidly be shared between workers in the field.

9.6 ILLUSTRATIVE EXAMPLE

As an illustration of the application of spectral fatigue analysis a brief example is presented. The main feature is to illustrate the importance of including dynamic effects in the calculation of stresses. This is particularly true of deep water jackets with fundamental periods at which there is significant wave energy. The structure being considered is intended for 150 m of water and has a fundamental frequency 0·43 Hz. Figure 9.11 shows stress range standard deviation versus significant wave height curves for a typical brace stress example. One curve shows the quasi-static response only while the other includes dynamic effects. The magnification is greatest when the sea state

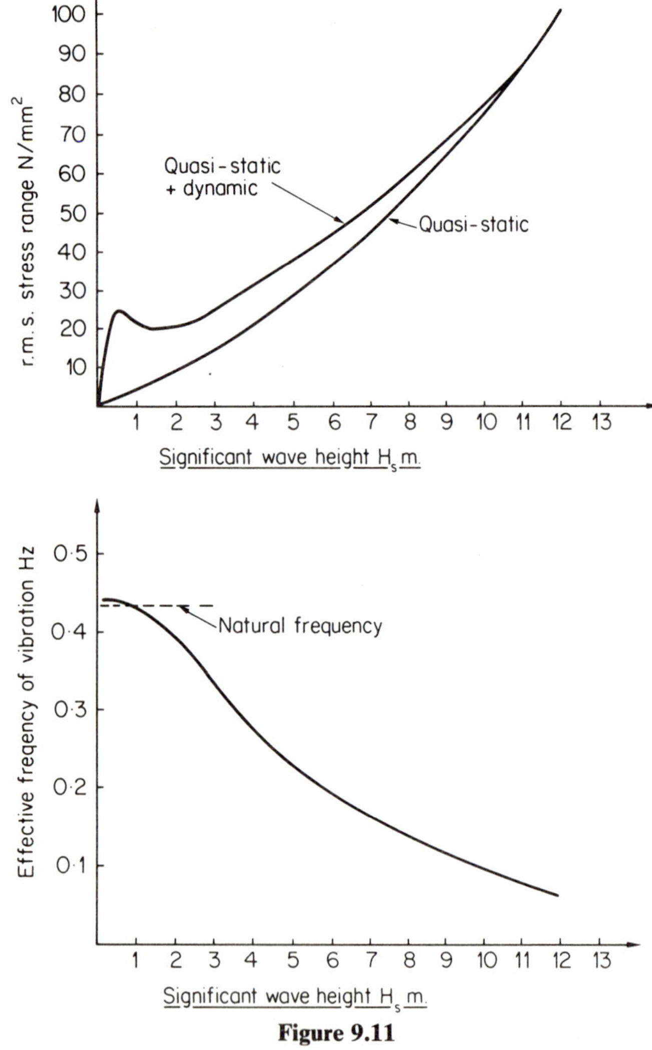

Figure 9.11

dominant frequency coincides with the structure natural frequency and the degree of magnification will depend on the damping in the system. In this case a structural proportion critical damping of $0 \cdot 015$ was used. Figure 9.11 shows the corresponding graph of effective frequency of vibration versus significant wave-height. Damage summation based on these curves gives the following results.

$$\begin{array}{ll} \text{Quasi-static} & \text{Life} = 55 \text{ years} \\ \text{Quasi-static} + \text{Dynamic} & \text{Life} = 12 \text{ years} \end{array}$$

The design life of offshore platforms is commonly around 50 years. The above example illustrates a case where the detail would be considered satisfactory in the basis of a quasi-static analysis only. Inclusion of dynamic effects has a dramatic effect on the life and indicates that a redesign of the joint would be required.

Acknowledgements

The Authors wish to thank Halcrow–Ewbank Petroleum and Offshore Engineering Company for permission to write this chapter and for the resources they have made available for the work described therein.

APPENDIX A

9A.1   FORCE SPECTRAL DENSITY MATRIX

One of the basic concepts in the study of random vibration is that the Spectral Density function and Autocorrelation functions form a Fourier Transform Pair. The crosscorrelation function is given by:

$$R_{ij}^{xx}(\tau) = \langle X_i(t)X_j(t+\tau) \rangle \tag{A9.1}$$

where $\langle \; \rangle$ denotes time average.

The Fourier Pair relationship is as follows:

$$S_{ij}^{xx}(f) = \int_{-\infty}^{+\infty} R_{ij}^{xx}(\tau)\, e^{-i2\pi fr}\, dr \tag{A9.2}$$

The linearized force vector given by Equation (9.17) is:

$$\{P(t)\} = [C^{MV}]\{\ddot{u}(t)\} + [\mathbf{C}^{DA}]\{\dot{u}(t)\} \tag{A9.3}$$

the ith term being:

$$P_i(t) = C_{ii}^{MV}\ddot{u}_i(t) + \mathbf{C}_{ii}^{DA}\dot{u}(t) \tag{A9.4}$$

The crosscorrelation function is therefore:

$$\begin{aligned}
R_{ij}^{p}(\tau) = \;& C_{ii}^{MV}\langle \ddot{u}_i(t)\ddot{u}_j(t+\tau) \rangle C_{jj}^{MV} \\
& + C_{ii}^{MV}\langle \ddot{u}_i(t)\dot{u}_j(t+\tau) \rangle \mathbf{C}_{jj}^{DA} \\
& + \mathbf{C}_{ii}^{DA}\langle \dot{u}_i(t)\ddot{u}_j(t+\tau) \rangle C_{jj}^{MV} \\
& + \mathbf{C}_{ii}^{DA}\langle \dot{u}_i(t)\dot{u}_j(t+\tau) \rangle \mathbf{C}_{jj}^{DA}
\end{aligned} \tag{A9.5}$$

The two central terms in this expression will cancel out for structures of this type as shown by Borgman,[19] therefore:

$$R_{ij}^{pp}(\tau) = C_{ii}^{MV} R_{ij}^{\ddot{u}\ddot{u}} C_{jj}^{MV} + \mathbf{C}_{ii}^{DA} R_{ij}^{\dot{u}\dot{u}} \mathbf{C}_{jj}^{DA} \tag{A9.6}$$

Taking the Fourier Transform of both sides the corresponding spectral density matrix follows:

$$S_{ij}^{pp}(f) = C_{ii}^{MV} S_{ij}^{\ddot{u}\ddot{u}} C_{jj}^{MV} + \mathbf{C}_{ii}^{DA} S_{ij}^{\dot{u}\dot{u}} \mathbf{C}_{jj}^{DA} \tag{A9.7}$$

The complete matrix may now be written:

$$[S^{pp}] = [C^{MV}][S^{\ddot{u}\ddot{u}}][C^{MV}] + [\mathbf{C}^{DA}][S^{\dot{u}\dot{u}}][\mathbf{C}^{DA}] \tag{A9.8}$$

Finally transforming to generalized co-ordinates:

$$[S^{p^*p^*}] = [\phi]^T[C^{MV}][S^{\ddot{u}\ddot{u}}][C^{MV}][\phi] + [\phi]^T[\mathbf{C}^{DA}][S^{\dot{u}\dot{u}}][\mathbf{C}^{DA}][\phi] \tag{A9.9}$$

This is Equation (9.36) in the text.

The expression just derived is applicable to any number and location of points in a plane. In the case of offshore structures it is usual to determine the modes and frequencies from a 'single column' lumped mass model while the areas and volumes are lumped in a number of columns. This is illustrated in Figure 9.9. The development here is carried out for two columns initially.

The essential point to note is that the modal matrix is identical for both columns of forces. This is a consequence of the assumed deformation behaviour of the structure.

The vertical spacing of lumped areas and volumes need not be the same as for the lumped masses. In general the spacing of the force parameters will be chosen so as to more accurately represent the loading on the platform. It is necessary then to create an interpolated modal matrix in which modal values at the lumped area and volumes have been interpolated in relation to their elevation. This new modal matrix will be denoted by $[\Phi_I]$.

The generalized force matrix may be partitioned as follows:

$$[S^{p^*p^*}] = [\phi_I \,|\, \phi_I]^T \begin{bmatrix} S_{LL}^{pp} & | & S_{LR}^{pp} \\ \hline S_{RL}^{pp} & | & S_{RR}^{pp} \end{bmatrix} \begin{bmatrix} \phi_I \\ \hline \phi_I \end{bmatrix}$$

$$= [\phi_I]^T ([S_{LL}^{pp}] + [S_{LR}^{pp}] + [S_{RL}^{pp}] + [S_{RR}^{pp}])[\phi_I] \qquad (A9.10)$$

Typical term $[S_{MN}^{pp}]$ is given by:

$$[S_{MN}^{pp}] = [C_M^{MV}][S_{MN}^{\ddot{u}\ddot{u}}][C_N^{MV}] + [C_M^{DA}][S_{MN}^{\ddot{u}\ddot{u}}][C_N^{DA}] \qquad (A9.11)$$

Summarizing Equation (A9.10) and extending to any number of columns gives:

$$[S^{p^*p^*}] = [\phi_I]^T \left(\sum_{i=1}^{NCOL} \sum_{j=1}^{NCOL} [S_{ij}^{pp}] \right)[\phi_I] \qquad (A9.12)$$

APPENDIX B

9B.1   SELECTION OF REPRESENTATIVE WAVE HEIGHT

Consider the solution for the linear system, i.e. the range of stress response S is related to the wave height H by (say)

$$S = C_1 H \tag{B9.1}$$

For the simplest practical fatigue S/N curve, namely the straight line with slope $-m$ on the double logarithmic plot with no cut off (i.e. no 'endurance limit', or limiting stress range for damage to occur), the contribution to Miner's damage quotient by a cycle of oscillation of range S can be written as $C_2 S^m$ or $C_2(C_1 H)^m$. The total damage by N cycles of oscillation in which the wave heights have a Rayleigh distribution with significant wave height (alternatively defined by the rms sea surface displacement $\sigma_\eta = H_s/4$) can be found by integration taking account of the probability density p_H (say) of occurrence of height H, giving an additional factor denoted C_3.

$$\text{Damage} = N \int C_2 S^m (H) p_H \, \mathrm{d}H = N C_2 C_3 (C_1 H_s)^m \tag{B9.2}$$

For many practical examples it is possible to take $m = 3$, for which there is an analytic solution

$$C_3 = \tfrac{3}{16}\sqrt{(2\pi)} = 0 \cdot 468 \tag{B9.3}$$

As indicated above, it is desired to use the linear relationship, substituting a fictitious 'representative' value C_{1R} for the constant in Equation (B9.1), derived from a careful analysis of the maximum and minimum values of stress (or moment, or any other appropriate response parameter) caused by passage of a suitable 'representative' wave, height H_R. The corresponding range of response is thus $S_R = C_{1R} H_R$. A suitable value of H_R defined as a multiple of H_s which will give satisfactory accuracy in widespread application of this procedure can be established from check analysis of a number of possible cases of non-linearity.

For any defined range of response—wave height relationship, the damage corresponding to Equation (B9.2) above can be evaluated, in general by numerical integration. The numerical factor will now be denoted C_{3R}

$$\int S^m (H) p_H \, \mathrm{d}H = C_{3R} S_s^m \tag{B9.4}$$

where S_s is the response range given by the pre-determined height H_s. The aim is clearly to replace Equation (B9.4) by Equation (B9.2) with C_{1R} in place of

C_1; comparing these equations, the error expressed as the ratio of the correct estimate due to the estimate from the replacement equation is

$$\text{Error} = \frac{C_{3R}S_s^m}{C_3(C_{1R}H_s)^m} = \frac{C_{3R}}{C_3}\left(\frac{S_sH_R}{S_RH_s}\right)^m \tag{B9.5}$$

For example, if $S(H) = (H/H_s)^2$, with $m = 3$, numerical integration gives $C_{3R} = 0.75$, and it can readily be shown that the error becomes zero if H_R is taken as $1.17H_s$. This clearly corresponds to a case which is abnormally sensitive to drag, and for most cases where there is no endurance limit it is sufficiently accurate within the practical limitations on the analysis to use H_s itself as the reference wave height, to obtain the quasi-static response range S_s. The rms value of this part of the response process is then taken as $\sigma_R = S_{s/4}$, which is combined with the dynamic contribution by summing squares (variances) in the usual way.

Where a damage cut-off extends to eliminate a substantial part of the wave height distribution for the given sea state, the result is more sensitive to non-linearity, because the effect of the upper tail of the Rayleigh distribution clearly becomes a larger part of the total. Study of a number of examples, notably with cut-off at the level corresponding to H_s, suggests that the 'representative wave' should be $H_R = 1.25H_s$, leading to $\sigma_R = \frac{1}{4}(S_R/1.25)$.

REFERENCES

1. Williams, A. K. and Rinne, J. E. (1976). 'Fatigue analysis of steel offshore structures', *Proc. Instn. Civ. Engrs.*, Part 1, 1976, 60, Nov. 635–654.
2. Draper, L. and Driver, J. S. (1971). 'Winter waves in the northern North Sea at 57° 30′N. 3° 00′E. Recorded by M. V. Famita', *N.I.O. Report A48*.
3. Longuet-Higgins, M. S. (1952). 'On the statistical distribution of the height of sea waves', *Journ. Marine Research*, Vol. II, No. 3.
4. Draper, L. (1976). 'Revisions in wave data presentation', *Proc. 15th Conf. Coastal Eng.*, Honolulu.
5. Wu, S. C. (1976). 'The effects of current on dynamic response of offshore platforms', *OTC Paper 2540*.
6. 'Recommended practice for planning, designing and constructing fixed offshore platforms', *American Petroleum Institute Code RP2A*.
7. Marshall, P. W. (1974). 'General consideration for tubular joint design', *W.R.C. Bulletin 193*.
8. Gurney, T. R. and Maddox, S. J. (1972). 'A re-analysis of fatigue data for welded joints in steel', *Welding Institute Report E/44/72*.
9. Miner, M. A. (1945). 'Cumulative damage in fatigue', *Journ. Appl. Mech.*, **12**, A-159.
10. Palmgren, A. (1924). 'Die Lebensdouen von Kugelhagern', *VDIZ*, **68**, 14, 339.
11. Maddox, N. R. and Wildenstein, A. W. (1975). 'A spectral fatigue analysis for offshore structures', *OTC Paper 2261*.

12. Cartwright, D. E. and Longuet-Higgins, M. S. (1956). 'The statistical distribution of the maxima of a random function', *Proc. Royal Soc.*, A**237**, 212–232.
13. Holmes, P. and Tickell, R. G. (1976). 'Spectral and probabilistic descriptions of wave loading', *BOSS Conference 1976 WTH Trondheim*/Pergamon Press.
14. Strating, J. (1973). *Fatigue and stochastic loadings*, Delft University.
15. Malhotra, A. K. and Penzien, J. (1970). 'Non-deterministic analysis of offshore structures', *ASCE Journ. Eng. Mech. Div.*, Dec. 1970, 985–1003.
16. Robson, J. D. (1963). *Random Vibrations*, Edinburgh Univ. Press..
17. Davenport, A. G. (1961). 'The application of statistical concepts to the wind loading of structures', *Proc. Inst. Civ. Engrs.*, **19**, Aug. 1961, 449–472.
18. Houmb, D. G. and Overik, T. 'Parameterisation of wave spectra and long term joint distribution of wave height and period', '*Boss 76*', N.T.H. Trondheim (Pergamon Press).
19. Borgman, L. E. (1965). 'The spectral density for ocean wave forces', *Coastal Eng. Speciality Conference*, Santa Barbara, Oct. 1965 (ASCE).
20. Aruliah, P. (1976). 'Fatigue durability under wave action', *Proc. Instn. Civ. Engrs.*, **60**, Part 1, Feb. 1976, 149–152.

ADDITIONAL READING

Pook, L. P. (1974). 'Fracture mechanics analysis of the fatigue behaviour of welded joints', *N.E.L. Report 561*.
Pook, L. P. and Greenan, A. F. (1974). 'Fatigue crack growth threshold for mild steel, a low alloy steel and a grey cast iron', *N.E.L. Report 571*.
Marsh, K. J., Martin, T., and McGregor, J. (1975). 'The effect of random loading and corrosive environment on the fatigue strength of fillet welded lap joints', *N.E.L. Report 587*.
Draper, L. (1966). 'The analysis and presentation of wave data—a plea for uniformity', *Proc. 10th Conf. Coastal Engineering*.
Vughts, J. H. and Kinra, R. K. (1976). 'Probabilistic fatigue analysis of fixed offshore structures', *OTC Paper 2608*.
Hurty, W. C. and Rubenstein, M. F. (1965). *Dynamics of Structures*, Prentice-Hall, New Jersey.

Chapter 10

The Effects of Foundation Radiation Damping on the Response of Gravity Platforms

R. Dungar

10.1 INTRODUCTION

Calculation of the response of gravity platform structures to time varying wave forces and possible earthquake forces is of concern to both the structural and foundation engineer. The former is more accustomed than the latter to treating this force as truly dynamic in, for example, the calculation of the fatigue life of structural members. The foundation engineer is traditionally accustomed to calculating the stability of foundations for given 'peak loading conditions' without regard to the dynamic nature of the applied force. In many cases this comparatively simple 'equivalent-static approach' is indeed justifiable and the foundation may be analysed with little reference to the structure itself. However with certain classes of foundation-structure a true dynamic analysis of the complete system must be performed to facilitate a true representation of the component parts.

The aim of this chapter is to explore the possibility that the gravity structure is part of a truly dynamic system when excited by realistic wave forces and earthquake ground accelerations. To this end particular attention is given to the effect of different foundation conditions, such as foundation modulii, which vary with depth and hard layers at particular depths below the platform. Attention is given to the ability of the foundation to radiate energy away from the structure in terms of compression, shear and Rayleigh Waves.[1] It is well known that this radiation of energy can contribute significantly to the total dissipation of the energy, and is usually known as 'radiation damping'. Other forms of energy dissipation are present in a real system, such as damping due to non-linear material constitutive behaviour, but 'radiation damping' is unique in that its presence transforms the system from a resonant to a non-resonant system, as will be fully explained later. Firstly let us consider the non-radiating system.

317

10.2 THE NON-RADIATING SYSTEM

Consider now the structural system illustrated in Figure 10.1. Here a linear elastic structure is considered which rests on a rigid foundation. The equations of motion for this particular system may be written in the form

$$KX + D\dot{X} + M\ddot{X} = F(t) \tag{10.1}$$

where the stiffness, damping and mass matrices, K, D and M respectively, are for an assumed number of generalized degrees of freedom n. Here the matrix D contains terms due to internal energy dissipation only, and M contains terms due to both the structural inertia and the fluid add-mass, as will be explained later.

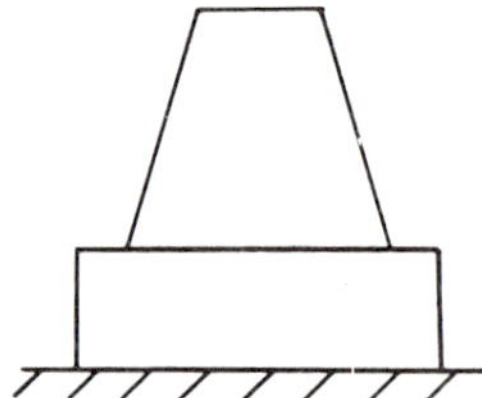

Figure 10.1   Structure on a rigid foundation

In general the forces $F(t)$ are due to a general time-varying loading. However it is convenient to consider the particular case of a set of forces which are all in phase and which vary as a sinusoidal function of time, i.e.

$$F(t) = \bar{F} \sin (\omega t) \tag{10.2}$$

If we now consider the particular case

$$X(t) = \bar{X} \sin (\omega t - \phi) \tag{10.3}$$

where the displacement components are all in phase and all have a phase lag ϕ with respect to the applied forces, it may be shown, by resolving the components of force into the real and imaginary direction, that Equation (10.1) may be rewritten as

$$(K - \omega^2 M)\bar{X} \cos \phi - \omega D\bar{X} \sin \phi + F(t) = 0 \tag{10.4}$$

and

$$(K - \omega^2 M)\bar{X} \tan \phi + D\bar{X} = 0 \tag{10.5}$$

From Equation (10.5), for any given value of ω, there exist n solutions of $\bar{X}$ and corresponding angles ϕ. The vectors $\bar{X}$ are known as the 'characteristic modes' and the corresponding angles ϕ the 'characteristic phase lags'.[2] In the special case when $\phi = \pi/2$ the Equations (10.4) and (10.5) become

$$\omega D\bar{X} = F(t) \tag{10.6}$$

and

$$(K - \omega^2 M)\bar{X} = 0 \tag{10.7}$$

The above Equation (10.7) corresponds to the equation of undamped or free vibration, from which the natural frequencies ω_c^0 and corresponding mode shapes $\bar{X}_i$ may be obtained. Notice that these frequencies and mode shapes may be calculated for all forms of the linear damping matrix D and are independent of the form of D (viscous, hysteretic, etc.).

In general the relative components of any characteristic mode shape $\bar{X}_r$ are dependent upon the frequency ω. That is the 'mode shape' is frequency dependent. This renders the characteristic modes of little use in the calculation of structural response. However if the assumption is made that the shape of all characteristic modes are independent of frequency then it follows that they are also equal to the 'normal modes' or 'natural modes' given from the solution of Equation (10.7). It also follows that the matrix D obeys the orthogonality relationship

$$\bar{X}_i' D\bar{X}_j = 0 \tag{10.8}$$

and Equation (10.1) may be rewritten in the form of 'n' dependent equations of the form

$$\ddot{z}_i + 2c\omega_i \dot{z}_i + \omega_i^2 z_i = \frac{\bar{X}_i' F(t)}{M_i} \tag{10.9}$$

where

$$\bar{M}_i = \bar{X}_i' M\bar{X}_i \tag{10.10}$$

and

$$X(t) = [\bar{X}_1 \bar{X}_2 \cdots \bar{X}_n] \begin{vmatrix} z_1 \\ z_2 \\ \cdot \\ \cdot \\ \cdot \\ z_n \end{vmatrix} \tag{10.11}$$

where c_i is the critical damping coefficient for mode i.

From the above several important results are worthy of note. Firstly the assumption that the characteristic mode shapes are independent of frequency enables these shapes to be calculated using standard eigenvalue-vector techniques.[3] Secondly the calculation of structural response is reasonably straightforward, being given in terms of individual modal contributions as in Equations (10.9) and (10.11). Thus having obtained the natural frequencies and mode shapes a large number of force conditions may be handled with relatively little extra cost and indeed only certain modes need be assumed to contribute to the response, these being within the frequency range of the exciting force.[4] Thirdly let us consider the nature of the system illustrated in Figure 10.1. This system can dissipate energy only by internal damping and there is a possibility that the energy may 'build-up' in the system due to the application of forces in sympathy with one or more of the resonant or natural frequencies ω_i to produce a quasi-resonant situation. From Equation (10.6) it is seen that, at a true resonance, the absolute value of the response is determined by the magnitude of the damping terms D and not simply upon the stiffness of the system. It is well known that as the system approaches resonance the peak response may build up to several times the value of the static response. The condition is known as dynamic magnification and this will now be considered in detail.

10.2.1 Dynamic magnification

The extent of the build up of response amplitude at resonance or quasi-resonance can be measured by the 'dynamic magnification factor' (D.M.F.) defined as

$$\text{D.M.F.} = \frac{\text{Peak dynamic 'response' due to } F(t)}{\text{'response' due to } |F(t)|_{\text{max}} \text{ applied statically}} \qquad (10.12)$$

Thus the D.M.F. provides an indication of the ability of the system to respond 'dynamically' to a given force input $F(t)$. If the D.M.F. is unity, or close to unity, the system may be regarded as 'quasi-static' for the calculation of response to the given forces. That is the displacements and stresses may be calculated from consideration of the stiffness terms only, and the maximum response is given for the minimum force conditions.

Strictly speaking a D.M.F. must be assigned to each mode of vibration and the generalized force, $\bar{X}_i'F(t)$ of Equation (10.9), should replace $F(t)$ in Equation (10.12). However from a practical point of view if the system is considered over a limited frequency range then the response from one mode of vibration will predominate and the expression of Equation (10.12) will suffice, the term 'response' being taken as a typical displacement or stress component within the system.

It is well known that the peak D.M.F. for a system acting as a single degree of freedom system is given by $1/2c$. Thus for example for a concrete structure with typical values of c within the range 0·02 to 0·04, the corresponding range of D.M.F. is 25 to 12·5. That is such a system can achieve magnifications in the displacement and stress response of between 12·5 to 25 times, due to a sinusoidal force being applied at the resonant frequency.

The ability of a non-sinusoidal forcing function to produce a quasi-resonance within a given system depends upon the frequency content of that force compared with the natural frequencies of the system. This will be considered in detail for both wave and earthquake forces later in this chapter.

10.3   THE RADIATING SYSTEM

All structural systems, other than free-free systems like aircraft, are attached to the ground and are therefore able to dissipate energy in the form of radiation damping. The system of Figure 10.2 for example is one in which the structure is regarded as resting on a semi-infinite elastic half-space. Many response calculations are based upon the assumption of a non-radiating system, as illustrated in Figure 10.1. For example much work has been done on the calculation of the response of dams[4] by neglecting the radiation effect, the calculation of response being based upon the use of Equations (10.9) and (10.11). In practice however some radiation damping is present in all non-free-free prototype situations.

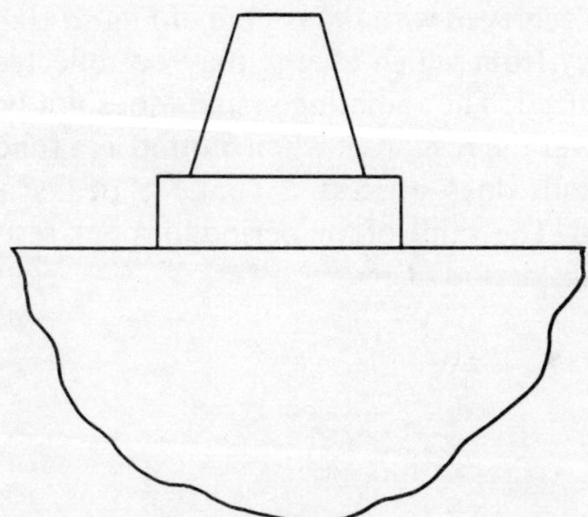

Figure 10.2   Structure on a semi-infinite
elastic half-space

Consider next the idealized system illustrated in Figure 10.3. This system is geometrically symmetrical and possesses modes of vibration which may be classified as either radiating or non-radiating. The symmetrical mode, for example, with $u_1 = -u_2$ is non-radiating in that the mode exerts zero net moment and horizontal force on the foundation, the motion being restricted simply to the bending of the vertical members and extension of the horizontal spring. Thus a true resonance will occur within this mode and the peak D.M.F.

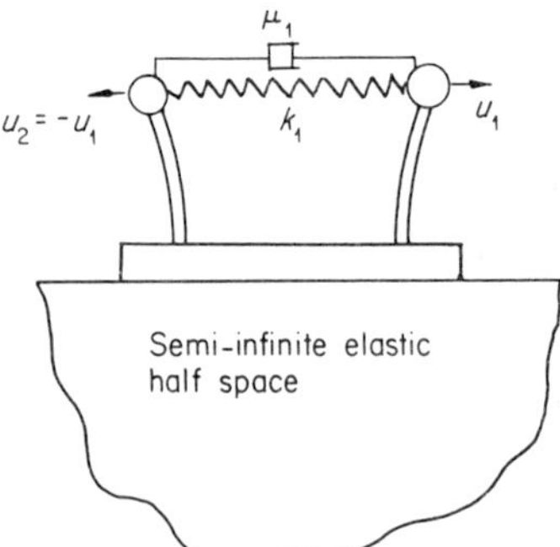

Figure 10.3 Structure with
an assumed rigid base

is simply a function of the damping μ_1. In contrast an anti-symmetrical mode, $u_1 = u_2$, will radiate energy as a net moment and force which will act on the rigid base and cause motion of this base and so dissipate energy in the foundation.

Consider now the system illustrated in Figure 10.4. Here the base is allowed flexibility and thus energy will be radiated in all modes. A truly resonant situation cannot be achieved with the system of Figure 10.4 because resonance requires a boundary from which energy may be reflected so that a standing wave may be produced. The radiating system does not possess true resonant frequencies. However the response, when plotted as a function of frequency of excitation, or period, does possess a peak or peaks at certain values of frequency or period. The value of the period for peak response will be termed 'the ground resonant period' Tg.[5]

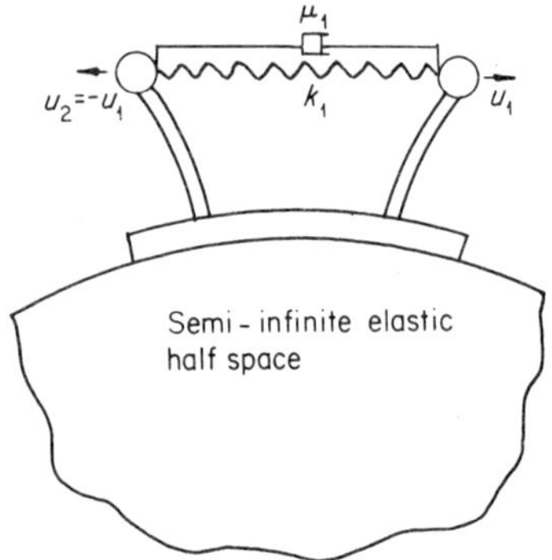

Figure 10.4 Structure with an
assumed non-rigid base

From a practical point of view it is often possible to classify a system as resonant if the radiation of energy is small compared with the energy dissipated due to internal damping. Take for example an arch dam located on relatively rigid foundations. Here it has been demonstrated that both model tests[6] and prototype tests[7] are able to locate resonant conditions, although in the case of the model tests using equipment especially designed to isolate the condition $\phi = \pi/2$ of Equation (10.3), it is notable that although very close agreement to this condition can be obtained for the majority of the model, in the region of the abutments the angle diverges from the desired $\pi/2$ condition. This divergence is attributed to radiation of energy but does not seriously affect the measured resonant properties, as compared with theoretical predictions based upon the solution of Equation (10.7). In practice each structure must be treated on its own merits and the all important 'structural stiffness-foundation stiffness ratio' must be assessed. There is currently an urgent need for positive classification of structural types into the resonant or radiating category.

Returning to Figure 10.3 this of course is an idealized system and, in practice, the decision of whether or not to assume a rigid base as an interface condition must be made with care.[6] From a practical point of view the base may be regarded as 'rigid' only in relation to the stiffness of the spring k and the two vertical members.

In the following sections no such rigid interface condition is assumed but the structure and the foundation are idealized using a descritized approach by employing finite elements in both of these solid regions. The idealization will now be described in detail.

10.3.1 Idealization of the radiating system

The representation of the gravity platform structure and its associated foundation and fluid regions can be accomplished by the use of axisymmetrical finite elements of the 'harmonic type', capable of representing motion in the horizontal and vertical degrees of freedom associated with non-axisymmetrical forces. Thus for example the effects of wave inertia forces and horizontal earthquake motion may be fully represented by the use of such an approach. The details involved in the formulation of the stiffness and inertia of such elements have been presented elsewhere[8,9,10] and will not be repeated here. However a typical mesh representing the structure-foundation is presented in Figure 10.5. Details of the material properties will be presented later.

In the fluid region the effects of compressibility of the fluid, surface waves and viscosity are neglected to enable the add-mass of this fluid to be calculated.[11] This fluid add-mass is included, together with the structural inertia, to form the mass matrix M for the system. Again axisymmetrical fluid elements are employed and are coupled to the structural elements by a transfer matrix

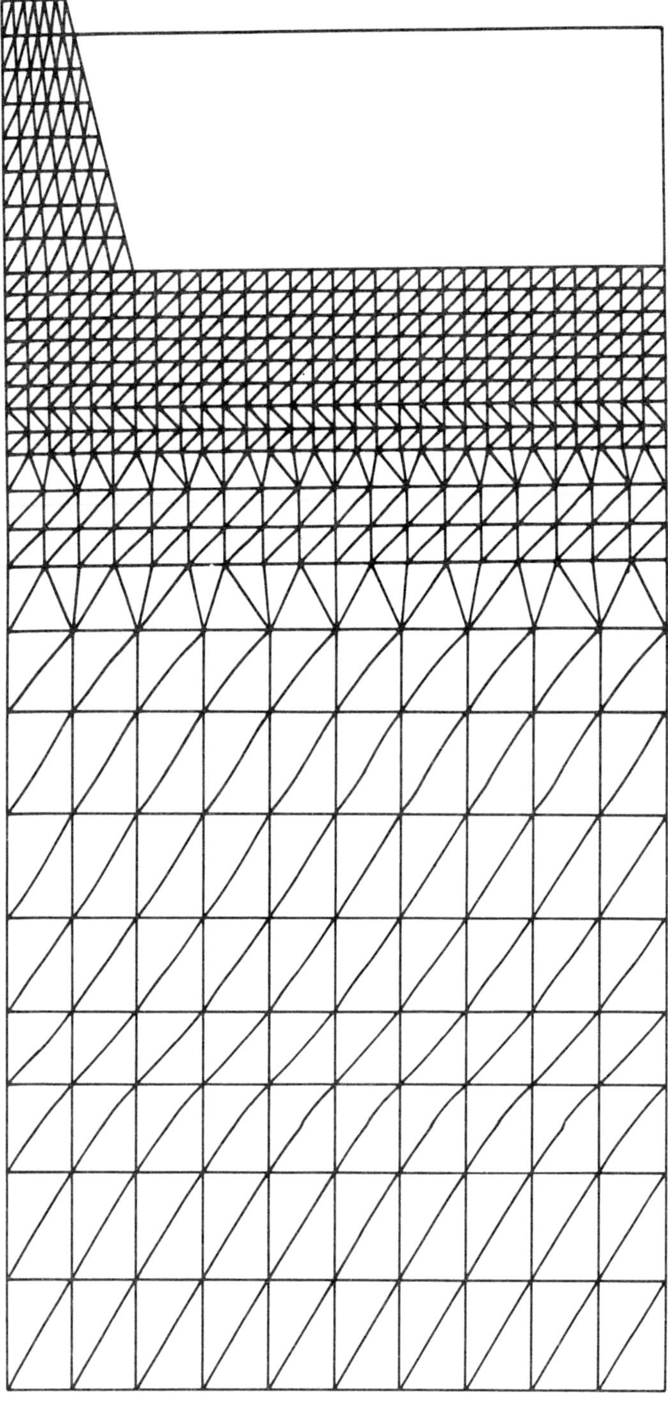

Figure 10.5 Typical structure-foundation-fluid mesh (individual fluid elements not shown)

method, the principles of which are given in Reference 12 and the details, for this type of axisymmetrical idealization, in Reference 13.

Turning now to the problem of representing the radiation damping, much work has been completed for this topic by Lysmer and his co-workers. The basic principles are explained in Reference 14 and the latest techniques including those associated with the representation of a quasi-three-dimensional formulation, such as a plane problem of given out-of-plane thickness, are presented in Reference 15. Essentially the assumed foundation mesh size and size of elements must be chosen with care, so that adequate representation of elastic wave motion is allowed within the considered finite volume, and the dissipation of wave energy at the boundary of the mesh is represented by damping terms.[14] The problems of choosing a suitable mesh size are briefly dealt with in Reference 5.

The equations that result from the above idealization are as given in Equation (10.1) above, where the matrix D now includes both internal damping and the above-mentioned 'radiation damping' terms. The Equation (10.1) may be solved in this case by considering a sinusoidally applied force of the form of Equation (10.2) and obtaining corresponding displacements of the form

$$X(t) = \lceil \sin (\omega t - \alpha_i) \rfloor \psi \tag{10.13}$$

where the brackets $\lceil \rfloor$ indicate a diagonal matrix, the angle α_i refers to the ith component of the vector X and ψ is an amplitude vector. Note that in this case the solution of Equation (10.1) is performed for a particular value of ω and the resulting amplitude vector ψ has no specific phase relationship with the applied force vector $\bar{F}$, as is the case for the 'characteristic modes' $\bar{X}(\omega)$ of Equation (10.3). That is in general the angles α_i are different for all degrees of freedom.

As an alternative to the above solution Equation (10.1) may be solved for a time varying force by employing a step-by-step solution technique. In this case material non-linearity may also be included in the solution by either a full time varying approach or by an equivalent linear approach.[15]

For the following results the solution of the form of Equation (10.13) above is employed and selected components of the response amplitude vector ψ are plotted as a factor of the equivalent component calculated on the basis of applying the vector $\bar{F}$ as a static load. That is the D.M.F. value of Equation (10.12) is plotted, as a function of frequency of excitation ω, for selected structural degrees of freedom.

10.3.2 Steady-state response for seven platform systems

In this section the results for the response of seven platform systems are considered in detail. As previously stated, of particular interest in this chapter is the question of how different foundation conditions can alter the ability of the system to respond dynamically to both wave and earthquaking loading. Thus particular attention is given to the details of the foundation. The results of this section are extracted from Reference 5 and Table 10.1 summarizes the main parameters for the seven cases considered. It is notable from this table that six of the platforms are of the same height, 120 m, whilst one platform of 200 m is considered for comparison. The main difference between the systems is in the assumed foundation material.

As stated in the previous section, the response results are presented as the D.M.F. values of Equation (10.12), plotted for various frequencies of sinusoidal vibration. Absolute response values are not considered in this section, thus a knowledge of the absolute force level is not required. The force $F(t)$ was in fact applied as a series of discrete forces which were vertically distributed to represent a given ratio of horizontal force to base moment. This ratio is presented in Table 10.1. All forces were applied at the same phase, which is reasonable if the sea state is long crested and produces forces on the structure which are of the same phase along a given vertical line, as discussed in Reference 5.

Consider now the results of System A, the system with a uniform foundation. These are presented in Figure 10.6 and represent the response of three selected degrees of freedom,

(i) Horizontal displacement at the edge of the base of the structure, indicating the contribution from sliding motion.

(ii) Vertical displacement at the edge of the base of the structure, indicating the contribution from rocking motion.

(iii) Horizontal displacement at the centre of the top of the structure, indicating a combination of (i) and (ii).

This pattern of presenting results is also followed for the other cases considered. It will also be noted from Figure 10.6 that both the amplitude $\psi_i(\omega)$ and the phase angle $\alpha_i(\omega)$ of Equation (10.13) are plotted. This type of presentation is in contrast to that of plotting the real and imaginary components of the response.[14]

It is notable from Figure 10.4 that the response contains contributions from both sliding motion, curve (i), and rocking motion, curve (ii). This is in contrast to the usual assumptions made in the text of Richart *et al.*[1] It is also notable that the three presented curves are similar in shape and peak amplitude, and that one distinct peak is indicated for the range of frequencies considered. Thus, on the evidence presented, it is reasonable to assume that this 'mode' of vibration is similar to the mode of deformation for a static application of the maximum force $\bar{F}$. Further use of this fact will be made later in this chapter.

The peak response of the system occurs at a value of period known as the 'ground resonant period' Tg which, for system A, is approximately 5·3 s. It is of

Table 10.1   Platform system details

| System | Structure | | | | | | Foundation | | |
	Height	Base diameter	Total mass	Elastic modulus	Base moment to horizontal force ratio	Water depth	Density	Elastic modulus	Poissons ratio
	m	m	kg	kN/m^2	m	m	kg/m^3	kN/m^2	
A	120	120	8×10^8	2×10^7	70	105	2×10^3	3×10^4	0·45
B	120	120	8×10^8	2×10^7	70	105	2×10^3	varying from 3×10^4 at 0 m to 3×10^5 at 400 m	0·45
C	120	120	8×10^8	2×10^7	70	105	2×10^3	3×10^4 with a layer of 1×10^5 from 60 m to 80 m	0·45
D	120	120	8×10^8	2×10^7	70	105	2×10^3	3×10^4 with a layer of 3×10^5 from 60 m to 80 m	0·45
E	120	120	8×10^8	2×10^7	70	105	2×10^3	3×10^4 down to 60 m, thereafter solid	0·45
F	120	120	8×10^8	2×10^7	70	105	2×10^3	3×10^4 with a layer of 3×10^5 from 30 m to 50 m	0·45
G	200	200	37×10^8	2×10^7	117	175	2×10^3	3×10^4	

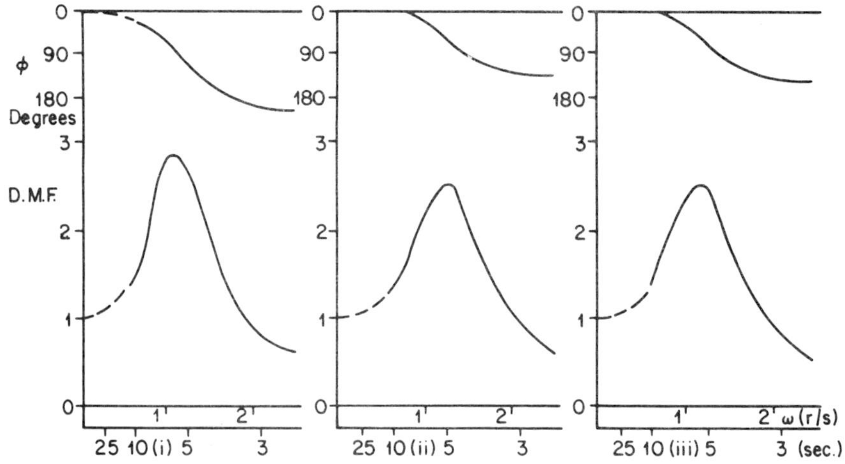

Figure 10.6 Results for system A

interest to note that this peak is approximately identified by the phase relationship of $\alpha = \pi/2$, which corresponds with the classical resonant phase angle $\phi = \pi/2$ of Equation (10.3). It is emphasized however that this situation is only approximate and in fact it would be impossible in practice to excite a classical resonance of such a system because, as can be seen from Equation (10.6), the forces required to procure this situation must contain components in all degrees of freedom associated with damping forces.

The value of Tg is important because, if this value corresponds with the frequency content of maximum wave energy, or earthquake energy, there will be a tendency for a quasi-resonant situation to develop. Also of importance is the ability of the system to accumulate or store this energy and this is indicated by the peak D.M.F. In the case of system A a peak of 2·8 is indicated for degree of freedom 1. This peak may be associated with the value $1/2c$, where c is the critical damping value of a single degree of freedom system. This approach has been used in the text of Richart *et al.*[1] and provides a useful means of assessing the system response in terms of an equivalent single degree system. This equivalent damping $\bar{c}$ will be referred to again later in the chapter and, for system A, a value of approximately 19 per cent critical is indicated.

Having identified the main features of the results of system A we are now able to consider the comparison of this system, of uniform foundation with those of varying foundation condition. The results of Figure 10.7 are presented for system B, which is for a Gibson type foundation, where the value of Young's Modulus is assumed to vary linearly with depth, a situation that is commonly found to be a good approximation in practice. It is notable that the peak response is larger in this case than in the case of system A. This effective

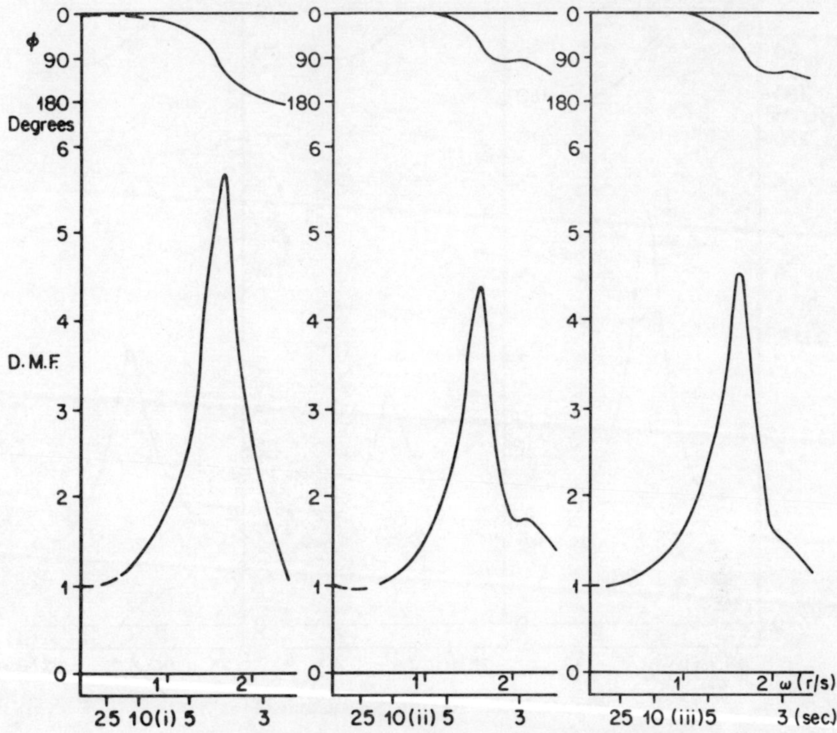

Figure 10.7 Results for system B

decrease in the value of $\bar{c}$, to approximately 10 per cent critical, is also associated with a shift in the value of Tg to the lower value of 3·7 s compared with the value of 5·3 s of system A. The reason for these trends is that the foundation becomes increasingly stiff with depth, thus lowering the period but providing a 'blanket area' for waves to be reflected, so increasing the tendency for a quasi-resonant situation to occur in conjunction with forces of appropriate frequency content.

In contrast to system B the results of system C, with a stiff layer at a depth of half the base diameter (60 m) below the foundation surface, produces results so similar to those of system A that they are not presented here. These results may be found in Reference 5 if required. However when the modulus of this layer is increased to a factor equal to ten times that of the remaining foundation the peak response is significantly increased in the case of sliding motion but slightly reduced in terms of its rocking contribution, as can be seen from Figure 10.8. These results are important as they indicate a trend in significant modification of structural and foundation response which can occur with the presence of

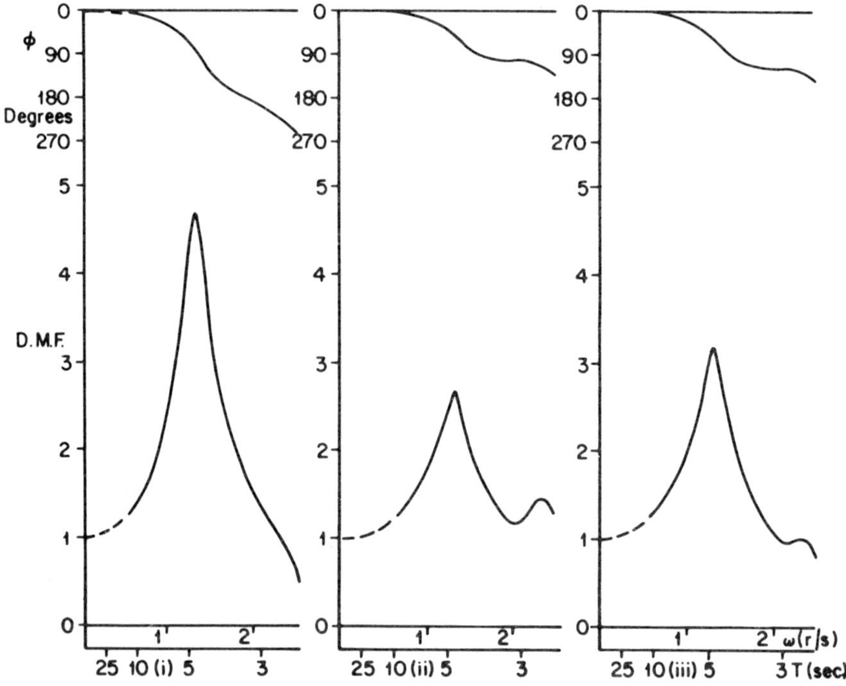

Figure 10.8   Results for system D

dense layers relatively close to the surface of the foundation. The fact that the sliding-shear type of deformation is significantly modified could also result in significant modification in the rate and type of permanent deformation experienced during storm action as compared with a uniform foundation. Very similar results are obtained for system F, the case of a layer of the same modulus and same thickness but in closer proximity to the foundation. This indicates that the stiffness of the layer is far more significant than its proximity to the structure.

Consider now the system E with an assumed solid strata at a depth of 60 m. The results for the sliding and rocking degrees of freedom indicate peak D.M.F. values of 446 and 165 respectively at a period of 4·2 s. These peak values indicate equivalent damping which is far lower than the material damping. In fact no internal damping was taken in this or any other solution in order to isolate the effect of radiation damping. In practice the presence of material damping will limit the response of this system, rather than the radiation of energy in the horizontal direction through the limited depth of relatively soft material.

Finally the results for the 200 m high structure of system G are presented in Figure 10.9. These indicate that there is a shift in the value of T_g to 7·9 s and an

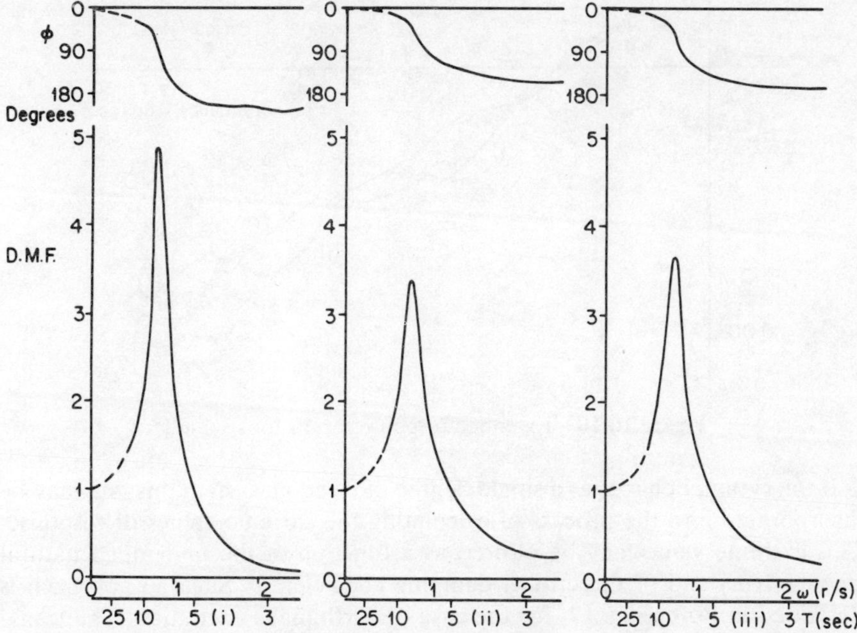

Figure 10.9 Results for system G

increase in the peak D.M.F. value to approximately 5, corresponding to an equivalent damping of 10 per cent. The significance of these results will be discussed later.

Before turning to the problem of calculating the response of the above systems to realistic wave and earthquake forces it is of interest to investigate the above response results in terms of a large range of frequency. Such a plot is presented in Figure 10.10 for the system F and here the frequency range covers that of the range of significant earthquake frequencies. This plot is presented in terms of displacement components (i) and (ii) only.

10.3.3 Frequency characteristics for North Sea storms and earthquakes

The frequency characteristics of an ergodic linear random forcing function may be expressed in terms of the Fourier transform $E(\omega_j)$ of an ensemble.[16] A knowledge of the transfer function $H(\omega_j)$ for the response quantity $x(t)$ enables this response to be defined in the frequency domain $X(\omega_j)$ i.e.

$$X(\omega_j) = H(\omega_j)E(\omega_j) \tag{10.13}$$

Statistics of maxima can be employed to obtain the extreme value of the response.[16]

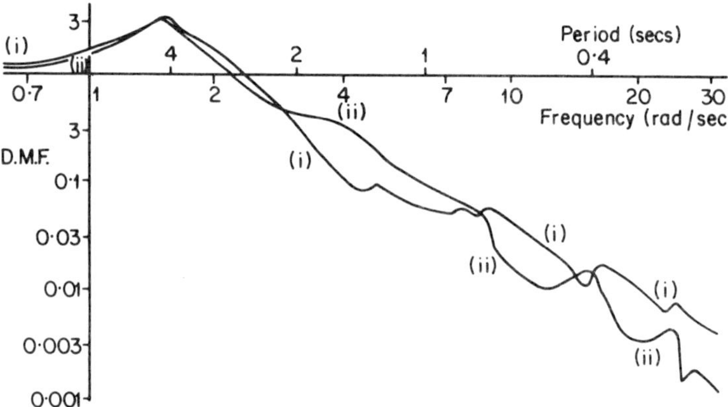

Figure 10.10   Extended response results for system F

If the system behaves as a single degree of freedom system this fact may be incorporated into the process of calculating the extreme values of response. This extreme value may be plotted as a function of the undamped natural frequency ω_i and of the critical damping coefficient c. Such an approach is described in Reference 17, for the case of earthquake excitation. The transform so obtained may be directly compared with that obtained directly from the ensemble in terms of spectral displacement, spectral velocity or spectral acceleration. The conclusion of Reference 17, for the case of earthquake excitation, is 'the computation of the response spectra based on the statistics of maxima of stationary functions represents only a first approximation when compared with the actual response to real earthquakes'. However 'accelerograms are far from ergodic, as they generally represent bursts of energy arriving with the various wave phases'. In the circumstances differences indicated in Reference 17 may be regarded as small, even for earthquake loading.

For the results of this chapter it was decided to use the direct method of 'spectral displacement which involves the implicit assumption that the structure does behave as a single degree of freedom system. This assumption is reasonable, for the structural systems considered, in view of the results of the previous section. This gives a simple method of establishing ψ_i and enables the system to be understood without the need for large numbers of computer runs to investigate the effects of different parameters, provided that spectral displacement curves are available.

Consider now the case of 'North Sea Storm' conditions. Here forty records taken from the North Sea (Famita, 57·5W 13E, 1969–72) have been analysed to establish their corresponding spectral displacement curves.[18] The steps taken to obtain these will now be repeated for completeness. It is assumed that the forces acting on the structure are proportional to either the wave amplitude h or to the wave amplitude squared h^2. That is

$$F(t) = Dh^2(t) + Gh^2(t) + Ih(t + \lambda/4) \qquad (10.14)$$

where D, G, I are generalized force vectors corresponding to unit wave amplitude conditions calculated for a particular characteristic wave period τ and λ is a second characteristic wave period, as will be explained later. The vectors D, G and I may be identified as drag, gravity and inertia vectors respectively. The precise form of these vectors is not required for the following discussion as the concern of this paper is with the frequency content of the wave motion and not with the spatial distribution of forces. However it will be noted from the previous section that a particular relationship between the base moment and horizontal force has already been assumed to enable the structural transfer functions to be obtained (Figure 10.6–Figure 10.10).

Turning now to the wave periods τ and λ, if the wave amplitude is sinusoidal, with a wave period τ, the forces D, G and I may be calculated by using standard wave theory and, in this case, $\lambda = \tau$. Difficulty arises when the amplitude h is a non-sinusoidal function of time, as is the case for realistic storm conditions. Here it will be assumed that the vectors D, G and I may still be calculated for some reference or characteristic wave period τ and by considering a unit wave amplitude. In this case the period λ may or may not be related to τ but in the following discussion the precise value of λ is unimportant as the three components of the force vector $F(t)$ will be later considered separately.

The dynamic displacement of the structure, in terms of the mode $\bar{X}_1$ (assuming a single degree system), may be obtained by substituting Equation (10.14) into Equation (10.9). Considering for example the contribution to the response from the inertia term and dropping the λ term,

$$\ddot{z} + 2c\omega_1\dot{z} + \omega_1^2 z = \frac{\bar{X}_1'}{\bar{M}_1} I . h(t) \qquad (10.15)$$

from which $z(t)$ may be obtained from $h(t)$, except for the unknown scalar factor $k_1 = \bar{X}_1' I / \bar{M}_1$. Also the peak response $\bar{z}_1$ may be obtained for the particular wave record considered. This value is referred to as the spectral displacement. For this paper the value of $\bar{z}_1$, the extreme value of the dynamic displacement, is presented as a proportion of the displacement,

$$z = \frac{\bar{X}_1' I}{\bar{M}\omega_1^2} . \bar{h} \qquad (10.16)$$

which is the displacement response due to applying the force obtained for the minimum wave amplitude $\bar{h}$ as a static force. In this way the factor k_1 is eliminated from this ratio. The ratio so obtained will be referred to as 'the spectral displacement magnification factor' S.D.M.F.

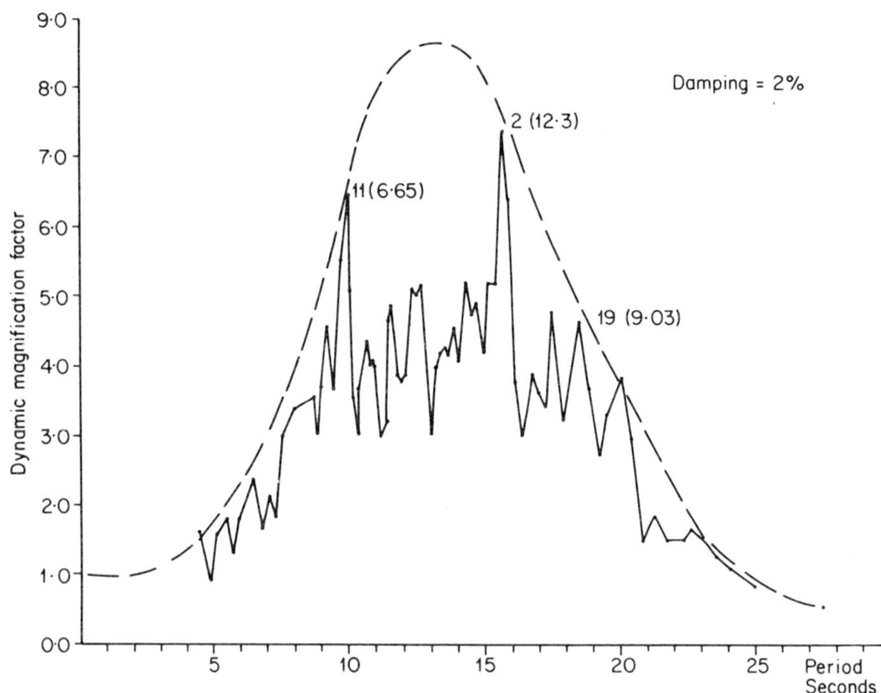

Figure 10.11 Spectral displacement magnification factors for North Sea storms—$h(t)$

As with all spectral displacement plots the S.D.M.F. curve must be plotted as a function of the assumed frequency of the single degree of freedom system ω_1 and different curves are obtained for different values of damping c. If a value of S.D.M.F. of unity is indicated for a particular value of ω_1 this indicates that a single degree of freedom system with resonant frequency ω_1 would react as a quasi-static system. This condition is of particular interest to the design engineer as it indicates that a calculation simply based upon statics and upon forces corresponding to peak force conditions is sufficient to obtain the extreme values of displacement and stress. However if a value in excess of unity is indicated then the system will experience magnification in response, over and above the static response, due to the structural frequency being within the range of the predominant wave frequency.

A typical S.D.M.F. curve, for 2 per cent critical damping, is presented in Figure 10.11. In fact the forty wave records as previously mentioned were used but only the peak response for all records are plotted. The dotted line is an attempt at proposing an envelope of those maxima expected if other storm records were available. Similar curves are obtained for other values of damping and for the case of $h^2(t)$. A complete set of these curves will be found in

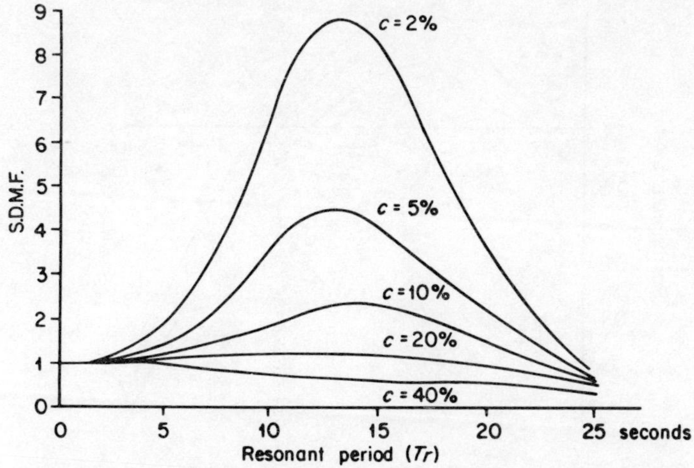

Figure 10.12 Spectral displacement magnification factors for North Sea storms—$h(t)$

Reference 18. A collection of the envelope curves, for the case of $h(t)$, are presented in Figure 10.12 and these will be referred to later.

Turning now to the case of earthquake loading; the forces $F(t)$ imposed upon a structure due to an earthquake are proportional to the free field acceleration of the earthquake $a(t)$. Here it is assumed that the foundation would act as a rigid body if the structure were not present[19] and the difference between the free-field motion and the motion with the structure may be obtained from Equation (10.1) where

$$F(t) = M_s \delta a(t) \tag{10.17}$$

The matrix M_s is the mass matrix due to the structure and the add-mass of the fluid but does not include foundation terms. The vector δ is a direction vector which indicates those degrees of freedom which are in the direction of the acceleration component $a(t)$.

The use of Equation (10.17) with Equation (10.9) again enables the response of the system z to be obtained for a given value of frequency ω_1 and damping value c. As for the case of wave loading this response is plotted as a proportion of the equivalent static response. In this case the 'static' response is that which would be calculated if the maximum acceleration $\bar{a}$ were to be applied continuously to the structure. The resulting ratio of dynamic to static response will be called the 'spectral acceleration magnification factor' and is simply obtained from any spectral acceleration plot by dividing by the peak acceleration of the particular earthquake record. For this chapter the average curves presented by Clough and Penzien[20] hav been replotted in the required form in

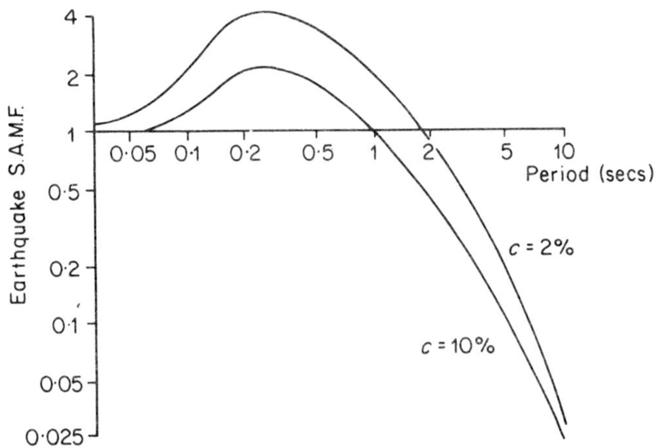

Figure 10.13 Spectral acceleration magnification factors for
average earthquake conditions

Figure 10.13 for 2 per cent and 10 per cent critical damping. These curves indicate the ability of the system with a given value of ω_1 and c to respond in a quasi-static way to an earthquake acceleration, in a similar way to the wave loading curves of Figure 10.12.

10.4 THE QUASI-STATIC/QUASI-RESONANT NATURE OF GRAVITY PLATFORMS

The frequency characteristics of wave and earthquake loading presented in the previous section, in the form of S.D.M.F. and S.A.M.F. curves, form a convenient basis on which to assess the potential for gravity platforms to behave as either quasi-static or quasi-resonant systems. These curves may now be used together with the dynamic characteristics presented earlier for the assumed structural systems. Of particular interest in this chapter is the extreme value of response, denoted by $\bar{\psi}_i$, and this will be expressed as a proportion of the response calculated by applying the peak force as a static force. Thus the value of $\bar{\psi}_i$ so calculated may be related to the D.M.F. factor of Equation (10.12) and represents the ability of the structural system to respond dynamically to the prescribed loading Thus if $\bar{\psi}_i \simeq 1$ then the motion may be regarded as quasi-static as previously explained. Alternatively if $\bar{\psi}_i$ is significantly greater than unity the system may be regarded as quasi-resonant under the action of the prescribed loading.

In a previous section it was explained that the structural system considered may be approximately idealized as single degree system, with ground resonant period T_g and critical damping coefficient c. The results for T_g and c, corresponding to some of the considered systems, are presented in Table 10.2 for the case of combined rocking-sliding motion. These results for T_g and c may now be used to estimate the value of $\bar{\psi}_i$, as described above, by simple reference to the S.D.M.F. and S.A.M.F. curves of Figures 10.12 and 10.13 respectively. The resulting values of $\bar{\psi}_i$ are also presented in Table 10.2 for both wave and earthquake loading. In the case of wave loading only those curves for the forces proportional to the wave amplitude h have been considered as the forces proportional to h^2 give correspondingly lower values of $\bar{\psi}_i$ in each case.

It will be seen from consideration of the results for $\bar{\psi}_i$ presented in Table 10.2 that, for the idealized gravity platforms considered, these structures behave as quasi-static structures in most cases when loaded by waves of storm intensity. The largest value of $\bar{\psi}_i$ for any structure is $1\cdot6$ and this is for the 200 m platform with the largest ground resonant period T_g of value $7\cdot9$ s. The reason for the values of $\bar{\psi}_i$ being approximately equal to unity, in all cases for wave loading conditions, is that the periods T_g are all lower than that required to cuase significant magnification in the response. It can be seen from Figure 10.12 that a period equal to approximately 12 s would be required, together with those damping values indicated, to cause significant amplification in the response over and above the equivalent static response. However if the damping due to radiated energy were to significantly decrease, as in the case of system E with the hard bedrock at a given depth, then the overall damping factor could well be small enough to cause large values of $\bar{\psi}_i$ even for moderately small values of period T_g. In fact, for wave loading, the period range of between 6 and 22 s is the range in which a quasi-resonant situation could develop given a low value of radiation damping c.

In the case of earthquake loading the values of period T_g calculated for the structural systems considered are in excess of the range of values which could possibly cause a quasi-resonant situation to develop. This range is $0\cdot1$ to 1 s, as can be seen from Figure 10.13. The considered structures would be over designed if the calculations for earthquake loading were based upon a static calculation with a peak force level corresponding to the force for peak acceleration. However when using the curves of Figure 10.13 it must be remembered that these are based upon the average curves taken for a large number of earthquakes. The effect of this averaging process does tend to filter out the large values of response that are indicated at certain frequencies when considering only one earthquake record.

The approach adopted in this chapter for calculating the response of gravity structures, in terms of the ratio between the dynamic and the equivalent static response, may be extended to obtain the absolute extreme values of displacement and stress provided that the actual wave or earthquake loading forcing conditions are known. This has been done for prototype conditions for wave

Table 10.2 Summary of response information for five platform systems

System	Ground resonant period			Equivalent damping			ψ_i for North Sea storm conditions			ψ_i for average earthquake conditions (relative to free field conditions)
	(i)	(ii)	(iii)	(i)	(ii)	(iii)	(i)	(ii)	(iii)	
	sec	sec	sec	per cent	per cent	per cent				
A	5·7	5·0	5·2	17·5	19·6	20·0	1·1	1·1	1·1	0·09
B	3·7	3·7	3·7	8·9	11·4	11·1	1·1	1·0	1·0	0·18
D	4·7	4·8	4·8	10·8	18·5	15·8	1·1	1·0	1·0	0·10
E[a]	4·2	4·2	4·2	2·0	2·0	2·0	1·3	1·3	1·3	0·19
G	7·9	7·9	7·9	10·2	14·7	13·7	1·6	1·4	1·4	0·045

[a] Value of damping assumed to be 2 per cent, and not extracted from peak D.M.F. values for the structural response of Figure 8.

loading and the results may be found in Reference 21. Also corresponding results for earthquake loading may be found in Reference 19.

10.5 CONCLUSIONS

Gravity platforms, together with associated foundation and fluid regions, have been idealized using finite elements of the axisymmetrical type. The effect of energy radiated away from the foundation is included in the solution, termed radiation damping, and special attention has been given to the effect of varying conditions in the foundation material. It is shown that the presence of a hard material layer within the foundation tends to increase the response of the platform due to a reduction in the effective damping provided by radiated energy. The magnitude of the stiffness of this layer is more significant than its proximity to the surface of the foundation. In the limit when the stiffness is such that the hard layer acts as a rigid boundary the effect of radiation damping may be neglected in relation to realistic values of internal damping. In this extreme case large values of response may be expected for forces with frequency or period content matching that of the ground resonant period of the system.

For the idealized systems considered with prototype height of 120 m except for one system of 200 m, and predominant foundation modules of $3 \times 10^4 \text{ kN/m}^2$ it is shown that values of ground resonant period T_{vg} are in the range of 3·7 to 7·9 s. The fundamental mode of vibration contains contributions from both rocking and sliding motion and, within the range of periods considered, the system behaves approximately as a single degree system. The use of this fact enables the response of the system to be calculated by reference to spectral displacement and spectral acceleration curves for wave and earthquake loading respectively.

It is shown that for realistic North Sea storm conditions the considered systems may be regarded as quasi-static, in the main, with the departure to a quasi-resonant situation being for the larger structure of 200 m height and for the case of a rigid layer at a given depth below the base of the platform. For the case of earthquake loading the ground resonant periods are above the predominant periods of average earthquake data and the response is correspondingly considerably less than that due to a quasi-static analysis, that is, due to the application of the maximum acceleration applied continuously to the system.

REFERENCES

1. Richart, F. E., Hall, J. R., and Woods, R. D. (1970). *Vibrations of Soils and Foundations*, Prentice-Hall.
2. Bishop, R. E. D. and Gladwell, G. M. (1963). 'An investigation into the theory of resonance testing', *Proc. Roy. Soc.*, **255**, A 1055.

3. Wilson, E. L. (1977). Numerical Methods for Dynamic Analysis, *Int. Symposium on Numerical Methods: Offshore Engineering*, Swansea.

4. Back, P. A. A., Cassell, A. C., Dungar, R., Gaukroger, D. R., and Severn, R. T. (1969). 'The seismic design study of a double curvature arch dam', *Proc. Instn. Civ. Engrs.*, **43**, 217–248.

5. Dungar, R. and Eldred, P. J. L. 'The dynamic response of gravity platforms', to be published by *International Journal of Earthquake Engineering and Structural Dynamics*.

6. Dungar, R. and Severn, R. T. (1971). 'The experimental vibration and response analysis of water retaining structures', *Dynamic Waves in Civil Engineering*, D. A. Howells *et al.* (eds.), Wiley.

7. Rouse, G. C. and Bowkamp, J. G. (1964). 'Vibration study of the Monticello Dam', *Water Research Publication, Report No. 9*, U.S. Bureau of Reclamation.

8. Wilson, E. L. (1965). 'Structural analysis of axisymmetric solids', *AIAA Journ.*, **3**, No. 12, Dec. 1965.

9. Dungar, R. (1972). 'The Three-dimensional analysis of structural foundations during strong motion earthquakes', *4th European Symposium on Earthquake Engineering*, London.

10. Dungar, R. (1972). 'The dynamic wind stresses in the clay foundation of a tall building', *Application of Experimental and Theoretical Structural Dynamics*, Southampton.

11. Zienkiewicz, O. C. (1977). 'Dynamic fluid-structure interaction–numerical modelling of coupled problems', *Int. Symp. on Num. Methods: Off-shore Structures*, Swansea.

12. Dungar, R. and Severn, R. T. (1975). 'A resume of experience gained in the numerical analysis of arch dams during the period 1964-74', *Int. Symposium on Criteria and Assumptions for Numerical Analysis of Dams*, Swansea, 364–386.

13 Dungar, R. and Jackson, E. A. (1975). 'The seismic analysis of the Bellmouth Spillway and value tower for an earth dam', *Int. Symp. on Criteria and Assumptions for Numerical Analysis of Dams*, Swansea, 603–624.

14. Lysmer, J. and Kuhlemeyer, R. L. (1969). 'Finite dynamic model for infinite media', *J. Eng. Mech. Div. (ASCE)*, EM4, 859–877, Aug. 1969.

15. Lysmer, J., Udaka, T., Tsai, C. F., and Seed, H. B. (1975). ' "Flush": a computer program for approximate 3-D analysis of soil-structure interaction problems', Earthquake Engineering Research Centre, Berkeley, California, *No. EERC 75-30*.

16. Penzien, J. and Tseng, W. S. (1977). 'Three-dimensional dynamic analysis of fixed offshore platforms', *Int. Symp. on Num. Methods in Offshore Engineering*, Swansea.

17 Udwadia, F. E. and Trifunac, M. D. (1973). 'The fourier transform, response spectra and their relationship through the statistics of oscillator response', *Report EERL 73-01*, Cal. Tech. Pasadena.

18. Dungar, R., Eldred, P. J. L., and Severn, R. T. (1976). 'Dynamic interaction of the structure-fluid-foundation system of gravity platforms', *Int. Conf. on Behaviour of Off-shore Structures*, Trondheim, Norway.

19. Dungar, R., and Eldred, P. J. L. (1977). 'The effect of earthquakes on the foundation stability of gravity oil platforms', *Sixth World Conference on Earthquake Engineering*, Delhi.

20. Clough, R. W. and Penzien, J. (1973). *Dynamics of Structures*, McGraw-Hill.

21. Dungar, R., Eldred, P. J. L., and Haws, E. T. (1977). 'The dynamic analysis of the foundation of a North Sea gravity platform', *Ninth Int. Conf. of Int. Soc. for Soil Mech. and Foundation Engineering*, Tokyo.

Chapter 11

Non-linear Response of Structure-Fluid-Foundation Systems to Earthquake Excitation

C. T. Chang, E. Hinton and O. C. Zienkiewicz

SUMMARY

The finite element method is employed in the study of non-linear response of structure-fluid-foundation systems subjected to earthquake excitation. An explicit central difference time marching scheme is adopted for the solution of the dynamic equilibrium equations. An elasto/viscoplastic constitutive relationship is used to model the material behaviour. Elastic and elastoplastic behaviour can be obtained from the elasto/viscoplastic formulation by adopting suitable values of the fluidity parameter. Finally, non-linear seismic analyses of an embankment dam and an offshore structure are presented.

11.1 INTRODUCTION

A full study of the non-linear response of a structure-fluid-foundation system, subjected to earthquake excitation is still a matter of research. While, in principle, solutions can be achieved by various step by step procedures, to date few solutions have been published and the procedures widely used in U.S.A. follow the work of Seed and Lysmer[1,2] involving a sequence of linear analyses in which the secant moduli of the materials are amended as the computation proceeds.

While such methods give reasonable predictions of the possible areas of liquefaction or overstress, they are incapable of giving any indication of permanent deformation or the extent of the damage which may occur after the passage of an earthquake.

Such procedures cannot, therefore, be considered conservative and a more precise evaluation by a finite element non-linear solution may be necessary for important structures. Preliminary studies indicate that this is possible and an explicit numerical procedure appears to be reasonably effective in this context.

In this paper we shall describe some applications of such a process to hypothetical structures with various degrees of soil viscoplasticity and observe their response to the passage of an earthquake.

A viscoplastic rather than purely plastic model is postulated to allow for rate effects which are undoubtedly important but for which appropriate physical constants are not yet available. Purely plastic behaviour can be obtained from the viscoplastic model as a limiting case.

11.2 NOMENCLATURE

It is useful at this point to lay out the notation which is employed in this paper. This notation follows closely that of Bathe *et al.*[3]

The motion of the body is considered in a fixed Cartesian co-ordinate system in which all kinematic and static variables are defined.

The co-ordinates describing the configuration of the body are ${}^{0}x_1$ and ${}^{0}x_2$, initially a time 0, ${}^{t}x_1$ and ${}^{t}x_2$ at time t and ${}^{t+\Delta t}x_1$ and ${}^{t+\Delta t}x_2$ at time $t+\Delta t$ where the left superscripts refer to the configuration of the body and the right subscripts to the co-ordinate axes.

The notation for the relative displacements of the body is similar to the notation for the co-ordinate; at time t the displacements are ${}^{t}u_i$, $i=1,2$ and at time $t+\Delta t$ the displacements are ${}^{t+\Delta t}u_i$, $i=1,2$, therefore,

$$\left.\begin{array}{c}{}^{t}x_i = {}^{0}x_i + {}^{t}u_i \\ {}^{t+\Delta t}x_i = {}^{0}x_i + {}^{t+\Delta t}u_i\end{array}\right\} i=1,2$$

The unknown increments in the displacements from time t to $t+\Delta t$ are denoted as

$$u_i = {}^{t+\Delta t}u_i - {}^{t}u_i \qquad i=1,2$$

During motion of the body, its volume, surface area, mass density, stresses and strains are assumed independent of the configuration. The specific mass, area and volume of the body at time 0 are ${}^{0}\rho$, ${}^{0}A$ and ${}^{0}V$ respectively.

The static body force components per unit mass at time $t+\Delta t$ are measured in configuration 0, ${}^{t+\Delta t}f_k$, $k=1,2$.

Considering stresses, the Cartesian components of the Cauchy stress tensor at time $t+\Delta t$ are denoted by ${}^{t+\Delta t}\sigma_{ij}$.

Considering strains, the Cartesian components of Cauchy's infinitesimal strain tensor referred to the configuration at time $t+\Delta t$ are denoted by ${}^{t+\Delta t}e_{ij}$.

In the formulation of the governing equilibrium equations, derivatives of displacements and co-ordinates are considered. In the notation adopted, a comma denotes differentiation with respect to the co-ordinate following. Thus, for example

$$u_{i,j} = \frac{\partial u_i}{\partial x_j}$$

11.3 FORMULATION OF EQUATIONS OF MOTION

Since the displacement based finite element procedure is to be adopted in conjunction with an explicit time integration scheme, the principle of virtual displacements is used to express the equilibrium of the body in the configuration at time t. Assuming that the direction and magnitude of the body loading and inertia loading is independent of the configuration (i.e. path-independent loading only is considered) and assuming infinitesimal displacements, the principle of virtual displacements requires that[4,5,6]

$$\int_{^0V} {^0\rho}\,{^t\ddot{u}_k}\delta u_k\,{^0}\mathrm{d}V + \int_{^0V} {^t\sigma_{ij}}\delta\,{^t e_{ij}}\,{^0}\mathrm{d}V = \int_{^0A} {^t f_k}\delta u_k\,{^0}\mathrm{d}A + \int_{^0V} {^0\rho}\,{^t\ddot{u}_k}\delta u_k\,{^0}\mathrm{d}V$$

$$k = 1, 2, \; i, j = 1, 2, 3 \quad (11.1)$$

In Equation (11.1) δu_k is a virtual variation in the current displacement components ${^t u_k}$ and $\delta\,{^t e_{ij}}$ are the corresponding virtual variations in strains, i.e.

$$\delta\,{^t e_{ij}} = \delta\tfrac{1}{2}({^t u_{i,j}} + {^t u_{j,i}}) \qquad (11.2)$$

It should be noted that in Equation (11.2) the summation convention of tensor notion is implied. The component of acceleration in direction k due to the earthquake is given as $\ddot{u}_k$.

The equilibrium Equation (11.1) is now used in the development of an explicit central difference time integration scheme in which a finite element spatial discretization is adopted. However before the discretized equation of motion is considered, the constitutive relations assumed are briefly described.

11.4 STRESSES, STRAINS AND CONSTITUTIVE RELATIONSHIPS

It is well known that the saturated soils encountered in many geotechnical problems such as those connected with the foundations of offshore structures or earthdams can be treated as two phase continua with a matrix of solid particles (the soil skeleton) and with water filling the space not occupied by solids. The water is considered to be an incompressible liquid. Although porewater may flow through the soil under the influence of excess pore pressures,[7] in the present studies it is assumed that no seepage flow occurs due to short term loading, i.e. the undrained condition is considered. Since water cannot resist a static shear stress, at any point within the soil the pore pressure p

acts uniformly in all directions and its effect is to put all solid components into a state of uniform hydrostatic compression. Thus, the total stress components $^{t+\Delta t}\sigma_{ij}$ present in the soil are divided into two parts; the effective stress components $^{t+\Delta t}\sigma'_{ij}$ and the pore pressure $^{t+\Delta t}p$ and can be presented as follows:

$$^{t+\Delta t}\sigma_{ij} = {}^{t+\Delta t}\sigma'_{ij} + \delta_{ij}{}^{t+\Delta t}p \tag{11.3}$$

Deformations or strains of the soil skeleton arise only if there is a change in the effective stress which acts on the soil skeleton. (In certain circumstances the pore pressure may reach a value equal to the total stress and lead to zero effective stress and the soil may liquefy, i.e. the soil flows like a liquid. This effect is not shown in the present study.) In the elasto/viscoplastic model adopted it can be assumed that the total strain $^{t}e_{ij}$ can be resolved into an elastic part $^{t}e_{ij}^{E}$ and a viscoplastic part $^{t}e_{ij}^{VP}$,[8,9] viz.

$$^{t}e_{ij} = {}^{t}e_{ij}^{E} + {}^{t}e_{ij}^{VP} \tag{11.4}$$

The viscoplastic strain which represents combined viscous and plastic effects is discussed later in this section. The elastic part is related to the elastic stress by the material property tensor E_{ijkl}. If the constants of this tensor are specified in terms of the effective stress moduli only, then this relationship can be used to define the effective stress and elastic strain by the following expression:

$$^{t}\sigma'_{ij} = E_{ijkl}{}^{t}e_{kl}^{E} \tag{11.5}$$

If the soil is subjected to any initial effective stress $^{0}\sigma'_{ij}$ the complete effective stress state can be written as

$$^{t}\sigma'_{ij} = E_{ijkl}({}^{t}e_{kl} - {}^{t}e_{kl}^{VP}) + {}^{0}\sigma'_{ij} \tag{11.6}$$

As pointed out previously, for a saturated soil under earthquake loading, the exciting force is applied so suddenly that it can be assumed no flow of pore fluid takes place in the interstitial pores. Consequently a complete undrained condition applies and the two phases deform and produce a common strain in both solid and fluid phases. In other words, the change in effective stress and the change in excess pore water pressure can therefore both be related to the same common change in strain if the soil is assumed to be fully saturated. Furthermore, the change in excess pore water pressure can be defined as

$$p_{ij} = k_a \delta_{ij}\delta_{kl}e_{kl} \tag{11.7}$$

in which k_a represents the apparent bulk modulus of compressibility of the pore fluid. This change in excess pore water pressure results in a change of volumetric strain e_{kk} in the soil skeleton which can be expressed as[10,11]

$$e_{kk} = \frac{n}{3k_f}p_{kk} + b + \frac{1}{3k_b}p_{kk} \tag{11.8}$$

where k_f and k_b are respectively the bulk moduli of the pore fluid and the soil skeleton and n is the porosity of the soil matrix. From Equations (11.7) and (11.8), k_a can readily be obtained as

$$k_a = \frac{1}{(n/k_f)+(1/k_b)} \tag{11.9}$$

If the pore fluid is assumed incompressible, i.e. $k_f \to \infty$ and porosity is small, then Equation (11.9) can be expressed simply as

$$k_a = k_b \tag{11.10}$$

in which k_b is known as the rebound modulus of the soil.[12] If the initial pore water pressure is ${}^0p$ then the total pore water pressure at $t+\Delta t$ is given as

$${}^{t+\Delta t}p_{ij} = {}^t p_{ij} + k_a \delta_{ij}\delta_{kl}e_{kl} + \delta_{ij}{}^0p \tag{11.11}$$

It is of interest to note that for drained condition the pore pressure remains unchanged, i.e.

$${}^{t+\Delta t}p_{ij} = \delta_{ij}{}^0p \tag{11.12}$$

Upon substitution of Equations (11.6), (11.11) and (11.12) into Equation (11.3) the following expression is obtained

$${}^{t+\Delta t}\sigma_{ij} = C_{ijkl}{}^{t+\Delta t}e_{kl} - E_{ijkl}{}^{t+\Delta t}e_{kl}^{VP} + {}^0\sigma_{ij}' + \delta_{ij}{}^0p \tag{11.13}$$

in which

$$C_{ijkl} = E_{ijkl} + k_a\delta_{ij}\delta_{kl}$$

for the undrained condition

$$C_{ijkl} = E_{ijkl}$$

for the drained condition.*

In the analysis of elasto/viscoplastic material behaviour, the total incremental strain can be written in the form

$$e_{ij} = e_{ij}^E + e_{ij}^{VP} \tag{11.14}$$

where the instantaneous response is elastic and the viscoplastic component is considered to be a delayed effect and includes both the viscous and plastic effects lumped together. The delayed viscoplastic strain can be treated in the same manner as prescribed initial strain. The viscoplastic flow rule implies that viscoplastic strain rates occur only when the state of stress exceeds a certain threshold static yield surface. In associative form, the flow rule is given by[9,13,14,15,16,17]

$${}^t\dot{e}_{ij}^{VP} = \gamma\langle\phi(F)\rangle\frac{\partial F}{\partial{}^t\sigma_{ij}'} \tag{11.15}$$

---

* For a full discussion of soil behaviour see Chapter 12.

where γ is a material fluidity parameter, F a scalar yield function and the notation $\langle\,\rangle$ implies that

$$\langle\phi(F)\rangle = \begin{cases} \phi(F) & \text{when } \phi(F)>0 \\ 0 & \text{when } \phi(F)\leq 0 \end{cases}$$

It may be noted that the viscoplastic strain rate exists only when $\phi(F)$ exceeds zero and depends on the amount by which the static yield surface is exceeded. Different functional forms have been introduced for $\phi(F)$ to describe the process. The choice of function $\phi(F)$ must be in accordance with the rate dependent properties of the materials considered. The simplest form is used in this study, viz.,

$$\phi(F)=F \tag{11.16}$$

while $F>0$ constitutes an inadmissible state in the theory of inviscid plasticity, the plastic state being identified by $F=0$, it is admissible in viscoplasticity and causes viscoplastic deformations to take place. However, elastoplastic problems can be solved by the elasto/viscoplastic algorithm as a limiting case in which $\gamma\rightarrow\infty$.

For non-associative laws $(\partial Q/\partial^t\sigma'_{ij})$ replaces $(\partial F/\partial^t\sigma'_{ij})$ in Equation (11.15) where Q is a separate plastic potential.[18]

Various yield functions and plastic potentials can be incorporated in the formulation depending on the nature of the materials used.[19,20] Mohr–Coulomb and a simplification due to Drucker and Prager[21] are employed here. Details are described by Nayak and Zienkiewicz.[22,23] For Mohr–Coulomb

$$F = \sigma_m \sin\phi + \left(\cos\theta - \frac{\sin\theta\sin\phi}{\sqrt{3}}\right)\sqrt{(J_2)} - c\,\cos\phi \tag{11.17}$$

For Drucker–Prager

$$F = \frac{\sqrt{(3)}\sin\phi}{(3+\sin^2\phi)^{1/2}}\sigma_m + \sqrt{(J_2)} - \frac{\sqrt{(3)}c\,\cos\phi}{(3+\sin^2\phi)^{1/2}} \tag{11.18}$$

in which

$$\sigma_m = \tfrac{1}{3}\sigma_{ii}$$

$$S_{ij} = \sigma_{ij} - \delta_{ij}\sigma_m$$

$$J_2 = \tfrac{1}{2}S_{ij}S_{ij}$$

$$J_3 = \tfrac{1}{3}S_{ij}S_{jk}S_{ki}$$

θ, the Lode angle, is defined as

$$-\frac{\pi}{6}\leq\theta = \frac{1}{3}\sin^{-1}\left(-\frac{3\sqrt{(3)}}{2}\frac{J_3}{J_2^{3/2}}\right)\leq\frac{\pi}{6}$$

c is the cohesion and ϕ is the angle of internal friction respectively.

It is well known that Tresca and Von Mises yield surfaces can be readily obtained by simply setting $\phi = 0$ in Equations (11.17) and (11.18) respectively. These yield surfaces are shown in Figure 11.1.

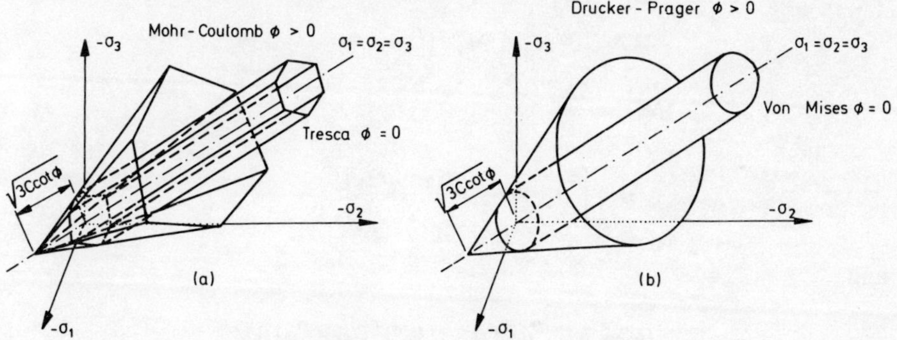

Figure 11.1   Yield surfaces

Using the viscoplastic strain rate given by Equation (11.15), the total viscoplastic strain at time $t + \Delta t$ can be obtained as

$$^{t+\Delta t}e_{ij}^{VP} = {}^{t}e_{ij}^{VP} + \int_{t}^{t+\Delta t} {}^{\tau}\dot{e}_{ij}^{VP} \, d\tau \tag{11.19}$$

If an assumption of constant viscoplastic strain rate during each time step is made, then the second term of the RHS in Equation (11.19) can be written as ${}^{t}\dot{e}_{ij}^{VP}\Delta t$.

11.5 FINITE ELEMENT SPATIAL DISCRETIZATION

An isoparametric finite element solution is adopted in which the co-ordinates and displacements at any point can be represented as

$$^{0}x_i = N_i^m \, {}^{0}x_i^m$$

$$^{t}u_i = N_i^m \, {}^{t}u_i^m \qquad i = 1,2 \qquad m = 1,r \tag{11.20}$$

in which $^{0}x_i^m$ is the co-ordinate of nodal point m corresponding to direction i at time 0, $^{t}u_i^m$ is the displacement at notal point m corresponding to direction i at time t, N_i^m is the interpolation function of nodal point m corresponding to direction i and r is the total number of nodal points.

The finite element discretization implied by Equation (11.20) can be used together with Equation (11.1) to obtain the following equations:

$$M_{ij}^{mn}\,{}^{t}\ddot{u}_j^n + \int_{{}^{0}V} [\mathbf{B}_i^m]^{Tt}\boldsymbol{\sigma}\,dV = {}^{t}R_i^m \qquad i,j = 1,2 \qquad m,n = 1,r \tag{11.21}$$

where

$$M_{ij}^{mn} = \delta_{ij} \int_{^0V} N_i^m N_j^n \, {}^0\rho \, {}^0\mathrm{d}V$$

$$\mathbf{B}_1^m = [N_1^{m,1}, 0, N_1^{m,2}]^T$$

$$\mathbf{B}_2^m = [0, N_2^{m,2}, N_1^{m,1}]^T$$

$$N_i^{m,j} = \frac{\partial N_i^m}{\partial x_j}$$

$${}^t\boldsymbol{\sigma} = \left[{}^t\sigma_{11}, {}^t\sigma_{22}, \frac{{}^t\sigma_{12} + {}^t\sigma_{21}}{2} \right]^T$$

and

$${}^t R_i^m = M_{ij}^{mn} \, {}^t\ddot{u}_j^n + \int_{^0V} [\mathbf{B}_i^m]^T ({}^0\boldsymbol{\sigma} + {}^0\mathbf{u}) \, \mathrm{d}V$$

If a special mass lumping scheme[24] is adopted so that

$$M_{ij}^{mn} = M_{(i)}^{(m)} \delta_{ij} \delta_{mn} \qquad \text{(implying no sum on } m \text{ or } i) \qquad (11.22)$$

then the first term in Equation (11.22) can be written as

$$M_{ij}^{mn} \ddot{u}_j^n = M_{(i)}^{(m)} \delta_{ij} \delta_{mn} \ddot{u}_j^n \qquad (11.23)$$

or

$$M_{ij}^{mn} \ddot{u}_j^n = M_{(i)}^{(m)} \ddot{u}_i^m$$

Equation (11.21) can thus be rewritten as

$$M_{(i)}^{(m)t} \ddot{u}_i^m + \int_{^0V} [\mathbf{B}_i^m]^{Tt} \boldsymbol{\sigma} \, {}^0\mathrm{d}V = {}^t R_i^m \qquad i = 1, 2; \, m, n = 1, r \quad (11.24)$$

Thus there is a separate uncoupled equation associated with each nodal variable u_i^m and a central difference time stepping algorithm can be adopted. Using Equation (11.23) ${}^t R_i^m$ can now be rewritten in the form

$${}^t R_i^m = M_{(i)}^{(m)t} \ddot{u}_i^m + \int_{^0V} [\mathbf{B}_i^m]^T ({}^0\boldsymbol{\sigma} + {}^0\mathbf{u}) \, {}^0\mathrm{d}V \qquad i = 1, 2, \qquad m, n = 1, r$$
$$(11.25)$$

It should be noted that the effect of the fluid (water) on the dynamic behaviour of the structure-foundation system can be modelled using finite elements to model the fluid directly[7,25] or, as in the case of the present work, by use of the 'added mass' concept.[6,26,27] Damping effects can be incorporated in the standard way.[25] All integration is performed numerically element by element in the usual manner.

11.6   CENTRAL DIFFERENCE TIME STEPPING ALGORITHM

Equation (11.24) can be solved using the central difference time stepping scheme[25,28,29] which can be expressed in the present case as

$$M_{(i)}^{(m)}(^{t-\Delta t}u_i^m - 2\,^t u_i^m + \,^{t+\Delta t}u_i^m)/(\Delta t)^2 + \int_{^0V}[\mathbf{B}_i^m]^T\,{}^t\boldsymbol{\sigma}\,dV - \,^t R_i^m = 0$$

$$i = 1, 2;\ m = 1, r \quad (11.26)$$

If values of $^t u_i^m$ and $^{t-\Delta t}u_i^m$ are known then the value of $^{t+\Delta t}u_i^m$ can be readily obtained as

$$^{t+\Delta t}u_i^m = [(\Delta t)^2(^t R_i^m - \int_{^0V}[\mathbf{B}_i^m]^T\,{}^t\boldsymbol{\sigma}\,dV) + 2M_{(i)}^{(m)}\,{}^t u_i^m - M_{(i)}^{(m)}\,{}^{t-\Delta t}u_i^m]/M_{(i)}^{(m)}$$

$$(11.27)$$

To initiate the solution of Equation (11.27), values of $^0 u_i^m$ and $^{-1}u_i^m$ are required. At the start of the excitation, the system is generally assumed to be at rest, i.e. $^0 u_i^m = \,^0\dot{u}_i^m = 0$ and since

$$^0\dot{u}_i^m \approx (^0 u_i^m - \,^{-\Delta t}u_i^m)/\Delta t \quad (11.28)$$

a value of $^{-\Delta t}u_i = 0$ is assumed.

A brief flow chart of the central difference scheme is given in Figure 11.2.

1. Initialization. Set $t = 0$, $^{-\Delta t}u_i = 0$, $^0 u_i = 0$. Select $\Delta t <$ critical time step lengths given by Equation (11.29) and Equation (11.30).

Evaluation of displacements $^{t+\Delta t}u_i^m$

2. Form $\int_{0V}[\mathbf{B}_i^m]^T\,{}^t\boldsymbol{\sigma}\,dV$ and $^t R_i^m$.
3. Compute $^{t+\Delta t}u_i^m$ from known values of $^{t-\Delta t}u_i^m$ and $^t u_i^m$ using Equation (11.27).

Evaluation of $^{t+\Delta t}\boldsymbol{\sigma}$ *and* $^{t+\Delta t}\dot{\mathbf{e}}^{VP}$ (required for next time step)

4. Calculate the strains $^{t+\Delta t}\mathbf{e} = [\mathbf{B}_i^m]^T\,{}^{t+\Delta t}u_i^m$ where $^{t+\Delta t}\mathbf{e} = [^{t+\Delta t}e_{11}, \,^{t+\Delta t}e_{22}, \,^{t+\Delta t}e_{12} + \,^{t+\Delta t}e_{21}]$.
5. Calculate the elastic strains $^{t+\Delta t}\mathbf{e}^E = \,^{t+\Delta t}\mathbf{e} - \,^{t+\Delta t}\mathbf{e}^{VP}$ where $^{t+\Delta t}\mathbf{e}^{VP}$ is formed from $^{t+\Delta t}e_{ij}^{VP}$ given by Equation (11.19).
6. Calculate the effective stresses $^{t+\Delta t}\boldsymbol{\sigma}'$ from $^{t+\Delta t}\sigma_{ij}'$ given by Equation (11.6).
7. Calculate the pore pressures $^{t+\Delta t}\mathbf{p}$ from $^{t+\Delta t}p_{ij}$ given by Equation (11.11).
8. Calculate the total stresses $^{t+\Delta t}\boldsymbol{\sigma}$ from $^{t+\Delta t}\sigma_{ij}$ given by Equation (11.13).
9. Calculate the viscoplastic strain rates $^{t+\Delta t}\dot{\mathbf{e}}^{VP}$ from $^{t+\Delta t}\dot{e}_{ij}^{VP}$ using Equation (11.15).
10. $t \leftarrow t + \Delta t$, if $t \leq t_{max}$ go to step 2; otherwise stop.

Figure 11.2   Flowchart of central difference time stepping scheme

11.7   CRITICAL TIME STEP LENGTH

The stability of the explicit central difference time marching schemes is conditional, i.e. it is unstable if a time step exceeding a critical value is used. If static computations with the elasto/viscoplastic model of an explicit (Euler)

scheme is used another critical time step length is found. Both these critical times are found to control the central difference scheme.

The critical time step in the dynamic elastic algorithm is given by[25,28,30]

$$\Delta t = \beta L \sqrt{\frac{\rho(1+\nu)(1-2\nu)}{E(1-\nu)}} \tag{11.29}$$

where L is the smallest distance between any two adjacent nodes and β is a parameter less than unity. Its value depends on the element types used and the coarseness of element mesh. Generally, for linear elements β ranges between 0·9 to 1, while for quadratic element β ranges between 0·2 to 0·6.[31] In Equation (11.29) ν and E are Poisson's ratio and Young's modulus of the material respectively.

Numerical stability in the quasi-static viscoplastic computation has been derived by Cormeau[32] and for a Mohr–Coulomb yield criterion as used in this study, it is given by the expression

$$\Delta t_{M-C} \leq \frac{4(1+\nu)(1-2\nu)c \cos \phi}{\gamma E(1-2\nu+\sin^2 \phi)} \tag{11.30}$$

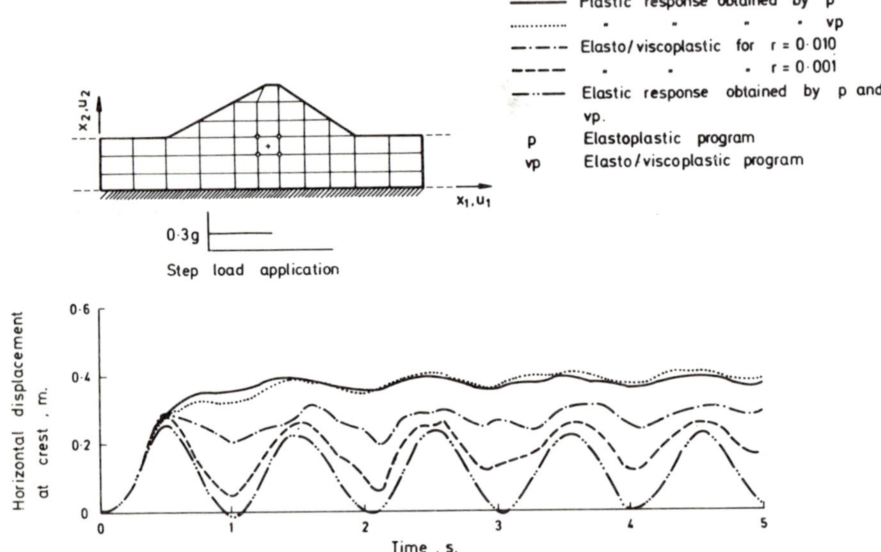

Figure 11.3 Responses with various values of fluidity parameter, γ (computer time on a CDC 7600 is 56 s for each analysis)

The practical time step length for stability must be selected from the value given by Equations (11.29) and (11.30) whichever is more critical.

Theoretically, the elastoplastic solution can be obtained from the elasto/viscoplastic algorithm as a limiting case in which $\gamma \to \infty$. However the time step length requirement of Equation (11.30) will become critical as γ increases. Therefore to obtain an elastoplastic solution critical time step Δt is calculated using Equation (11.29).

For a given time step length, the largest value of γ which can be used is given as

$$\gamma = \beta_1 \frac{4(1+\nu)(1-2\nu)c \cos \phi}{\Delta t\, E(1-2\nu+\sin^2 \phi)} \qquad (11.31)$$

The reduction factor β_1 is usually less than unity. For uniform dynamic load as shown in Figure 11.3 a value of $\beta_1 = 1$ was found to give a reasonable approximation to the elastoplastic response using the elasto/viscoplastic model.

The elastic solution can easily be obtained from the elasto/viscoplastic solution by simply putting $\gamma = 0$. The response obtained with various values of γ are presented in Figure 11.3. To illustrate the general application of the procedures and indeed the validity of the stability criteria two examples are given.

11.8 EXAMPLES

Example 1. An Earth dam (plain strain) subjected to NS-component of El Centro Earthquake.

An earth dam with dimensions and material properties shown in Figure 11.4 is subjected to the base accelerations of the first 10 seconds of the N-S component of 1940 El Centro earthquake as given in Figure 11.5. The results of this study are presented in terms of the time history of displacements as well as stresses and the $^t u_1$ displacement contours and the principal stress distributions. Figures 11.6 and 11.7 show the displacement time history of the dam crest and the shear stress time history at the Gauss point marked by X. Figure 11.8 shows the displacement contours with the principal stress distributions every second through the passage of the earthquake. In Figure 11.6, it can be seen that the elastic solution leads to cyclic response while the plastic solution yields permanent deformation, and the elasto/viscoplastic solution lies between elastic and plastic responses. Although the effect of water in the reservoir has not been incorporated in this study directly using 'fluid' finite elements, this effect can be incorporated without any difficulties.

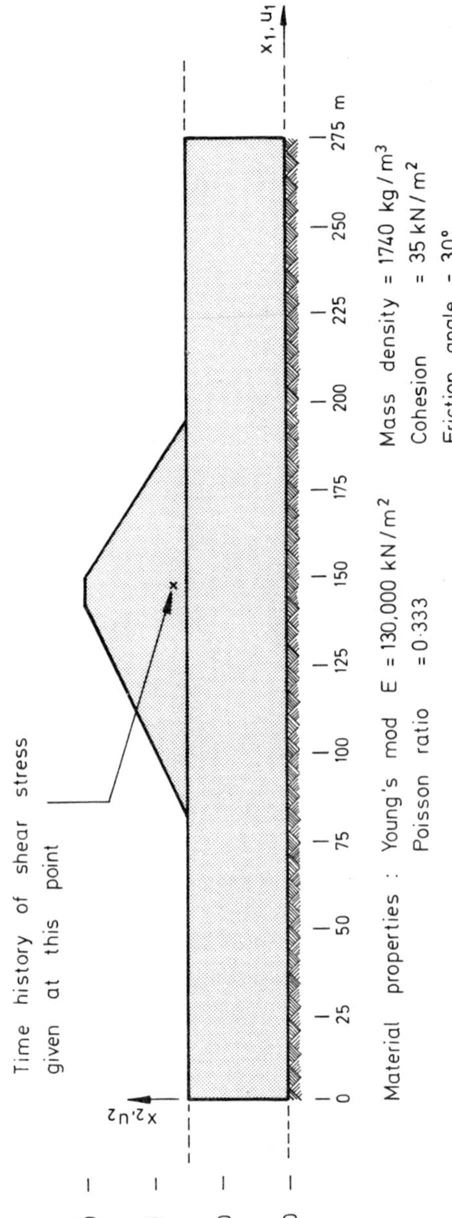

Figure 11.4 Earthdam—Example 1

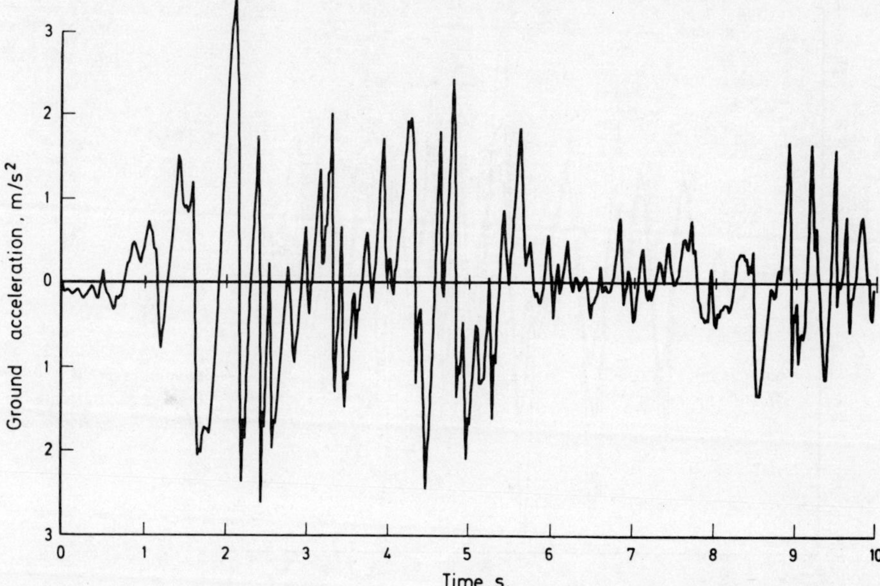

Figure 11.5   Accelerogram from El Centro earthquake (May 1940), NS component

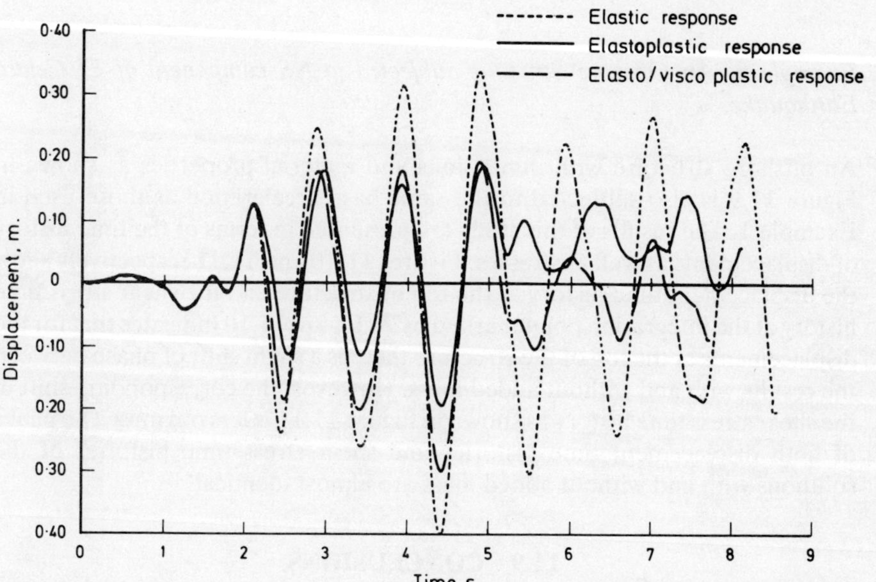

Figure 11.6   Horizontal displacement, $'u_1$, time-history at dam crest (computer time on a CDC 7600 is 160 s for each analysis)

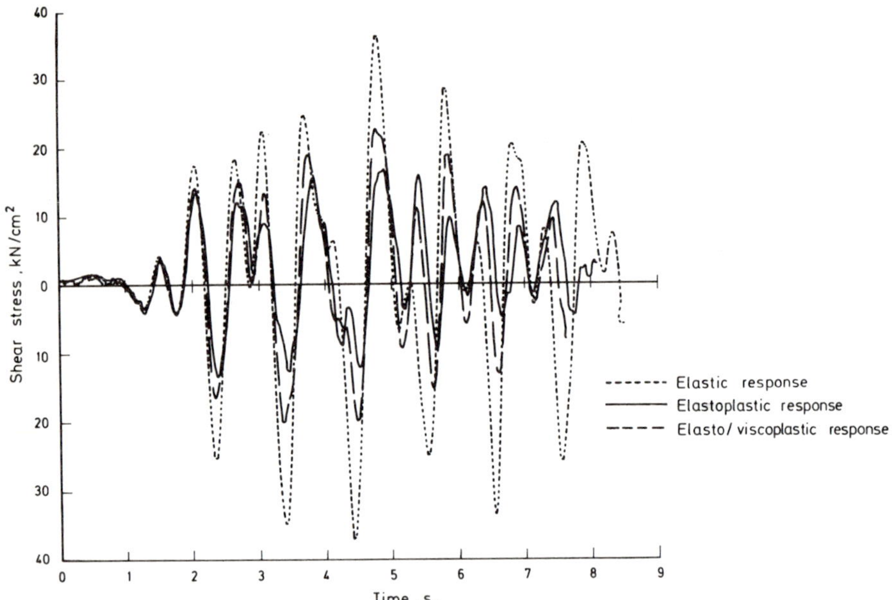

Figure 11.7 Shear stress time history at the point given in Figure 11.4

Example 2. An Offshore Structure subjected to NS-component of El Centro Earthquake.

An offshore structure with dimensions and material properties as shown in Figure 11.9 is also subjected to the same base acceleration as those used in Example 1. The results of this study are presented in terms of the time history of displacements as well as stresses. Figures 11.10 and 11.11 respectively show the displacement time history at the top of structure and the shear stress time history at the integration point marked by X. Figure 11.10 indicates that for the displacements at the top of the structure there is a slight shift of phase between the results with and without added mass. However, the corresponding shift in the shear stress time history as shown in Figure 11.11 is less obvious. The peaks of both displacement time histories and shear stress time histories of the solutions with and without added mass are almost identical.

11.9 CONCLUSIONS

The dynamic non-linear seismic response of a dam section or an offshore structure, with a segment of the foundation included for interaction purposes

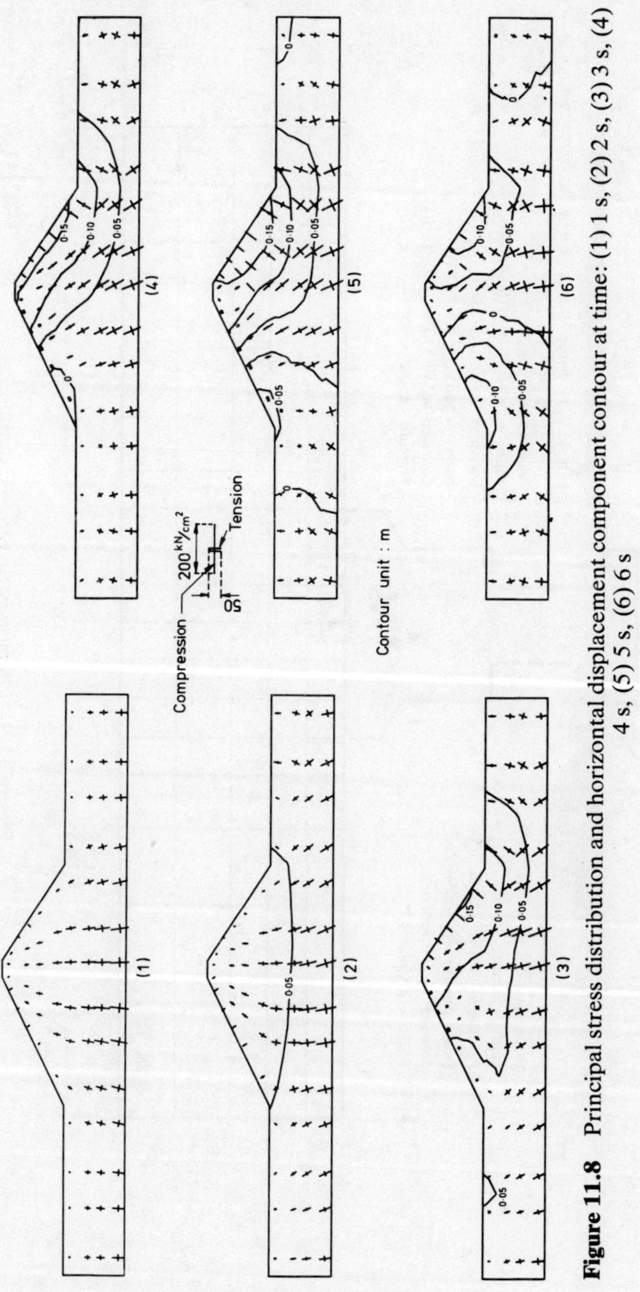

Figure 11.8   Principal stress distribution and horizontal displacement component contour at time: (1) 1 s, (2) 2 s, (3) 3 s, (4) 4 s, (5) 5 s, (6) 6 s

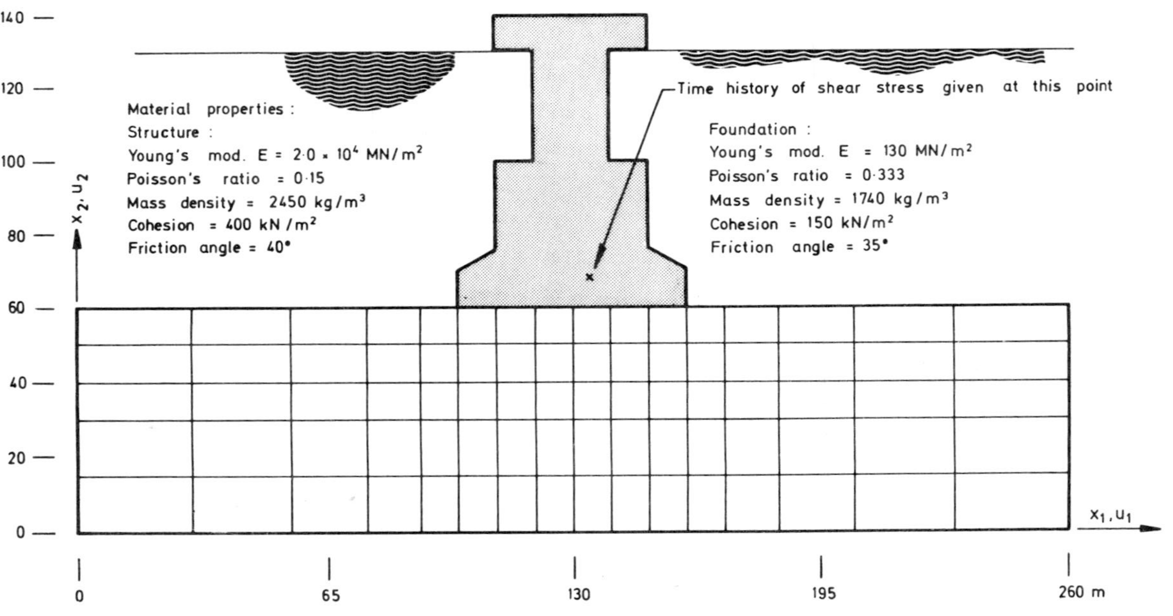

Figure 11.9   Offshore structure—Example 2

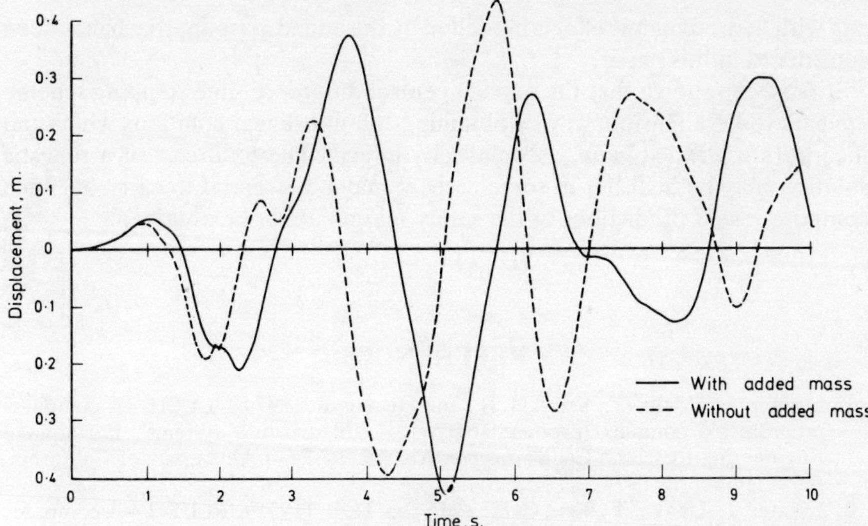

Figure 11.10   Horizontal displacement time-history at the top of structure (computer time on a CDC 7600 is 288 s each)

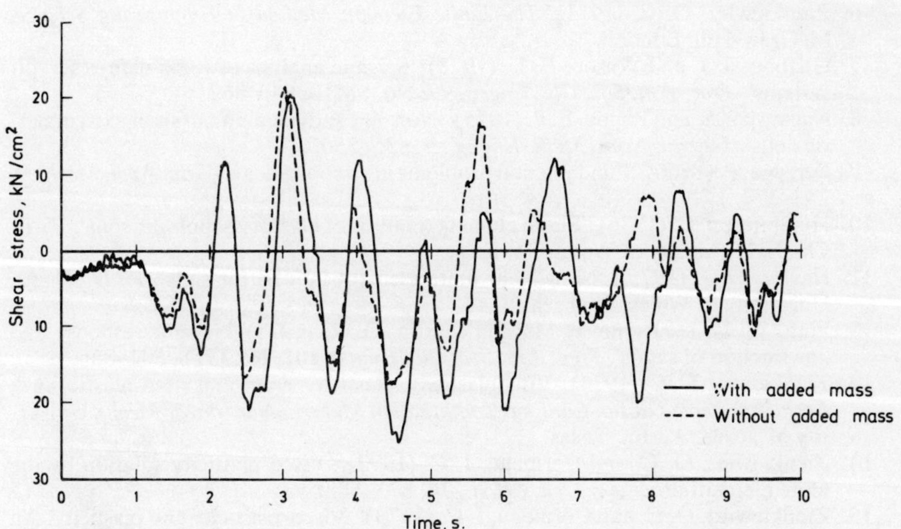

Figure 11.11   Shear stress time history at the point given in Figure 11.9

and with hydrodynamic effects modelled by the added mass approach, has been considered in this paper.

It has been shown that the explicit central difference time stepping scheme appears to be a feasible way of obtaining full non-linear solutions which can incorporate plasticity and viscoplasticity effects. The total cost of a realistic solution may be high but in some cases it may be essential to carry out such computations if predictions of the safety margin are to be obtained.

REFERENCES

1. Lysmer, J., Udaka, T., Seed, H. B., and Hwang, R. (1974). 'LUSH—A computer program for complex response analysis of soil-structure systems', Earthquake Engineering Research Centre, *Report No. EERC 74–4*, University of California, Berkeley, April, 1974.
2. Lysmer, J., Udaka, T., Tsai, C. F., and Seed, H. B. (1975). 'FLUSH—A computer program for approximate 3-D analysis of soil-structure interaction problems', Earthquake Engineering Research Centre, *Report No. EERC 75–30*, University of California, Berkeley, November, 1975.
3. Bathe, K. J., Ramm, E., and Wilson, E. L. (1975). 'Finite element formulations for large deformation dynamic analysis', *Int. Jnl. Num. Meth. Engng.*, **9**, 353–386.
4. Clough, R. W. and Penzien, J. (1975). *Dynamics of Structures*, McGraw-Hill, New York.
5. Newmark, N. M. and Rosenblueth, E. (1971). *Fundamentals of Earthquake Engineering*, Prentice-Hall, Englewood Cliffs, H.J.
6. Zienkiewicz, O. C. (1971). *The Finite Element Method in Engineering Science*, McGraw-Hill, London.
7. Ghaboussi, J. and Wilson, E. L. (1973). 'Seismic analysis of earth dam-reservoir systems', *Proc. Am. Soc. Civ. Engrs.*, **99**, No. SM10, 849–862.
8. Nagarajan, S. and Popov, E. P. (1975). 'Non-linear dynamic analysis of axisymmetric hells', *Int. Jnl. Num. Meth. Engng.*, **9**, 535–550.
9. Perzyna, P. (1966). 'Fundamental problems in viscoplasticity', *Adv. Appl. Mech.*, **9**, 243–377.
10. Humpheson, C. (1976). 'Finite element analysis of elasto/viscoplastic soils', *Ph.D. Thesis*, University of Wales, Swansea.
11. Naylor, D. J. (1975). 'Non-linear finite element models for soils', *Ph.D. Thesis*, University of Wales, Swansea.
12. Finn, W. D. L., Byrne, P. M., and Martin, G. R. (1975). 'Seismic response and liquefaction of sands', *Proc. Am. Soc. Civ. Engrs.*, **102**, No. GT8, 841–856.
13. Zienkiewicz, O. C. (1974). 'Viscoplasticity, plasticity, creep and visco-plastic flow', *Lecture presented at Int. Conf. on Computation Meth. in Non-linear Mech.*, University of Texas, Austin, Texas.
14. Zienkiewicz, O. C. and Cormeau, I. C. (1972). 'Visco-plasticity solution by the finite element process', *Arch. Mech.*, **24**, 873–888.
15. Zienkiewicz, O. C. and Cormeau, I. C. (1973). 'Visco-plasticity and plasticity. An alternative for finite element solution of material Non-Linearities', *Proc. Colloque Méthodes Calcul. Sci. Tech.*, 171–199, Paris: IRIA.

16. Zienkiewicz, O. C. and Cormeau, I. C. (1974). 'Visco-plasticity—plasticity and creep in elastic solids, a unified numerical solution approach', *Int. Jnl. Num. Meth. Engng.*, **8**, 821–845.
17. Zienkiewicz, O. C., Humpheson, C., and Lewis, R. W. (1975). 'Associated and non-associated visco-plasticity and plasticity in soil mechanics', *Géotechnique*, **25**, No. 4, 671–689.
18. Hill, R. (1950). *The Mathematical Theory of Plasticity*, Oxford, Clarendon Press.
19. Kalev, I. and Gluck, J. (1977). 'Cyclic elastic-plastic dynamic analysis by the finite element method', *Computer Methods in Applied Mechanics and Engineering 10*, 63–74.
20. Prevost, J. H. and Høeg, K. (1975). 'Mathematical model for static and cyclic undrained clay behaviour', *Reports 52412, NGI.*
21. Drucker, D. C. and Prager, W. (1952). 'Soil mechanics and plastic analysis on limit design', *Q. Appl. Math.*, **10**, 157–165.
22. Nayak, G. C. and Zienkiewicz, O. C. (1972). 'Elasto-plastic stress analysis. A generalization for various constitutive relations including strain softening', *Int. Jnl. Num. Meth. Engng.*, **5**, 113–135.
23. Nayak, G. C. and Zienkiewicz, O. C. (1972). 'Convenient form of stress invariants for plasticity', *Proc. Am. Soc. Civ. Engrs.*, **98**, ST4, 949–954.
24. Hinton, E., Rock, T. A., and Zienkiewicz, O. C. (1976). 'A note on mass lumping and related processes in the finite element method', *Int. J. Earthquake Eng. and Struct. Dyn.*, **4**, 245–249.
25. Shantaram, D., Owen, D. R. J., and Zienkiewicz, O. C. (1976). 'Dynamic transient behaviour of two- and three-dimensional structures including plasticity, large deformation effects and fluid interaction', *Int. J. Earthquake Eng. and Struct. Dyn.*, **4**, 561–578.
26. Liaw, C. Y. and Chopra, A. K. (1973). 'Earthquake response of axisymmetric tower structures surrounded by water', Earthquake Engineering Research Centre, *Report No. EERC*, 73–25, University of California, Berkeley, October, 1973.
27. Westergaard, H. M. (1933). 'Water pressures on dams during earthquakes', *Transactions, ASCE*, **98**.
28. Rock, T. A. (1975). 'Dynamic analysis of civil engineering structures', *Ph.D. Thesis*, University of Wales, Swansea.
29. Thakkar, S. K. and Stagg, K. G. (1976). 'Non-linear dynamic analysis of stress-wave-propagation problems using parabolic isoparametric finite elements', *Report of the Science Research Council Project*, University of Wales, Swansea.
30. Belytschko, T., Holmes, N., and Mullen, R. (1975). 'Explicit integration-stability, solution properties, cost', *Finite Element Analysis of Transient Non-linear Structural Behaviour presented at The Winter Annual Meeting of the ASME*, Houston, Texas.
31. Deffense, G. V. (1976). 'Non-linear seismic analysis of earthdams by the finite element method', *M.Sc. Thesis*, University of Wales, Swansea.
32. Cormeau, I. C. (1975). 'Numerical stability in quasi-static elasto-viscoplasticity', *Int. Jnl. Num. Meth. Engng.*, **9**, 109–127.

Chapter 12

A Unified Approach to the Soil Mechanics Problems of Offshore Foundations

O. C. Zienkiewicz, V. A. Norris, L. A. Winnicki, D. J. Naylor and R. W. Lewis

12.1 INTRODUCTION

The design of offshore structures has introduced new problems in foundation analysis, the solution of which can rely to only a limited extent on past experience from onshore work. Theoretical studies must therefore be employed, using the best models and solution techniques available, so that the uncertainty is reduced to a minimum. Exact prediction of soil behaviour cannot be achieved in the present state of understanding of soil properties, but it is important to be able to model accurately those properties which are essential in the solution of a problem. For this purpose the Finite Element Method has been used together with the assumption that soil behaviour may be represented by an elasto-plastic model. Various forms of the plastic relationships are considered here and their effects shown for both drained and undrained conditions, so that their correctness may be judged.

The essential foundation problem for gravity platforms is, of course, stability, for which previous onshore experience is of utmost importance. However, since the consequences of failure would be catastrophic it is important to know the ultimate bearing capacity as precisely as possible. A detailed investigation has therefore been made of the differences between the results of standard formulae and those of a finite element model, which does not make assumptions concerning the location of a slip surface.

Also important is the question of settlement, both initially and after a period of time has allowed dissipation of excess pore pressures. Excessive uneven settlement may cause structural damage, whilst excessive vertical movement or tilting may cause operational difficulties. An elasto-plastic analysis of consolidation is compared with a purely elastic analysis to show the importance of taking non-linearity into account.

Affecting both stability and settlement is the 'fatigue' phenomenon caused by cyclic loading. For drained conditions repeated loading/unloading causes a gradual accumulation of permanent strain, whilst under undrained conditions a build-up of pore pressure occurs, positive for normally or lightly over-consolidated clays and loose sands, possibly negative for heavily over-consolidated clays and dense sands. In the case of a pore pressure rise the effective stresses* are increased and the soil strength decreases, whilst the opposite is true for a fall in pore pressure. The magnitude of such changes of pore pressure will depend upon the rate of dissipation, which will eventually bring about their cancellation. The modelling of this difficult problem is only partially resolved and much work has yet to be done here. The phenomenon of 'shake down' or continuing plastic deformation under load reversal occurs, however, independently of the above considerations and can be adequately modelled using isotropic hardening or even ideal plasticity. Offshore structures are particularly sensitive to such effects due to frequent load cycling and some predictions of this kind will be given in examples.

Other subjects of offshore research which will not be dealt with here are the cyclic loading of platform foundations by earthquakes, the dynamic analysis of platforms, the cyclic loading of piles, and the strength of anchor systems used for floating structures. Some aspects of these problems are discussed in other chapters, but much research is still needed there to assess such factors as the influence of rate effects, etc.

Before dealing with the actual problems of offshore foundations it is necessary to make clear what are generally considered to be universal aspects of soil behaviour, upon which any analysis may be founded. The way in which a finite element analysis is developed from these is indicated.

12.2 BASIC CHARACTER OF SOIL AND FORMULATION IN TERMS OF FINITE ELEMENTS

In the present context it may be assumed that the soil involved is a two phase medium—a skeleton of particles surrounded by a fluid in which shear stresses are small and which exerts a pressure p on the solid phase. Biot,[1-4] Terzaghi,[5] and others[6-7] have formulated the essential behaviour relation of such materials with a linear elastic skeleton. Here we shall extend such a formulation to a general non-linear case introducing simultaneously the finite element discretization processes.[8]

12.2.1 Effective stress constitutive relation

Defining the total stress vector $\boldsymbol{\sigma}$ as

$$\boldsymbol{\sigma} = [\sigma_{xx}, \sigma_{yy}, \sigma_{zz}, \sigma_{xy}, \sigma_{yz}, \sigma_{zx}]^T \tag{12.1}$$

---

* Compressive stresses are regarded as negative, in the normal convention of mechanics.

(with tensions as $+\mathrm{ve}$) we can decompose this into a hydrostatic component $\mathbf{m}p$ and an *effective* stress:

$$\boldsymbol{\sigma} = \boldsymbol{\sigma}' - \mathbf{m}p \tag{12.2}$$

in which $\mathbf{m} = [1, 1, 1, 0, 0, 0]^T$. It will be noted that this is a generalized statement of the principle of effective stress, and that p is the pore pressure (compression $+\mathrm{ve}$). As the pore pressure places all the solid phase in a purely hydrostatic stress state (assuming on the average homogeneous and isotropic properties) its effect is merely to introduce a general volumetric strain,

$$\varepsilon_v^p = -\frac{p}{K_s}$$

or equivalent increments

$$\mathrm{d}\varepsilon_v^p = -\frac{\mathrm{d}p}{K_s} \tag{12.3a}$$

This can be written as a generalized strain

$$\boldsymbol{\varepsilon}^p = -\mathbf{m}\frac{p}{3K_s}$$

or its increments

$$\mathrm{d}\boldsymbol{\varepsilon}^p = -\mathbf{m}\frac{\mathrm{d}p}{3K_s} \tag{12.3b}$$

in which K_s is the average bulk modulus of the solid phase. In soil mechanics this strain is generally considered negligible but as its importance in rocks is significant we shall retain it here.

By the above argument we can conclude that it is the 'effective *stress*', $\boldsymbol{\sigma}'$, which is responsible for all major deformations, linear or non-linear, and indeed that the failure states can only be adequately expressed in terms of such effective stresses. With this generally agreed assumption we can write the constitutive skeletal relations. Whatever the instantaneous or time based response such relations can be written in the incremental form

$$\mathrm{d}\boldsymbol{\sigma}' = \mathbf{D}_T(\mathrm{d}\boldsymbol{\varepsilon} - \mathrm{d}\boldsymbol{\varepsilon}^c - \mathrm{d}\boldsymbol{\varepsilon}^p) \tag{12.4}$$

in which $\mathbf{D}_T$ is the tangent modulus matrix, $\mathrm{d}\boldsymbol{\varepsilon}^p$ is the strain increment due to pressure changes and $\mathrm{d}\boldsymbol{\varepsilon}^c$ is due to creep stresses which with some generality can be written as

$$\mathrm{d}\boldsymbol{\varepsilon}^c = \mathbf{g}(\boldsymbol{\sigma}')\,\mathrm{d}t \tag{12.5}$$

where $\mathbf{g}(\boldsymbol{\sigma}')$ is a stress independent vector defining the strain rate. Both $\mathbf{D}_T$ and $\mathbf{g}$ are obtained from appropriate tests of either the *drained* material or the undrained material with pore pressure measurement.

Introducing at this stage the finite element discretization in its *displacement form* with $\bar{\mathbf{u}}$ and $\bar{\mathbf{p}}$ defining the nodal displacement and pore pressure parameters, we can write for the displacements and strains (using the notation of Reference 8)

$$\mathbf{u} = \mathbf{N}\bar{\mathbf{u}}; \qquad \boldsymbol{\varepsilon} = \mathbf{B}\bar{\mathbf{u}}; \qquad p = \mathbf{N}\bar{\mathbf{p}} \tag{12.6}$$

and for the *incremental* equilibrium equation in the total stress state

$$\int_{\Omega} \mathbf{B}^T \, d\boldsymbol{\sigma} \, d\Omega - d\mathbf{f} = 0 \tag{12.7}$$

by invoking the principles of virtual work.[8] Here $d\mathbf{f}$ stands for any *changes* of external forces due to boundary or body force loading. (Note that *total* stress boundary traction has to be specified.) Again the detailed form of such loads is calculated in the standard finite element manner as in Reference 8, i.e.

$$d\mathbf{f} = \int_{\Omega} \mathbf{N}^T \, d\mathbf{b} \, d\Omega + \int_{\Gamma} \mathbf{N}^T \, d\mathbf{t} \, d\Gamma \tag{12.8}$$

where $\mathbf{b}$ and $\mathbf{t}$ are body forces and boundary tractions respectively.

Substitution of Equations (12.2, 4–6) into (12.7) gives the incremental equilibrium conditions as

$$\mathbf{K}_T \frac{d\bar{\mathbf{u}}}{dt} - L \frac{d\bar{\mathbf{p}}}{dt} - \mathbf{c} - \frac{d\mathbf{f}}{dt} = 0 \tag{12.9}$$

where $\mathbf{K}_T$ is the tangential stiffness matrix,

$$\mathbf{K}_T = \int_{\Omega} \mathbf{B}^T \mathbf{D}_T \mathbf{B} \, d\Omega \tag{12.10a}$$

and

$$L = \int_{\Omega} \mathbf{B}^T \left(\mathbf{m} - \frac{\mathbf{D}_T \mathbf{m}}{3K_s} \right) d\Omega \tag{12.10b}$$

$$\mathbf{c} = \int_{\Omega} \mathbf{B}^T \mathbf{D}_T \mathbf{g} \, d\Omega \tag{12.10c}$$

Equation (12.9) is the fundamental one from which, if pressures, $\bar{\mathbf{p}}$, are known, the displacements and stresses in the system can be obtained. In general, the pressure development is coupled with the strain changes and hence these have to be found for the flow conditions.

At this stage it is important to note that Equation (12.9) is generally highly non-linear with all matrices dependent on stresses and strains which occur at a particular increment.

12.2.2 The flow continuity relation

With the hydraulic head h defined by

$$\gamma h = \gamma z + p \tag{12.11}$$

where γ is the fluid density and z the vertical co-ordinate, flow velocities can be obtained by Darcy's law as

$$\mathbf{v} = -\mathbf{k}\nabla(\gamma h) = -\mathbf{k}\nabla(\gamma z + p) \tag{12.12}$$

In the above, $\mathbf{k}$ is the permeability matrix (which in isotropic situations can be replaced by a simple scalar value k). The permeability will in general be a quantity strongly dependent on ε_v, the total volumetric strain reached.

The basic statement of continuity of flow requires that the divergence of the flow velocity plus the rate of fluid accumulation per unit volume of space, must be equal to zero:

$$\nabla^T \mathbf{v} + \text{rate of fluid accumulation} = 0 \tag{12.13a}$$

Now, various factors contribute to the accumulation term:

(1) change of total strain resulting in a rate of accumulation,

$$\frac{\partial \varepsilon_v}{\partial t} = \mathbf{m}^T \frac{\partial \mathbf{\varepsilon}}{\partial t} \tag{12.13b}$$

(2) the changes of grain volume due to pressure changes yielding

$$(1 - \eta)\frac{1}{K_s}\frac{\partial p}{\partial t} \tag{12.13c}$$

where η is the porosity, K_s is the skeleton bulk modulus,
(3) fluid compressibility defined by a bulk modulus K_f contributes

$$\eta \frac{1}{K_f}\frac{\partial p}{\partial t} \tag{12.13d}$$

and finally
(4) compression of solid grains due to effective stress changes $\partial \mathbf{\sigma}'/\partial t$. Although this strain is small its importance in rock flow associated with oil technology is such that it should not be neglected. Thus, due to above change of effective stress an *average* hydrostatic compression of magnitude

$$-\frac{1}{3}\mathbf{m}^T \frac{\partial \mathbf{\sigma}'}{\partial t}\frac{1}{(1-\eta)}$$

is developed. This acting on a total volume of solid $(1-\eta)$ produces a volume accumulation,

$$-\frac{1}{3K_s}\mathbf{m}^T\frac{\partial\boldsymbol{\sigma}'}{\partial t}$$

Substituting Equations (12.3(b) and 12.4) we have for the rate of accumulation due to this change

$$-\frac{1}{3K_s}\mathbf{m}^T\mathbf{D}_T\left(\frac{\partial\boldsymbol{\varepsilon}}{\partial t}+\frac{1}{3K_s}\mathbf{m}\frac{\partial p}{\partial t}-\mathbf{g}\right) \tag{12.13e}$$

The combined differential equation governing the flow becomes on substitution of (12) (with no source terms)

$$-\boldsymbol{\nabla}^T\mathbf{k}\boldsymbol{\nabla}(\gamma z+p)+\left(\mathbf{m}^T-\frac{\mathbf{m}^T\mathbf{D}_T}{3K_s}\right)\frac{\partial\boldsymbol{\varepsilon}}{\partial t}+\frac{\mathbf{m}^T\mathbf{D}_T\mathbf{g}}{3K_s}$$

$$+\left[(1-\eta)/K_s+\eta/K_f-\frac{1}{(3K_s)^2}\mathbf{m}^T\mathbf{D}_T\mathbf{m}\right]\frac{\partial p}{\partial t}=0 \tag{12.14}$$

Again a finite element discretization of the above can be performed by using the Galerkin procedure in a standard way incorporating any prescribed flow boundary conditions. Details of such discretizations are standard (see Reference 8) and will result, using the three expansions of Equation (12.6), in

$$\mathbf{H}\bar{\mathbf{p}}-\mathbf{S}\frac{d\bar{\mathbf{p}}}{dt}-\mathbf{L}^T\frac{d\bar{\mathbf{u}}}{dt}-\mathbf{f}=0 \tag{12.15}$$

with

$$\mathbf{H}=\int_\Omega (\boldsymbol{\nabla}\mathbf{N})^T\mathbf{K}\boldsymbol{\nabla}\mathbf{N}\,d\Omega \tag{12.16a}$$

$$\mathbf{S}=\int_\Omega \mathbf{N}^T s\,\mathbf{N}\,d\Omega \quad\text{where}\quad s=\left[\frac{1-\eta}{K_s}+\frac{\eta}{K_f}-\frac{\mathbf{m}^T}{(3K_s)^2}\mathbf{D}_T\mathbf{m}\right] \tag{12.16b}$$

$$\mathbf{L}=\int_\Omega \left(\mathbf{m}^T-\mathbf{m}^T\frac{\mathbf{D}_T^T}{3K_s}\right)\mathbf{B}\,d\Omega \tag{12.16c}$$

$$\mathbf{f}=\int_\Gamma \mathbf{N}^T\mathbf{q}\,d\Gamma+\int_\Omega \frac{1}{3K_s}\mathbf{m}^T\mathbf{D}_T\mathbf{g}\,d\Omega-\int_\Omega (\boldsymbol{\nabla}\mathbf{N})^T\mathbf{K}\boldsymbol{\nabla}\gamma\mathbf{Z}\,d\Omega \tag{12.16d}$$

where $\mathbf{q}$ is the outflow across the boundary Γ. In the above s denotes the compressibility terms contained in the square bracket of Equation (12.14). Although the last term is generally insignificant no computational advantage accrues due to its omission.

Summarizing, the complete *coupled behaviour* of the two phase skeleton-fluid state is governed by a system of ordinary differential equations which we can write concisely as

$$\begin{bmatrix} 0 & 0 \\ 0 & \mathbf{H} \end{bmatrix} \begin{Bmatrix} \bar{\mathbf{u}} \\ \bar{\mathbf{p}} \end{Bmatrix} + \begin{bmatrix} \mathbf{K}_T & -\mathbf{L} \\ -\mathbf{L}^T & -\mathbf{S} \end{bmatrix} \frac{d}{dt} \begin{Bmatrix} \bar{\mathbf{u}} \\ \bar{\mathbf{p}} \end{Bmatrix} = \begin{Bmatrix} \dfrac{d\mathbf{f}}{dt} + \mathbf{c} \\ \mathbf{f} \end{Bmatrix} \tag{12.17}$$

The symmetry of this general equation is assured providing $\mathbf{K}_T$ is symmetric (as is the case in associated plasticity) and a general solution can be obtained provided initial conditions of $\boldsymbol{\sigma}_0$, $\mathbf{p}_0$ etc. are known. Solution of such transient equations can be accomplished by various processes of time stepping and provide a complete consolidation history. With linear elastic assumptions such solutions have been obtained by Sandhu and Wilson,[9] Ghaboussi and Wilson,[10] Morgenstern[11] and many others. Elastoplastic consolidation has been modelled by Small[12] and also by the current writers, as illustrated in Section 12.7. Here we shall be concerned with either long term fully drained behaviour or alternatively with instantaneous, undrained conditions. Both these conditions are special cases of the above discretization and can be derived from Equation (12.17).

12.2.3 Fully drained, steady state conditions

If the second of Equations (12.17) (or Equation (12.15)) is examined we note that provided pressures and displacements reach constant values, it reduces to

$$\mathbf{H}\bar{\mathbf{p}} = \mathbf{f} \tag{12.18}$$

and the pressures can be independently determined, assuming of course that the permeability is known and independent of strain.

The first of Equations (12.17) (or Equation (12.9)) results then in

$$\mathbf{K}_T \, d\bar{\mathbf{u}} = d\mathbf{f} + \mathbf{L} \, d\bar{\mathbf{p}} + [\mathbf{c} \, dt] \tag{12.19}$$

with time intervening only in the creep term which is assumed to tend to zero as steady state is approached. (Indeed, the time here can be considered in a fictitious manner as shown later.) Equation (12.19) represents essentially an incremental form of a non-linear mechanics problem with forces $\mathbf{L}d\bar{\mathbf{p}}$ due to pore pressures added in. Standard programs and methods used in such solutions can be directly applied once the constitutive relations are known. For plastic or viscoplastic constitutive laws such analyses are derived in Reference 8 and other publications.[13–16] For a simple elastic medium we have on integration

$$\mathbf{K}_T\bar{\mathbf{u}} = \mathbf{f} + \mathbf{L}\bar{\mathbf{p}} \tag{12.20}$$

showing the 'forces' due to pore pressure.*

---

* The above formulation is a slight variant on that derived in Reference 8 where forces due to pressure are computed in a different way. The equivalence of both forms has been proved elsewhere.[17]

12.2.4.1 Instantaneous load (undrained condition)

If Equations (12.17) are multiplied by dt and $dt \rightarrow 0$ we note that all finite terms disappear. Reforming, however, a creep term which may appear in a different time scale, we reduce the general equations to a system

$$\begin{bmatrix} \mathbf{K}_T & -\mathbf{L} \\ -\mathbf{L}^T & -\mathbf{S} \end{bmatrix} \begin{Bmatrix} d\bar{\mathbf{u}} \\ d\bar{\mathbf{p}} \end{Bmatrix} = \begin{Bmatrix} d\mathbf{f} + [\mathbf{c}\, dt] \\ 0 \end{Bmatrix} \tag{12.21}$$

Again this is an incremental form of the well known nearly incompressible (or, if $S = 0$, incompressible) solid formulation, and once again for an elastic linear behaviour, on integration, will yield

$$\begin{bmatrix} \mathbf{K}_T & -\mathbf{L} \\ -\mathbf{L}^T & -\mathbf{S} \end{bmatrix} \begin{Bmatrix} \bar{\mathbf{u}} \\ \bar{\mathbf{p}} \end{Bmatrix} = \begin{Bmatrix} \mathbf{f} \\ 0 \end{Bmatrix} \tag{12.22}$$

a form developed originally by Herrmann[18] and used subsequently by others[19] for solution of incompressible elastic solids.

12.2.4.2 An alternative form of instantaneous load solution (undrained conditions)

Equations (12.21) or (12.22) represent the classical approach to the study of incompressible or nearly incompressible solids. In the former case $\mathbf{S}$ is zero and $\bar{\mathbf{p}}$ is in effect the Lagrangian multiplier inserted to ensure incompressibility. When $\mathbf{S} \neq 0$ other possible approaches exist. As the pressures can, in such a case, be determined from the strains (or displacements) it is possible to eliminate these from the equations and deal solely with displacement as the basic unknown.

For instance, by inversion of matrix $\mathbf{S}$, $\bar{\mathbf{p}}$ can be found from the second of Equations (12.21) explicitly and elimination from the first set of equations results in

$$[\mathbf{K}_T + \mathbf{L}\mathbf{S}^{-1}\mathbf{L}^T]\, d\bar{\mathbf{u}} = d\mathbf{f} + [\mathbf{c}\, dt] \tag{12.23}$$

Here a new 'stiffness matrix' replaces the original one, based on skeletal (effective stress) properties.

Clearly the procedure implied by Equation (12.23) is inconvenient, involving the inversion of a matrix $\mathbf{S}$, which in many cases is still defined, and it is more convenient to return directly to the constitutive relations and reformulate the problem. We shall now find that a direct non-singular relationship can be written between the *total stresses* and *strains*. This total stress relation is generally one of near-incompressibility and corresponds precisely to the type of formulation used in solid mechanics when a Poisson's ratio with a value approaching but not equal to 0·5 is inserted in an ordinary displacement formulation. Naylor[20] has shown that such a formulation can be very effectively used.

To derive the direct relationship we examine the differential form of the continuity Equation (12.14) and note that if no flow occurs

$$\left[\mathbf{m}^T - \frac{\mathbf{m}^T \mathbf{D}_T}{3K_s}\right] \mathrm{d}\boldsymbol{\varepsilon} = -s\,\mathrm{d}p - \frac{\mathbf{m}^T \mathbf{D}_T}{3K_s}\mathrm{d}\boldsymbol{\varepsilon}^c \tag{12.24}$$

Inserting the above in Equations (12.2) and (12.4) we have

$$\mathrm{d}\boldsymbol{\sigma} = \mathrm{d}\boldsymbol{\sigma}' - \mathbf{m}\,\mathrm{d}p = \mathbf{D}_T(\mathrm{d}\boldsymbol{\varepsilon} - \mathrm{d}\boldsymbol{\varepsilon}^c) - \mathbf{m}\left(1 - \frac{\mathbf{D}_T}{3K_s}\right)\mathrm{d}p$$

$$= \left[\mathbf{D}_T + \mathbf{m}\left(1 - \frac{\mathbf{D}_T}{3K_s}\right)\mathbf{m}^T\left(1 - \frac{\mathbf{D}_T}{3K_s}\right)\frac{1}{s}\right]\mathrm{d}\boldsymbol{\varepsilon}$$

$$- \left[\mathbf{D}_T - \mathbf{m}\left(1 - \frac{\mathbf{D}_T}{3K_s}\right)\frac{\mathbf{m}^T \mathbf{D}_T}{3K_s}\frac{1}{s}\right]\mathrm{d}\boldsymbol{\varepsilon}^c$$

$$= \check{\mathbf{D}}_T\,\mathrm{d}\boldsymbol{\varepsilon} - \hat{\mathbf{D}}_T\,\mathrm{d}\boldsymbol{\varepsilon}^c \tag{12.25}$$

The above relation gives essentially a constitutive relation in terms of total stress which, in the absence of creep, is entirely determined by tests on the undrained material as

$$\mathrm{d}\boldsymbol{\sigma} = \check{\mathbf{D}}_T\,\mathrm{d}\boldsymbol{\varepsilon} \tag{12.26}$$

It is common to establish such laws and deal with the analysis explicitly in total stress terms. We note that the skeletal matrix $\mathbf{D}_T$ is related to the total matrix $\check{\mathbf{D}}_T$ by 'penalty' terms which incorporate approximate incompressibility.[21]

In terms of the total stress the analysis procedure follows from Equation (12.7) yielding

$$\int_\Omega \mathbf{B}^T \check{\mathbf{D}}_T \mathbf{B}\,\mathrm{d}\bar{\mathbf{u}}\,\mathrm{d}\Omega - \mathrm{d}\mathbf{f} - \int \mathbf{B}^T \hat{\mathbf{D}}_T\,\mathrm{d}\boldsymbol{\varepsilon}^c\,\mathrm{d}\Omega = 0 \tag{12.27}$$

As already indicated the equations developed are highly non-linear. The main factors are the variability of the permeability[22] and of the stress–strain relationship. The latter will now be considered in detail, on the assumption that it may be modelled adequately in terms of elastoplasticity.

12.3 PLASTICITY AND VISCOPLASTICITY

Before attempting to construct models for predicting the stress–strain behaviour of soils a brief simple description of plasticity will be given. Viscoplasticity (meaning creep of the solid, skeleton phase, not consolidation of a two phase medium) is also considered and even though its true constants are not often available due to lack of physical data it is always a very convenient alternative for computation of plastic deformation. In future research into cyclic loading (when the change in load is too rapid to allow full plastic flow) the true nature of viscoplastic behaviour will doubtless be of importance.

12.3.1 Uniaxial stress

The basic nature of plasticity is most easily explained in terms of a uniaxial stress–strain relationship, as shown in Figure 12.1. It is impossible to determine whether the behaviour is elastic or plastic until an unloading test is carried out. If such a test shows a complete reversal of the stress–strain behaviour from any stress point, the material can be described as a (non-linear) elastic one. In reality, such reversibility is present only in the small (linear) segment, and return paths such as (c) are usually noted, showing permanent deformation on removal of load. Such behaviour is characteristic of plasticity and indeed is its definition.

A further feature of Figure 12.1(a) is that on subsequent reloading a return along the path (d) is obtained and plasticity begins only when the 'yield' stress condition, $\sigma' = Y$, is reached. This condition is given at the same point A.

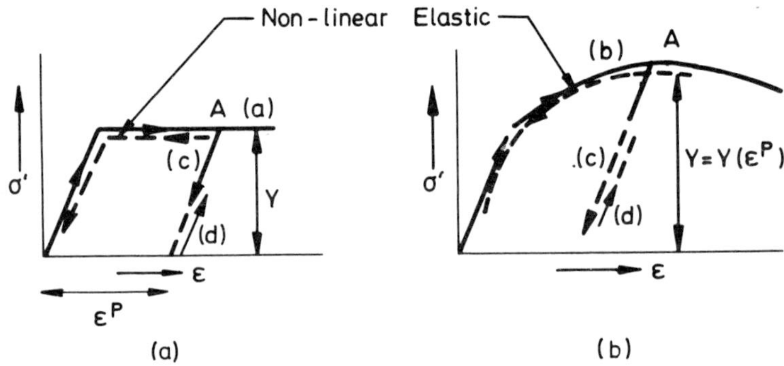

Figure 12.1 Ideally plastic and strain dependent (hardening or softening) behaviour in uniaxial stress

In Figure 12.1(a) the yield stress Y remains constant for all loading paths, and such a material model is termed ideally plastic. In Figure 12.1(b) the yield stress Y changes with the accumulation of plastic strain, ε_p, and such a material may be called strain-hardening (or softening).

The model of elasto-plastic behaviour shown in Figure 12.1 can be visualized mechanically as a combination of a spring and a slider as shown in Figure 12.2.

Soil-like materials show behaviour patterns similar to those indicated in Figure 12.1(b); i.e. exhibit both hardening and softening which is present to varying degrees. As will be discussed in Section 12.6 plasticity may be present at very low stresses, and may occur during unloading and reloading. The maximum load that a sample can carry is called the *failure* load. In analysis of complex forms it is convenient to replace at times the strain hardening behaviour by an ideal plastic model which yields at a constant stress corresponding to the maximum failure conditions. This is shown in Figure 12.3.

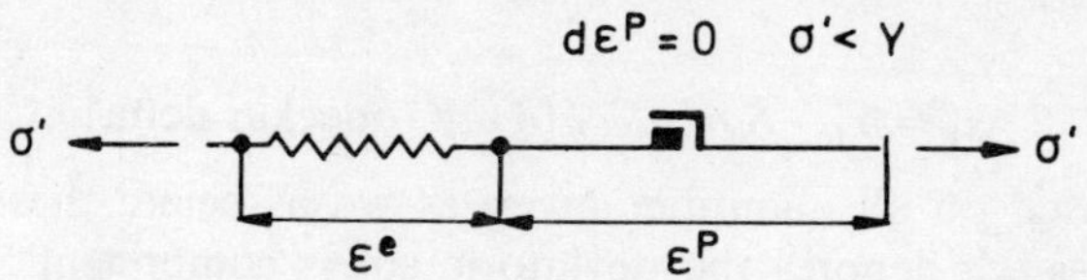

Figure 12.2 Uniaxial model of elasto-plastic material

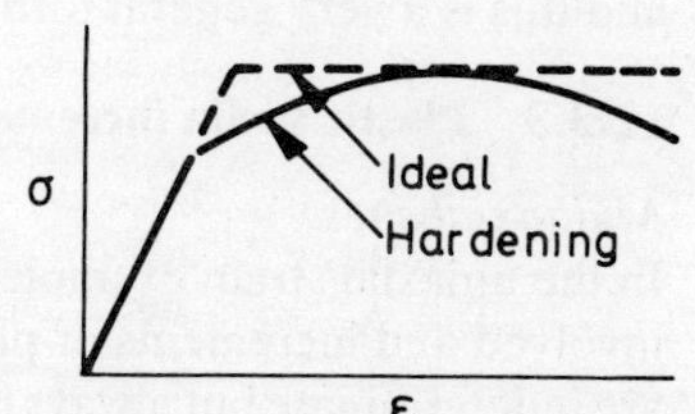

Figure 12.3 Replacement of strain hardening plasticity by an ideal plastic approximation

Clearly such a model will not give a very good representation of strains in a sample, *but the collapse or failure load of a structure should be well approximated.* Indeed, a non-linear load deformation characteristic before failure will be obtained for a structure although this would not be present for homogeneous stresses in a simple sample. As we shall see later, such models replacing strain hardening by ideal plasticity can be used very successfully in real problems.

12.3.2　Yield under multiaxial stress

In the uniaxial behaviour we can conclude that plastic deformation takes place only if

$$F(\sigma') = \sigma' - Y = 0; \tag{12.28}$$

if $F(\sigma') < 0$ only elastic strains are permitted.

For multiaxial stress plasticity will be caused by some combination of effective stresses, and a surface

$$F(\sigma'_{ij}) = 0 \tag{12.29}$$

defined in the stress (hyper) space will give the onset of plastic yield. This is the well known *yield surface.*

In an isotropic material the six independent stress components σ'_{ij} can be described fully by three stress invariants

$$
\begin{aligned}
J_1 &= \tfrac{1}{3}\sigma'_{ii} = \sigma'_m \\
J_2 &= \tfrac{1}{2}s'_{ij}s'_{ij} = \bar{\sigma}^2 \\
J_3 &= \tfrac{1}{2}s'_{ij}s'_{jk}s'_{ki}
\end{aligned}
\tag{12.30}
$$

with

$$s'_{ij} = \sigma'_{ij} - \delta_{ij}\sigma'_m \qquad (\delta = \text{Kronecker delta})$$

Primes are used on all quantities given as we associate these further with effective stress. s'_{ij} denotes the deviatoric stress components. For isotropic materials, therefore, the yield surface can be written as

$$F(J_1, J_2, J_3) = 0 \qquad\qquad (12.31)$$

and this is a very general form.

12.3.3   Plastic strain increments

Multiaxial stress

In the uniaxial strain example of Figures 12.1–3 only one strain component is involved and increments of plastic strain in the case of an ideally plastic state are indeterminate but always in the direction of this component. What happens in an equivalent ideal plasticity under multiaxial stress conditions needs now to be specified. When the material reaches the plastic yield surface of say the type sketched in Figure 12.4 the strain magnitudes are indeterminate, but observation shows that under such conditions the ratios of various strain increments will be fully defined. If, for instance, we plot the direction of the strain components associated with particular stresses such as 1 and 2 shown in the figure, we can describe this direction by a set of vectors shown in Figure 12.4.

Such directions can be uniquely specified by requiring that the ratios of strains be associated with some plastic potential Q, requiring that

$$\frac{\mathrm{d}\varepsilon^p_1}{\mathrm{d}\varepsilon^p_2} = \frac{\partial Q/\partial \sigma'_1}{\partial Q/\partial \sigma'_2} \text{, etc.;} \qquad Q = Q(\sigma_1, \sigma_2 \cdots) \qquad (12.32)$$

The surface Q is a unique function of stress components and is termed the plastic potential. Equation (12.32) can be interpreted as a requirement that the strain vector be normal to the surfaces of *the constant Q*. It must be emphasized here that the introduction of plastic potential is nothing else but a convenient way of describing a physical law in a simple and unique manner but that other descriptions could equally well be used.

In a general situation we can therefore write for any yield surface, F, that the plastic strains are given by

$$\mathrm{d}\varepsilon^p_{ij} = \lambda \frac{\partial Q}{\partial \sigma_{ij}}; \qquad Q = Q(\sigma_{ij}) \qquad (12.33)$$

where λ is a constant as yet indeterminate.

If the surface of $Q = \text{constant}$ corresponds with the yield surface $F = 0$, the plasticity law is known as *associated*. If this is not the case, the law is termed *non-associated*.

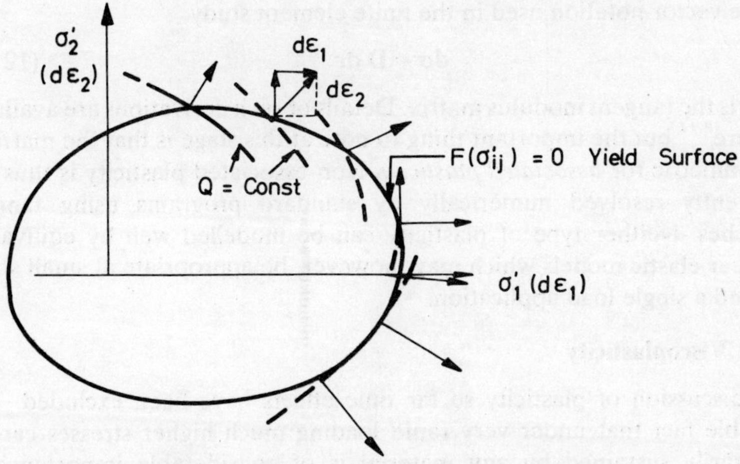

Figure 12.4　Yield surface, plastic; strain increment directions; and plastic potential surfaces in stress hyper-space

For associated plasticity certain well known bound theorems can be proved for collapse[23]; e.g. the lower static bound on the collapse load, which states that if any equilibrating set of stresses can be found which does not exceed the plastic limit, the corresponding load is less than the collapse one. This theorem is further supported by a kinematic upper bound theorem.

As such bounds are only available for associated plastic behaviour they need not (and indeed are not) generally valid and much misuse of such bounding has occurred in soil mechanics and other applications.

The classical plasticity work does not *prove* association but simply asserts this as a consequence of certain postulates,[23] which need not be true in a frictional material such as soil. Further, uniqueness proofs are not available for non-associated plasticity and, indeed, on occasions non-unique solutions can be obtained. This does not, however, invalidate the fact that non-associated materials do in fact exist.

With a specified plastic yield surface F and plastic potential Q and the requirement that the total strain ε_{ij} be the sum of elastic and plastic components,

$$d\varepsilon_{ij} = d\varepsilon_{ij}^{e} + d\varepsilon_{ij}^{p} \tag{12.34}$$

it is possible to derive a unique relationship between strain and stress increments,

$$d\sigma_{ij} = D_{ijkl}\, d\varepsilon_{kl} \tag{12.35}$$

or in the vector notation used in the finite element study

$$d\boldsymbol{\sigma} = \mathbf{D}\,d\boldsymbol{\varepsilon} \tag{12.36}$$

Here $\mathbf{D}$ is the tangent modulus matrix. Details of such derivations are available elsewhere[8,14] but the important thing to note at this stage is that the matrix is only symmetric for *associated plasticity*. Non-associated plasticity is thus not conveniently resolved numerically by standard programs using tangent approaches. Neither type of plasticity can be modelled well by equivalent non-linear elastic models which may, however, be appropriate at small stress levels and a single load application.

12.3.4 Viscoplasticity

In the discussion of plasticity so far time effects have been excluded. The observable fact that under very rapid loading much higher stresses can be momentarily sustained by any material is of considerable importance in dynamic (earthquake) analysis, and models capable of reproducing this should be considered.

The model of Figure 12.2 can be extended to give the desired time response by placing a viscous dashpot in parallel with the plastic element, as shown in Figure 12.5. Immediately, two phenomena are noted.

1. The stresses can exceed the plastic limit for rapidly applied load.
2. Creep (time dependent deformation) proceeds for all stresses exceeding the plastic limit. When this creep stops, the stresses must be at or below the yield surface and, therefore, the plastic solution is obtained as a by-product.

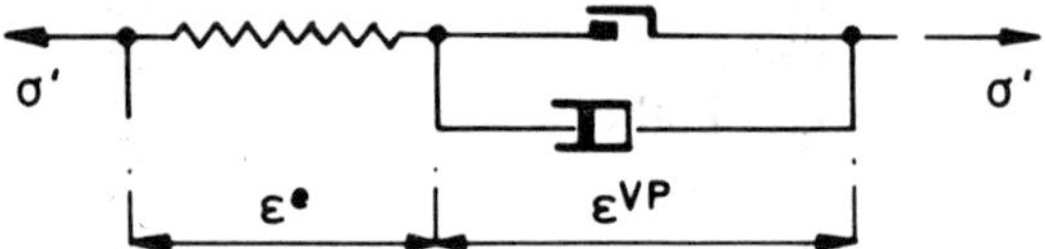

Figure 12.5 Uniaxial model of a visco-plastic material

For a general case of viscoplastic flow this may thus be written as

$$\varepsilon_{ij} = \varepsilon_{ij}^{e} + \varepsilon_{ij}^{vp} \tag{12.37}$$

and

$$\frac{d\varepsilon_{ij}^{vp}}{dt} = \gamma \langle F \rangle \frac{\partial Q}{\partial \sigma_{ij}'} = h(\sigma_{ij}') \tag{12.38}$$

where

$$\langle F \rangle = 0 \quad \text{if} \quad F \leq 0$$

$$\langle F \rangle = F \quad \text{if} \quad F > 0$$

and

$$\sigma'_{ij} = D_{ijkl}\varepsilon_{kl}$$

Here the form of the general expressions (12.4 or 5) is once again obtained. This model is very effective not only for true viscoplastic solution, but for the numerical computation of simple plasticity by studying the converged solution of steady state. Here it supercedes the previously widely used initial stress process as it embraces its advantages—in particular, the ability to deal with the non-symmetric matrix resulting from non-associated plasticity.

12.4 NON-LINEAR MODELS FOR SOILS

Much work has been done on the development of non-linear elastic models based on a generalized Hooke's law.[24,25] According to Duncan[26] his model 'gives reasonably accurate results for many problems involving movements in earth masses which are not close to failure,' and 'the parameters required to characterize the soil behaviour may be determined readily from the results of conventional triaxial compression tests with volume change measurements.' On the other hand, his model 'does not provide accurate predictions of the strains for some strain paths,' 'it represents the soil as non-dilatant, i.e. shear stresses cause no volume changes,' and 'it provides a very poor representation of the properties of soil at and after failure.'

Such models should be suitable for the prediction of settlement so long as zones of near failure stress are limited. In fact, in such cases they may well be preferable to plastic models which often assume linear elasticity up to the point of yielding.

In the context of offshore structures, however, we are mainly concerned with problems of complex stress states and widespread yielding, especially in the estimation of ultimate loads for the determination of factors of safety. The use of plasticity models is therefore necessary, except perhaps for the computation of simple settlement. However, if this is accompanied by cyclic loading, a plastic analysis appears to be essential. No further attention will be given to elastic models, therefore, though it should be borne in mind that they do have the advantage of incorporating directly strains based on experimental evidence, and perhaps thus representing more accurately the pre-failure deformation states.

12.4.1 Plasticity models

The most well established feature of soil deformation is the failure condition, which for an isotropic material is generally accepted as being given by a Mohr–Coulomb type of criterion based on an envelope of principal stress circles, as shown in Figure 12.6. The shape of this envelope is dependent on the intermediate principal stress (as discussed in Section 12.4.2) and also the overconsolidation history if effective stresses are used.

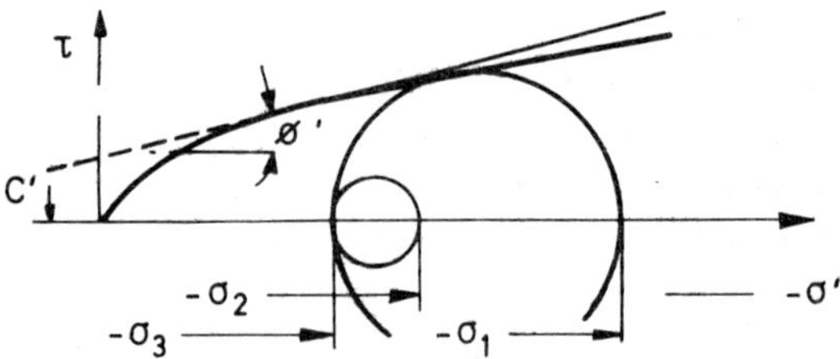

Figure 12.6 Mohr circle envelope as failure criterion

In the case of ideal plasticity the failure surface is also the yield surface, whilst in the case of strain-hardening/softening the yield surface may lie inside and/or outside the Mohr–Coulomb type surface for a normally consolidated soil. Inside this surface yielding causes expansion of the yield surface (hardening) until the stress path touches the Mohr–Coulomb type surface. Further yielding causes no further expansion of the yield surface and hence the term 'critical state' is applied, and the Mohr–Coulomb type surface is called the 'critical state surface.' Yielding outside this surface causes contraction of the yield surface (softening) until the 'critical state' is again achieved. Such a yield surface thus has the characteristic of predicting a higher maximum load for overconsolidated soil states than does the ideal plasticity model, as well as having the potential for determining pre-failure plastic strains. A more detailed description of both types of model, involving the use of plastic potential surfaces for the determination of plastic strain direction, is given in Section 12.4.3.

12.4.2 The failure surface

The failure surface for an isotropic soil is assumed to be symmetric about planes through each principal stress axis and the $\sigma_1 = \sigma_2 = \sigma_3$ axis. It is therefore convenient to express the surface in terms of the three invariants σ'_m, $\bar{\sigma}$ and θ, i.e. the effective mean stress, the deviatoric invariant and the Lode angle, defined as

$$-\frac{\pi}{6} \le \theta = \frac{1}{3}\sin^{-1}\left(\frac{-3\sqrt{3}}{2}\frac{J_3}{\bar{\sigma}.3}\right) \le \frac{\pi}{6}$$

where J_3 is the normal third invariant, defined in Equation (12.30). The geometrical interpretation of this set of invariants is shown in Figure 12.7(a). In

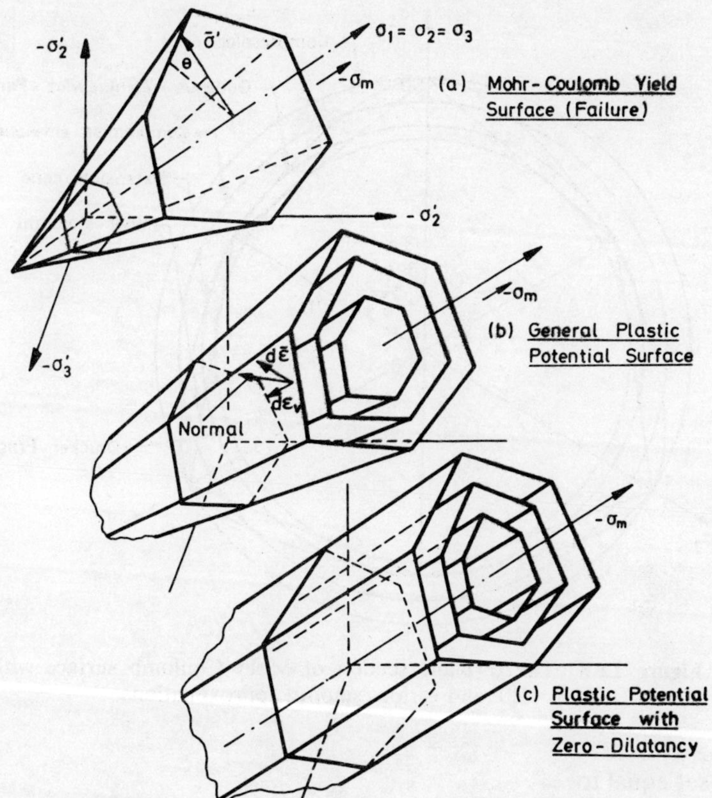

Figure 12.7　Idealized yield plastic potential surface

a π-section $(\sigma'_m = \text{constant})$ the value of $\bar{\sigma}$ is determined from triaxial compression tests for planes through the principal stress axes and the σ'_m axis. Using the Mohr–Coulomb rule and defining $\bar{\sigma}$ as unity in this case $(\theta = \pi/6)$ its value for varying θ is given by

$$\bar{\sigma}(\theta) = \frac{\cos(\pi/6) - \sin(\pi/6)\sin\phi/\sqrt{3}}{\cos\theta - \sin\theta\sin\phi/\sqrt{3}} \tag{12.39}$$

which defines a straight line between $\theta = (\pi/6)$ and $\theta = (-\pi/6)$, as shown in Figure 12.8. Also shown are the radially symmetric surfaces which are often used as approximate failure surfaces, and a smoothed version of the Mohr–Coulomb surface, due to Gudehus and Zienkiewicz[27,28]:

$$\bar{\sigma}(\theta) = \frac{2q}{(1+q) - (1-q)\sin 3\theta} \tag{12.40}$$

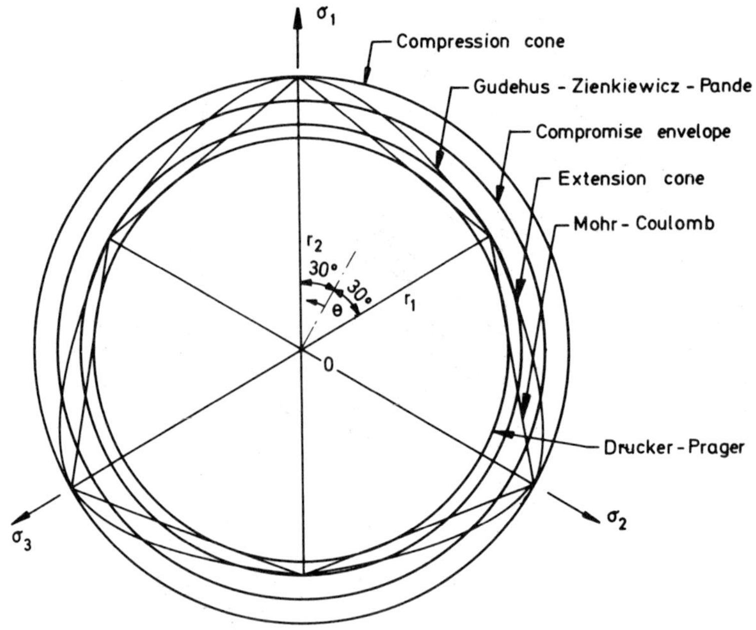

Figure 12.8   The π plane section of Mohr–Coulomb surface with $\phi = 20°$ and various smooth approximations

If q is set equal to

$$\frac{3 - \sin \phi}{3 + \sin \phi}$$

then $\bar{\sigma}(\theta)$ agrees with the Mohr–Coulomb surface for $\theta = (-\pi/6)$, as shown in Figure 12.8. However, some experimental findings of Lade[29] have given results for the form of $\bar{\sigma}(\theta)$ with $\bar{\sigma}$ greater than that given by the Mohr–Coulomb rule for $\theta = (-\pi/6)$.

The various surfaces of failure have been used as yield surfaces for a finite element analysis of a drained non-dilatant non-strain-hardening soil subjected to a flexible strip load. The results are compared in Figure 12.9 which shows a very wide range of ultimate loads. Since the results for the Mohr–Coulomb surface and the surface of Equation (12.40) are substantially different and as the latter appear to be nearer experimental findings,[29] there would appear to be a need for investigation, as the Mohr–Coulomb surface is often used on the assumption that it is quite accurate. We have, however, in further work adopted the Mohr–Coulomb criterion as being a sufficiently good approximation.

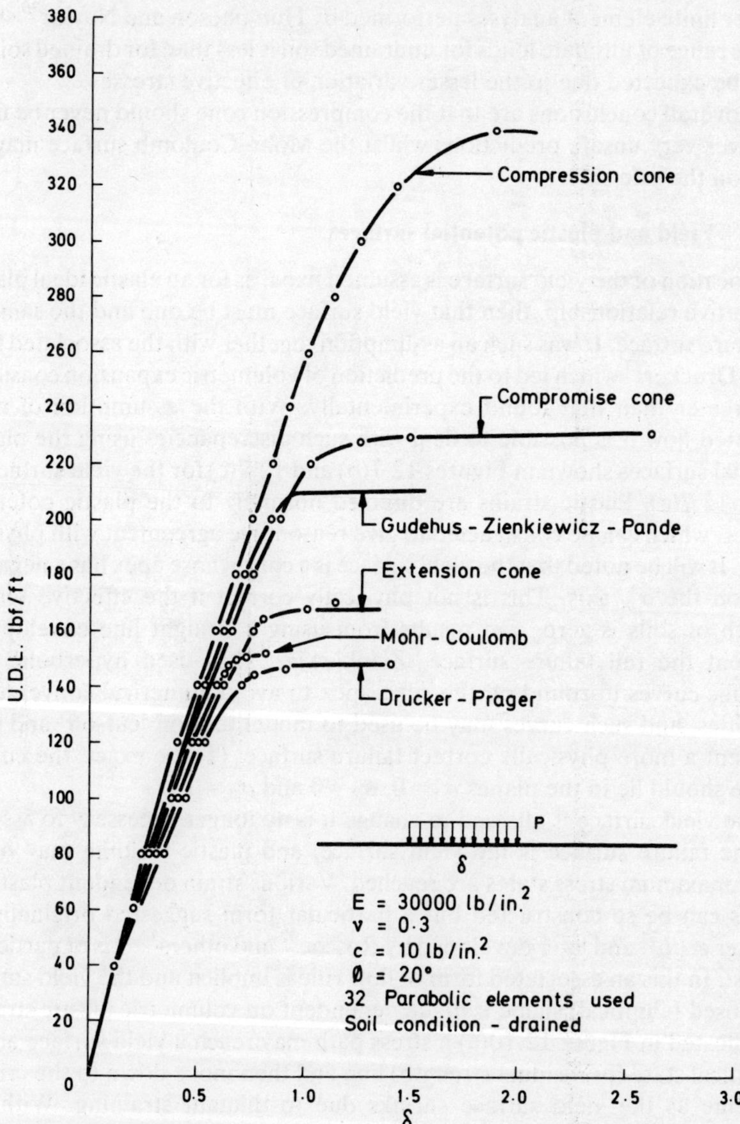

Figure 12.9 Load-deformation curves for ideal associated plasticity for various forms of Mohr–Coulomb approximation

Other finite element analyses performed by Humpheson and Naylor[30] show that the range of ultimate loads for undrained soil is less than for drained soil, as would be expected due to the lesser variation of effective stresses.

The overall conclusions are that the compression cone should never be used as it gives very unsafe predictions whilst the Mohr–Coulomb surface may be erring on the safe side.

12.4.3　Yield and plastic potential surfaces

If the location of the yield surface is assumed fixed, as for an elastic ideal plastic constitutive relationship, then that yield surface must be one and the same as the failure surface. It was such an assumption together with the associated flow rule of Drucker[23] which led to the prediction of volumetric expansion considerably greater than that found experimentally. With the assumption of non-associated flow it is possible to deal with such discrepancies using the plastic potential surfaces shown in Figures 12.7(b) and 12.7(c) for the yield surface of Figure 12.7(a). Plastic strains are directed normally to the plastic potential surfaces, which can be constructed to give reasonable agreement with physical results. It will be noted that the yield surface is a cone whose apex has a negative value on the σ'_m axis. This is not physically correct if the effective tensile strength of soils is zero, and results from using a straight line envelope to represent the full failure surface. Zienkiewicz[28] has used hyperbolic and parabolic curves to round off the cone apex to avoid numerical convergence difficulties, and such curves may be used to model tension 'cut-off' and thus represent a more physically correct failure surface. (To be exact, the cut off surface should lie in the planes $\sigma_1 = 0$, $\sigma_2 = 0$ and $\sigma_3 = 0$.)

If the yield surface is allowed to change it is no longer necessary to assume that the failure surface is the yield surface, and plastic yielding may occur before maximum stress states are reached. Various strain dependent plasticity models can be so constructed but a particular form suggested originally by Drucker *et al.*[31] and later developed by Roscoe[32] and others[33,34] is of particular interest. In this an associated form of flow rule is implicit and the yield surface of a closed (elliptical) shape is made dependent on volumetric plastic strain.* As indicated in Figure 12.10(a) a stress path may reach a yield surface above the critical state (or residual strength) line and then move down to the critical state line as the yield surface shrinks due to dilatant straining. With the direction of straining assumed normal to the yield surface (associated flow)

---

* The model used in later examples assumes that the diameter along the mean effective stress axis is given by

$$p = p_0 \exp \left(X[\varepsilon^p - \varepsilon_0^p] \right)$$

where $[\varepsilon^p - \varepsilon_0^p]$ is the plastic strain increment corresponding to an increase in mean stress $[p - p_0]$ under isotropic stress conditions, and X is a hardening constant.

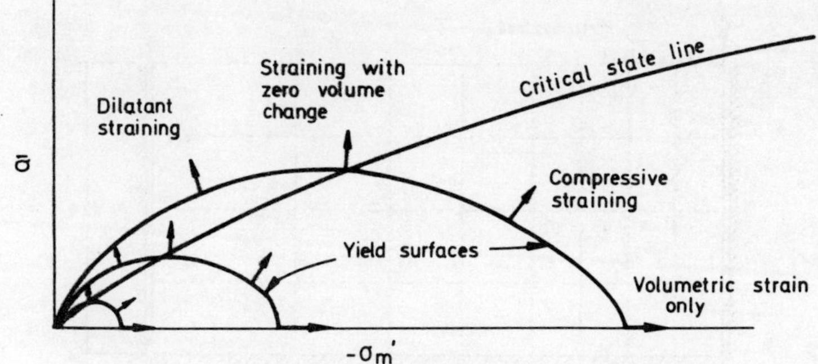

Figure 12.10(a)　Strain-hardening/softening model

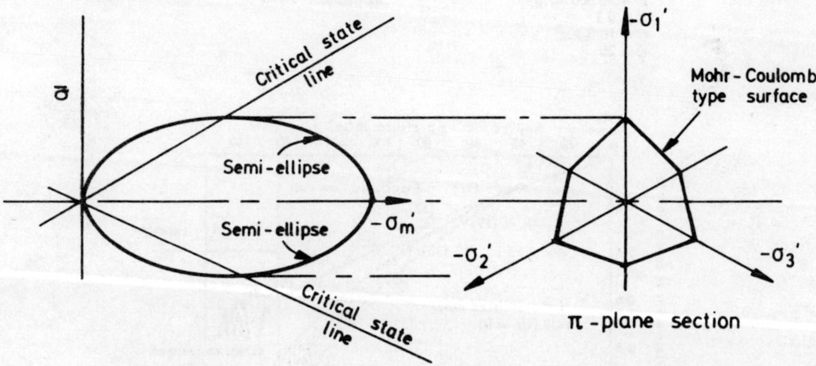

Figure 12.10(b)　Elliptical strain-hardening/softening model

predictions in reasonable agreement with physical evidence are obtainable because the normal becomes more and more parallel to the deviatoric stress axis as the stress path approaches the critical state line. By definition, the volumetric strain must be zero when the stress state lies on the yield surface and on the critical state line, and no further change of the yield surface will then take place.

The above arguments have their counter-part below the critical state line where strain-hardening is found to occur. Yielding causes compression, leading to expansion of the yield surface. Again, the yield surface may be used as a plastic potential surface since its normal approaches the direction of the $\bar{\sigma}$ axis as the stress path approaches the critical state line [see Figure 12.10(a)].

Whilst the assumption of associated plasticity may not be correct it may be used in conjunction with suitable strain-hardening, strain-softening rules[32,33] to ensure that the volumetric strains are reasonably correctly predicted, even though the deviatoric strains are not. Thus ultimate loads, which are dependent

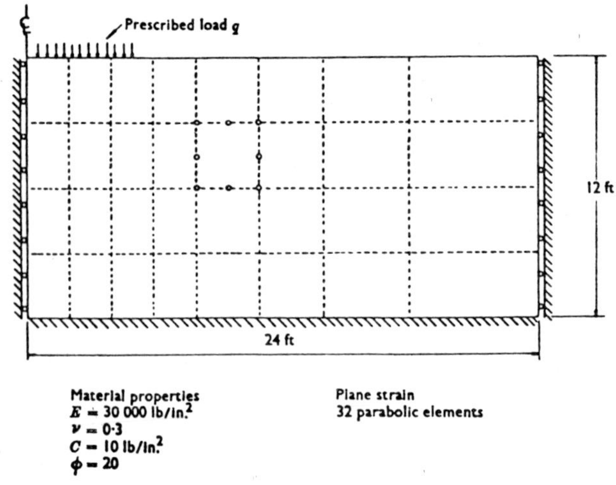

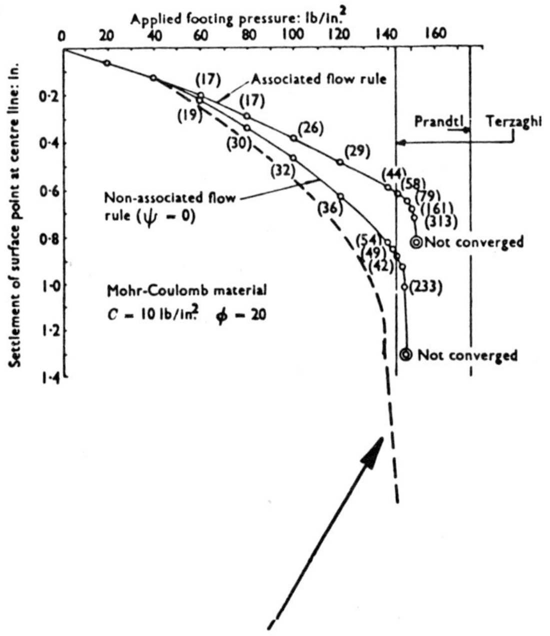

---- Strain Hardening Model ($E = 70\,000$ lb/in.^{2})

(a) Mesh and load-deformation behaviour

Figure 12.11 Strip load on a foundation of a weightless c–ϕ material; ideal plasticity with associated and non-associated (non-dilatant) flow rules and strain-hardening plasticity

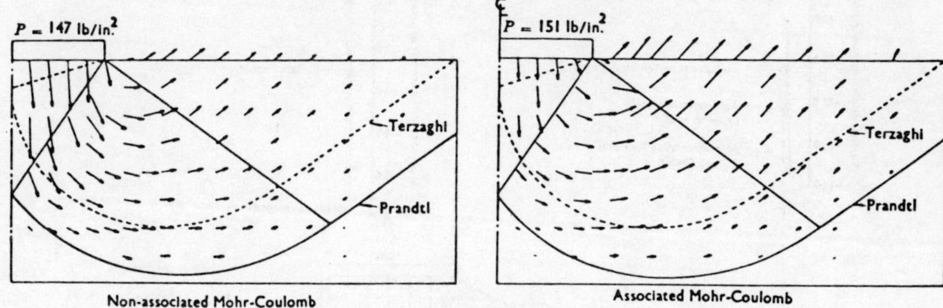

(b) Flow patterns

Figure 12.11 (*continued*)

on plastic dilatancy and contraction, especially in the undrained case, may be estimated with reasonable correctness even though the displacements may not be totally satisfactorily modelled. In later examples such assumptions have been made in using a model (Figure 12.10(b)) based on the modified cam clay ellipse of Roscoe and Burland,[34] though without placing too much reliance on the elliptical shape for stress states above the critical state line. An oversafe model may be obtained by introducing a non-associated Mohr–Coulomb cut-off yield surface for the super-critical zone. Experimental evidence is not at present sufficient to warrant greater refinement.

12.4.4　Numerical modelling of plasticity

A general outline of the numerical modelling of soil behaviour has been given in Section 12.2, but without the derivation of the non-linear stiffness matrix D_T. This has been developed elsewhere[8,35] and takes the form

$$\mathbf{D}_T = \mathbf{D}_e\left[\mathbf{I} + \frac{\partial Q}{\partial \boldsymbol{\sigma}} \cdot \frac{\partial F}{\partial \boldsymbol{\sigma}} \cdot \frac{\mathbf{D}_e}{A} + \left(\frac{\partial F}{\partial \boldsymbol{\sigma}}\right)^T \mathbf{D}_e \frac{\partial Q}{\partial \boldsymbol{\sigma}}\right] \tag{12.41}$$

where $\mathbf{I}$ is an identity matrix, and A is a strain hardening parameter, equal to zero for ideal plasticity. It will be observed that if $F \neq Q$ (non-associated plasticity) the matrix will not be symmetric, causing numerical difficulties in a tangential stiffness matrix solution.

One further point concerning numerical problems is that corners in yield surfaces need special attention. As mentioned in Section 12.4.3 special devices may be used to round off the apices of conical Mohr–Coulomb type yield surfaces, as well as the corners in the π-sections; they may even be more representative of physical behaviour. In the cases where the Mohr–Coulomb

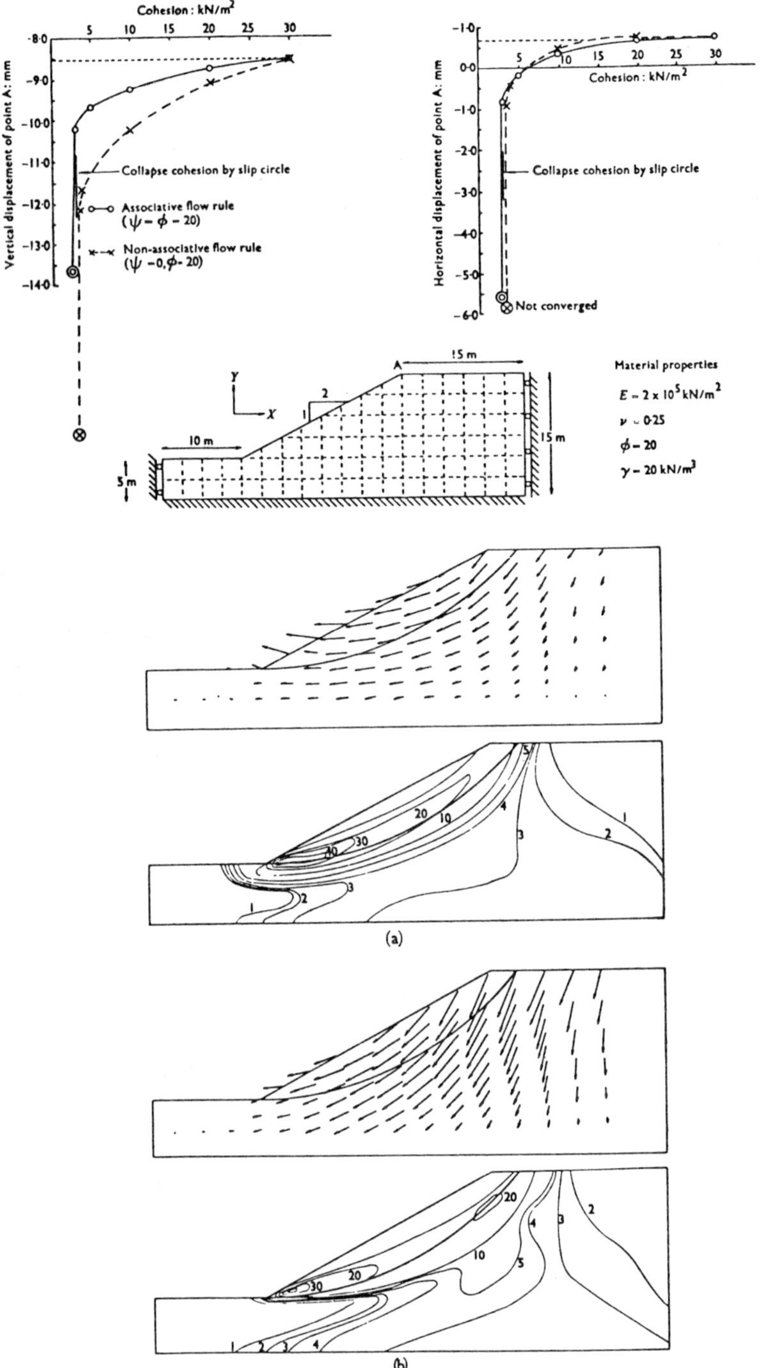

Figure 12.12 Embankment deformation flow patterns and maximum effective shear strain contours (gravity constant; progressive decrease of cohesion)

surface has been used without smoothing in later examples the technique used is simply to assume that if θ is very close to $\pm(\pi/6)$ the direction of plastic strain is in the direction of the corresponding principal stress axis; i.e. in the direction of the resultant of the two unit normals to the faces of the Mohr–Coulomb surfaces which meet at the corner ($\sin\theta > 0{\cdot}49$).

To avoid difficulties with non-symmetrical tangent matrices and, indeed, to obtain maximum efficiency the authors advocate the use of the *viscoplastic procedure*[15,16] even if purely plastic situations are to be solved. Such models were used, in fact, throughout the solution procedure in this chapter, except in Section 12.7.

12.4.5 The effect of volumetric strains on ultimate load

Although experiments indicate some dilation of the soil before failure, a plastic potential parallel to the failure surface would give excessive and continuing volume changes, which do not in fact occur. A completely associated ideally plastic model of soil thus does not appear justified, and in some recent publications of the authors the effect of comparing associated and non-associated models has been fairly comprehensively investigated.[36,37]

The general conclusions of such investigations can be summarized as follows.

(1) For relatively unconfined deformations such as occur in drained behaviour of embankments or foundations the results of fully associated and zero volume change (non-dilatant) plasticity give virtually identical collapse loads although the deformation history is apparently different. Figure 12.7(c) illustrates such non–dilatant potential surfaces associated with the Mohr–Coulomb criterion and Figures 12.11 and 12.12 illustrate appropriate examples.

(2) Where the degree of confinement is moderate the associated and non-associated models will give substantially different collapse loads and displacements. This is illustrated in the case of a drained test piece placed between rigid plattens, the results of which are shown in Figure 12.13.

(3) When the degree of restraint is large, such as occurs in the fully undrained behaviour where the presence of the fluid virtually restrains all volume deformation, *the fully associated ideally plastic material is incapable of failure as plastic strains are almost prevented.*

(4) For strain-hardening/softening models the ultimate load will be close to that given by the Mohr–Coulomb line under drained conditions, as shown in Figure 12.11, but may be very different under undrained conditions, as shown in Figure 12.14. The *plastic* volumetric compression must here be balanced by *elastic* volumetric expansion, since the undrained conditions allow little total volumetric change. Consequently, the mean effective stress, σ'_m, must

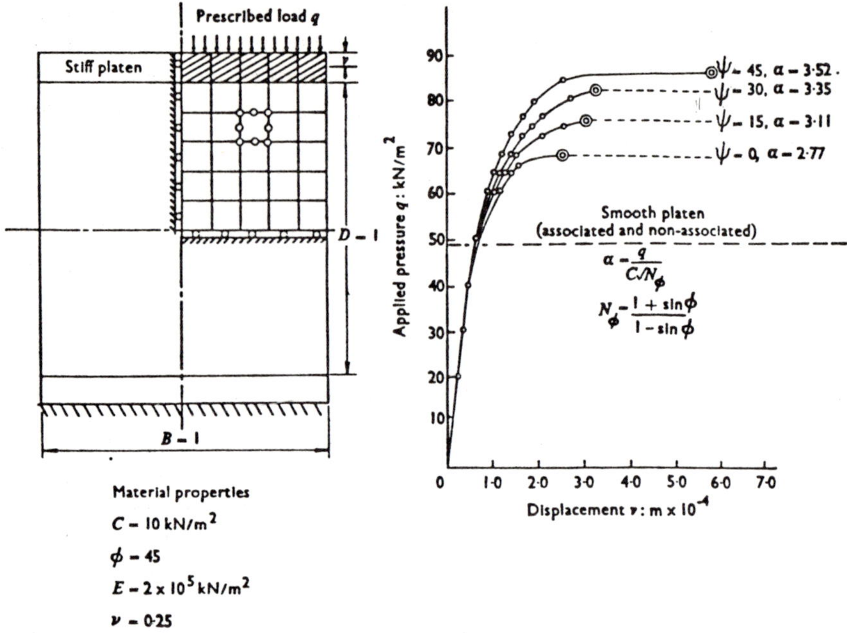

$$\alpha = \frac{q}{C\sqrt{N_\phi}}$$

$$N_\phi = \frac{1 + \sin\phi}{1 - \sin\phi}$$

Figure 12.13 Axisymmetric sample between rough platens. Effect of degree of dilatancy

$\psi = 0$ Zero dilatancy
$\psi = \phi$ 45° Fully associated flow rule

increase,* as shown in Figure 12.15, leading to a lower ultimate load than that given by the Mohr–Coulomb yield criterion.

Three important conclusions can be drawn from these results. If ideally elastic-plastic models are used, the zero dilatancy model is capable of predicting the failure load under most conditions reasonably well. The fully associated ideally plastic model should not be used in cases of confined deformation, and *never* for undrained analyses. The strain-hardening/softening model predicts ultimate loads reasonably well under all conditions, as well as having the potential for improved calculation of displacements. However, it may give considerably lower failure loads in the case of *undrained* analyses of *normally* consolidated soils, suggesting that the use of a Mohr–Coulomb criterion may be unsafe in such cases.

A basis for the numerical modelling of soil has now been established, which, whilst not claiming to give accurate estimates of displacements, does claim to suffice for the prediction of ultimate loads, at least to the degree of accuracy

---

* Compressive stresses are regarded as negative, in the normal convention of mechanics.

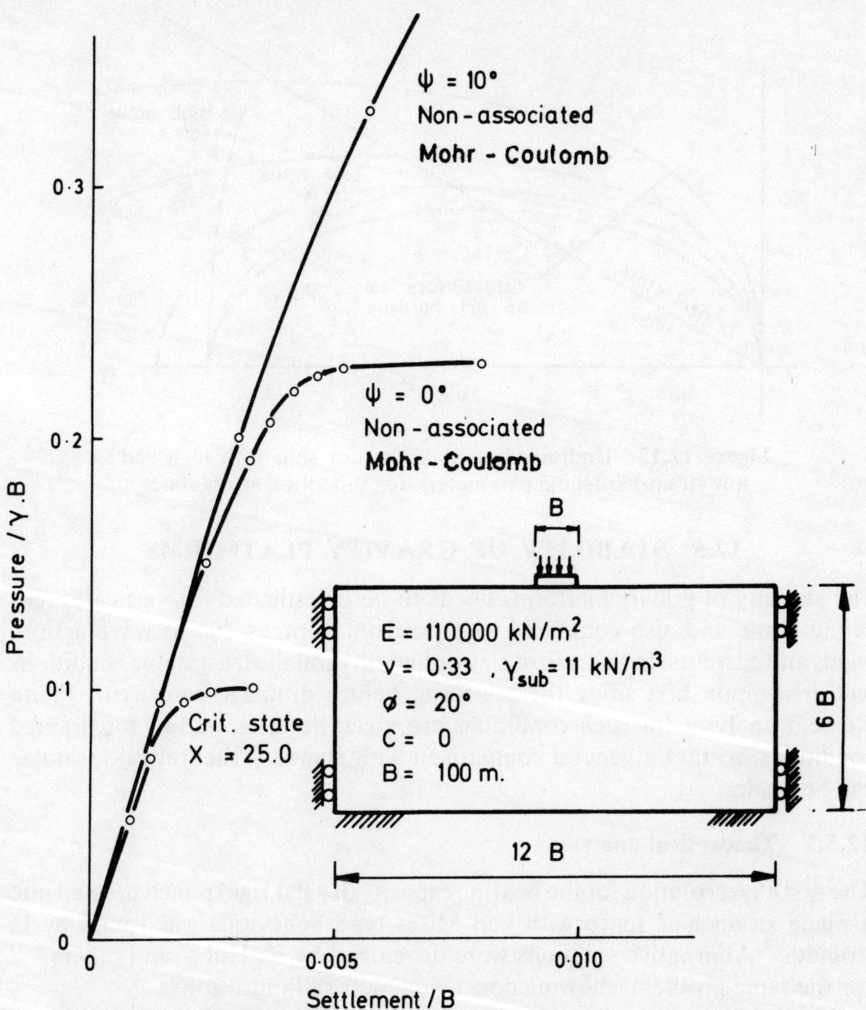

Figure 12.14 Comparison of load/settlement curves for:
(i) Dilatant Mohr–Coulomb yielding $\psi = 10°$
(ii) Non-dilatant Mohr–Coulomb yielding $\psi = 0°$
(iii) Critical state yielding
for soil with self-weight, under undrained conditions

warranted by the normal procedures of soil sampling and testing. The application of the procedures so far described will now be shown for some problems of offshore structures; in the case of cyclic loading further refinement of the numerical models will be discussed for the prediction of 'fatigue' phenomena not explained by the models discussed so far.

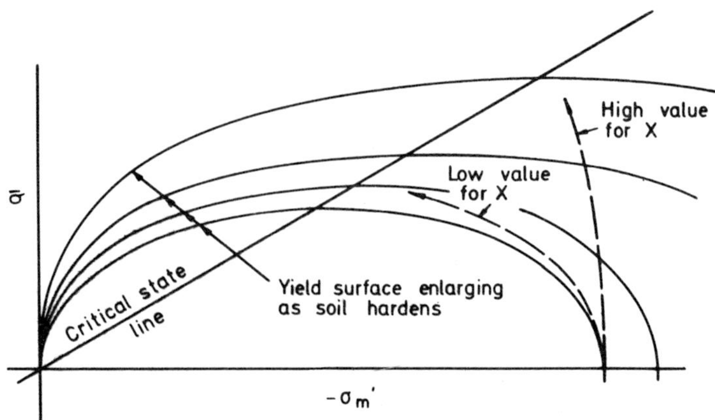

Figure 12.15 Undrained stress paths, for soils with high and low strain-hardening parameters, for subcritical stress states

12.5 STABILITY OF GRAVITY PLATFORMS

The stability of gravity platforms needs to be investigated for vertical forces acting alone and also combined with horizontal forces due to wave action, winds and currents. In the case of clay or fine silty soil the least stable conditions will arise upon first siting the platform, before drainage can occur. Finite element analyses for such conditions are given here, as well as for drained conditions, so that a general comparison with standard theoretical formulae can be made.

12.5.1 Theoretical analyses

The first exact solution for the bearing capacity of a flat rigid punch pressed into a plane strain half space with von Mises type behaviour was found by L. Prandtl.[38] Alternative solutions were developed by R. Hill,[39] and others[40,41] for the same problem, showing non-uniqueness of failure zones.

When exact solutions are not available as in many engineering problems stability solutions can be obtained by the application of the limit (bounding) theorems developed by Drucker and Prager.[42] For example, Caquot's[43] method of determination of the bearing capacity gives an upper bound (kinematical) solution whereas Bell's[44] and Chen's[45] solutions are cases of lower bound (statical) solutions. The case of a weightless medium with internal friction was considered by H. Reissner[46] and also by R. T. Shield[47,48] who dealt with stress determination and velocity fields using the method of characteristics and the associated flow rule.* A numerical solution of bearing capacity and

---

* Velocity fields for non-associated flow rules have also been considered.[49,50]

stability problems for soil with self weight was presented by W. W. Sokolovski.[51,52] In the plane strain case, and for given boundary conditions, it reduces to the solution of hyperbolic partial differential equations which can be derived from the conditions of static equilibrium and the Mohr–Coulomb failure criterion, as given by

$$\left.\begin{array}{c}\dfrac{\partial\sigma_x}{\partial x}+\dfrac{\partial\tau_{xy}}{\partial y}=0\\[2mm]\dfrac{\partial\sigma_y}{\partial y}+\dfrac{\partial\tau_{xy}}{\partial x}-\gamma=0\end{array}\right\}\tag{12.42}$$

and

$$((\sigma_y-\sigma_x)^2+4\tau_{xy}^2)^{1/2}+(\sigma_x+\sigma_y)\sin\phi-2c\cos\phi=0\tag{12.43}$$

12.5.1.1 *Prandtl-type, upper-bound solutions*

Starting from the collapse mechanism proposed by L. Prandtl (shown in Figure 12.16) it is assumed that the shear strength of the soil is developed along the slip surfaces. The middle triangle AA′B is moving down as a rigid body with the strip foundation whereas the triangular wedge ACD is moving in a direction orthogonal to the line AC. The transit shear area ABC is bounded by a logarithmic spiral BC and straight lines AB, AC. The angle of internal shearing resistance ϕ governs the size of the mobile area. The length and depth of distortion can be obtained from geometric relationships:

$$h=[B\cos\phi/2\sin(\pi/4-\phi/2)]\exp[(\pi/4+\phi/2)\tan\phi]\tag{12.44}$$

$$l=[B\cos(\pi/4-\phi/2)/\cos(\pi/4+\phi/2)]\exp(\pi/2\tan\phi)\tag{12.45}$$

Values of h and l for different values of internal friction are shown in Table 12.5.1.

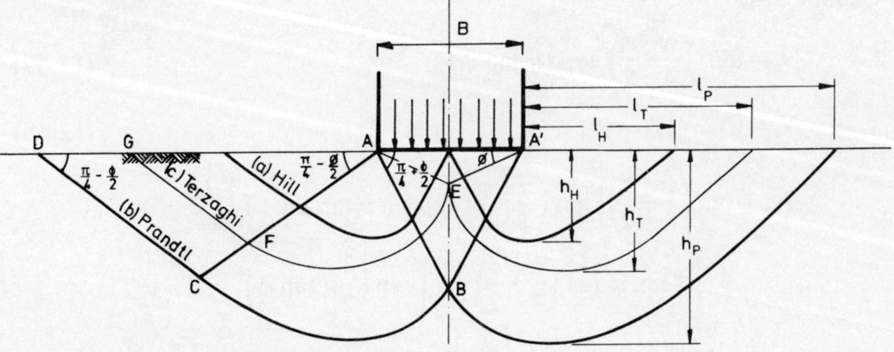

Figure 12.16　Slip surfaces assumed for theoretical prediction of bearing capacity

Table 12.5.1 Size of failure zone for varying angle of internal friction ϕ°

ϕ°	h/B			l/B		
	Hill	Terzaghi	Prandtl	Hill	Terzaghi	Prandtl
0.	0·354	0·500	0·707	0·500	0·707	1·000
5.	0·396	0·574	0·793	0·626	0·906	1·252
10.	0·447	0·660	0·893	0·786	1·161	1·572
15.	0·507	0·762	1·014	0·993	1·491	1·985
20.	0.581	0·886	1·162	1·265	1·929	2·530
25.	0·673	1·040	1·347	1·633	2·522	3·265
30.	0·793	1·238	1·585	2·145	3·351	4·290
35.	0·952	1·502	1·904	2·885	4·552	5·770
40.	1·174	1·868	2·348	4·006	6·375	8·012
45.	1·500	2·405	3·001	5·807	9·308	11·614

These limits of the plastic zone form a useful guide for the mesh size necessary for solution of problems using a finite element approximation. Comparing external load power with power dissipated internally leads to the evaluation of ultimate pressure. This can be presented in a form introduced by Terzaghi:

$$p = cN_c + gN_g + 0\cdot5\gamma N_\gamma B \tag{12.46}$$

where

p = ultimate pressure,

c = cohesion,

g = pressure due to overburden at base of footing,

γ = soil self-weight (submerged weight of a saturated soil in an effective stress analysis),

B = width of strip footing.

Dimensionless parameters N_c, N_g and N_γ are found to depend only on the value of internal friction, ϕ, and for a Prandtl-type slip surface are given by Caquot[43] as

$$N_g = \tan^2\left(\frac{\pi}{4}+\frac{\phi}{2}\right)\exp\left(\pi\tan\phi\right) \tag{12.47}$$

$$N_c = (N_g - 1)\cot\phi \tag{12.48}$$

$$N_\gamma = \tan\left(\frac{\pi}{4}+\frac{\phi}{2}\right)\left[\tan\left(\frac{\pi}{4}+\frac{\phi}{2}\right)\exp\left(\frac{3}{2}\pi\tan\phi\right)-1\right]\bigg/2$$

$$+\left\{\left[3\tan\phi\tan\left(\frac{\pi}{4}+\frac{\phi}{2}\right)-1\right]\exp\left(\frac{3}{2}\pi\tan\phi\right)\right.$$

$$\left.+\left[3\tan\phi+\tan\left(\frac{\pi}{4}+\frac{\phi}{2}\right)\right]\right\}\bigg/4(1+9\tan^2\phi)\sin^2\left(\frac{\pi}{4}-\frac{\phi}{2}\right) \tag{12.49}$$

Table 12.5.2   Bearing capacity coefficients for varying angle of internal friction, ϕ

Bearing Capacity formula		ϕ								
		0°	5°	10°	15°	20°	25°	30°	35°	40°
Terzaghi	N_c	5·7	7·0	9·5	13·0	17·0	24·0	37·0	58·0	98·0
	N_g	1·0	1·6	2·7	4·5	7·5	13·0	23·0	42·0	77·0
	N_γ	0·0	0·14	0·7	2·0	4·8	9·8	20·0	43·0	98·0
Prandtl–Caquot [Equations (12.47)–(12.49)]	N_c	5·14	6·49	8·34	10·98	14·80	20·72	30·14	46·12	73·31
	N_g	1·00	1·57	2·47	3·94	6·40	10·66	18·40	33·30	64·19
	N_γ	0·00	0·50	1·44	3·28	6·90	14·32	30·38	67·74	163·50
Hill type		0·00	0·25	0·72	1·64	3·45	7·16	15·19	33·87	81·75
Sokolowski	N_γ*	0·00	0·16	0·46	1·40	3·16	6·92	15·32	35·18	86·46
B. Hansen [Equation (12.52)]		0·00	0·08	0·39	1·19	2·95	6·76	15·07	33·93	79·53
Meyerhof [Equation (12.53)]		0·00	0·07	0·37	1·13	2·87	6·74	15·67	37·16	93·68

* N_c and N_g as for Prandtl–Caquot.

Values of N_c, N_g and N_γ for different values of ϕ are shown in Table 12.5.2, contrasted with alternative solutions.

The above method gives an upper bound solution employing a Prandtl-type failure mechanism which should agree well with experiment in the case of a perfectly rough footing. For a weightless soil, this solution is identical with the upper bound solution obtainable by the characteristics method.[47]

In contrast, the kind of velocity field proposed by Hill[39] is reasonable for a smooth footing (Figure 12.16). This type of slip surface has given good agreement with experiment.[53] Adopting the same procedure as before to determine the ultimate load again results in Equation (12.46) where the dimensionless parameters N_g and N_c are identical to those given by Equations (12.47) and (12.48) respectively, whereas parameter N_γ is one half that given by Equation (12.49) for Prandtl's mechanism. Similarly, the depth and length of the distortion area are reduced to half those given by Equations (12.44) and (12.45). As for the Prandtl-type mechanism the solution is upper bound.

12.5.2   Engineering applications

Since exact theoretical solutions for different types of load and shapes of footing are not available approximate formulae are employed. Historically, the first bearing capacity solution was developed by Terzaghi[5] but at present, the Brinch Hansen,[54] and Meyerhof[55] formulae are widely used.

12.5.2.1 Terzaghi solution (1943)

The solution was originally developed for rigid, rough, shallow footings under plane strain conditions but it has since been modified to cover rectangular and circular footings. Terzaghi assumed wedge AA′E (Figure 12.16), compressed by the foundation load, to be in a state of limiting equilibrium with passive resistance of the soil within the zone AEFG and cohesive resistance along the plane AE. Evaluating the passive resistance of the soil by minimization of foundation load Terzaghi expressed the ultimate load as a function of c, γ, g and dimensionless parameters N_c, N_γ, N_g [Equation (12.46)]. Values of these parameters, which are different from those of Prandtl's solution for $\phi = 0$ are shown in Table 12.5.2. Terzaghi's solution does not fully satisfy the requirements of an upper bound limit theorem.

The length and depth of the area of distortion are now given by

$$l = B \,.\, \exp\left[(3\pi/4 - \phi/2)\tan\phi\right]\cos(\pi/4 - \phi/2)/\cos\phi$$
$$h = (B/2)\exp(\pi/2\tan\phi). \tag{12.50}$$

Values calculated thus are shown in Table 12.5.1.

12.5.2.2 Brinch Hansen and Meyerhof formulae

More general formulae which apply to inclined loads and different shapes of foundation have been introduced by Brinch Hansen, and Meyerhof. The bearing capacity parameters N_c and N_g are the same as in Prandtl or Hill solutions, whilst the parameter N_γ is given by a simpler approximation. Both of their formulae can be expressed in the form

$$\frac{Q}{B'L'} = P_r = cN_c d_c s_c i_c + g'N_g d_g s_g i_g + \tfrac{1}{2}\gamma'B'N_\gamma d_\gamma s_\gamma i_\gamma \tag{12.51}$$

where

$\qquad P_r$ = ultimate bearing capacity for the effective foundation area,
$\qquad \gamma'$ = effective unit weight of soil,
$\qquad B'$ = effective foundation width,
$\qquad L'$ = effective foundation length,
$\qquad g'$ = effective overburden pressure at base level of foundation,
$\quad d_c, d_g, d_\gamma$ = depth factors,
$\quad s_c, s_g, s_\gamma$ = shape factors,
$\quad i_c, i_g, i_\gamma$ = load inclination factors,
$\qquad c$ = drained cohesive shear strength.

The bearing capacity factors, N_g, N_c, are taken from the upper bound Prandtl solution [Equations (12.47) and (12.48)], but the factor N_γ is assumed to be as given below.

Brinch Hansen:
$$N_\gamma = 1\cdot5(N_g - 1)\tan\phi \tag{12.52}$$

Meyerhof:
$$N_\gamma = (N_g - 1)\tan(1\cdot4\phi) \tag{12.53}$$

where $N_g = \tan^2(\pi/4 + \phi/2)\exp(\pi\tan\phi)$. Values of N_γ for different ϕ values are shown in Table 12.5.2. The depth factors are determined by approximate formulae.

Brinch Hansen:
$$d_c = 0\cdot4D/B, \qquad d_g = 1 + 2\tan\phi(1 - 2\sin\phi)^2 D/B$$
$$\text{for } D \leq B \quad \text{and for } D > B$$

$$d_c = 0\cdot4 \operatorname{arc}\tan D/B$$
$$d_g = 1 + 2\tan\phi(1 - \sin\phi)^2 \operatorname{arc}\tan D/B$$

Meyerhof:
$$d_c = 1 + 0\cdot2\tan(\pi/4 + \phi/2)\,.\,D/B$$
$$d_g = d_\gamma = 1 + 0\cdot1\tan(\pi/4 + \phi/2)\,.\,D/B \text{ for } \phi > 10°$$
$$d_g = d_\gamma = 1\cdot0 \text{ for } \phi < 10°$$

where D is the depth of the foundations below the soil surface.

The shape factors depend on the effective width and effective length of foundations as well as the eccentricity and inclination of the load.

Brinch Hansen:
$$s_g = 1 + i_g B'/L' \sin\phi$$
$$s_\gamma = 1 - 0\cdot4i_\gamma B'/L'$$
$$s_c = 0\cdot2\,(1 - 2i_c)B'/L' \qquad (\phi > 0)$$
$$s_c = 0\cdot2i_c B'/L' \qquad (\phi = 0)$$

The effective width for a strip foundation is determined by the eccentricity of the load:
$$B' = B - 2e$$

where e is the eccentricity.

Meyerhof:

$$s_g = s_\gamma = 1 + 0\cdot1 \tan^2 (\pi/4 + \phi/2)B/L$$

$$s_c = 1 + 0\cdot2 \tan^2 (\pi/4 + \phi/2)B/L \qquad (\phi > 10°)$$

$$s_g = s_\gamma = 1 \qquad (\phi = 0)$$

Finally, the load inclination factors are calculated as follows:

Brinch Hansen:

$$i_g = (1 - 0\cdot5H/(V + Ac \cot \phi))^5,$$

$$i_\gamma = (1 - 0\cdot7H/(V + Ac \cot \phi))^5,$$

$$i_c = 1 - H/2Ac \qquad (\phi > 0),$$

$$i_c = 0\cdot5[1 + \sqrt{(1 - H/Ac_u)}] \qquad (\phi = 0)$$

Meyerhof:

$$i_c = i_g = [1 - (\text{arc } \tan H/V)/90°]^2$$

$$i_g = [1 - (\text{arc } \tan H/V)/\phi]^2$$

where

$$V = \text{vertical load,}$$
$$H = \text{horizontal load,}$$
$$A = \text{effective foundation area,}$$
$$c_u = \text{undrained cohesive shear strength.}$$

12.5.2.3 *Other methods and formulae*

The bearing capacity formulae introduced by Terzaghi,[5] Meyerhof,[55] Brinch Hansen[54] and others are based on the development of critical shear zones, but the geometry of these zones is fixed. To improve results empirical factors had to be employed. Methods that do allow for a chosen failure surface have been published by Janbu,[56] Morgenstern and Price,[57] and Karal.[58] The disadvantage of these methods is that they are both complicated to use and time consuming.

A simplified way to get quick and reliable answers for the ultimate load and the location of the failure zone is to adopt the slip surface method used by the Norwegian Geotechnical Institute.[59] In this method the resistance to sliding for various slip surfaces is evaluated simply by changing one parameter (which can be performed by a computer). An advantage of the method is that the soil need not be homogeneous and shear strength variation with depth is accounted for. The results obtained by this method give good agreement with those of the Brinch Hansen, and Meyerhof methods for the case of homogeneous soils.

12.5.3 Numerical analyses

The theoretical formulae are derived for plane strain conditions, and empirical factors are used to derive approximate bearing capacities for rectangular and circular footings. Also, they relate to homogeneous soils. The finite element method, on the other hand, allows direct analysis of three-dimensional conditions as well as for the variation of soil properties, including the dependence of undrained strength on depth, and changing pore water pressure.

In this section, for the purpose of comparison, only plane strain conditions for a single layered soil will be investigated, this alone involving a wide range of variables covering drainage conditions, soil stress–strain behaviour and boundary conditions.

12.5.3.1 Drained analyses

Drained analyses are considered first, being the more likely to show agreement between theoretical ultimate load predictions and those of finite element analyses. Volumetric changes due to dilatancy cannot cause pore pressure changes and consequent changes of effective stress, so that the degree of dilatancy has little influence on failure stress states, though it may more seriously affect settlement. Weightless soils are considered separately so that the relative influence of N_c and N_γ in Equation (12.46) may be studied. The influence of N_g will not be considered here.

Weightless Soil (Influence of N_c). Before any comparisons can be made between limit and numerical displacement analyses it is essential to determine the influence of the boundaries used in the finite element analyses. The effect of changing from a small domain 'A' to a large domain 'B' is shown in Figure 12.17. It would appear that the bearing capacity for a flexible footing decreases

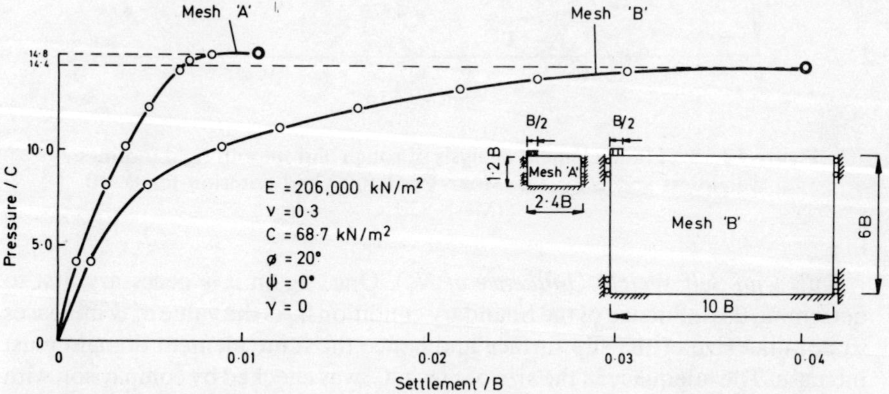

Figure 12.17 Comparison of meshes used for the analysis of weightless soils, for the case of a flexible footing

only slightly as the mesh size increases beyond that of mesh A, in agreement with the shear zone limits for $\phi = 0$ given by Table 12.5.1.*

Using an intermediate mesh size (as shown in Figure 12.18) analyses were performed to determine the soil response to

(i) a rigid footing with a rough surface for $\phi > 0°$ (allowing no sliding),
(ii) a rigid footing with a smooth surface (allowing sliding).

The results are compared in Figure 12.18 for a value of internal friction angle ϕ equal to 0° and in Figure 12.19 for ϕ equal to 20°, using the associated Mohr–Coulomb yield criterion. The results show an increase in bearing capacity as the footing/soil-interface conditions become more constraining on displacements. In all cases the bearing capacity is above the slip line solution which is an exact solution in the case of ϕ equal to 0.

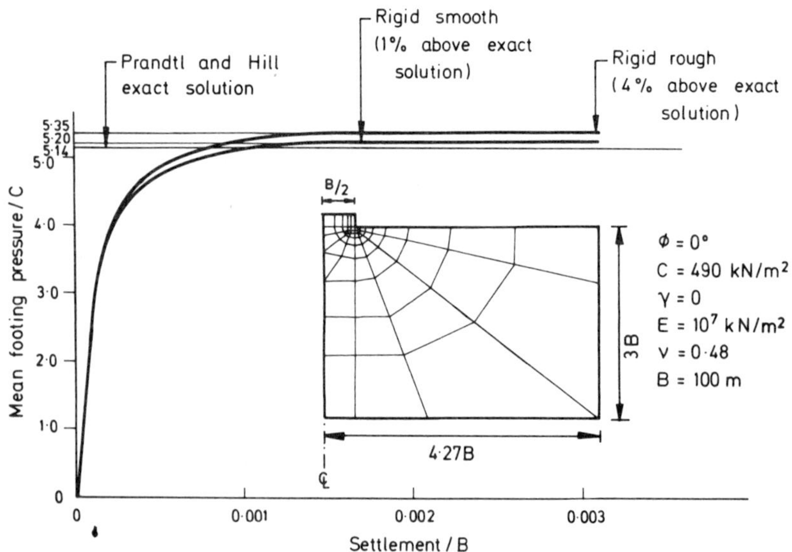

Figure 12.18   Finite element analysis of rough and smooth rigid footings on a weightless soil using the Mohr–Coulomb yield criterion for $\phi = 0$. (Mesh 'C')

Soils with Self Weight (Influence of N_γ). Once again it is necessary first to determine the influence of the boundary conditions. As the value of ϕ increases so does the size of the slip surface and hence the finite element domain must increase. The adequacy in the size of mesh 'C' was checked by comparison with a mesh 'D' of size 7.2B breadth by 4.3B depth for ϕ equal to 20°, using a

---

* The widely different displacements are simply due to the well known non-uniqueness of displacements if two-dimensional analysis is used.

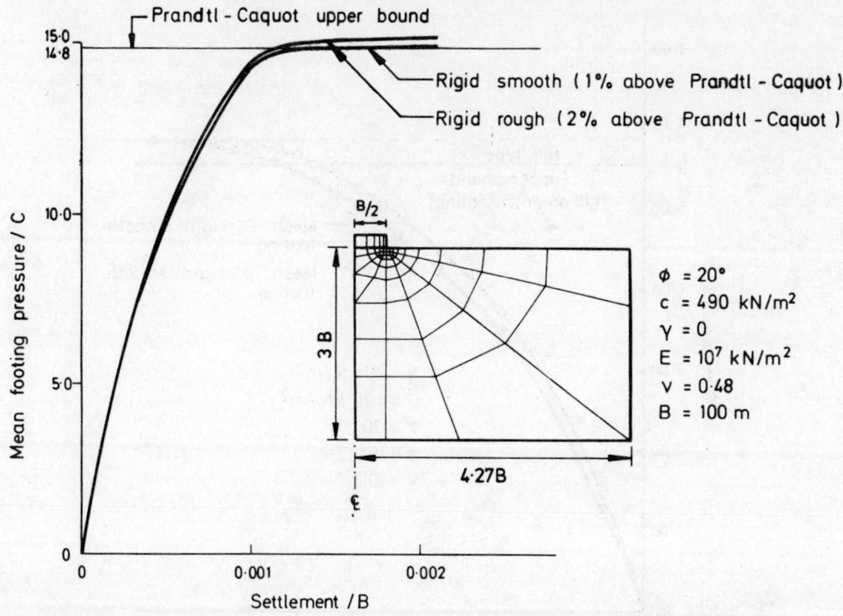

Figure 12.19 Finite element analysis of rough and smooth rigid footings on a weightless soil using the Mohr–Coulomb yield criterion (associated), for $\phi = 20°$

Mohr–Coulomb flow rule. The similarity of results may be seen in Figure 12.20.

Figure 12.21 compares finite element analyses for rigid footings for ϕ equal to 20°. It may be seen that the rigid smooth footing results agree with Hill's formula, whilst those for a rigid rough footing lie below the upper bound Prandtl–Caquot solution but close to that of Terzaghi.

12.5.3.2 Undrained analyses

In order to obtain a value for the undrained bearing capacity by means of Equation (12.46) it is necessary to assume that $\phi = 0$ and $c = a$ constant, averaged throughout the soil. The results of a finite element analysis for $\phi = 0$ have already been compared with those given by cN_c (Figure 12.18). For a soil which has variable strength due to consolidation under its own weight, an effective stress analysis may be carried out. The results for non-associated ($\phi = 0$) and fully-associated Mohr–Coulomb flow rules, as well as the strain-hardening/softening critical state ellipse model are described in Section 4.5 and shown in Figure 12.14. It will be repeated that the results show that *associated* Mohr–Coulomb flow rules should not be used in practice; also that the strain-softening/hardening model predicts lower ultimate loads than does the Mohr–Coulomb model.

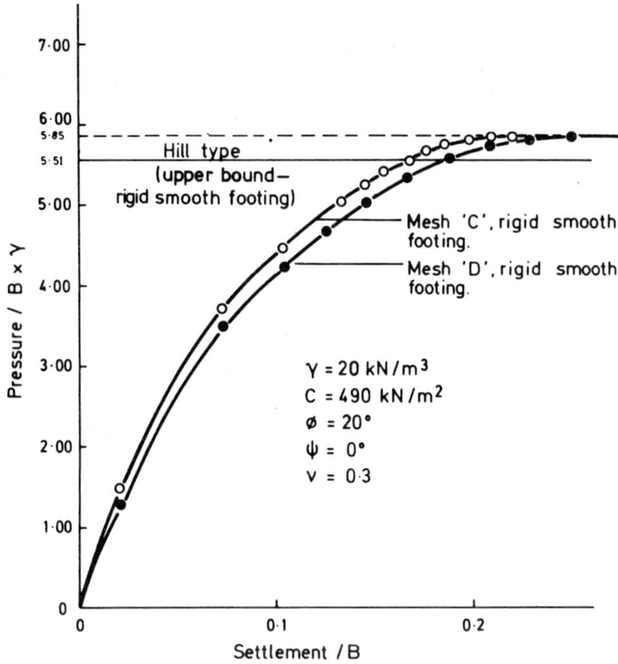

Figure 12.20 Comparison of meshes used for the analysis of soils with self-weight

12.5.3.3 *Stability under the action of inclined forces*

Finite element analyses have been carried out for both vertical and inclined eccentric loads for comparison with the Meyerhof and Brinch Hansen formulae. The results for drained conditions using the Mohr–Coulomb yield criterion are shown in Figures 12.22 and 12.23. For vertical eccentric loading of a *rough* rigid plate the bearing capacity is about 40 per cent higher than that given by the Hansen and Meyerhof formulae. In partial explanation of this large difference it should be noted that the Hansen and Meyerhof values for N_γ are close to those of Hill, which are based on the assumption of a *smooth* rigid plate. In order to perform a more comparable F.E.M. analysis it would be necessary to constrain the element nodes under the footing to remain on a line of rotation whilst being free from constraint by the plate elements in moving along that line.

In the case of an inclined eccentric load for which the Meyerhof and Hansen predictions differ by 100 per cent the F.E.M. gives results which are yet another 100 per cent higher. Again the F.E.M. difference may be partially explained by the roughness of the footing, but further investigation is obviously required.

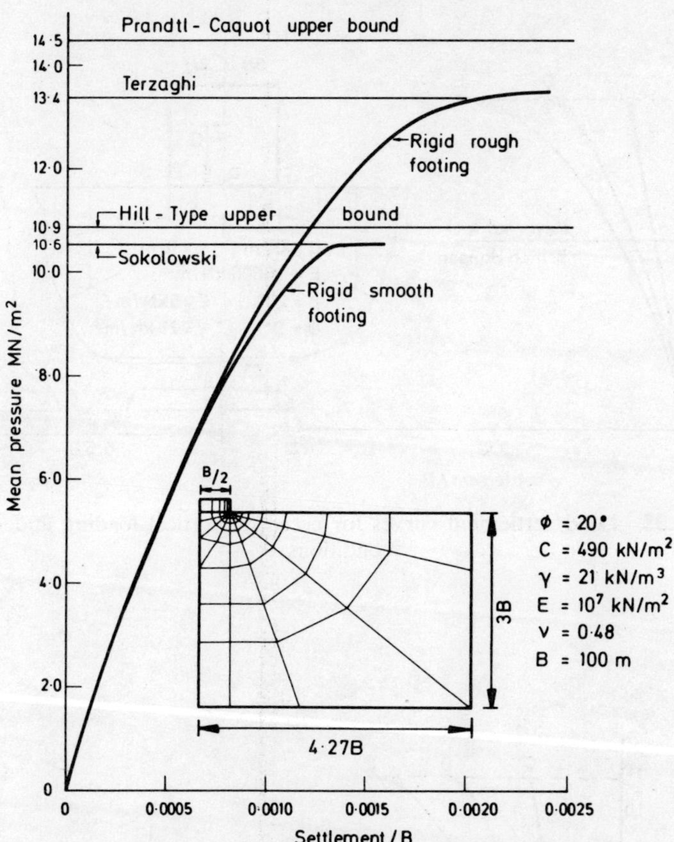

Figure 12.21 Finite element analysis of rough and smooth rigid footings on a soil with self-weight, using the Mohr–Coulomb yield criterion (associated) for $\phi = 20°$

Experimental evidence has shown[55] that the Meyerhof and Hansen formulae are over-safe.

12.6 REDUCTION OF BEARING CAPACITY AND PROGRESSIVE DISPLACEMENT DUE TO CYCLIC LOADING

In the case of offshore structures the applied loads are not all static and corresponding allowances must be made with respect to both stability and settlement. Figure 12.24 shows how wave action changes the moment and horizontal and vertical forces acting on a gravity platform, as well as the pore

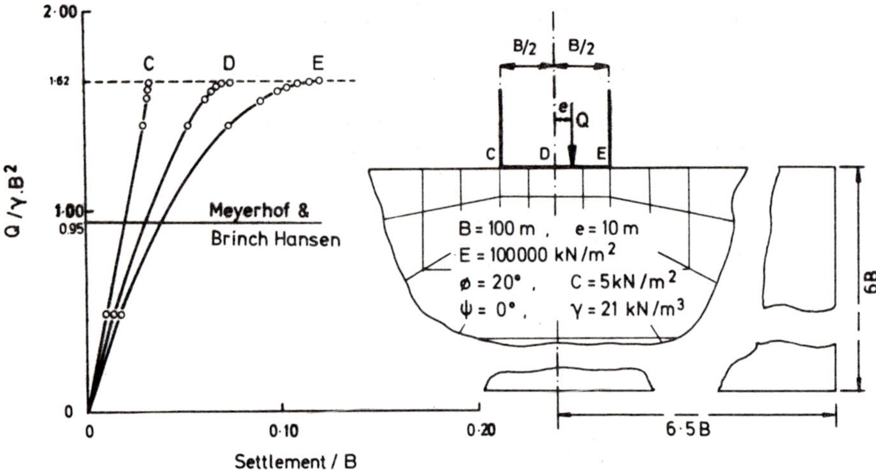

Figure 12.22 Load/settlement curves for eccentric vertical loading under drained conditions

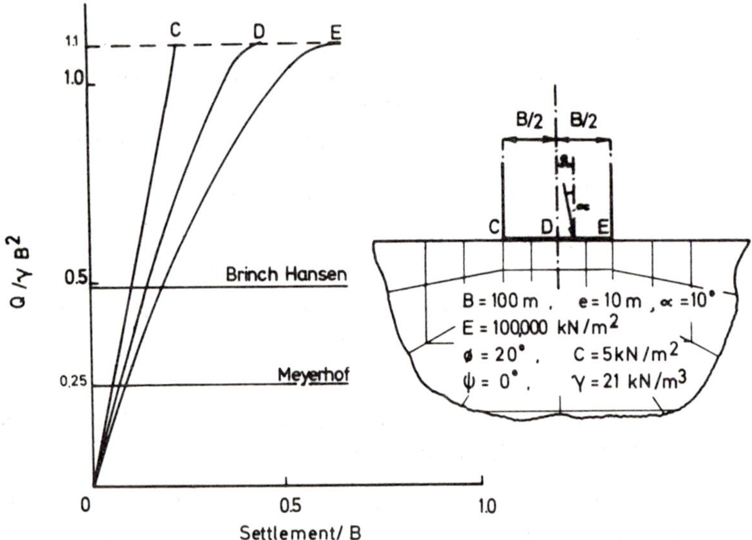

Figure 12.23 Load/settlement curves for eccentric inclined loading under drained conditions

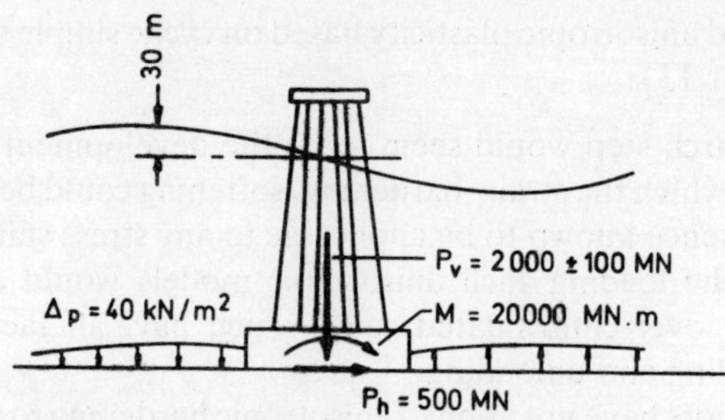

Figure 12.24 Maximum forces acting on platform due to wave action and current

pressures in adjacent soil. It is known that such cyclic loading causes accumulative strain and, in the case of undrained, saturated soils, changes in effective stresses, leading to a change in strength.

A practical approach to cyclic loading analysis is to determine the stress distribution due to static loads and then find out experimentally the effect of cyclic application of such stress states to test samples. From such tests the pore pressure changes due to cyclic loading equivalent to the varying wave forces of a storm may be estimated, and used for an approximate re-analysis of the soil structure. However, such analysis procedures are incapable of predicting the simple 'shakedown' phenomenon which occurs *without* any cumulative material damage.

To try to perform such an analysis more directly by numerically applying varying cyclic loads has been thought by some to be too complex and expensive of computing time, if not impossible, and unwarranted due to inadequate understanding of soil behaviour. Whilst such a viewpoint may be partly justified the authors feel that it is worth pursuing the fully numerical approach for the following reasons.

(1) The experimental approach uses simple shear and triaxial tests, which are very limited in their ability to reproduce the true stress states predicted for field conditions.

(2) Numerical analysis may be simplified by representing random cycling as a series of wave sets, each of constant amplitude.[60,61]

(3) Although much has still to be understood concerning soil behaviour, it is known that the main feature leading to strength reduction during cyclic loading under undrained conditions is the build-up of pore pressure due to the tendency towards overall compression. Such compression is not modelled by isotropic yield surfaces but can be modelled approximately by means of anisotropic hardening, as carried out by Prevost and Hoeg[62] using the technique of Mroz.[63] Basically this is a semi-empirical approach involving a total

stress analysis and anisotropic plasticity based on cyclic simple shear or triaxial tests (see Chapter 13).

The next research step would seem to be the development of an effective stress analysis in which the strain-hardening/softening could be more theoretically based and hence known to be applicable to any stress states. Apart from dealing with cyclic loading such anisotropic models would also deal more realistically with over-consolidated soils, which have in fact already been subjected to loading and unloading.

Whilst the authors have not applied anisotropic hardening to the problems of cyclic loading, nevertheless, results have been obtained which predict progressive displacement. The explanation of this phenomenon lies in the fact that in the application of a moment some forces must increase whilst others decrease. Hence for cyclic moments the reloading stress path for a point in the soil may not be the reverse of its previous unloading path, due to the effect thereon of plastic straining in other parts of the soil simultaneously undergoing loading. This effect would not be present for a simpler form of loading such as that of the isotropic compression test for which stresses at all points increase in unison.

The behaviour of soil under cyclic loading may thus be separated into two parts: that due to permanent cyclic deformation of the material, and that caused by irreversibility of the stress paths in the actual structure in which residual stresses develop. The remainder of this section will be concerned with the latter, which is a phenomenon of 'shakedown' or 'ratcheting', and in an examination of the factors contributing to its importance.

Figure 12.25 shows this 'ratcheting' phenomenon for an effective stress analysis under undrained conditions. Figure 12.25(c) compares displacements for two different cyclic moments using the critical state/Mohr–Coulomb flow rule of the type described in Section 12.4.3. It may be seen that progressive settlement is negligible for realistic loading ($M = 200$ kN m) for a value of internal friction given by $\phi = 30°$, though there is significant settlement in the first cycle. For a moment of twice the magnitude, however, the progressive settlement is considerable. Figure 12.25(d) compares displacements for different values of the hardening parameter defined in Section 12.4.3 and for two values of the vertical load (applied under undrained conditions). The magnitude of the settlement is seen to increase with the magnitude of the vertical load, whereas it has been found previously[64] that for a vertical load applied under drained conditions the settlement decreases with increase of vertical load. The settlement is also seen to increase as the value of X decreases.

The impression given so far is that progressive effects are small for realistic conditions. However, if ϕ is reduced to $20°$ a far different situation may arise, as shown in Figure 12.25(e). Here a value of X equal to $25\cdot0$ (which is realistic) is used for a comparison of the results using $\phi = 20°$ and $\phi = 30°$. Also included is the result for the non-hardening zero-dilatant Mohr–Coulomb flow rule. It is obvious that the reduction of ϕ from 30° to 20° has a very important effect in the case of a strain-hardening material. The pore pressure distribution and build-up of pore pressures are shown in Figures 12.25(f) and (g). The pore pressure build-up in this case has been shown to be of a *structural* kind rather than due to material behaviour. This point is worth making as in some triaxial tests uniform stresses can not be imposed and at least some of the measured effects may well be due to this same cause.

It must be emphasized that the examples shown are not intended to give a complete picture of the soil behaviour due to cyclic loading, as the constitutive relationship used is not adequate for that purpose. They are, however, intended to illustrate some of the factors involved. It has been shown that progressive settlement under undrained conditions becomes much more significant as the tendency towards strain-hardening increases, or the vertical loading increases. Perhaps most important is the marked increase in settlement as the angle of internal friction falls in the range 30° to 20°. It remains to be seen how much influence anisotropic hardening will have on progressive displacements, and the reduction of bearing capacity.

The values of moments and horizontal forces used in the above examples are as high as or higher than any likely to occur in the North Sea. For a more realistic analysis account should be taken of the fact that such forces occur seldom, accompanied by many lesser forces, and storms are separated by periods of time sufficient to allow partial dissipation of pore pressures and hence changes in strength. A more exact analysis would therefore include a consolidation stage such as is illustrated in Section 12.7.

12.7 CONSOLIDATION

Consolidation or dissipation of pore pressures is of importance in most problems of soil mechanics. It controls the rate of settlement which may be important in nonhomogeneous conditions if tilting of a platform is to be controlled by eccentric ballasting; it decides the rate at which soils change their bearing capacity; perhaps most important of all it determines whether the build-up of pore pressure in one storm will remain large enough to influence the shear strength in the next storm. For these reasons it is important to be able to model the consolidation phenomenon accurately. The method of analysis should therefore take into account the non-linear stress–strain relationship of the soil skeleton as well as the variation of the permeability. Figure 12.26 shows the different results given by using linear elastic and elasto-plastic stress–strain laws, the latter being based on the critical state ellipse as described in Section 12.4.3 and using realistic parameters ($X = 25$, $\phi = 25°$). No claims are made for

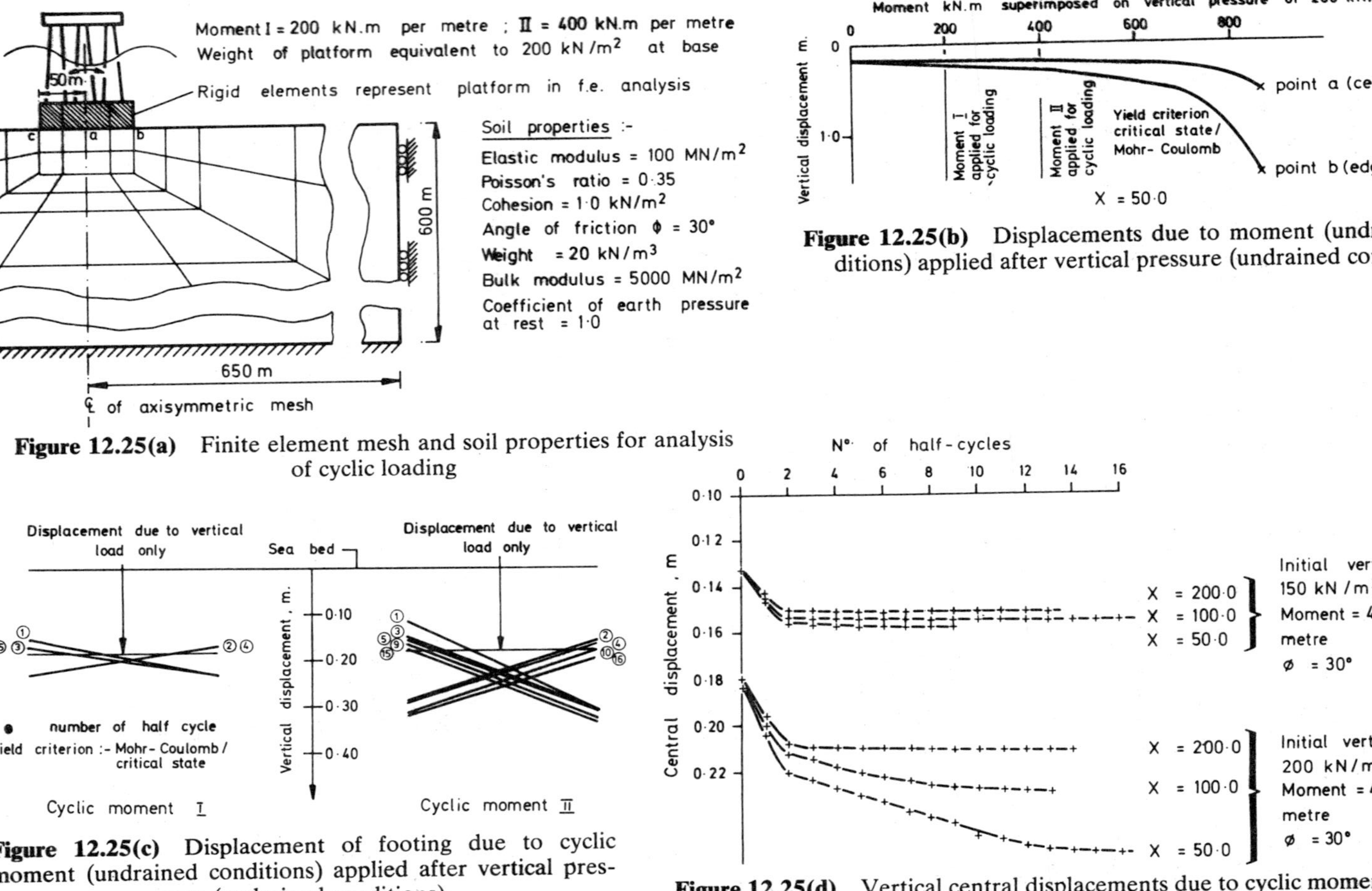

Figure 12.25(b) Displacements due to moment (undrained conditions) applied after vertical pressure (undrained conditions)

Figure 12.25(a) Finite element mesh and soil properties for analysis of cyclic loading

Figure 12.25(c) Displacement of footing due to cyclic moment (undrained conditions) applied after vertical pressure (undrained conditions)

Figure 12.25(d) Vertical central displacements due to cyclic moments for varying hardening factor (X) and two initial vertical loadings (undrained conditions). ($\phi = 30°$)

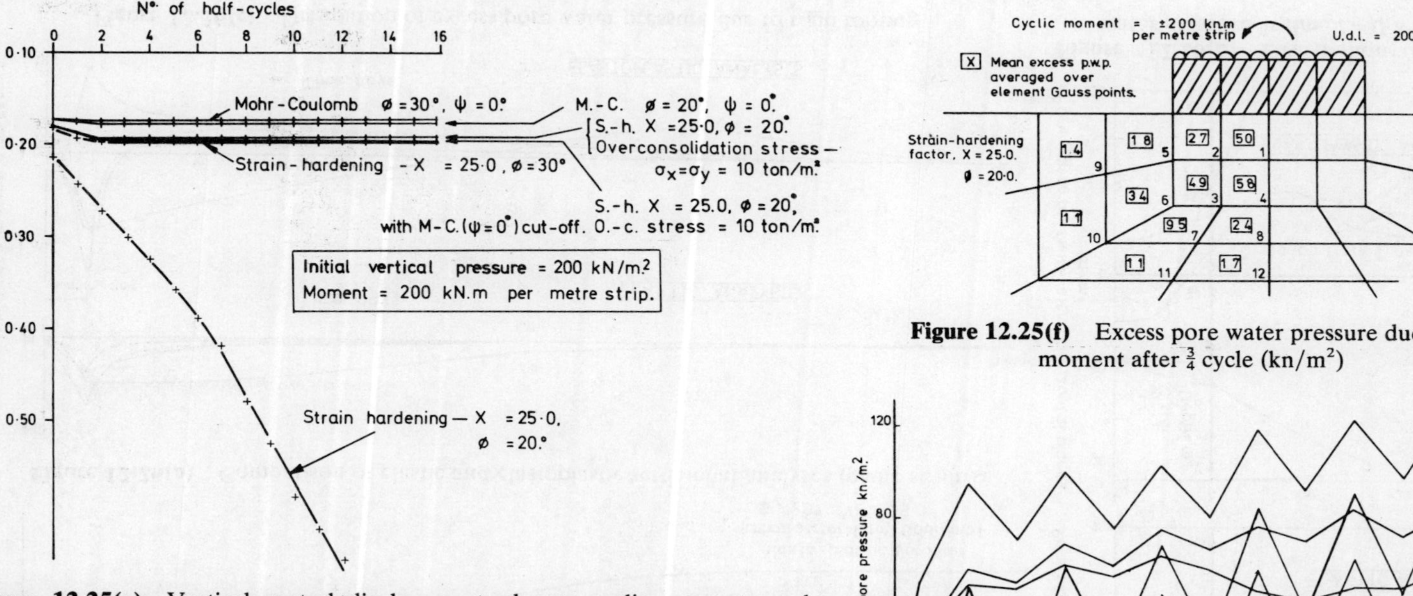

Figure 12.25(e)   Vertical central displacements due to cyclic moments under undrained conditions. ($\phi = 20°$ and $30°$)

Figure 12.25(f)   Excess pore water pressure due to moment after $\frac{3}{4}$ cycle (kn/m²)

Figure 12.25(g)   Variation of excess pore pressures due to cyclically applied moment

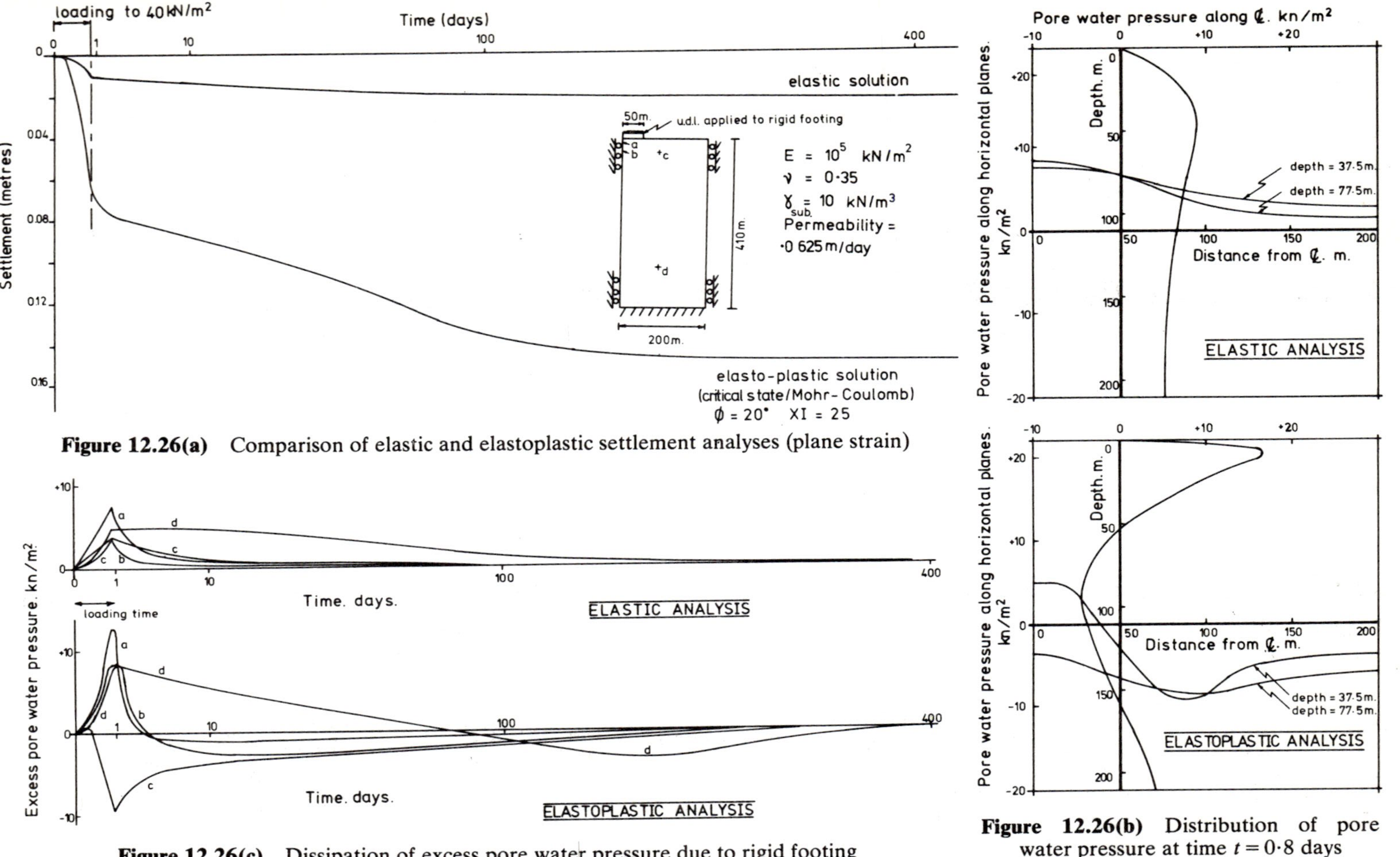

Figure 12.26(a) Comparison of elastic and elastoplastic settlement analyses (plane strain)

Figure 12.26(c) Dissipation of excess pore water pressure due to rigid footing

Figure 12.26(b) Distribution of pore water pressure at time $t = 0·8$ days

the accuracy of the latter model, except that it is nearer the correct model than is the elastic version. In actual fact the settlement should probably be even greater than that given by the plastic model as experimentally shear strains are found to occur even before the stress path touches the yield surface of overconsolidated soils, and on the yield surface itself are greater than those predicted by an associated flow rule for both normal and overconsolidated soils.[34]

The comparison of pore pressure distributions for elastic and critical state analyses, shown in Figures 12.26(b) and (c), is of interest: it shows a rapid tendency for the excess pore pressures to fall and even become negative, resulting in increased effective compressive stresses and greater strength than would be predicted by an undrained analysis. The foundation soil has been assumed normally consolidated; a more realistic analysis of overconsolidated soil would probably show less marked behaviour.

12.8 CONCLUSIONS

An attempt has been made to summarize the various factors involved in the application of the finite element method to the solution of elasto-plastic static and transient soil mechanics problems. Much attention has been given to the actual form of the yield surface. For drained analyses, or those in which the soil exhibits little strain-hardening, it is safe to use a Mohr–Coulomb yield surface combined with a zero dilatancy flow rule. For undrained analyses of soils showing definite strain-hardening a model such as the critical state ellipse should be used, as plastic compressive straining may lead to an overall pore pressure increase, resulting in failure loads lower than those given by the Mohr–Coulomb model. Such strain-hardening models are not, however, sufficiently well developed to deal with problems in which stresses rising above the critical state line have a marked influence on the failure load.

Both the Mohr–Coulomb type and strain-hardening models have been used with various approximations to the true Mohr–Coulomb section in the π-plane. Experimentally, it has been shown that the true Mohr–Coulomb section may be oversafe, but the compression cone approximation can be very unsafe and should never be used.

Using the non-associated Mohr–Coulomb model the finite element method has been employed to predict the bearing capacity of a footing under plane strain conditions for comparison with standard formulae. The finite element results for vertical loading are in good agreement with exact and upper bound standard solutions. In the case of eccentric and inclined loading, for which there are no exact or upper bound solutions available, the finite element results are much higher than those given by the Brinch Hansen and Meyerhof formulae. It is known from experimental evidence that these formulae are oversafe, but further investigation using finer element meshes is obviously required. A further subject for research is the comparison of three-dimensional finite element analyses with the use of current approximations based on the plane strain formulae.

Other offshore applications dealt with are the effects of cyclic loading and consolidation. The former effects may be due to both irreversible material properties and irreversible structural changes. It has been shown that the result of structural changes may be significant for lightly consolidated soils of low friction angle. (The results of irreversible material behaviour are described in Chapter 13.)

The example of elasto-plastic consolidation using a critical state model shows a more rapid decrease in excess pore water pressure than would have been expected from the results of a simple elastic analysis. Further investigation would obviously be useful; in particular, using material properties for *overconsolidated* clay.

Much work remains to be done in the field of numerical analysis of soil behaviour. In particular, plasticity models need to be improved in their prediction of displacements; even in the case of the critical state model, which may predict volumetric plastic strains reasonably well during loading below the critical state line, provision needs to be made for more accurate modelling above the critical state line and during unloading, reloading, etc. Then, perhaps, it would be possible to make a more precise analysis of cyclic loading than that given by superimposing test results from simple laboratory stress states on the more complex stress states of real foundations.

REFERENCES

1. Biot, M. A. (1941). 'General theory of three-dimensional consolidation', *Journal of Applied Physics*, **12**.
2. Biot, M. A. (1955). 'Theory of elasticity and consolidation for a porous anisotropic solid', *Journal of Applied Physics*, **26**.
3. Biot, M. A. (1956). 'Theory of deformation of a porous viscoelastic anisotropic solid', *Journal of Applied Physics*, **27**.
4. Biot, M. A. (1956). 'General solutions of the equations of elasticity and consolidation for a porous material', *Journal of Applied Mechanics, Proc. ASME*.
5. Terzaghi, K. (1943). *Theoretical Soil Mechanics*, Wiley, New York.
6. Schiffman, R. L., Chen, A. T. F., and Jordan, J. C. (1969). 'An analysis of consolidation theories', *Journal of Soil Mechanics and Foundation Div.*, ASCE, **95**, No. SM1.
7. Gibson, R. E. and McNamee, J. (1963). 'A three-dimensional problem of the consolidation of a semi-infinite clay structure', *Quarterly Journal of Mechanics and Applied Mathematics*, **XVI** (2).
8. Zienkiewicz, O. C. (1977). *The Finite Element Method in Engineering Science*, Third edition, McGraw-Hill, London/New York.

9. Sandhu, R. S. and Wilson, E. L. (1969). 'Finite element analysis of flow in saturated porous media', *Proc. Am. Soc. Civ. Engs.*, **95**, EM. 641–652.

10. Ghaboussi, J. and Wilson, E. L. (1973). 'Flow of compressible fluid in porous elastic media', *Int. J. Num. Meth. in Eng.*, **5**, 419–442.

11. Hwang, C. T., Morgenstern, N. R., and Murray, D. W. (1971). 'On solutions of plane strain consolidation problems by finite element methods', *Canadian Geotechnical Journal*, **8**, 109–118.

12. Small, J. C., Booker, J. R., and Davies, E. H. (1977). 'Elasto plastic consolidation of soil', *Research Report No. R268*, School of Civ. Eng., University of Sidney.

13. Zienkiewicz, O. C., Valliappan, S., and King, I. P. (1969). 'Elasto-plastic solutions of engineering problems. Initial stress finite element approach', *Int. J. Num. Meth. in Eng.*, **1**, 75–100.

14. Nayak, G. C. and Zienkiewicz, O. C. (1972). 'Elasto-plastic stress analysis. A generalization for various constitutive relations including strain softening', *Int. J. Num. Meth. in Eng.*, **5**, 113–135.

15. Zienkiewicz, O. C. and Cormeau, I. C. (1972). 'Visco-plasticity solution by the finite element process', *Arch. Mech.*, **24**, 873–888.

16. Zienkiewicz, O. C. and Cormeau, I. C. (1974). 'Visco-plasticity—plasticity and creep in elastic solids. A unified numerical solution approach', *Int. J. Num. Meth. Eng.*, **8**, 821–845.

17. Zienkiewicz, O. C. (1974). *Lecture Notes in Mathematics*, 363, Springer-Verlag.

18. Herrmann, L. R. (1965). 'Elasticity equations for incompressible, or nearly incompressible materials by variational theorem', *J.A.I.A.A. No. 3*, 1896.

19. Christian, J. T. (1968). 'Undrained stress distribution by numerical methods', *J. Soil Mech. Found. Div.*, A.S.C.E., **94**, 1333–1345.

20. Naylor, D. J. (1974). 'Stresses in nearly incompressible materials by finite elements with applications to the calculation of excess pore pressures', *Int.J.N.M.E.*, **8**, 443–460.

21. Zienkiewicz, O. C. (1974). 'Constrained variational principles and penalty function methods in the finite element analysis', *Lecture Notes in Mathematics*, Springer-Verlag, New York, 207–214.

22. Lewis, R. W., Roberts, G. W., and Zienkiewicz, O. C. (1976). 'A non-linear flow and deformation analysis of consolidation problems', *2nd Int. Conf. on Num. Meth. in Geomech.*, Blacksburg, Virginia.

23. Drucker, D. C. (1951). 'A more fundamental approach to plastic stress–strain solutions', *Proc. 1st U.S. National Congress of Applied Mechanics*, 487–491.

24. Duncan, J. M. and Chang, C. Y. (1970). 'Non-linear analysis of stress and strain in soils', *Proc. A.S.C.E.*, **96**, No. SM 5, 1629–1653.

25. Nelson, J. and Baron, M. L. (1971). 'Application of variable moduli to soil behaviour', *Int. J. of Solids and Struct.*, **7**, 399–417.

26. Duncan, J. M. (1973). Written discussion, *Proc. of Symp. on the Role of Plasticity in Soil Mechanics*, Cambridge.

27. von Gudehus, G. (1973). 'Elastoplastische Stoffgleichungen fur trockenen Sand Ingenieur', *Archiv.* **42**.

28. Zienkiewicz, O. C. and Pande, G. N. (1975). 'Some useful forms of isotropic yield surfaces for soil and rock mechanics', *C/R/248/75*, Karlsruhe, September (1975).

29. Lade, P. V. and Duncan, J. M. (1975). 'Elastoplastic stress–strain theory for cohesionless soils', *Proc. A.S.C.E.*, **101**, GT10, 1037–53.

30. Humpheson, C. and Naylor, D. J. (1975). 'The importance of the form of the failure criterion', *Int. Symp. on Num. Methods in Soil Mech. and Rock Mech.*, Karlsruhe.

31. Drucker, D. C., Gibson, R. E., and Henkel, D. J. (1957). 'Soil mechanics and work-hardening theories of plasticity', *Trans. A.S.C.E.*, **122**, 338–346.
32. Roscoe, K. H., Schofield, A. N., and Wroth, C. P. (1958). 'On the yielding of soils', *Geotechnique*, No. 8, 22–53.
33. Schofield, A. N. and Wroth, C. P. (1968). *Critical State Soil Mechanics*, London: McGraw-Hill.
34. Roscoe, K. H. and Burland, J. B. 'On the generalized stress–strain behaviour of "wet" clay'. In *Engineering Plasticity* (ed. J. Heyman and F. A. Leckie), 535–609, Cambridge, University Press.
35. Nayak, G. C. and Zienkiewicz, O. C. (1972). 'A convenient form of invariants and its application in plasticity', *Proc. A.S.C.E.*, **98**, 949–954.
36. Zienkiewicz, O. C., Humpheson, C., and Lewis, R. W. (1975). 'A unified approach to soil mechanics including plasticity and viscoplasticity', *Int. Symp. on Num. Methods in Soil Mech. and Rock Mech.*, Karlsruhe.
37. Humpheson, C. (1976). *Ph.D. Thesis*. 'Finite element analysis of elasto/viscoplastic soils', University College of Swansea.
38. Prandtl, L. (1920). 'Über die Härte plasticher Körper', Nachr. Kgl. Ges. Wiss. Göttingen, *Math. Phys.*, **K1**, 74–85.
39. Hill, R. (1950). *The Mathematical Theory of Plasticity*, Clarendon Press, Oxford.
40. Prager, W., Hodge, P. G. (1951). *Theory of Perfectly Plastic Solids*, Wiley, London.
41. Bykovlev, G. I. (1961). 'O Pole Skorostiev pri Vdavlevaniy ploskovo shtampa', *PMM*, **25**, 552–553.
42. Drucker, D. C. and Prager, W. (1952). 'Soil mechanics and plastic analysis or limit design', *Quart. Appl. Math.*, **10**, 157–165.
43. Caquot, A. and Kerisel, J. (1949). *Traite de Mechanique des Sols*, Gauthier-Villars, Paris.
44. Bell, A. L. (1915). 'The lateral pressure and resistance of clay and the supporting power of clay foundations', *Min. Proc. Inst. Civil Engrs. (London)*, Paper No. 4131.
45. Chen, W. F. (1969). 'Soil mechanics and theorems of limit analysis', *J. Soil Mech. Found. Division, Proceed. ASCE*, **95**, No. SM2, 493–518.
46. Reissner, H. (1924). 'Zum Erddruckproblem', Proceed. First *Int. Conf. Appl. Mech., Delft*, The Netherlands, 295–311.
47. Shield, R. T. (1953). 'Mixed boundary value problems in soil mechanics', *Q. Applied Math.*, **11**, 61–75.
48. Shield, R. T. (1953). 'Plastic potential theory and Prandtl bearing capacity solution', *J. Applied Mech.*, **21**, 193–194.
49. Davis, H. H. (1968). 'Theories of plasticity and the failure of soil masses', *Soil Mechanics, Selected Topics*, ed. I. K. Lee, Butterworths, London, 371–380.
50. Drescher, A. 'Some remarks on plane flow of granular media', *Arch. Mech. Stosowanej*, **24** (72), 837–848.
51. Sokolovski, W. W. (1958). *Statyka osrodkow sypkich*, PWN Warsaw.
52. Sokolovski, W. W. (1960). *Statistics in Soil Media*, Butterworth, London.
53. Tien Hsing Wu (1976). *Soil Mechanics*, Allyn and Bacon, Inc., 229.
54. Hansen, J. B. (1970). 'A revised and extended formula for bearing capacity', The Danish Geotechnical Institute, Copenhagen, *Bulletin No. 28*, 5–11.
55. Meyerhof, G. G. 'Some recent research on the bearing capacity of foundations', *Canadian Geotechnical Journal*, **1**, No. 1, 16–26.
56. Janbu, N. (1973). 'Slope stability computation, embankment dam engineering', *Casagrande Volume*. Ed. R. C. Hirschfeld and S. J. Poulos, New York, Wiley, 47–86.

57. Morgenstern, N. R. and Price, V. E. 'The analysis of the stability of general slip surfaces', *Geotechnique*, **15**, No. 1. 79–93.
58. Karal, K. (1973). 'En energimetode for geotekniske stabilitetsanalyser' (in Norwegian). *Ph.D. Thesis*, The Norwegian Institute of Technology.
59. Lauritzsen, R. A. and Schjetne, K. (1976). 'Stability calculations for offshore gravity structures', *Offshore Technology Conference, 8*, Houston, Texas, May 1976.
60. Schjetne, K. (1976). 'Foundation engineering for gravity structures in the North Sea', *SPE/DUT—European Spring Meeting*, Amsterdam.
61. Lee, K. L. and Focht, J. A. (1975). 'Cyclic testing of soil for ocean wave loading problems', *7th Offshore Technology Conference*, Houston, May 1975.
62. Prevost, J. H. and Høeg, K. (1975). *Mathematical Model for Static and Cyclic Undrained Clay Behaviour*, Norwegian Geotechnical Inst.
63. Mroz, Z. (1967). 'On the description of anisotropic workhardening', *Journal of the Mech. and Phys. of Solids, London*, **15**, 163–175.
64. Zienkiewicz, O. C., Lewis, R. W., Norris, V. A. and Humpheson, C. (1976). 'Numerical analysis for foundations of offshore structures with special reference to progressive deformation', *SPE/DUT—European Spring Meeting*, Amsterdam.

Chapter 13

Soil Deformations due to Cyclic Loads on Offshore Structures

K. H. Andersen, O. E. Hansteen,
K. Høeg, and J. H. Prévost

13.1 INTRODUCTION

The objective of this contribution is to describe, interpret and analyse the effects of cyclic loading on the soil beneath offshore structures. The understanding of soil stress–strain and strength behaviour during cyclic loading is essential for the evaluation of foundation stability and deformations as well as for a reasonable analysis of the dynamic soil-structure interaction during wave- and earthquake loading.

The type of fixed offshore platform to be installed at a given site is governed to a significant extent by the foundation conditions, and the use of concrete gravity structures as alternatives to piled steel platforms has greatly increased the engagement of geotechnical engineers in the overall analysis and design. The foundation engineer is now involved in all aspects of the cyclic soil-foundation-structure interaction analyses and the field instrumentation and performance observation programs for these structures.[1] Some significant advances have been made in theory and practice over the last couple of years, and most importantly, one now has the experience and recorded data from the installation phase and the later performance of fixed platforms in the North Sea (Figures 13.1 and 13.2).

In general, when a structure is placed on a soil foundation, there is an immediate settlement, followed by consolidation and secondary creep settlements. For an offshore structure there are in addition the cyclic motions and permanent deformations that occur in storm periods, and the consolidation settlements that occur during and subsequent to each storm in connection with the dissipation of the excess pore pressures that build up during the cyclic loading.

For an offshore structure the foundation behaviour under cyclic loading may be separated into a short-term undrained condition with generation of excess average pore pressures due to the cyclic wave loading, and a long-term condition in which drainage occurs. For a clay, the short-term condition may

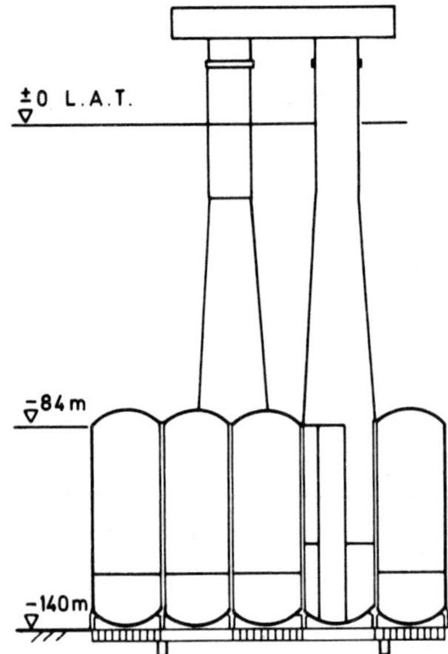

Figure 13.1   Field instrumentation of the Condeep structure at the Brent B site

contain several succeeding storms. On the other hand, for a sand significant drainage may occur during one single storm, and the undrained short-term condition is limited to the first part of the storm.

In the undrained short-term condition the excess pore pressure leads to a reduction in effective stresses with accompanying reductions in soil stiffness and strength. For the platform this means an increased natural period of vibration with increasing cyclic displacements, gradually increasing vertical settlements and a reduction in safety against ultimate bearing capacity failure during the storm.

The undrained cyclic loading combined with the drainage which occurs afterwards, will influence the soil behaviour during subsequent undrained cyclic storm loading. For platforms on sand, or normally consolidated clay, the soil stiffness and strength will increase. This leads to reduced cyclic displacements and improved stability compared to previous storms. For a platform on an overconsolidated clay the opposite may be the case, at least for the first few storms until the originally overconsolidated clay starts to behave more like a normally consolidated deposit. Only the behaviour of the soil during the undrained short-term condition will be discussed in this contribution.

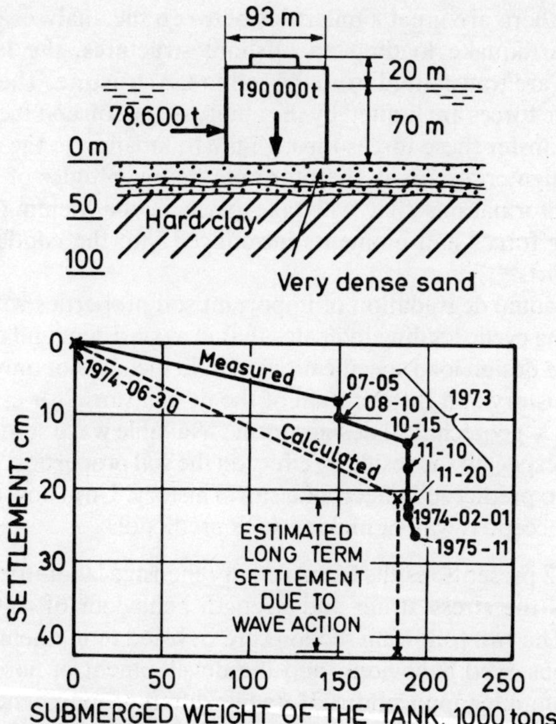

Figure 13.2   Observed settlements of the Ekofisk tank

In the design of offshore structures, the following points should be considered:

(i) Since the soil strength decreases with time during the undrained short-term condition, the overall foundation stability is usually checked for the situation of a long storm followed by the 100-year wave.

(ii) The soil deformations introduce significant stresses in the conductors under the platform as well as in the risers. Axial stresses arise due to negative skin friction along the conductors, and bending stresses develop due to horizontal displacements and shear deformations in the soil.

(iii) The wave forces acting on the platform are dynamic forces. Hence, both the maximum response during a severe storm and the fatigue life of the structure are studied through dynamic analyses including the inertia effects. Parametric studies[2] show that the soil properties, in particular the stiffness, are among the more important factors influencing the dynamic behaviour of typical offshore gravity platforms.

(iv) While there are great similarities between the analyses performed for wave- and earthquake loading on offshore structures, the latter case has 'forces' which are transmitted from the soil to the structure. The magnitude of the earthquake forces are limited by the ability of the soil and the soil-structure interface to transfer these forces through the foundation to the structure. The governing design criteria may therefore be the magnitudes of the cyclic and permanent deformations that will take place in the yielding soil, and the corresponding forces and moments introduced into the conductors and the foundation skirts.

(v) The gradual degradation of important soil properties with the number of cycles during cyclic loading indicates that in a consistent and rational design procedure, the design load specifications should include not only the intensity, but also the history and the duration of the design storm (or earthquake). To establish such a 'geotechnical design storm', available wave statistics should be studied with respect to the resulting effect on the soil properties. This requires a rational way to predict soil effects of a storm history. Until now there has been no generally accepted way of making such predictions.

Section 13.2 presents results from a comprehensive laboratory test program which studied the stress–strain and strength behaviour of clay subjected to cyclic loads. The two following sections are devoted to mathematical descriptions of the observed behaviour, and the development of numerical models suitable for computer applications. Here two different approaches are pursued. In Section 13.3 a simplified approach based directly on the test results is used, while Section 13.4 gives a mathematical model within the framework of the flow theory of plasticity. It is believed that both approaches will contribute to a more thorough understanding of the effects of cyclic loading on the foundation soil.

13.2   RESULTS FROM LABORATORY TESTS

The effect of cyclic loading on soil behaviour is illustrated by means of results from the laboratory testing of triaxial and simple shear samples of Drammen clay. However, the behaviour of sand and silt under cyclic loading is governed by the same fundamental laws as clay, and may be explained in terms of effective stresses. In principle the description in this section is therefore valid for sand and silt as well as for clay.

A description of Drammen clay and the laboratory testing techniques are given elsewhere.[3,4,5] These references also form the basis for the description of soil behaviour under cyclic loading which is presented in this section.

13.2.1   Stress–Strain behaviour

Figure 13.3 shows typical stress–strain curves for a soil element which is subjected to undrained cyclic loading. Cyclic shear strains increase with number of cycles, and the equivalent secant deformation modulus decreases from one cycle to another. In the cycle with large cyclic strains ($N = 884$), however, the slope of the stress–strain curve increases at large strains. This is an effect which is caused by a tendency for the soil to dilate when reaching large strains.

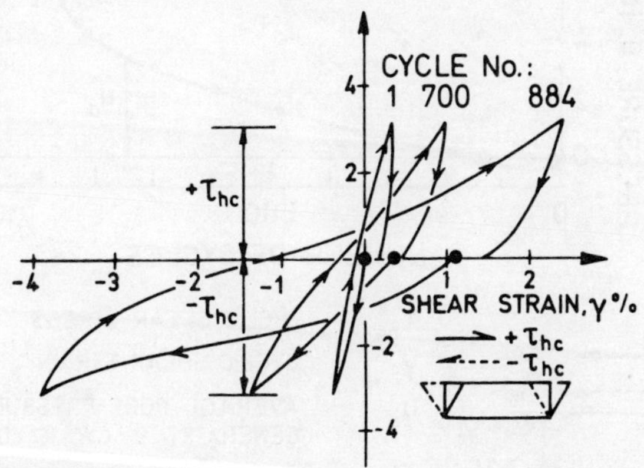

Figure 13.3   Stress–strain curves from simple shear
tests with undrained two-way cyclic loading

The stress–strain curves describe hysteresis loops, indicating that damping takes place in the soil. This damping is of importance in connection with a dynamic response analysis.

13.2.2   Development of strains and pore pressures

Figure 13.4 shows the development of strains and pore pressures with the number of cycles for the simple shear test with two-way symmetrical cyclic loading in Figure 13.3. The average pore pressure increases with increasing number of cycles. This average pore pressure is a permanent pore pressure increase with only small variations due to dilatancy effects within each cycle. The increasing average pore pressure causes the sample to move towards the failure line as indicated by the stress paths in Figure 13.5. As the failure line is approached, the cyclic strains increase as shown in Figure 13.4, and the behaviour of soil in cyclic loading may thus be explained in terms of effective stresses. Failure under conditions of cyclic loading is in this case reached after about 800 cycles even though the cyclic shear stress is only 45 per cent of the

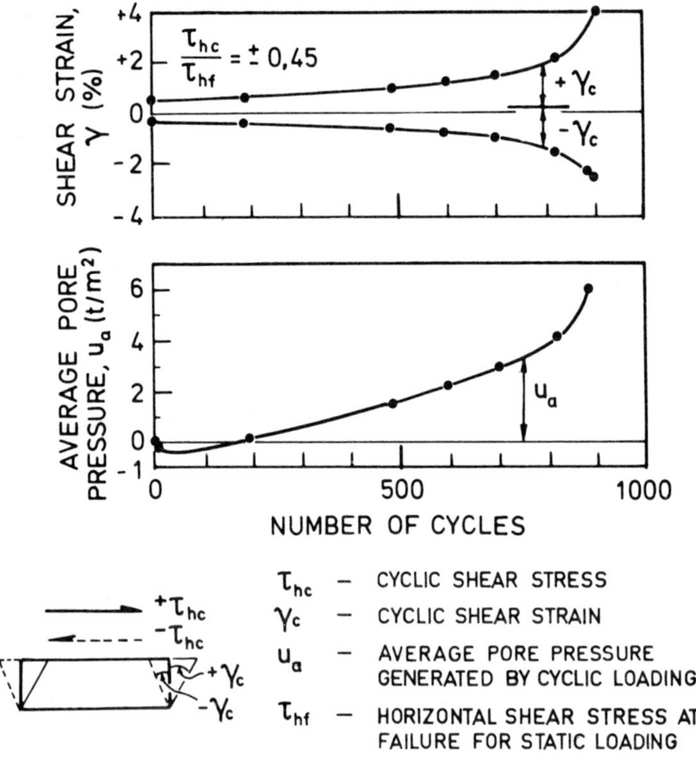

Figure 13.4   Development of shear strains and average pore pressure in simple shear tests with undrained two-way cyclic loading

undrained static shear strength of the clay. Failure is here defined as a cyclic shear strain of ±3 per cent. The symmetrical two-way loading of this sample gives approximately symmetrical cyclic strains, and the average strains are close to zero.

Figure 13.6 shows the development of strains and pore pressures in a clay element subjected to one-way cyclic triaxial loading. Since the cyclic loading is not symmetrical, the sample will experience an increasing average strain in addition to the cyclic strain. Both the average and the cyclic strain components will increase with increasing number of cycles. In the triaxial sample there will also be a cyclic pore pressure component in addition to the gradual increase in average pore pressure. The cyclic pore pressure is mainly caused by the variation in total stresses on the sample during a cycle.

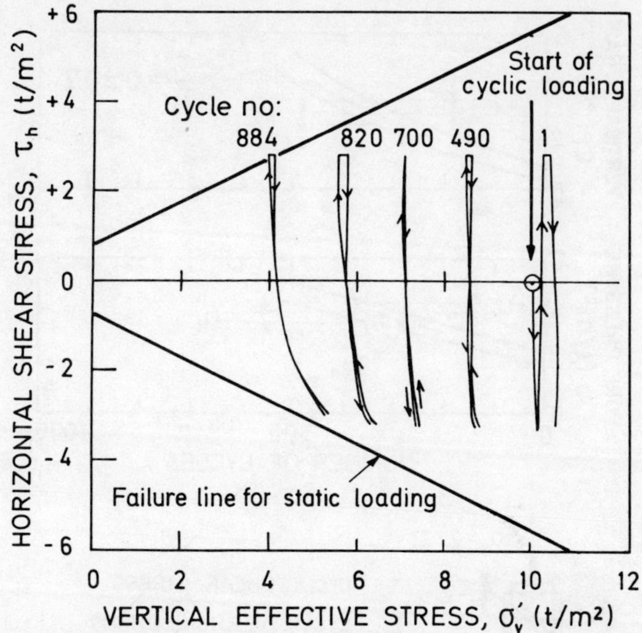

Figure 13.5   Variation in vertical effective stress in a simple
shear test with undrained two-way cyclic loading

13.2.3   Effects of stress conditions

The stress conditions in the foundation beneath a gravity platform are complex, and as indicated by the examples in Figures 13.4 and 13.6, the behaviour of a clay subjected to cyclic loading is highly dependent upon the stress conditions. This is further illustrated in Figure 13.7, where the results from three triaxial tests with cyclic loading on Drammen clay with an overconsolidation ratio (OCR) of 4 are shown. The samples are all loaded to the same maximum shear stress of 5 t/m^2 during cyclic loading, but the cyclic shear stress amplitudes, τ_c, are different. The results which are indicated in Figure 13.7 show that the behaviour of these three samples is very different. This means that the maximum shear stress cannot be used as the only parameter to define the effects of cyclic loading. In general, both the cyclic and the average shear strains are functions of both cyclic and average shear stresses as well as the number of cycles, soil type, overconsolidation ratio and principal stress directions.

For the tests in Figure 13.7, the cyclic variation in shear stress, τ_c, seems to be a more dominant parameter than the maximum or average shear stresses. For the special case of Drammen clay (OCR = 4) Figure 13.8 shows the cyclic shear

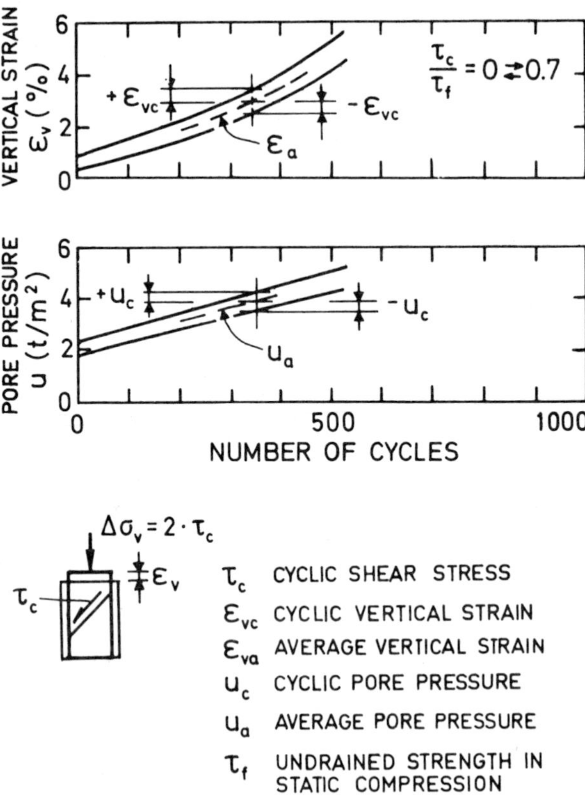

Figure 13.6   Development of strains and pore pressures in a triaxial test with undrained one-way cyclic loading

stress, τ_c, plotted against the cyclic shear strain amplitude, γ_c, for 10, 100 and 1000 cycles. The plot is based on triaxial and simple shear tests, including both one-way and two-way cyclic loading at various stress levels. All the tests were stress-controlled with a constant cyclic shear stress in each test. The vertical consolidation stress was $10\ t/m^2$. Undrained shear strengths for static loading was found to be $9\cdot8\ t/m^2$ for triaxial compression, $5\cdot5\ t/m^2$ for triaxial extension and $7\cdot1\ t/m^2$ for simple shear loading. There is a scatter in the plots shown in Figure 13.8, but the difference from one type of test to another seems to be smaller than the scatter within each specific test type. It thus seems that for Drammen clay with OCR = 4, the cyclic shear strain may be expressed as a function of the cyclic shear stress, τ_c, and the number of cycles, N. It must be emphasized, however, that the simple shear samples in this case (OCR = 4), have isotropic consolidation stresses ($K_0' = (\sigma_h'/\sigma_v') = 1\cdot0$). The majority of the

TEST No.	TYPE OF LOADING	τ_{max} CONSOLI- DATION	τ_{max} LOADING	RESULTS		
A	Two-way	0.0	$-5 \rightleftarrows +5$	Failure after 10 cycles		
B	One-way	0.0	$0 \rightleftarrows +5$	$\gamma_c = \pm 0.3\%$ $\gamma_a = 0.8\%$	After	
C	One-way	3.5.	$+3.5 \rightleftarrows +5$	$\gamma_c = \pm 0.02\%$ $\gamma_a = 0.03\%$	2500 cycles	

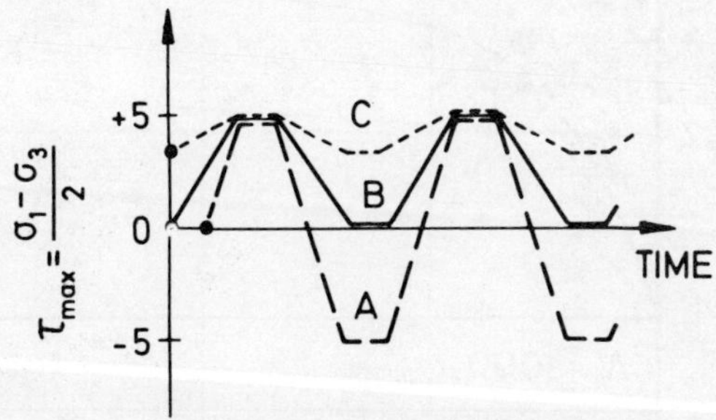

Figure 13.7 Comparison of triaxial tests with various types of undrained cyclic loading

triaxial tests are also isotropically consolidated. Those which are not have shear stresses during consolidation of $3 \cdot 5\,\text{t/m}^2$. These tests seem to compare well with the isotropically consolidated specimens, but some effect is likely for larger consolidation shear stresses.

For other values of overconsolidation ratios or for different soil types, the cyclic shear strains may not be so well related to the cyclic shear stress and number of cycles.

13.2.4 Strain contour diagrams

For the special case of Drammen clay with OCR = 4, the diagrams in Figure 13.8 may be transformed into a strain-contour diagram as presented in Figure 13.9. The number of cycles to 'failure' increases rapidly with decreasing cyclic shear stress, and below a cyclic shear stress of $\pm 2 \cdot 25\,\text{t/m}^2$, the cyclic shear strains are essentially independent of the number of cycles.

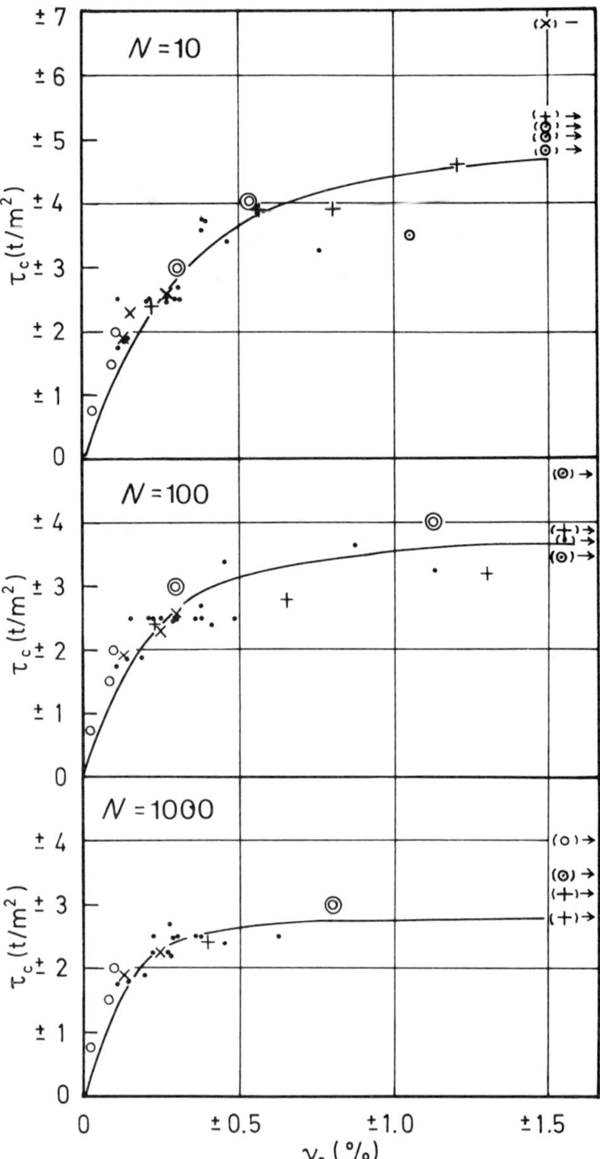

X one-way simple shear
+ two-way simple shear
o triax $\tau_{cons} = 0.35 \cdot s_u$, one-way
◎ — " — $= 0.35 \cdot s_u$, two-way
⊙ triax, $\tau_{cons} = 0$, two-way
• triax, $\tau_{cons} = 0$, one-way

Figure 13.8 Relationship between cyclic shear stress and cyclic shear strain after 10, 100 and 1000 cycles. The plots include triaxial and simple shear tests with one-way as well as two-way undrained cyclic loading. Valid for Drammen clay with OCR = 4

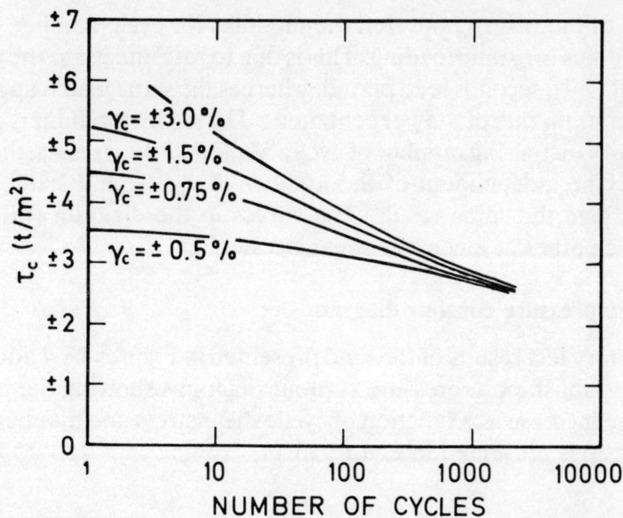

Figure 13.9 Strain-contour diagram for Drammen clay with
OCR = 4

The diagram is based on stress-controlled tests with constant cyclic shear stress, but may be used to predict cyclic strains for soil elements with varying cyclic shear stresses as well. The procedure for doing this is described in Section 13.3.

13.2.5 Deformation modulus

The variation in secant shear modulus, $G = (\tau_c/\gamma_c)$, with cyclic shear stress level and number of cycles is shown in Figure 13.10. The diagram is established from Figure 13.8. The corresponding modulus variation for static loading is included as a reference.

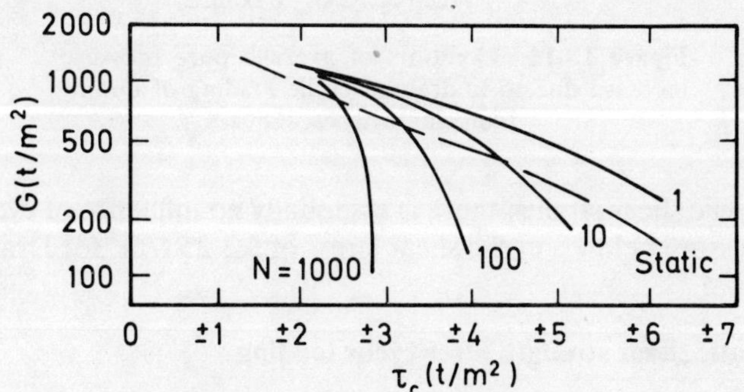

Figure 13.10 Secant moduli for cyclic and static loading of
Drammen clay with OCR = 4

It may be of interest to note that the modulus for cycle number 1 is higher than the modulus for static loading. This is due to rate effects, as the cyclic load is applied with a 10 seconds load period, whereas the static load is applied much slower with a strain rate of 4·5 per cent/hour. The secant modulus is decreasing drastically with increasing number of cycles. At low cyclic stresses, the modulus will probably be independent of the number of cycles, and static and cyclic loading will give the same result. The curves in the diagram will therefore approach each other at low cyclic shear stresses.

13.2.6 Pore pressure contour diagram

From laboratory test results of the kind presented in Figures 13.4 and 13.6, it is possible to establish pore pressure contour diagrams showing the permanent pore pressure increase as a function of cyclic shear stress and number of cycles. Such a diagram is presented in Figure 13.11.

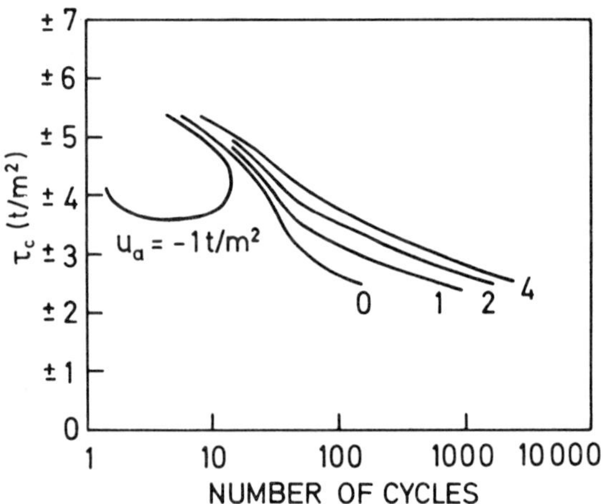

Figure 13.11 Contours of average pore pressure increase due to undrained cyclic loading of Drammen clay with OCR = 4

As for cyclic shear strains, there is essentially no influence of cyclic loading on pore pressure below a cyclic shear stress of $\pm 2\cdot25$ t/m² for Drammen clay with OCR = 4.

13.2.7 Static shear strength after cyclic loading

Figure 13.12 shows the results of some samples subjected to undrained static loading. All but one of the samples were first subjected to undrained cyclic

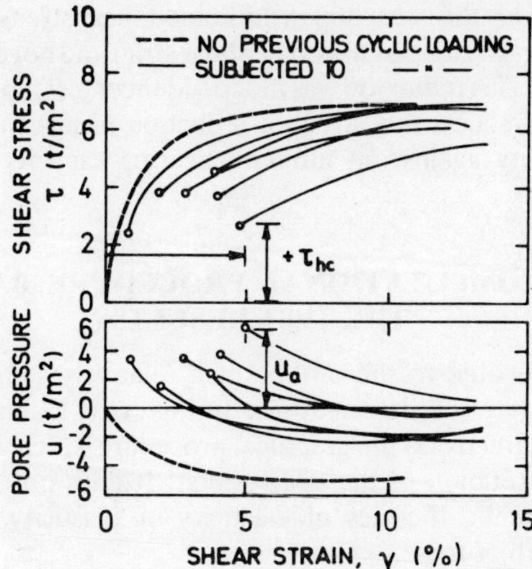

Figure 13.12 Results from undrained, static simple shear tests on samples with and without previous undrained cyclic loading

loading. It appears that the samples which were subjected to cyclic loading underwent a shear strength reduction which increases with increasing cyclic shear strains. This is further illustrated in Figure 13.13, which shows the reduction in strength as a function of cyclic shear strains and number of cycles.

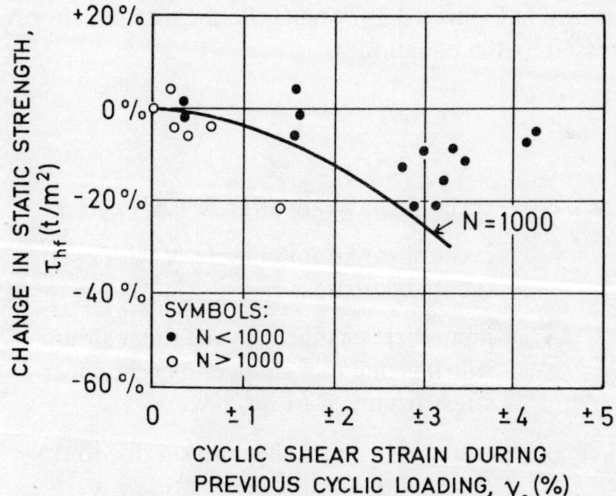

Figure 13.13 Reduction in undrained static shear strength due to undrained cyclic loading. (Simple shear tests on Drammen clay with OCR = 4)

For Drammen clay the reduction in undrained shear strength was found to be less than 25 per cent as long as the cyclic shear strain did not exceed ±3 per cent after 1000 cycles. This reduction was independent of test type and overconsolidation ratio. In practice, this strength reduction means that for an offshore platform the safety against an ultimate bearing capacity failure decreases during a storm.

13.3 A COMPUTATIONAL PROCEDURE BASED ON THE TEST RESULTS

This section is devoted to the outline of a relatively simple computational procedure to analyse soil deformations during cyclic loading. The numerical method is based directly on the graphical procedure which was first developed during the interpretation of the test results.[3] It does not rely on additional assumptions from the theories of elasticity or plasticity, but has features common with both of these.

At present the computational procedure is limited with regard to the type of soil for which it should be used, and to the aspects of cyclic behaviour covered. The prospects for extensions to more general conditions appear to be good.

13.3.1 Graphical procedure

(a) Stress-controlled test with varying stress level

The stress–strain curves for a soil element subjected to varying cyclic shear stresses are shown in Figure 13.14. The cyclic shear strain after $N + \Delta N$ cycles may be expressed by the equation:

$$\gamma_{c,N+\Delta N} = \gamma_{c,N} + \Delta\gamma_{c,i} + \Delta\gamma_{c,\Delta N} \tag{13.1}$$

where:

$$\gamma_{c,N+\Delta N} = \text{cyclic shear strain after } N + \Delta N \text{ cycles}$$

$$\gamma_{c,N} = \text{cyclic shear strain in cycle } N \text{ with a}$$
$$\text{cyclic shear stress } \tau_{c,N}$$

$$\Delta\gamma_{c,i} = \text{immediate change in cyclic shear strain}$$
$$\text{which is due to a change in cyclic shear}$$
$$\text{stress from } \tau_{c,N} \text{ to } \tau_{c,N+1}$$

$$\Delta\gamma_{c,\Delta N} = \text{increase in cyclic shear strain due to } \Delta N$$
$$\text{cycles with a cyclic shear stress } \tau_{c,N+1}$$

If the cyclic shear strain in the first cycle, $\gamma_{c,1}$, and expressions for $\Delta\gamma_{c,i}$ and $\Delta\gamma_{c,\Delta N}$ are known, the accumulated cyclic shear strain in a storm with varying

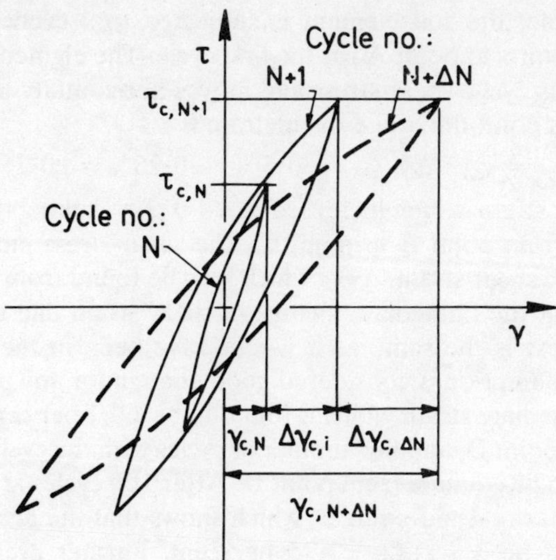

Figure 13.14 Stress–strain behaviour of soil element subjected to varying cyclic shear stresses

wave heights may be found by integrating Equation (13.1). A graphical method for doing this is illustrated by means of an example in Figure 13.15. Both diagrams in Figure 13.15 are based on the results presented in Figures 13.8 and 13.9, obtained from tests with constant cyclic shear stress.

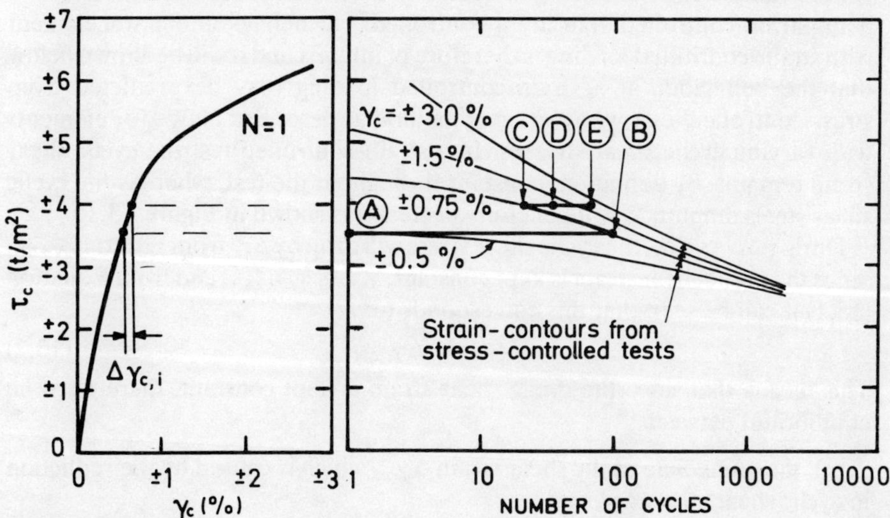

Figure 13.15 Prediction of cyclic shear strains for soil element subjected to varying cyclic shear stresses

In the example, the soil element is subjected to a cyclic shear stress of $\pm 3 \cdot 5$ t/m^2 and starts at point A for the first cycle. The element is subjected to 100 cycles at this cyclic shear stress and moves horizontally to point B in the diagram. At this point the cyclic shear strain is

$$\gamma_{c,N} = \gamma_{c,1} + \Delta\gamma_{c,\Delta N} = \pm 0 \cdot 5\% + \pm 0 \cdot 25\% = \pm 0 \cdot 75\%$$

The cyclic shear stress is then increased to $\pm 4 \cdot 0$ t/m^2, taking the soil along the strain contour from point B to point C. The shear stress increase causes an immediate cyclic shear strain, $\Delta\gamma_{c,i}$, which may be found from Figure 13.15(a) by assuming that the immediate change in shear strain due to the change in cyclic shear stress is the same as it would have been in the first load cycle ($N = 1$). This assumption is considered good enough for any practical computation. The immediate strain which is found to be $\pm 0 \cdot 1$ per cent, takes the soil from point C to point D, and the number of cycles with the cyclic shear stress of $\pm 4 \cdot 0$ t/m^2 are to be counted from point D. After 100 cycles at this cyclic shear stress, the soil has reached point E, which shows that the accumulated cyclic shear strain will be $\gamma_{c,N+\Delta N} = \pm 1 \cdot 75$ per cent. Further cyclic loading with increasing or decreasing cyclic shear stresses may be considered in the same manner, and the method may be used to evaluate the cyclic shear strains throughout a storm which causes arbitrary variations in cyclic loads on the foundation.

(b) Strain-controlled test

The foundation stresses beneath a gravity platform will redistribute during the storm. Many elements will be subjected to a cyclic loading which is closer to being strain-controlled than stress-controlled. The behaviour of a soil element with strain-controlled loading is therefore of interest and it will be shown below that the behaviour in a strain-controlled loading may be predicted from stress-controlled tests with the same method as described above for elements with varying cyclic shear stresses. In a strain-controlled test the cyclic shear strain remains, by definition, constant throughout the test, whereas the cyclic shear stress amplitude will generally decrease as shown in Figure 13.16.

During ΔN cycles, the cyclic shear stress will drop by $\Delta\tau_c$ from $\tau_{c,N}$ to $\tau_{c,N+\Delta N}$. Since the cyclic shear strain is kept constant, $\gamma_{c,N} = \gamma_{c,N+\Delta N}$, and from Equation (13.1) it can be seen that this corresponds to:

$$\Delta\gamma_{c,i} = -\Delta\gamma_{c,\Delta N} \tag{13.2}$$

This means that since the cyclic shear strain is kept constant, there must be equilibrium between

(a) the immediate cyclic shear strain $\Delta\gamma_{c,i}$, which is caused by the reduction in cyclic shear stress $\Delta\tau_c$, and

(b) the cyclic shear strain $\Delta\gamma_{c,\Delta N}$ which develops during ΔN cycles at the cyclic shear stress, $\tau_{c,N+\Delta N}$.

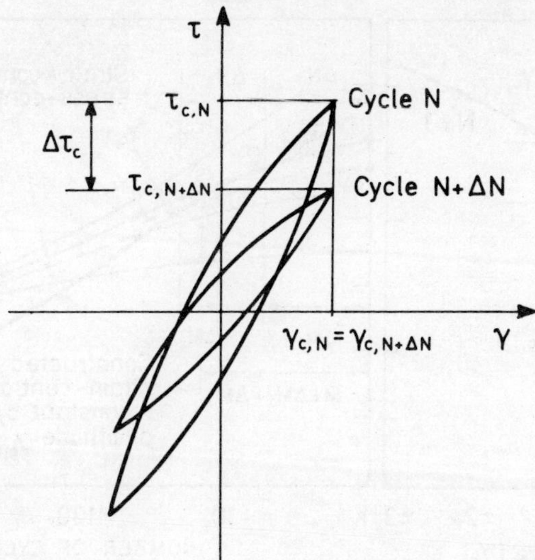

Figure 13.16 Stress–strain behaviour of soil element with two-way strain-controlled cyclic loading

The principle for analysis of strain-controlled loading is graphically illustrated in Figure 13.17. The diagram on the left describes the curve for the cyclic shear strain in the first cycle at different cyclic shear stress levels. This curve may be constructed from stress-controlled tests, as previously described for the same curve in Figure 13.15. The diagram on the right shows the development of cyclic strains with the number of cycles in stress-controlled tests and corresponds to the strain-contour diagram presented in Figure 13.9.

Consider a strain-controlled test with a cyclic shear strain of $\gamma_{c,1}$. In the first cycle, the sample will be at the strain-contour for $\gamma_{c,1}$ at point 1 in the diagram. After a few cycles, ΔN_1, the cyclic stress level will have dropped from point 1 to point 2 in the diagram. The immediate cyclic shear strain, $\Delta\gamma_{c,i,1}$, corresponding to this drop in cyclic stress, can be determined from the diagram on the left which shows that the cyclic shear strain is reduced to $\gamma_{c,2} = \gamma_{c,1} - \Delta\gamma_{c,i,1}$ corresponding to point 3 in the diagram on the right. Point 3 will plot on the axis for $N = 1$. Since the resultant cyclic shear strain remains constant in the strain-controlled test, the increase in cyclic shear strain due to cyclic loading, $\Delta\gamma_{c,\Delta N_1}$, must be equal to the immediate cyclic strain component, $\Delta\gamma_{c,i,1}$ (cf. Equation (13.2)). Cyclic loading must therefore bring the soil element out to point 2 again, and the number of cycles, ΔM_1, which is required to achieve this may be read directly from the right hand-side diagram in Figure 13.17.

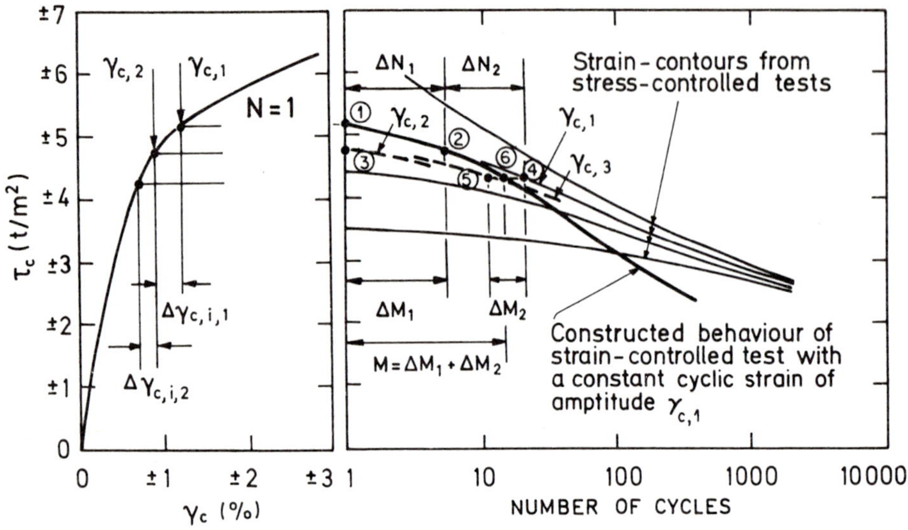

Figure 13.17   Principles for interpretation of test with strain-controlled two-way cyclic loading

When considering a second small number of cycles, ΔN_2, the cyclic stress level will drop from point 2 to point 4. The immediate cyclic shear strain, $\Delta\gamma_{c,i,2}$, corresponding to this drop in cyclic shear stress is found from the left hand-side diagram and takes the soil to point 5 which lies on the strain-contour for $\gamma_{c,3} = \gamma_{c,1} - \Delta\gamma_{c,i,2}$. The requirement that the resultant cyclic shear strain remains unchanged is again used to determine the number of load cycles, ΔM_2, which has to be applied in order to bring the soil back to point 4 on the strain-contour from $\gamma_{c,1}$.

When plotting results from strain-controlled tests, the cyclic shear stress is plotted against the actual number of cycles, M, applied to the sample, which will correspond to the sum of all the increments of actual load cycles, $M = \sum_i \Delta M_i$. After the second increment, the soil will therefore be located at point 6 which corresponds to $M = \Delta M_1 + \Delta M_2$ actual load cycles. The same procedure is continued, and the complete curve for a strain-controlled test with cyclic shear strain of $\gamma_{c,1}$ will be as indicated in Figure 13.17. This does not coincide with the strain-contour for $\gamma_{c,1}$ which was determined from stress-controlled tests.

Figure 13.18 shows a comparison between the constructed and measured behaviour for two strain-controlled simple shear tests on Drammen clay with OCR = 4. It appears that there is good agreement between constructed and measured behaviour.

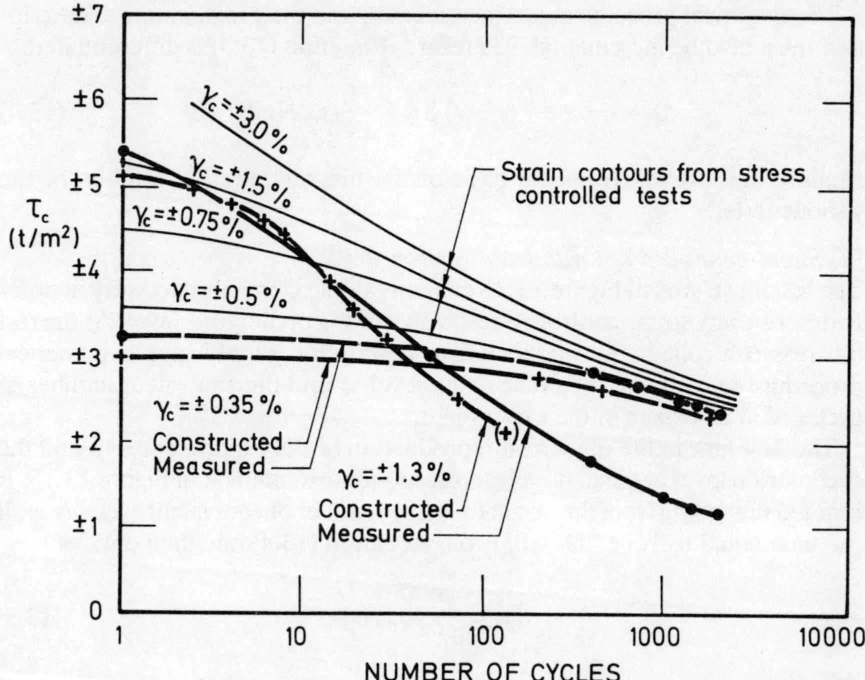

Figure 13.18   Comparison between constructed and measured behaviour of strain-controlled simple shear tests with two-way undrained cyclic loading

13.3.2   Mathematical formulation of the graphical procedure

In order to form a suitable basis for numerical computations, the graphical procedure explained above must be formulated in mathematical terms. The basis for the graphical procedures is a strain-contour diagram such as the one shown in Figure 13.9. Mathematically, the strain-contour diagram may be represented by the equation

$$F = \gamma_c - f(\tau_c, N) = 0 \tag{13.3}$$

In this equation, γ_c is the cyclic shear strain amplitude, and τ_c the amplitude of the maximum shear stress. N is the equivalent number of cycles, which together with τ_c represents the stress history of the test sample. N is called an 'equivalent' number of cycles because the properties of a sample with a current cyclic stress level τ_c (which has varied) and an equivalent number of cycles N, are the same as those of a sample which has experienced an actual number of cycles N of constant cyclic stress level τ_c.

The graphical procedures are incremental, and their mathematical description must also be incremental. Therefore, Equation (13.3) is differentiated:

$$dF = d\gamma_c - \frac{\partial f}{\partial \tau_c}(\tau_c, N)\, d\tau_c - \frac{\partial f}{\partial N}(\tau_c, N)\, dN = 0 \qquad (13.4)$$

Equation (13.4) will form the basis of the mathematical description of the various tests.

(a) Stress-controlled test with varying stress level
The test illustrated in Figure 13.15 contains all the elements necessary in order to describe any stress-controlled test with varying cyclic stress level. As the test is stress-controlled, the mathematical description should give a numerical procedure to compute the cyclic strain level γ_c and the equivalent number of cycles N at any stage in the experiment.

The first task in the numerical reproduction of the experiment is to find the cyclic strain level in the first cycle, corresponding to point A in Figure 13.15. It is noted that going from the origin to A, the number of equivalent cycles N is all the time equal to 1, i.e. $dN = 0$. From Equation (13.4) one then obtains

$$d\gamma_c = \frac{\partial f}{\partial \tau_c}(\tau_c, 1)\, d\tau_c \qquad (13.5)$$

The initial cyclic strain corresponding to the stress in point A can thus be found by dividing the cyclic stress amplitude in a number of sufficiently small finite increments, compute the strain increment for each stress increment from Equation (13.5), and add the strain increments.

The next stage in the experiment is to go from A to B. In this stage, the cyclic stress amplitude τ_c is constant, such that from Equation (13.4)

$$d\gamma_c = \frac{\partial f}{\partial N}(\tau_c, N)\, dN \qquad (13.6)$$

Therefore, the strain increase in this stage can be computed by dividing the applied number of cycles into a number of sufficiently small finite increments ΔN. For each increment, the corresponding strain increment can be computed from Equation (13.6), continually updating the equivalent number of cycles N and the derivative $\partial f/\partial N$.

In the third stage of the experiment, a cyclic stress increment $\Delta\tau_c$ is applied. Moving from point B to point C in Figure 13.15 it is seen that $d\gamma_c = 0$. Equation (13.4) then gives a change in the equivalent number of cycles

$$\Delta N = \frac{\partial f/\partial \tau_c(\tau_c, N)}{\partial f/\partial N(\tau_c, N)}\,\Delta\tau_c \qquad (13.7)$$

On moving from point C to point D in Figure 13.15 there is an increase in cyclic strain of

$$\Delta\gamma_c = \frac{\partial f}{\partial \tau_c}(\tau_c, 1)\ \Delta\tau_c \tag{13.8}$$

Simultaneously, the equivalent number of cycles is increased by

$$\Delta N = \frac{1}{\partial f/\partial N(\tau_c, N)}\ \Delta\gamma_c \tag{13.9}$$

For sufficiently small increments $\Delta\tau_c$, the two-step movement from B to D in Figure 13.15 may be summarized as:

(a) Compute strain increment from Equation (13.8)
(b) Compute increment in equivalent number of cycles as

$$\left.\begin{aligned}
\Delta N &= \frac{1}{\partial f/\partial N(\tau_c, N)}\left[\Delta\gamma_c - \frac{\partial f}{\partial \tau_c}(\tau_c, N)\cdot\Delta\tau_c\right] \\
&= \frac{1}{\partial f/\partial N(\tau_c, N)}\left[\frac{\partial f}{\partial \tau_c}(\tau_c, 1) - \frac{\partial f}{\partial \tau_c}(\tau_c, N)\right]\Delta\tau_c
\end{aligned}\right\} \tag{13.10}$$

It should be noted that Equation (13.8) is the same as Equation (13.5), which was used to compute the strain increments in the first cycle. Similarly, Equation (13.10) will yield $\Delta N = 0$ if the equivalent number of cycles $N = 1$. Thus, each increment in the initial loading during cycle 1 is just a special case of the more general increment from point B to point D described above.

The last stage in the experiment, moving from point D to point E in Figure 13.15, is exactly similar to the movement from A to B, and thus is handled the same way computationally.

The numerical procedure described here closely parallels the graphical procedure described previously. Given an adequate mathematical or numerical (interpolation tables) description of the function $f(\tau_c, N)$, the two procedures will therefore lead to the same results.

(b) Strain-controlled test

The same concepts that were used in the numerical simulation of the stress-controlled tests can also be used in predicting numerically the results of a strain-controlled test from the results of the stress-controlled tests.

Equation (13.4) forms the basis also for the simulation of the strain-controlled tests. However, it now proves useful to divide the cyclic strain increment into two parts:

$$d\gamma_c = d\gamma_{c,i} + d\gamma_{c,\Delta N} \tag{13.11}$$

where $d\gamma_{c,i}$ is the 'immediate' strain increment caused by a change in stress level, and $d\gamma_{c,\Delta N}$ is a strain increment caused by a number of stress cycles at a constant cyclic stress level.

The first task in the simulation of the strain-controlled test is to establish the cyclic stress level in the first cycle corresponding to the known cyclic strain. In the first cycle $dN = d\gamma_{c,\Delta N} = 0$, and Equation (13.4) reduces to

$$\Delta\tau_c = \frac{1}{\partial f/\partial\tau_c(\tau_c,\, 1)}\, \Delta\gamma_c \tag{13.12}$$

The initial cyclic stress level can thus be found by dividing the cyclic strain into a number of sufficiently small finite increments, then compute the stress increment for each strain increment from Equation (13.12), and add the stress increments. This procedure is analogous to the first stage in the simulation of the stress-controlled tests.

Next, an increment ΔN in the equivalent number of cycles is applied to the sample, while the cyclic strain is kept constant. With $d\gamma_c = 0$, Equation (13.4) gives

$$\Delta\tau_c = \frac{\partial f/\partial N(\tau_c,\, N)}{\partial f/\partial\tau_c(\tau_c,\, N)}\, \Delta N \tag{13.13}$$

This change (reduction) in cyclic stress level leads to a corresponding change in 'immediate' cyclic strain:

$$\Delta\gamma_{c,i} = \frac{\partial f}{\partial\tau_c}(\tau_c,\, 1)\, \Delta\tau_c = -\frac{\partial f/\partial\tau_c(\tau_c,\, N)}{\partial f/\partial\tau_c(\tau_c,\, N)}\, \frac{\partial f}{\partial N}(\tau_c,\, N)\, \Delta N \tag{13.14}$$

Because the total cyclic strain level does not change during the increment, Equation (13.11) gives

$$\Delta\gamma_{c,\Delta N} = -\Delta\gamma_{c,i} = \frac{\partial f/\partial\tau_c(\tau_c,\, 1)}{\partial f/\partial\tau_c(\tau_c,\, N)}\, \frac{\partial f}{\partial N}(\tau_c,\, N)\, \Delta N \tag{13.15}$$

The actual number of cycles necessary at the current point $(\tau_c,\, N)$ to achieve this increase in cyclic strain is

$$\Delta M = \frac{\Delta\gamma_{c,\Delta N}}{\partial f/\partial N(\tau_c,\, N)} = \frac{\partial f/\partial\tau_c(\tau_c,\, 1)}{\partial f/\partial\tau_c(\tau_c,\, N)}\, \Delta N \tag{13.16}$$

From Equation (13.16) it is seen that in the first increment, when $N = 1$, the number of actual constant strain cycles necessary to achieve a given reduction in cyclic stress level is the same as the increase in number of equivalent cycles. However, in the subsequent increments the number of cycles necessary to achieve a given reduction in cyclic stress level is different from (and generally smaller than) the corresponding increase in number of equivalent cycles. This explains why the stress curves from the strain-controlled tests (where the stresses are plotted against actual number of cycles) drop off more quickly than the strain contours from the stress-controlled tests.

Combining Equations (13.13) and (13.16) gives

$$\Delta \tau_c = -\frac{\partial f/\partial N(\tau_c, N)}{\partial f/\partial \tau_c(\tau_c, N)} \Delta M \tag{13.17}$$

and

$$\Delta N = -\frac{\partial f/\partial \tau_c(\tau_c, N)}{\partial f/\partial N(\tau_c, N)} \Delta \tau_c \tag{13.18}$$

Equations (13.17) and (13.18) may be used to compute the stress curves for the strain-controlled, constant cyclic strain tests, by repeated application of increments ΔM of the actual number of cycles. For each increment, the cyclic stress level τ_c and the equivalent number of cycles N is updated. It may be noted that Equation (13.18) agrees with Equation (13.10) when the cyclic strain increment $\Delta \gamma_c = 0$, as expected.

The numerical procedure described here is thus fully consistent with the numerical procedure for the stress-controlled tests, and again closely parallels the graphical procedure described previously.

13.3.3   Limitations and capabilities of the numerical model

Being in possession of a consistent numerical model which has been derived directly from the experimental results, and which can thus be assumed to reflect satisfactorily the significant elements of the true material behaviour, it is of considerable interest to explore the limitations and potential of incorporating the model in a computational procedure.

Let us consider a stiff gravity platform base, resting on the soil. Assuming wave action primarily in one direction, the forces acting on the foundation are a vertical force V, a horizontal force H and a moment M. All these forces are time-dependent. However, for any given wave-, current- and wind condition it seems reasonable to represent them as consisting of an average and a cyclic part:

$$V = V_a + V_c \, e^{i\omega t}$$
$$H = H_a + H_c \, e^{i\omega t} \tag{13.19}$$
$$M = M_a + M_c \, e^{i\omega t}$$

The average vertical force V_a is the submerged weight of the platform, while the average horizontal force H_a and the average moment M_a are results of unsymmetrical wave- and wind-forces, and of current forces. The cyclic forces are due to the symmetrical part of the wave- and wind-loads. In general V_c, H_c and M_c are complex, due to the phase differences between the different forces. If these phase differences can be disregarded, the cyclic parts of the forces may be taken as real and equal to the amplitude values.

At a given point in time during a storm, the vertical displacement of the foundation r_z, the horizontal displacement r_x and the rotation r_ϕ may also be assumed to consist of average and cyclic parts:

$$r_z = r_{za} + r_{zc}\, e^{i\omega t}$$

$$r_x = r_{xa} + r_{xc}\, e^{i\omega t} \tag{13.20}$$

$$r_\phi = r_{\phi a} + r_{\phi c}\, e^{i\omega t}$$

Similarly, at a given point in time during the storm, the stress components σ_{ij} in a soil element under the platform and the corresponding strain components ε_{ij} may also be assumed to consist of average and cyclic parts:

$$\sigma_{ij} = \sigma_{ija} + \sigma_{ijc}\, e^{i\omega t}$$

$$\varepsilon_{ij} = \varepsilon_{ija} + \varepsilon_{ijc}\, e^{i\omega t} \tag{13.21}$$

If the simplifications implied in Equations (13.19–13.21) are accepted, it follows that at any point in time during the storm, each of the two systems (average and cyclic) must separately be compatible and in equilibrium. This means that the cyclic stresses σ_{ijc} must be in equilibrium locally and with the external forces V_c, H_c and M_c, and that the cyclic strains ε_{ijc} must be compatible locally and with the foundation displacements r_{zc}, r_{xc} and $r_{\phi c}$. The same must hold for the average parts of the stresses, forces, strains and displacements. This separation tends to encourage a separate analysis of the two parts of platform behaviour, average and cyclic. Unfortunately, the experimental results indicate that such a simplification would not reflect all the significant elements of the real behaviour of the platform foundation and the soil underneath it.

Qualitatively, the tests show that in a soil sample subjected to unsymmetrical cyclic shear stresses, both the average and the cyclic parts of the shear strains develop gradually with the number of stress cycles, even when the average stresses are constant. This indicates that even during a period with constant cyclic and average forces acting on the foundation, the average part of the soil strains will change gradually with increasing number of cycles, and the average platform displacements will increase. In model tests simulating real platform behaviour, such average, irrecoverable displacements do occur.[6] Reliable predictions for this kind of displacements are of the utmost importance for the design of conductors, risers and riser connections to the platform. However, at present these effects have not been studied quantitatively in detail, and they are not incorporated in the graphical/numerical model of soil behaviour presented above. A computational procedure based on this model alone cannot, therefore, produce such predictions.

The present numerical model concentrates on the relationship between the cyclic parts of the soil stresses and strains. As long as the average shear stresses are moderate ($< \sim35$–50 per cent of undrained shear strength) and undrained conditions prevail, there is some experimental support for an assumption that this relationship is not significantly influenced by the average stresses and strains in the soil. Thus, Figures 13.8 and 13.9 are part of the basis for the numerical model. In these figures the results of tests with both one-way and two-way cyclic stresses are plotted, and no significant difference can be distinguished between the different types of loading. For the particular case represented by these diagrams (Drammen clay with OCR $= 4$), it therefore seems reasonable to proceed with the development of a computational procedure which assumes the relationship between cyclic soil stresses and cyclic soil strains to be independent of the average deviatoric soil stresses and strains. Possibly, such a model may later be validated also for more general conditions. Even if this turns out not to be the case, this simplified procedure is likely to have main features which also should be incorporated in a more general and elaborate computational procedure.

The numerical model gives a real relationship between the amplitudes of the cyclic stresses and strains in the soil. In reality there is also a phase lag between the stress- and corresponding strain components, resulting in a hysteresis loop with an area >0 (damping). This effect could be modelled by representing the cyclic parts of the stresses and strains as complex quantities, and introduce a complex relationship between them. Such a model would be fully consistent with the force representation given above. However, at present the non-linear, history-dependent damping properties have not been studied in detail. It is therefore natural at this stage to disregard the damping effect, and assume the stresses and strains to be in phase.

Even if of limited value, a computational procedure based on the present, simplified numerical model will still bring immediate and practical advantages. Not the least among these is that it will provide a more rational basis on which to assess the relationship between the cyclic forces on and displacements of the foundation, including the interacting effects of storm severity (non-linearity) and duration (history during a large number of cycles). Such a relationship can be converted into non-linear, history-dependent spring stiffnesses representing the soil in a dynamic analysis of the platform. At the present stage, the values of these spring stiffnesses are among the most uncertain and the most decisive parameters determining the dynamic behaviour of the entire platform.[2]

13.3.4 Finite element application

In this section, a brief outline is given of a finite element program based on the numerical model presented above. The main purpose is to demonstrate the principal features of a computational procedure which permits taking into account the non-linear, history-dependent soil stiffnesses through a storm consisting of a very large number of load cycles.

The storm is represented by a sequence of wave trains. Each wave train gives rise to a number of load cycles with constant load amplitudes, H_c, M_c and V_c. The load amplitudes will change from one wave train to the next.

For simplicity, all cyclic forces acting on the structure and the stresses and strains in the soil may be assumed to be in phase. This assumption is not strictly necessary, but appears reasonable in view of the simplifications in the numerical description of soil behaviour. It will also greatly simplify the computations, permitting the use of real arithmetic only.

In analogy with the numerical simulation of the various tests, the first stage in the computation is concerned with finding the displacement-, stress- and strain amplitudes in the first cycle of the first wave train. This may be done in an incremental way, in that the load amplitudes are divided into a number of finite increments ΔH_c, ΔM_c and ΔV_c. For each increment, the incremental displacement-, strain and stress amplitudes are computed and added to the total amplitudes. In formulating the stiffness matrix for such an incremental computation, a constant, high bulk modulus corresponding to an incompressible soil material should be used. The tangent shear modulus in a soil element is represented by the expression

$$G_t = \frac{1}{\partial f / \partial \tau_c (\tau_c,\, 1)} \tag{13.22}$$

The result of the first computational stage is the cyclic displacement-, stress- and strain amplitudes in the first load cycle. Therefore, the equivalent number of cycles N in all soil elements is 1.

Next, a number of actual cycles ΔM, representing a part of the first wave train, is applied to the structure. Initially, it is assumed that the displacement and strain amplitudes remain constant during these cycles. Therefore incremental cyclic shear stresses develop in the soil elements. The magnitude of these stress changes may be determined from Equation (13.17), and their directions are determined by the assumption that they appear as direct reductions of the existing cyclic shear stresses. Simultaneously, an increase in the equivalent number of cycles occurs in the soil. This increase is in general (except for the first increment) different for each soil element, and can be computed from Equation (13.18).

Because of the cyclic stress reductions, the structure is not in equilibrium at this stage. Treating the stress reductions as initial stresses, a statically equivalent vector of unbalanced, cyclic nodal point forces may be found. It is consistent with Equation (13.8) using the same high bulk modulus as before, and the shear modulus given by Equation (13.22), when forming the stiffness matrix for the linear set of equations that yields incremental cyclic displacements caused by the unbalanced cyclic forces. These incremental displacements will yield new incremental cyclic stresses restoring equilibrium of the cyclic forces. In addition, they will again cause changes in the equivalent number of cycles in all the soil elements. The change ΔN in number of equivalent cycles is different from element to element, and may be computed from Equation (13.10).

A new increment in actual number of cycles may now be applied to the structure, and treated in the same way. This procedure may be repeated as many times as necessary to account for the first wave train, and will result in a gradual redistribution of the cyclic stresses in the soil, accompanied by a gradual increase in the cyclic displacements and strain levels.

When the next wave train arrives, the only modification necessary to the procedure described above is that the unbalanced cyclic force vector, instead of representing the stress reductions caused by a number of cycles, should initially represent the incremental increase in the forces acting on the structure. Subsequently, the second wave train is treated exactly like the first one. In this manner, the computation can proceed through all the wave trains representing the storm.

In the computational procedure outlined above, a number of load cycles on the structure is effectively converted into an incremental cyclic force vector. The cyclic stress history of a soil element is taken into account by keeping track of the equivalent number of cycles on the element. The result is a procedure which has a great deal in common with a conventional, incremental elastic-plastic solution procedure. A noticeable difference is, however, that the present procedure always works with amplitudes of the forces, displacements, strains and stresses.

In incremental, non-linear computational procedures it is common practice to perform iterations within each increment, in order to maintain equilibrium as well as fulfillment of the stress–strain law after the increment. Such iterations should also be incorporated into the present computational procedure, in order to ensure that after each increment, cyclic equilibrium is maintained and all soil elements remain on the surface as defined by Equation (13.3).

13.3.5 Extension of the procedure to predict average displacements

As mentioned previously, there is also a need for a model which is able to give reasonably reliable predictions for the permanent, average displacements that develop during a storm. The present approach, which promises considerable economy in computer time compared to a strict application of the incremental flow theory of plasticity, may also lend itself to extensions in order to include the gradual development of the average stresses, strains and displacements during the storm. In this section, a preliminary discussion of some of the concepts and problems of such an extension is given.

In order to tie the discussion to a practical case, the prediction of vertical settlement of a platform during a storm is considered. The submerged weight of the platform V_a is the only non-zero average force acting on the foundation, while the cyclic forces are a horizontal force H_c and an overturning moment M_c. The cyclic forces are assumed to be in phase.

The average stresses under the platform consist of two parts. One part is the initial stresses that existed in the soil before the platform was installed, where the initial horizontal stresses are dependent upon the geological history at the site. The other part of the average stresses is due to the submerged weight of the platform. The distribution of these stresses changes gradually during settlement after installation of the structure, and is also affected by stress redistribution caused by storms.

The increase in vertical settlement discussed here takes place during the storm, and there is no time for drainage to occur. Hence, there will be no change in volume of the soil elements, and the settlements that develop are the result of gradually increasing average shear strains in the soil.

The tendency towards increasing average shear strains during the cyclic loading depends upon the angle between the directions of the average and cyclic principal stresses. Figure 13.19 shows schematically the maximum shear

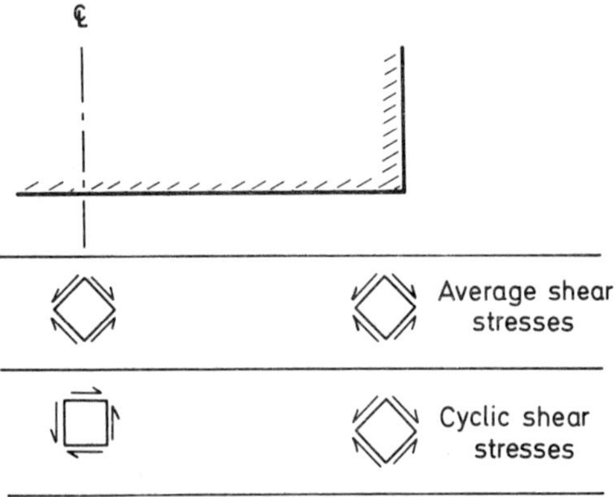

Figure 13.19 Average and cyclic maximum shear stresses on typical soil elements beneath a gravity platform

stresses in two typical soil elements, one beneath the centre and one beneath the edge of the platform. The stress components are shown on the planes on which they act.

For the element under the centre of the platform, it is seen that no cyclic shear stresses act on the planes where the principal average shear stresses act, as long as zero damping is assumed. In this case, it is assumed that the tendency towards increasing average shear strains is governed mainly by the excess average pore pressure generated during the cyclic loading. This pore pressure causes a reduction in effective stresses and stiffness in the soil ('pore pressure effects').

On the other hand, for the element under the edge of the platform it is seen that the planes where the maximum average shear stresses act may also experience considerable cyclic shear stresses. These planes are thus subjected to unsymmetrical, cyclic shear stresses. In this case, the average shear strains increase also when no increase in average pore pressure occurs. Thus, for soil elements beneath the platform periphery, the tendency towards increasing average shear strains consists of two parts: the reduction in soil stiffness caused by pore pressure build-up, and an additional effect which occurs irrespective of pore pressure build-up.

In every soil element beneath the platform there is a tendency towards an increase in the average shear strain during the storm. These strains however do not develop freely, but are constrained by the compatibility requirements within the soil. Aiming at an incremental finite element displacement formulation, it is therefore envisaged that the strains are kept constant, while the tendency towards increasing average strains is converted into a corresponding drop in average stresses. In further analogy with the procedure explained previously for cyclic strains, the reduction in average stresses is then represented by a statically equivalent, unbalanced force vector that is applied to the structure. Solution of the set of coupled incremental equilibrium equations gives an increment to the average displacement vector which will restore equilibrium of the average forces.

The procedure outlined briefly above is parallel to the one described previously for the cyclic part of the solution. It suggests that the drop in average shear forces caused by an incremental number of actual cycles ΔM can be formulated by an expression similar to Equation (13.17). In view of the discussion above, it is reasonable to postulate that

$$\Delta \tau_a = -g(\tau_c, \tau_a, \alpha, N_a) \, \Delta M \qquad (13.23)$$

where τ_c is the cyclic shear stress amplitude, τ_a the maximum average shear stress, α the angle between the principal axes of the cyclic and the average stresses, and N_a some measure of the recent stress history. Thus, there are similarities between N_a and the equivalent number of cycles N used previously.

13.4 MATHEMATICAL MODEL BASED ON PLASTICITY THEORY

Based on the experimental data presented above and elsewhere in the soil mechanics literature, a mathematical elastic-plastic model for undrained stress–strain behaviour is proposed below.

Soils in general undergo simultaneously both elastic and plastic deformations upon shearing, and the relative magnitudes of the two contributions depend on the stress level, the type of soil and the stress history. For stress paths involving loading, unloading and subsequent reloading in an arbitrary direction, the soil stress–strain properties depend greatly on the loading history. A comprehensive mathematical model should account for the anisotropic, elastic-plastic, and path-dependent stress–strain behaviour.

Only the special case of undrained behaviour with no volume change is modelled herein, and the reader is referred to the literature[11,12,13] for more general and detailed discussions.

It is assumed that the elasticity of the material is linear and isotropic, and that non-linearity and anisotropy result from its plasticity. In order to define the plastic strains, one needs:

(1) the yield condition specifying the states of stress for which plastic flow occurs;

(2) the flow rule connecting the plastic strain-increment tensor with the stress and stress-increment tensors; and

(3) the hardening rule specifying the modification of the yield condition in the course of plastic flow.

It is convenient and customary to picture the yield condition as a surface in stress space and to refer to it as a yield surface. The flow rule used is the normality rule of plasticity, which states that the plastic strain-increment vector lies along the exterior normal to the yield surface at the stress point. However, the rule of isotropic plastic hardening[7] used frequently in soil mechanics is not adequate for unloading–loading reversals as it implies exclusively elastic behaviour until the stress is fully reversed. It is observed experimentally that upon unloading, both elastic and plastic deformations occur well before the stress is fully reversed. This is similar to the so-called Bauschinger effect observed in metals. Hence, it is proposed to use a combination of the isotropic and kinematic[8] plastic hardening rules in which the yield surface is allowed to be translated in stress space by the stress point and to change in size simultaneously. Furthermore, the concept of a field of shear moduli[9,10] is introduced. This field is defined in stress space by the collection $f_0, f_1, \ldots, f_n$ of nested yield surfaces with respective sizes $k^{(0)} < k^{(1)} < \cdots < k^{(n)}$ which define the regions of constant shear moduli.

13.4.1 Yield surfaces

Since during undrained loading of saturated soil the yielding is independent of the imposed octahedral normal total stress component $p = \sigma_{ii}/3$, only deviatoric stresses $s_{ij} = \sigma_{ij} - \frac{1}{3}\delta_{ij}(\sigma_{kk})$ need appear in the yield functions. Thus, for this special case, von Mises type circular cylindrical yield surfaces may be used. The yield functions f_m are then represented by equations of the form

$$[\tfrac{3}{2}(s_{ij} - \alpha_{ij}^{(m)})(s_{ij} - \alpha_{ij}^{(m)})]^{1/2} - k^{(m)} = 0 \tag{13.24}$$

for all m, where $\alpha_{ij}^{(m)}$ represents the co-ordinates of the centre of the yield surface f_m in stress space. Because $\alpha_{ij}^{(m)}$ is not necessarily proportional to the isotropic tensor δ_{ij}, the Kronecker delta, the yielding of the material is anisotropic. Direction is therefore of great importance, and the physical reference axes (x, y, z) must be fixed with respect to the soil element. The translations $\alpha_{ij}^{(m)}$ of the yield surfaces in stress space may be thought of as indirect expressions of the material 'memory' of its past loading history associated with that direction of loading.

If the physical co-ordinate axes originally coincide with the principal axes of the material anisotropy, then $\alpha_{xy}^{(m)} = \alpha_{yz}^{(m)} = \alpha_{zx}^{(m)} = 0$ for all m initially. Moreover, if in addition $\alpha_x^{(m)} = \alpha_z^{(m)}$ for all m, then the anisotropy is initially rotationally symmetric about the y-axis. Finally, if $\alpha_x^{(m)} = \alpha_y^{(m)} = \alpha_z^{(m)}$ for all m, then there is complete spherical symmetry, and the soil is initially isotropic.

All yield surfaces may be translated in stress space without changing in form or orientation, and they consecutively touch and push each other but cannot intersect. When the stress point reaches the yield surface f_m, all the yield surfaces $f_0, f_1, \ldots, f_m$ are tangent to each other at the contact point, and

$$\frac{s_{ij} - \alpha_{ij}^{(m)}}{k^{(m)}} = \frac{s_{ij} - \alpha_{ij}^{(m-1)}}{k^{(m-1)}} = \cdots = \frac{s_{ij} - \alpha_{ij}^{(0)}}{k^{(0)}} \tag{13.25}$$

Figure 13.20a shows the set of yield surfaces in the octahedral plane for an element which was initially isotropic but has subsequently been subjected to one-dimensional (K_0) consolidation.

13.4.2 Shear strains

A plastic shear modulus H'_m is associated with each of the yield surfaces, and an associative flow is used to compute the plastic strains. The total strain is equal to the sum of the elastic and plastic strains.

When an infinitesimal stress increment is applied to the soil element such that the stress increment vector points out of the yield surface f_m (i.e. $(s_{ij} - \alpha_{ij}^{(m)})\,\mathrm{d}s_{ij} > 0$), the total strain increment is given by Equation (A1) in Appendix A. The yield surfaces $f_0, \ldots, f_m$ are then translated together by the stress point

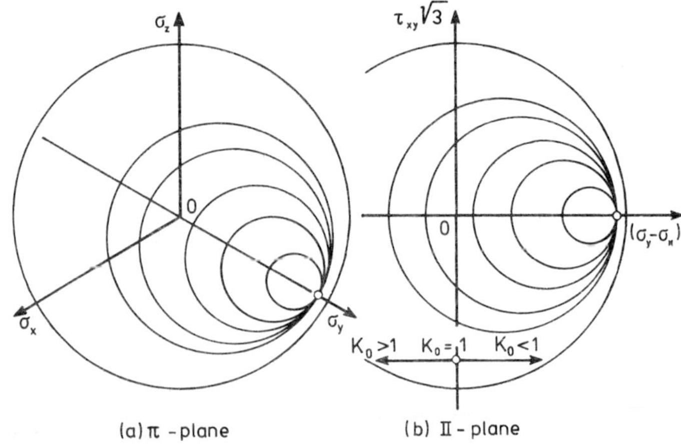

Figure 13.20   Initial set of yield surfaces for K_0-consolidation

towards the yield surface f_{m+1}, and they remain tangent to each other on the stress path. The instantaneous translation $d\alpha_{ij}^{(m)}$ is given by Equation (A3) in Appendix A. If on the other hand, the stress point retreats from the yield surface f_m (i.e. $(s_{ij} - \alpha_{ij}^{(m)})\, ds_{ij} < 0$), $H'_m = H'_0$ in Equations (A1–A3), and loading reversal occurs.

13.4.3   Plastic hardening and softening

The model is further generalized by assuming that the surfaces are allowed to change in size as well as to be translated. The associated shear moduli are also allowed to vary, and in general both $k^{(m)}$ and H'_m are functions of λ, where λ is a scalar parameter monotonically increasing during the deformation process and chosen equal to the length of the plastic deviatoric shear strain trajectory

$$\lambda = \int [\tfrac{2}{3}\, de_{ij}^p\, de_{ij}^p]^{1/2} \tag{13.26}$$

The integration is carried out over the strain path. A superscript p denotes a plastic strain, and the de_{ij} denote the incremental deviatoric strain vector components, $de_{ij} = d\varepsilon_{ij} - \tfrac{1}{3}\delta_{ij}(d\varepsilon_{kk})$.

The resulting mathematical formulation for the constitutive laws associated with the proposed model is presented in Appendix A together with the mathematical description of the motion of the yield surfaces. The equations are presented in a form suitable for finite element analyses and may be implemented in a general finite element program. H_m in Equations (A2) and

(A5) denotes the elastic-plastic shear modulus associated with the yield surface f_m and is defined as:

$$\frac{1}{H_m} = \frac{1}{2G} + \frac{1}{H'_m} \tag{13.27}$$

where G is the elastic shear modulus.

13.4.4   Model interpretation and application

Complete specification of the model parameters requires the determination of (1) the initial positions and radii of the yield surfaces together with their associated moduli, and (2) their size and shear modulus changes as loading proceeds.

For any loading or unloading history, the present configuration of the field of yield surfaces is determined by calculating the translation and contraction or expansion of each yield surface during successive changes in load.

The soils' anisotropy originally develops during consolidation, which in many practical cases occurs under K_0-conditions, i.e. no lateral strain. In the following the y-axis is vertical and coincides with the direction of the consolidation stress, and initially $\alpha_x^{(m)} = \alpha_z^{(m)} = \alpha_{xy}^{(m)} = \alpha_{yz}^{(m)} = \alpha_{zx}^{(m)} = 0$ for all m. Since the yield surfaces are allowed to change in position in stress space in order to follow the stress point [Equation (A3)], the yield functions [Equation (13.24)] initially reflect, but do not always conserve, the original anisotropy of the soil element. If, as in a triaxial test, the principal axes of stress remain fixed during shearing, it follows from Equation (A1) that the strain-increment tensor and thus also the strain tensor remain coaxial with the stress tensor. Furthermore, it follows from Equation (A3) that $d\alpha_{ij}^{(m)} = 0$ for $i \neq j$, and the initial anisotropy of the material is preserved. On the other hand, if, as in a simple shear test, the principal axes of stress rotate during shearing, it follows from Equations (A1) and (A3) that $d\varepsilon_{ij} \neq 0$ and $d\alpha_{ij}^{(m)} \neq 0$ for $i \neq j$. Therefore, due to the anisotropy caused by plastic flow, the strain tensor is not coaxial with the stress tensor. Hence the application of shearing stresses tends to erase the soil's memory of its previous history in that its state of initial anisotropy is disrupted by shear, and a new state of anisotropy is created.

Since the material is anisotropic, no aspect of its deformational behaviour, especially yielding, can be dealt with in principal stress space. When some stress components vanish, the set of yield surfaces can be represented in the subspace of the non-vanishing components, since during the subsequent motion of the yield surfaces, all components of $\alpha_{ij}^{(m)}$ not belonging to this subspace also vanish [Equation (A3)]. In particular, for axisymmetric stress states ($\tau_{zx} = \tau_{yx} = 0$) with equal horizontal normal stresses ($\sigma_x = \sigma_z$), the position in stress space of the set of yield surfaces is defined by the sole determination of two parameters

$\alpha^{(m)}$ and $\beta^{(m)}$ ($m = 0, \dots, n$). Equation (13.24) then simplifies to:

$$[(\sigma_y - \sigma_x) - \alpha^{(m)}]^2 + [\tau_{xy}\sqrt{(3)} - \beta^{(m)}]^2 - [k^{(m)}]^2 = 0 \qquad (13.28)$$

where $\alpha^{(m)} = \alpha_y^{(m)} - \alpha_x^{(m)} = 3\alpha_y^{(m)}/2$ and $\beta^{(m)} = \alpha_{xy}^{(m)}\sqrt{(3)}$ therefore represent the centre co-ordinates of the circular yield surfaces in a $(\sigma_y - \sigma_x)$ vs. $[\tau_{xy}\sqrt{(3)}]$-stress plane, which in the following is referred to as the II-plane. For soil elements with initial rotational symmetry about the y-axis with respect to anisotropy, the yield surfaces are initially centred along the $(\sigma_y - \sigma_x)$-axis. All the model parameters $\alpha_0^{(m)}$, $k^{(m)}(\lambda)$ and $H_m(\lambda)$ may then be determined based on the observed experimental behaviour in either triaxial or simple shear tests. The subscript zero refers to initial conditions.

(a) Interpretation of triaxial tests

Consider the yield surfaces in the octahedral plane (π-plane) as well as in the II-plane shown in Figure 13.20. Originally all the yield surfaces were concentric with the centre at the origin. During the one-dimensional consolidation during soil deposition in the field (K_0-consolidation), the yield surfaces were successively translated into their present locations. If the lateral stress coefficient K_0 increases in value subsequent to initial soil deposition, as it does in overconsolidated deposits, the inner circles will be translated back towards the origin. Figure 13.21(a) shows the yield surfaces before shearing for a triaxial specimen which went through the process described above and presently has $K_0 = 1$.

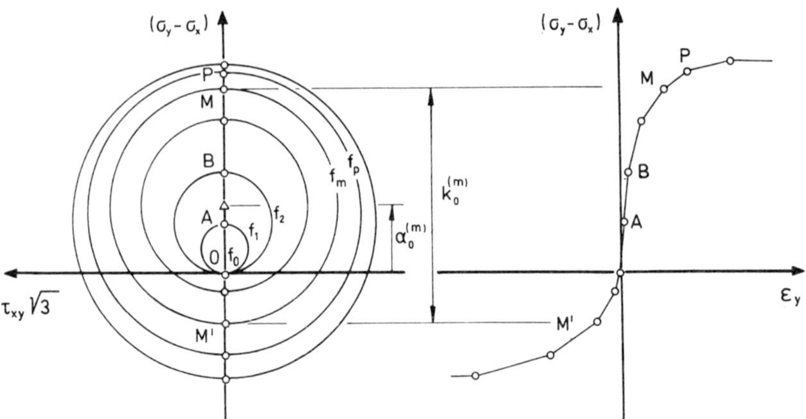

Figure 13.21　Monotonic triaxial compression and extension tests.
(a) Field of yield surfaces in II-plane
(b) Stress–strain curves

During both triaxial compression and extension tests, $\sigma_x = \sigma_z$ and $\tau_{xy} = 0$. The stress point moves along the $(\sigma_y - \sigma_x)$-axis, and when it touches f_m in triaxial compression, $(\sigma_y - \sigma_x) = \alpha_0^{(m)} + k_0^{(m)}$, whereas in triaxial extension $(\sigma_y - \sigma_x) = \alpha_0^{(m)} - k_0^{(m)}$ where $\alpha_0^{(m)}$ and $k_0^{(m)}$ denote the initial position and radius of the yield surface f_m [Figure 13.21(a)]. At this time the incremental vertical strain is given by Equation (A1) in both cases. This equation simplifies to

$$\mathrm{d}\varepsilon_y = \left(\frac{1}{3G} + \frac{2}{3H'_m}\right)\mathrm{d}(\sigma_y - \sigma_x) = \frac{2}{3H_m}\,\mathrm{d}(\sigma_y - \sigma_x) \qquad (13.29)$$

By comparing the experimental stress–strain curves obtained in monotonic triaxial compression and extension tests, one may determine the initial positions $\alpha_0^{(m)}$, radii $k_0^{(m)}$ and associated shear moduli H_m of the yield surfaces (Figure 13.21). The plastic shear moduli are obtained by using Equation (13.27).

Figure 13.22(a) presents the situation upon reaching point P on f_p in compression. Upon a loading reversal, the stress point leaves point P, inverse plastic flow occurs, and the stress point translates the surfaces downwards. The predicted reverse loading curve for the case when the yield surfaces do not change in size but just translate, i.e. $k^{(m)} = k_0^{(m)}$, is shown in Figure 13.22(b). Changes in size usually start to occur upon the first loading reversal, and the $k^{(m)}(\lambda)$ functions may be determined by comparing the differences between the segments PA'_1 and OA, $A'_1B'_1$ and AB, etc. $\cdots$ in Figures 13.21 and 13.22.

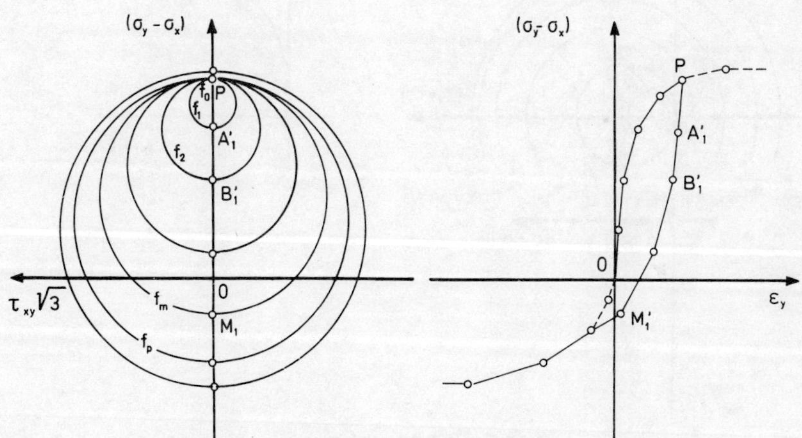

Figure 13.22   Triaxial compression test with reverse loading.
(a) Field of yield surfaces before reversing load
(b) Reverse loading stress–strain curve

When $k^{(m)} = k_0^{(m)}$, $\text{PA}_1' = k_0^{(1)}, \ldots, \text{PM}_1' = k_0^{(m)}$, etc., and experimental deviations from these equalities are attributed to the functions $k^{(m)}(\lambda)$. Under cyclic loading conditions, the stress–strain curves consist of consecutive hysteresis loops. The values of $k^{(m)}$ can be found for each branch of these loops, and the function $k^{(m)}(\lambda)$ can be determined. Once the yield surfaces reach their ultimate limiting sizes (Figure 13.9), their associated shear moduli H_m start to vary, and the functions $H_m(\lambda)$ are determined by comparing the shapes of the consecutive loops.

(b) *Interpretation of constant volume, simple shear tests*

In the case of simple shear loading ($d\varepsilon_x = d\varepsilon_y = d\varepsilon_z = 0$), the stress point moves in stress space in such a way that the resulting strain-increment vector does not have a normal strain component. For a soil specimen initially subjected to equal horizontal normal stresses, it is apparent from Equations (A1–A3) that the stress path occurs in the II-plane (i.e. $\sigma_x = \sigma_z$ at all times). If the magnitude of the plastic increment vector is large in comparison to the magnitude of the elastic strain-increment vector (i.e. $2/H'_m$ is large compared to $1/G$), the stress path goes through the apexes of the subsequent yield surfaces. Such a simplified stress path is shown in Figure 13.23(a). The stress point and the currently attached yield surfaces then move along together in the same direction, and

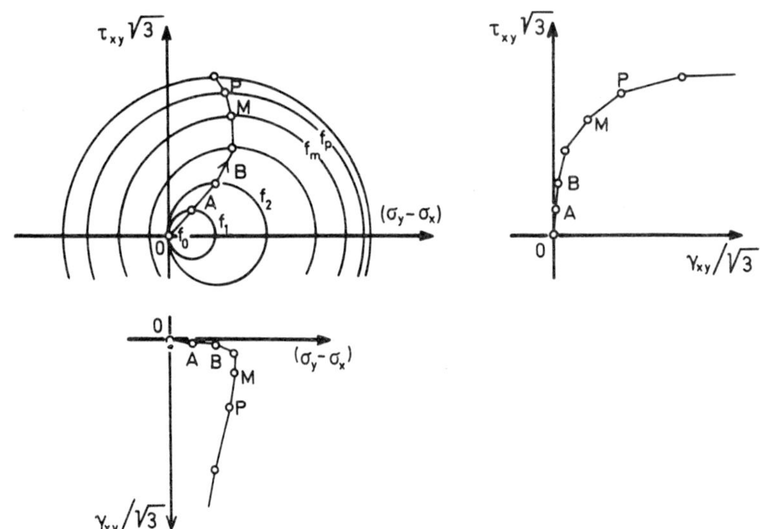

Figure 13.23 Monotonic simple shear test.
(a) Field of yield surfaces in II-plane
(b) Shear stress vs. shear strain
(c) Normal stress difference vs. shear strain

when the stress point reaches the yield surface f_m, $(\sigma_y - \sigma_x) = \alpha_0^{(m)}$ and $\tau_{xy}\sqrt{(3)} = k_0^{(m)}$. The stress–strain relation [Equation (A1)] then simplifies to:

$$d\gamma_{xy} = \left(\frac{1}{G} + \frac{2}{H_m'}\right) d\tau_{xy} = \frac{2}{H_m} d\tau_{xy} \tag{13.30}$$

By combining the experimental stress–strain curves (i.e. shear stress vs. shear strain and normal stress difference vs. shear strain) obtained in a monotonic simple shear test, one may determine the initial positions, sizes and associated shear moduli of the yield surfaces. This is illustrated schematically in Figure 13.23.

Figure 13.24(a) presents the situation upon reaching point P of f_p. Upon loading reversal, the stress point P leaves the yield surface f_p and travels vertically downwards in the II-plane, pushing back the yield surfaces until it reaches the yield surface f_p once more (Figure 13.24(a)). Thereafter, the stress path bends over, and its direction is governed by the current position of f_{p+1}. If the surfaces are translated without changing in size (i.e. $k^{(m)} = k_0^{(m)}$), the model predicts that the reverse loading curve (Figure 13.24(b)) is uniquely defined by the primary loading curve OABMP, since during loading reversal, the stress difference corresponding to any shear modulus H_m equals twice the difference observed during primary loading. Experimental deviations from these equalities are therefore due to the functions $k^{(m)}(\lambda)$. The functions $k^{(m)}(\lambda)$ and $H_m(\lambda)$ may be determined by using cyclic simple shear experimental test results, as explained previously for the triaxial tests.

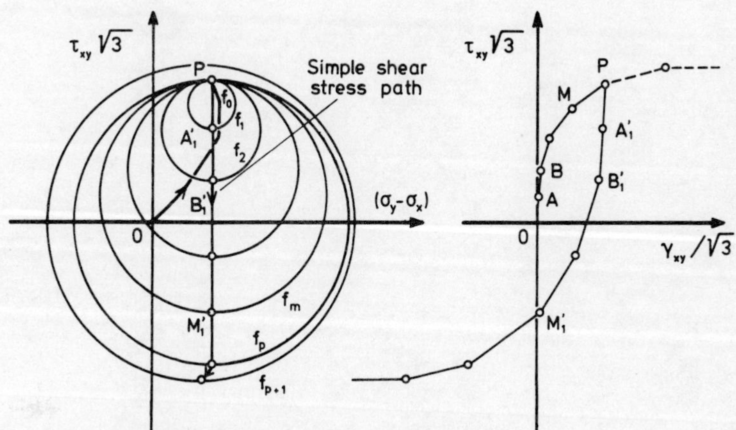

Figure 13.24   Simple shear test with reverse loading.
(a) Field of yield surfaces before reversing load
(b) Reverse loading stress–strain curve

(*c*) *Model application*

To test the mathematical models' ability to predict undrained behaviour, the model parameters were determined based solely on the laboratory results from monotonic and cyclic strain-controlled simple shear tests. The model was subsequently applied to predict the behaviour of the same soil in (1) monotonic triaxial compression and extension tests, (2) cyclic two-way strain- and stress-controlled triaxial tests, and (3) cyclic two-way stress-controlled simple shear tests. The theoretical predictions were found to agree very well with the experimental results as outlined elsewhere.[12,13]

13.5　CLOSING REMARKS

Two alternative approaches to the analysis of soil deformations under cyclic loading were discussed in Sections 13.3 and 13.4. Both approaches have been pursued far enough to allow development of finite element computer programs which will give cyclic platform displacements. The approach in Section 13.4 will also lead to calculated increases in permanent vertical settlements (shakedown), whereas the approach in Section 13.3 needs further development in order to give such results (see Section 13.3.5).

In order to check the validity of the material models and the computational procedures, computed displacements and settlements should be compared with performance observations from physical model tests and from existing offshore platforms.

APPENDIX: MATHEMATICAL DERIVATIONS FOR PLASTICITY MODEL

In view of the form adopted for the yield function f [Equation (13.24)] it is adequate to deal only with deviatoric components.

FLOW RULE

The plastic strain increment vector lies along the exterior normal to the yield surface at the stress point. The total strain is the sum of the elastic and plastic components. In tensor notation, it is represented by:

$$de_{ij} = \frac{ds_{ij}}{2G} + \frac{3}{2H'_m} \frac{s_{ij} - \alpha_{ij}^{(m)}}{(k^{(m)})^2}(s_{kl} - \alpha_{kl}^{(m)})\, ds_{kl} \tag{A1}$$

Equation (A1) may be inverted to give:

$$ds_{ij} = 2G\, de_{ij} - (2G - H_m)\frac{3}{2}\frac{s_{ij} - \alpha_{ij}^{(m)}}{(k^{(m)})^2}(s_{kl} - \alpha_{kl}^{(m)})\, de_{kl} \tag{A2}$$

HARDENING RULE

If the stress point lies on the yield surface f_m and moves towards the yield surface f_{m+1}, the instantaneous translation of f_m is given by[9]:

$$d\alpha_{ij}^{(m)} = \frac{d\mu}{k^{(m)}}[(k^{(m+1)} - k^{(m)})s_{ij} - (\alpha_{ij}^{(m)}k^{(m+1)} - \alpha_{ij}^{(m+1)}k^{(m)})] \tag{A3}$$

where $k^{(m)}$ and $k^{(m+1)}$ denote the instantaneous sizes of the yield surfaces f_m and f_{m+1} for a given λ. The parameter $d\mu$ is obtained from the condition that the stress point remains on the surface f_m, namely

$$\tfrac{3}{2}(s_{ij} - \alpha_{ij}^{(m)})(d\alpha_{ij}^{(m)} - ds_{ij}) + k^{(m)}dk^{(m)} = 0 \tag{A4}$$

Combining Equations (A2), (A3) and (A4), one finally gets:

$$d\mu = \frac{(3H_m/2)(s_{ij} - \alpha_{ij}^{(m)})\, de_{ij} - k^{(m)}\, dk^{(m)}}{k^{(m+1)}k^{(m)} - \tfrac{3}{2}(s_{ij} - \alpha_{ij}^{(m+1)})(s_{ij} - \alpha_{ij}^{(m)})} \tag{A5}$$

If the surface f_{m-1} previously touches f_m, it remains in contact during the active deformation process and the point of contact lies on the stress path. The position of f_{m-1} is given by

$$s_{ij} - \alpha_{ij}^{(m)} = \frac{k^{(m)}}{k^{(m-1)}}(s_{ij} - \alpha_{ij}^{(m-1)}) \tag{A6}$$

Similar relations occur for any number of surfaces which are tangential to each other on the stress path.

REFERENCES

1. Høeg, K. (1976). 'Foundation engineering for fixed offshore structures; state of the art', International Conference on Behaviour of Offshore Structures, 1. *BOSS' 76.* Trondheim. *Proceedings,* **1**, 39–69. Also publ. in: Norwegian Geotechnical Institute. Publication, 114.
2. Bell, K., Hansteen, O. E., Larsen, P. K., and Smith, E. K. (1976). 'Analysis of a wave-structure-soil system; case study of a gravity platform', International Conference on Behaviour of Offshore Structures, 1. *BOSS' 76.* Trondheim. *Proceedings,* **1**, 846–863. Also publ. in: Norwegian Geotechnical Institute. Publication, 114.
3. Andersen, K. H. (1976). 'Behaviour of clay subjected to undrained cyclic loading', International Conference on Behaviour of Offshore Structures, 1. *BOSS' 76.* Trondheim. *Proceedings,* **1**, p. 392–403. Also publ. in: Norwegian Geotechnical Institute. Publication, 114.
4. Andersen, K. H., Brown, S. F., Foss, I., Pool, J. H., and Rosenbrand, W. F. (1977). 'Effect of cyclic loading on clay behaviour', Conference on Design and Construction of Offshore Structures. London, 1976. Proceedings Publ. by the Institution of Civil Engineers, London, pp. 75–79. Also publ. in: Norwegian Geotechnical Institute. Publication, 113.
5. Brown, S. F., Andersen, K. H., and McElvaney, J. (1977). 'The effect of drainage on cyclic loading of clay', International Conference on Soil Mechanics and Foundation Engineering, **9**, Tokyo, Proceedings, Vol. 2, pp. 195–200.
6. Rowe, P. W., Craig, W. H., and Procter, D. C. (1976). 'Model studies of offshore gravity structures founded on clay', International Conference on Behaviour of Offshore Structures, 1. *BOSS' 76.* Trondheim. *Proceedings,* **1**, 439–448.
7. Hill, R. (1950). *The mathematical theory of plasticity,* Osford.
8. Iwan, W. D. (1967). 'On a class of models for the yielding behaviour of continuous and composite systems', *Journal of Applied Mechanics, ASME,* **34**, No. E3, 612–617.
9. Mróz, Z. (1967). 'On the description of anisotropic workhardening', *Journal of the Mechanics and Physics of Solids,* **15**, No. 3, 163–175.
10. Prager, W. (1955). 'The theory of plasticity; a survey of recent achievements', Institution of Mechanical Engineers, London. *Proceedings,* **169**, 41–57.
11. Prévost, J. H. and Høeg, K. (1975). 'Effective stress-strain-strength model for soils', American Society of Civil Engineers. *Proceedings,* **101**, No. GT3, 259–278. Also publ. in: Norwegian Geotechnical Institute. Publication, 107.
12. Prévost, J. H. (1977). 'Mathematical modelling of monotonic and cyclic undrained clay behaviour', *International Journal for Numerical Methods in Geomechanics,* **1**, 2, 195–216.
13. Prévost, J. H. and Høeg, K. (1977). 'Plasticity model for undrained stress-strain behaviour', International Conference on Soil Mechanics and Foundation Engineering, **9**, Tokyo, Proceedings, Vol. 1, pp. 255–261.

Chapter 14

Some Applications of Numerical Methods to the Design of Offshore Gravity Structure Foundations

R. Hobbs, P. J. George and G. G. W. Mustoe

14.1 INTRODUCTION

This chapter describes some of the techniques available to the industry and currently employed by Lloyd's Register of Shipping in their independent evaluation of offshore gravity structure foundation designs. It attempts to assess the usefulness and limitations of these techniques as engineering tools.

In addition to 'desk top' and computer studies, the evaluation of the foundation design of offshore structures by the Certifying Authority includes involvement in site investigation and laboratory testing, physical modelling and installation. The present text, however, is primarily concerned with the numerical methods used. To help put the computer studies in context, some of the desk-top studies are also briefly described.

The effects of cyclic loading on the strength and stiffness of soils are not discussed in this chapter.

14.2 DESK-TOP STUDIES

14.2.1 Overturning stability

The stability of gravity structures against overall overturning is assessed by use of the general static bearing capacity equation due to Brinch Hansen[1,2] (Table 14.1), this being programmed on a desk-top calculator.

Since the calculation is strictly only applicable to a foundation bearing on a single, semi-infinite, stratum of isotropic, homogeneous soil, two sets of hypothetical uniform soil models are analysed, viz. cohesionless material under drained conditions and purely cohesive material under undrained conditions. Because real situations involve a number of layers of different materials, and

---

Table 14.1   Equations used in Brinch Hansen overturning stability calculations

For drained cohesionless material:

$$Q/A' = 0\cdot5\gamma'B'N_\gamma s_\gamma i_\gamma$$

For undrained cohesive material:

$$Q/A' = (\pi+2)C_u(1+s_c-i_c)$$

where

$$s_\gamma = 1-0\cdot4i_\gamma B'/L'$$
$$i_\gamma = (1-0\cdot7H/V)^5$$
$$s_c = 0\cdot2(1-2i_c)B'/L'$$
$$i_c = 0\cdot5-0\cdot5\sqrt{(1-H/(A'C_u))}$$

$B', L' = $ 'effective' dimensions of base.
$A' = B'L'$
$N_\gamma = $ bearing capacity factor.
$H, V = $ horizontal, vertical loads.
$Q = $ Vertical bearing capacity.
$\gamma' = $ effective unit weight
$C_u = $ undrained shear strength

---

because skirts and grout injected below the base are not directly taken into account, the results of these analyses can only be applied in a subjective, qualitative manner. Despite this limitation, the calculation is of value inasmuch as it provides an inexpensive means of comparing the structure with others. Moreover, it may be regarded as a basic check on the structure's stability. The effects of varying parameters, such as loads and base dimensions can also be quickly assessed.

Figure 14.1 shows typical plots of material factor of safety, F, against ϕ, the angle of internal friction of granular soils, and C_U, the undrained shear strength of cohesive soils, obtained for the CDP1 gravity platform which is in the North Sea Frigg field. The two curves shown correspond to two different idealizations of the annular base of the platform—one a solid circular base of the same diameter as the annulus, the other a solid circular base with the same area. Generally, a range of load combinations is considered in order to find the most critical.

From these calculations uniform soil properties required to achieve a given factor of safety are found and compared with measured site values. This approach is more satisfactory than attempting to define a representative soil before any calculations are performed.

While it may be reasonable to assume that the overall behaviour of the actual soils can be represented by hypothetical uniform materials (although this is by no means proven), it is important to realize that the calculation makes no allowance for local failure of weaker materials, which could lead to deformations in excess of serviceability limits. This aspect of the problem is considered by elasto-plastic finite element analyses (Section 13.3.2) and, where necessary, by physical model tests.

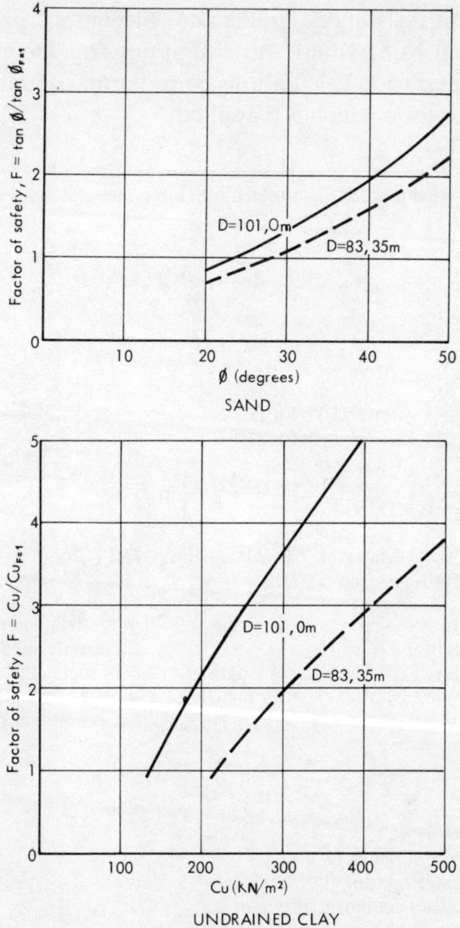

Figure 14.1 Stability based on Brinch Hansen bearing capacity equations. (Reproduced from Reference 7 by permission of the American Society of Civil Engineers)

14.2.2 Sliding stability

Calculations of static sliding stability are made in which account is taken of resistance to sliding on soil-base, soil-skirt and soil-soil interfaces.

14.2.3 Installation of structure

To ensure that the submerged weight of the structure (including ballast) is sufficient to cause skirts, etc. to penetrate, calculations are performed to estimate penetration resistance. Unlike the calculations performed for axial pile capacity (Section 14.5), where the assumptions tend to give a lower bound resistance, skirt penetration calculations consider maximum resistance. Table 14.2 indicates the criteria which are applied.

Table 14.2   Equations used to calculate skirt penetration characteristics

End Bearing

$$q_{\text{ult}} = 0 \cdot 5\gamma' N_\gamma B s_\gamma + \sigma'_v N_q s_q d_q$$

or for undrained clays:

$$q_{\text{ult}} = (\pi + 2)C_u(1 + s_c + d_c)$$

where

$$\begin{aligned}
s_\gamma &= 1 - 0 \cdot 4(B/L) \quad \text{for } L > B \\
s_q &= 1 + \sin(B/L) \quad \text{for } L > B \\
s_c &= 0 \quad \text{for } L \gg B \\
\left. \begin{aligned} d_q &= 1 + 2\tan\phi(1 - \sin\phi)^2 D/B \\ d_c &= 0 \cdot 4(D/B) \end{aligned} \right\} &\text{ If } D \leqslant B
\end{aligned}$$

or

$$\left. \begin{aligned} d_q &= 1 + 2\tan\phi(1 - \sin\phi)^2 \arctan(D/B) \\ d_c &= 0 \cdot 4 \arctan(D/B) \end{aligned} \right\} \text{ If } D > B$$

B is thickness of skirt.
D is depth of toe of skirt.
L is length of skirt (perimetral length).
N_γ and N_q are bearing capacity factors.

γ' is effective unit weight.
C_u is undrained shear strength.
ϕ is angle of internal friction.
σ'_v is vertical effective stress.
q_{ult} is unit end bearing.

Skin Friction

$$\tau_D = K_D D$$

where

τ_D is shear stress at depth D.
K_D is a parameter relating the skin friction to depth.
An alternative (but similar) expression is $\tau_D = K_D \sigma'_{VD}$.
σ'_{VD} is vertical effective stress at depth D.

For undrained clays:

$$\tau_D = \propto C_{uD}$$

where

C_{uD} is shear strength at depth D.

14.3   FINITE ELEMENT STUDIES

14.3.1   Linearly elastic finite element calculations

Three-(sometimes two-)dimensional, linearly elastic finite element calculations of soil-structure interaction are performed using the MSC/LR NASTRAN package, to determine base movements and soil stresses. The structure is represented by plate, membrane, bar and solid elements while the soil is represented by linear strain hexahedral and wedge elements.

Ideally, the soil and structure should be modelled together in detail. However, even allowing for space-saving cyclic and rotational symmetry options (which do not constrain the user to axisymmetric loading), the cost of such idealizations involving a three-dimensional finite element representation is prohibitive. On the other hand, a crude model of the full system would give only an overall picture of the behaviour of the structure. Thus, two separate series of three-dimensional static analyses are normally carried out:

(a) foundation analyses incorporating coarsely modelled structures (e.g. Figure 14.2);

(b) analyses of the structure founded on springs derived from the foundation analyses.

Numerical techniques such as the finite element method can give an appreciation of the behaviour of gravity structure foundations and identify possible

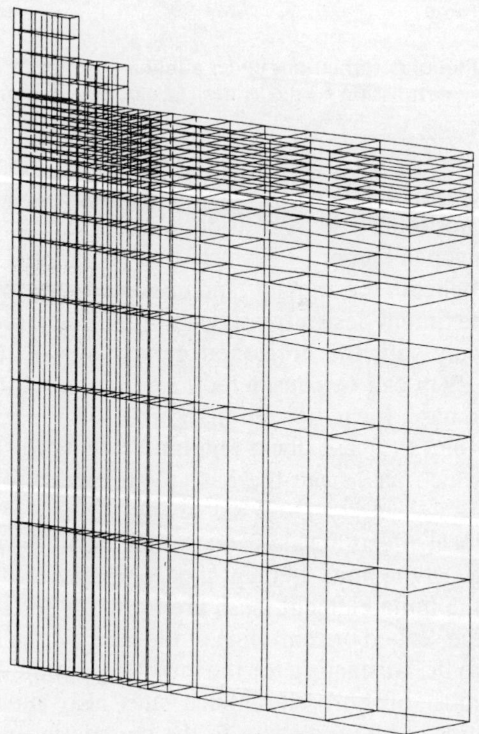

Figure 14.2 Three-dimensional finite element idealization of Frigg CDP1 foundation. (Reproduced from Reference 7 by permission of the American Society of Civil Engineers)

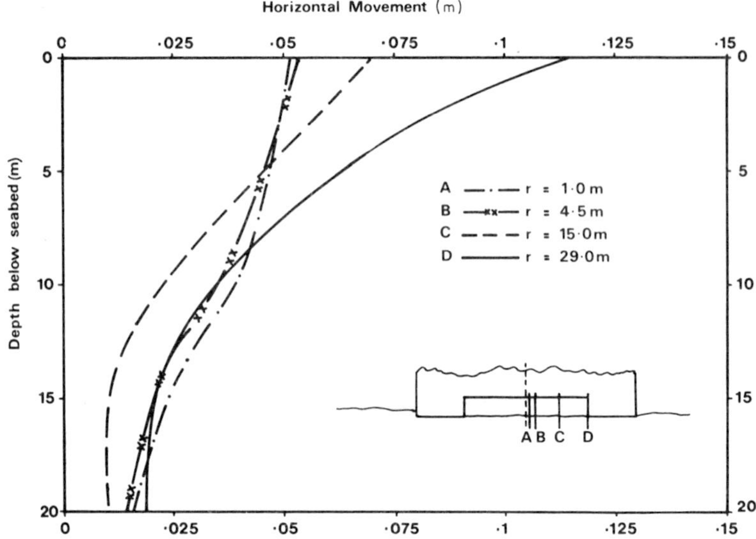

Figure 14.3   Plot of deformations under annular base. (Reproduced from Reference 7 by permission of the American Society of Civil Engineers)

failure modes (Section 14.3.2). They may also point to detailed design problems which might not otherwise be considered.

An example is given in Figure 14.3 which shows computed elastic horizontal soil deformations beneath the annular (and unskirted) base of the Frigg CDP1 platform, under maximum design (gravity plus environmental) loading. Since the base is relatively stiff, the horizontal movements of the seabed at the internal radius of 29 m can be considered to be representative of the movements of the whole base. It can be seen that soil below the void of the annulus moves much less than the base does. The implications for the design of the conductor tubes which pass from the base to the soil through the void are apparent.

In general, the local effects of skirts and conductor tubes (where these exist) are not represented in the finite element model.

Boundaries of the finite element mesh are normally set at about twice the base extent from the seabed or centreline of the base. These boundary extents have been found to be satisfactory for the increasing stiffness with depth soils encountered on most offshore sites. Such sites also allow relatively high element aspect ratios to be used, close to the boundaries, without numerical problems. Further comments on boundary extents will be found in Section 14.3.4.

14.3.2 Non-linear finite element calculations

There can be little doubt that soil is a non-linear material and that the constitutive non-linearity of soil is a plastic, rather than an elastic phenomenon. Thus, elasto-plastic analyses should model soil behaviour more realistically than linearly or non-linearly elastic analyses.

For static, linearly elastic-perfectly plastic calculations in plane strain or axisymmetry, the program FEEPHO[3] is used. These elasto-plastic analyses supplement the three-dimensional linearly elastic calculations described above, for predominantly cohesive sites. The program is an incremental, initial stress formulation based on the von Mises yield criterion:

$$\bar{\sigma}_c = \frac{1}{\sqrt{2}}[(\sigma_x - \sigma_y)^2 + (\sigma_y - \sigma_z)^2 + (\sigma_z - \sigma_x)^2 + 6\tau_{xy}^2 + 6\tau_{yz}^2 + 6\tau_{zx}^2]^{1/2}$$

$$(14.1)$$

and the Prandtl–Reuss equations.

A load increment is first solved elastically to give increments of strain $d\varepsilon$ and of stress

$$d\boldsymbol{\sigma}' = \mathbf{D}\,d\varepsilon \qquad (14.2)$$

where $\mathbf{D}$ is the elasticity matrix.

Since the material is actually non-linear after yield then, for the given increment of strain, the stress increment will not be correct. The stress increment which should have occurred be given by

$$d\boldsymbol{\sigma} = \mathbf{DPL}\,d\varepsilon \qquad (14.3)$$

where $\mathbf{DPL}$ is the 'plasticity' matrix, which may be written

$$\mathbf{DPL} = \mathbf{D} - \mathbf{P} \qquad (14.4)$$

For plane strain,

$$\mathbf{D} = \frac{E}{(1+\nu)(1-2\nu)}\begin{bmatrix} 1-\nu & & \text{SYMMETRICAL} \\ \nu & 1-\nu & & \\ 0 & 0 & \frac{1}{2}-\nu & \\ \nu & \nu & 0 & 1-\nu \end{bmatrix} \qquad (14.5)$$

$$\mathbf{P} = \frac{1}{A}\begin{bmatrix} \sigma_x'^2 & & \text{SYMMETRICAL} \\ \sigma_x'\sigma_y' & \sigma_y'^2 & & \\ \sigma_x'\tau_{xy} & \sigma_y'\tau_{xy} & \tau_{xy}^2 & \\ \sigma_x'\sigma_z' & \sigma_y'\sigma_z' & 0 & \sigma_z'^2 \end{bmatrix} \qquad (14.6)$$

$$A = \tfrac{2}{3}\bar{\sigma}^2(1 + H'/3G)(1+\nu)/E \qquad (14.7)$$

$$G = E/[2(1+\nu)] \tag{14.8}$$

$$\sigma'_x = \sigma_x - (\sigma_x + \sigma_y + \sigma_z)/3 \tag{14.9}$$

H' is a hardening parameter relating equivalent stress $\bar{\sigma}$ to equivalent plastic strain $\bar{\varepsilon}^p$ through

$$H' = \mathrm{d}\bar{\sigma}/\mathrm{d}\bar{\varepsilon}^p \tag{14.10}$$

and this term can be used to model strain softening or hardening, if required.

The corresponding stress and strain vectors are

$$\boldsymbol{\sigma} = [\sigma_x \sigma_y \tau_{xy} \sigma_z]^T \tag{14.11}$$

$$\boldsymbol{\varepsilon} = [\varepsilon_x \varepsilon_y \gamma_{xy} 0]^T \tag{14.12}$$

A set of nodal 'body forces' $\mathrm{d}\boldsymbol{R}$ are computed which equilibrate the initial stress

$$\mathrm{d}\boldsymbol{\sigma}'' = \mathrm{d}\boldsymbol{\sigma}' - \mathrm{d}\boldsymbol{\sigma} \tag{14.13}$$

required to bring Equation (14.3) and the quasi-elastic equation

$$\mathrm{d}\boldsymbol{\sigma} = \mathbf{D}\,\mathrm{d}\boldsymbol{\varepsilon} + \mathrm{d}\boldsymbol{\sigma}'' \tag{14.14}$$

into coincidence. These forces are given by

$$\mathrm{d}\boldsymbol{R} = \int_{VOL} \mathbf{B}^T \,\mathrm{d}\boldsymbol{\sigma}'' \,\mathrm{d}(VOL) \tag{14.15}$$

where $\mathbf{B}$ is the strain-displacement matrix. $\mathrm{d}\boldsymbol{R}$ are then added to the incremental load vector.

By allowing the continuum to deform elastically under the new system of loads, new strain and stress increments are found which again require the application of equilibrating body forces. The redistribution of these forces continues until convergence is attained, that is, until $\mathrm{d}\boldsymbol{R}$ becomes sensibly constant. Throughout this process the overall stiffness matrix is unchanged and thus has to be reduced only once.

The iterative initial stress calculations allow a definition of 'collapse' through lack of convergence.[4] However, while this type of analysis has been shown to give collapse loads consistent with classical theories for plane strain undrained saturated clay slopes,[3,5] it is unlikely to be able to predict realistic collapse loads for the three-dimensional, mixed strata conditions associated with gravity structure foundations. Collapse loads are, in any case, dependent upon such factors as mesh size, boundary conditions and load increment size. The analyses are useful, however, in providing an insight into possible development and modes of failure under increased load.[6,7]

Figure 14.4 shows the computed development of failure for the Frigg CDP1 platform while Figure 14.5 shows the corresponding acceleration of horizontal and vertical base movements. In this example the design loads have been incremented simultaneously. Normally, however, the gravity loads are first applied (incrementally, if necessary) and the environmental loads are then incremented to 'collapse'. Both soil and structure were represented by linear strain quadrilateral elements, these elements being preferred to those of higher order as soil stratification can more easily be modelled. A lateral stress coefficient K_0 of unity was used, so that the in situ stresses had no effect on yield.

For undrained saturated clays, the von Mises yield criterion employed by the program is identical to the Mohr–Coulomb criterion. For drained (or unsaturated) materials, on the other hand, the stress-dependency of the constitutive behaviour cannot be properly modelled. Therefore, for mixed strata sites, qualitative studies are made in which soil stiffnesses and strengths are varied on a stratum or zone basis. Alternatively, a non-linear elastic program,[8] based on a hyperbolic stress–strain law can be employed. NASTRAN also has a non-linear elastic capability. The differences between non-linear elastic and elasto-plastic formulations in representing soil behaviour[9] must, however, be realised.

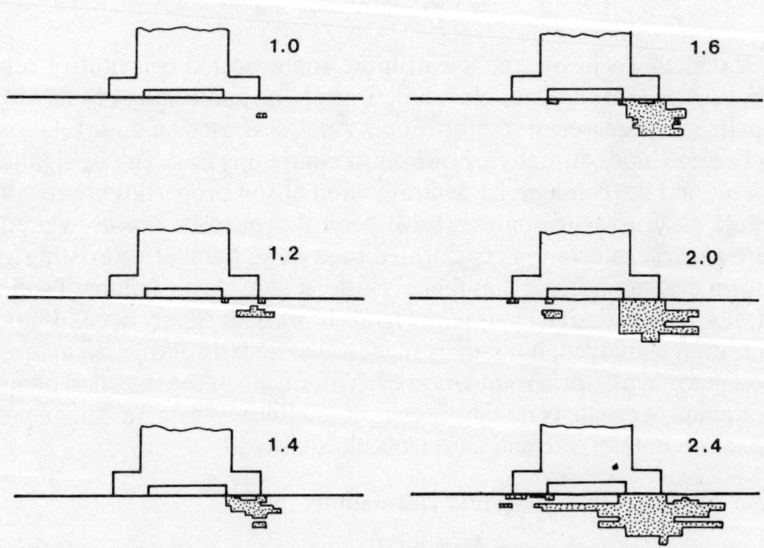

Figure 14.4   Development of plastic yield with increasing load

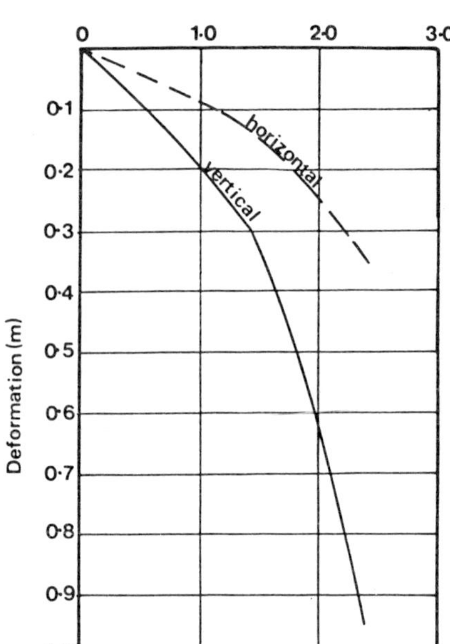

Figure 14.5 Load-deformation curves
for forward edge of base

It is difficult to justify the use of more sophisticated constitutive relationships; such as that for Cam-clay, unless sampling and testing of offshore soils
can match their degree of sophistication. Analyses based on these relationships
may become more attractive, however, if computing costs can be significantly
reduced, and the commercial determination of soil properties improved.

While elasto-plastic analyses have been shown to be useful in predicting
failure modes, calculations performed on a large number of existing gravity
platform designs indicate that there is little, if any, plastic failure of soil under
statically applied design loads (e.g. Figure 14.4). This clearly depends upon the
idealization employed, but it does suggest that linearly elastic calculations still
have a place in platform foundation evaluation, albeit the selection of appropriate moduli of elasticity, in the absence of suitable test data, remains a problem
common to both elastic and elasto-plastic analyses.

14.3.3 Pore pressure response calculations

Pore pressure response can be modelled using a two-dimensional finite element formulation based on Biot's equation.[3,10,11] Developments on this formulation are described in the following chapter. However, because of the lack of

data on consolidation parameters for offshore sites, analyses of this type, while important, are currently only of a qualitative nature.

14.3.4 Dynamic response calculations

Dynamic analyses of gravity platforms have, until recently, generally represented the soil by springs and dashpots based on elastic half-space theories for homogenous, isotropic materials. An obvious alternative, which allow for example soil stratification and variable damping to be modelled, is to represent the soil by finite elements. Comparisons between the two methods are currently being made using data from a number of instrumented platforms.[12]

Factors which need to be studied in an assessment of the use of finite elements to represent soil include the effects of mesh boundary positions and mesh element size. These can be studied relatively cheaply using a two-dimensional idealization, as discussed in the following section.

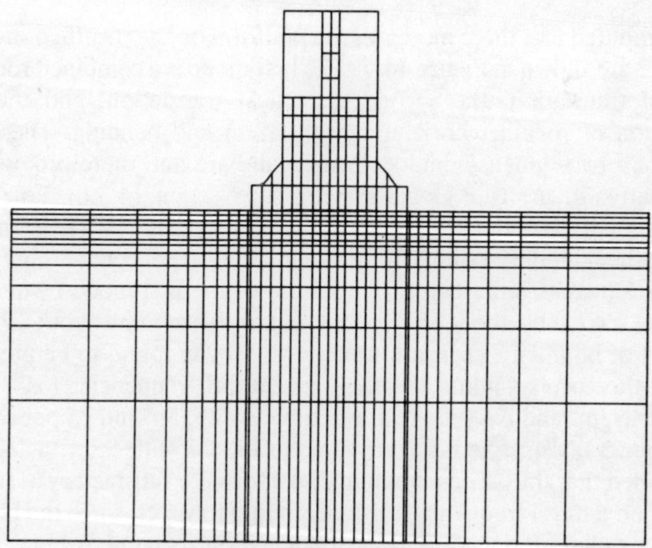

Figure 14.6 Two-dimensional finite element idealization of Frigg CDP1 platform

Figure 14.6 shows a plane strain finite element mesh for the Frigg CDP1 platform in which the structure has been idealized using membrane and bar elements. A normal modes analysis (inverse power method) for this idealization with a fixed base gives natural frequencies of the first three bending modes in close agreement with those for a three-dimensional model, suggesting that the structural stiffnesses are consistent.

Table 14.3   Soil profile for CDP1 platform. (Reproduced from Reference 12
by permission of the Society of Petroleum Engineers)

Depth below seabed (m)	Young's modulus $E(\text{kN/m}^2)$	Poisson's ratio ν
0–3	30,000	0·3
3–6	40,000	0·3
6–9	50,000	0·3
9–13·5	26,600	0·48
13·5–18	35,900	0·48
18–21	50,000	0·3
21–25·5	75,000	0·48
25·5–33	100,000	0·48
33–46	150,000	0·48
46–70	250,000	0·48
70–115	300,000	0·3
115–200	300,000	0·3

density $\rho = 2\cdot1\,\text{Mg/m}^3$ throughout.

The computed first three modes of the platform on the stratified site given in
Table 14.3 are shown in Figure 14.7. The first mode is a combined rocking and
horizontal translation, the second a vertical translation, and the third a
combination of rocking, horizontal translation and bending. These are, of
course, for a two-dimensional idealization and are not, therefore, necessarily
representative of the true global system (see Section 14.3.6). For the same
idealization and site, the effect of soil boundary positions has been studied.
Reducing the boundary extents by approximately 50 per cent of those shown in
Figure 14.6 increases the natural frequency of the first mode by about 5 per
cent of the second by about 10 per cent and of the third by about 20 per cent.
The effect of boundaries for uniform sites has been found to be greater—for
example, the corresponding increases in natural frequencies for a site with
$E = 100\,\text{MN/m}^2$ and $\nu = 0\cdot3$ are approximately 25, 50, and 75 per cent. Since
sites are generally stratified, and have increasing soil stiffness with depth, it can
be concluded that the chosen boundary positions are satisfactory as far as their
influence on natural frequencies is concerned. However, since the boundaries
also affect radiation damping, their influence on dynamic response must be
taken into account.

Finer meshes serve a number of purposes; they enable strain variations and
soil stratigraphy to be more closely followed and they allow higher frequency
vibrations to be picked up.

The effect of soil stiffness has been investigated for both stratified and
homogeneous sites. Figure 14.8 shows how natural frequencies increase with
Young's modulus, E, for a uniform site. The effect of changing Poisson's ratio
from 0·48 to 0·3 however, is small (e.g. $E = 100\,\text{MN/m}^2$, $\nu = 0\cdot48$
($G = 33\cdot8\,\text{MN/m}^2$) gives natural frequencies of the first three modes of 0·229,

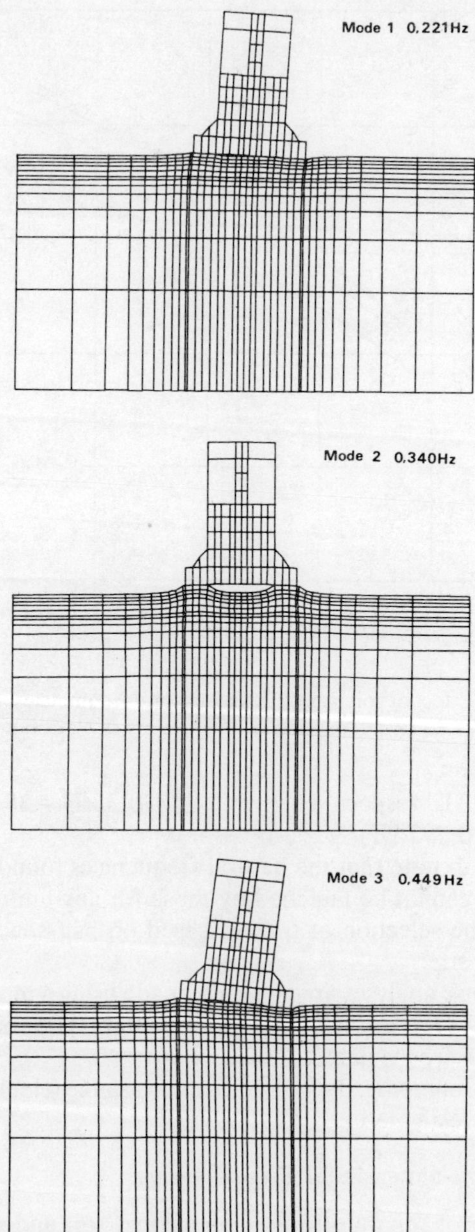

Figure 14.7 Modal deformations. (Reproduced from Reference 12 by permission of the Society of Petroleum Engineers)

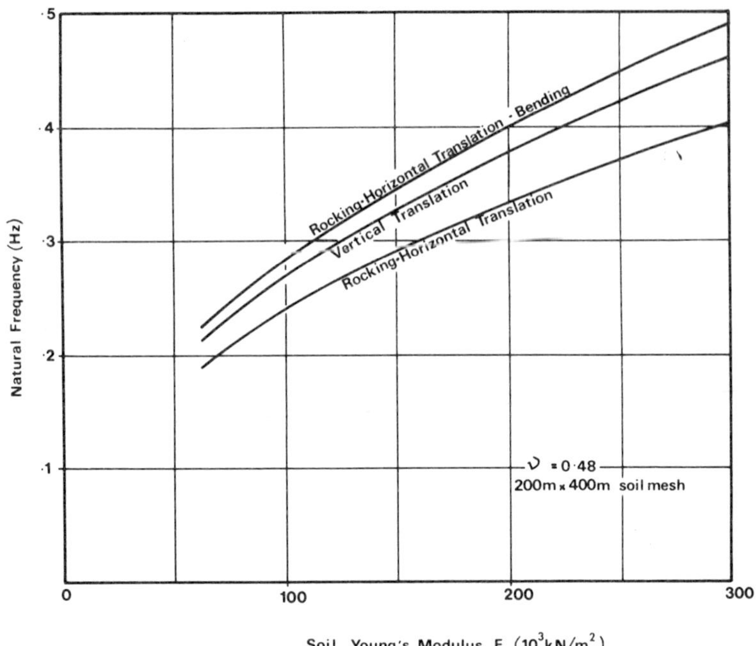

Figure 14.8 Effect of soil properties on natural frequencies. (Reproduced from Reference 12 by permission of the Society of Petroleum Engineers)

0·268 and 0·282 Hz respectively, while $\nu = 0.3$ $(G = 38.5\,MN/m^2)$ gives 0·228, 0·251 and 0·297 Hz).

It is interesting to note that the natural frequencies found for the stratified site (Figure 14.7) cannot be matched by those for any uniform site. This has implications for the selection of springs based on half-space theories for an averaged uniform site.

Dynamic response analyses are generally made using a modal superposition method,[12] but NASTRAN has a number of alternative methods and other in-house programs are available.

The effect of soil damping, both uniform and variable (element by element) is currently being investigated.

14.3.5 Use of two-dimensional discretizations

Although the use of separate three-dimensional soil and structural models reduces the computing cost sufficiently for one-off static linearity elastic finite element calculations, the desire to include soil non-linearity, its two-phase nature, and to do a range of parametric studies usually dictates the further expedient of a two-dimensional idealization (e.g. Sections 14.3.2 to 14.3.4).

Examples of studies which can be made economically using two-dimensional static analyses are those performed for the Frigg CDP1 platform[7] viz:

(a) varying the yield stress of sand layers in order to account for the limitations of the constitutive model;
(b) altering soil profiles within the bounds identified by site investigation;
(c) assuming in-filling of an annular base;
(d) simulating the possible effects of cyclic loading by reducing the stiffness and strength of elements which had yielded in a previous analysis;
(e) simulating centrifugal model tests.

The calculations for (e) were included to investigate the effects of the differences in soil stratigraphy, and of the position of the model container walls. The boundaries were found not to significantly affect the tests carried out; for example, the change in vertical displacements of the base was 1 per cent and in horizontal displacements 5 per cent.

14.3.6 Comparison of two- and three-dimensional solutions

As noted in previous sections it is often necessary to perform calculations in two- rather than three-dimensions. It cannot be expected that two-dimensional analyses will accurately reflect the true three-dimensional nature of wave-loaded gravity structures and therefore, to some extent, two-dimensional analyses must be viewed as giving a qualitative rather than quantitative picture of behaviour. However, there is very rarely any attempt to quantify the differences between the reduced degree-of-freedom system and the true global system. To indicate the magnitude of deviation between results from two- and three-dimensional static computations, two examples are considered:

The first is a comparison of computed vertical displacements and 'equivalent' stresses* under a typical gravity platform subjected to gravity load only for plane strain and axisymmetric elastic idealizations. Total loads were used for the axisymmetric case while loads for the plane strain analysis were calculated on an area basis. In both cases the loads were distributed on top of the base. K_0 was taken to be unity. It can be seen (Figure 14.9) that displacements agree extremely well (the maximum displacement of the base was 74 mm for axisymmetry and 72 mm for plane strain), while the stresses show similar agreement in upper layers. The differences between stresses at depth, which may be explained by load spread into the plane of the paper in the axisymmetric case, is not thought to be important in this instance as soil stiffness and strength increase with depth.

* See Section 14.3.2.

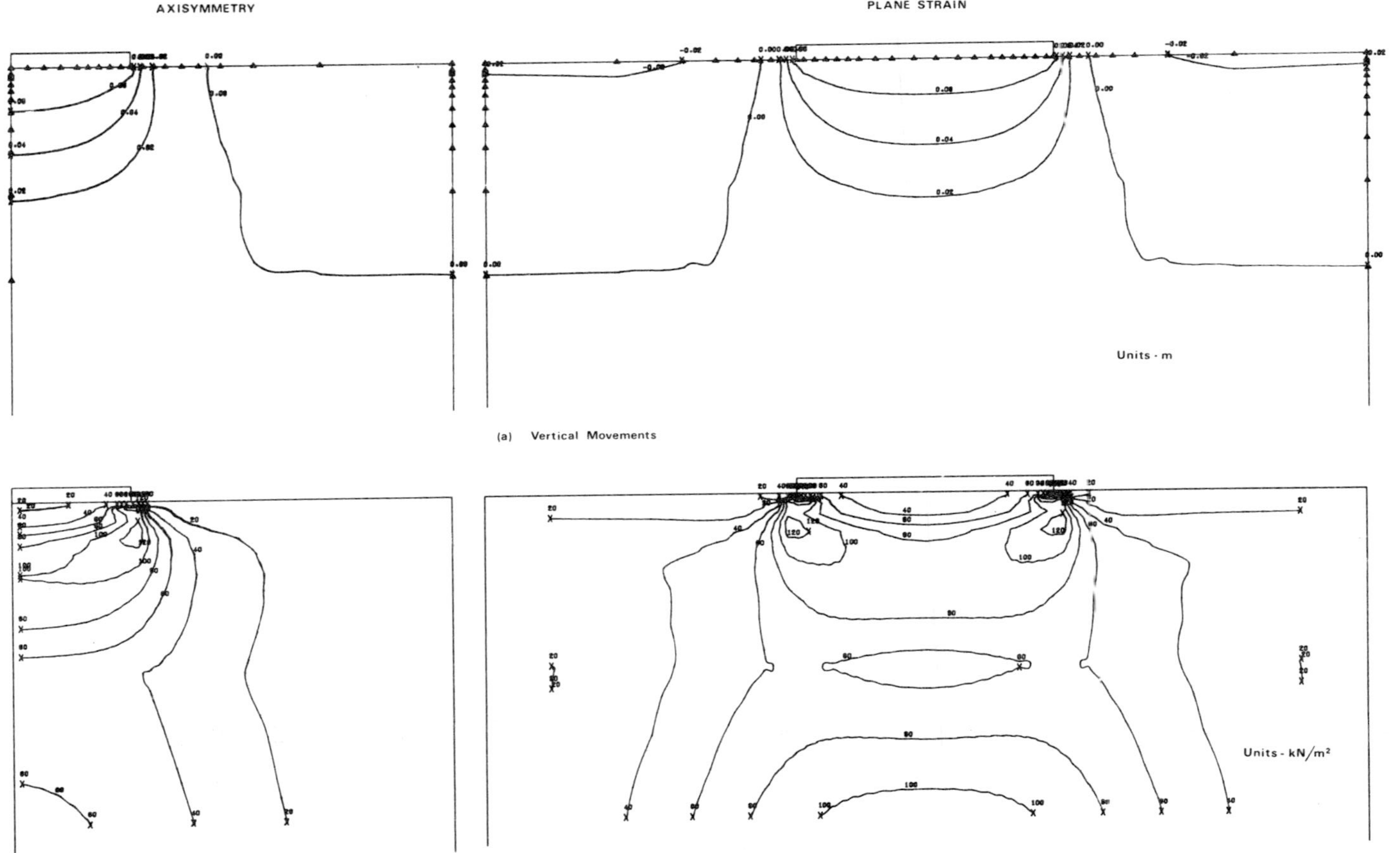

Figure 14.9 Comparison of deformations and stresses under gravity loading for plane strain and axisymmetric idealizations

The second example is a comparison of plane strain and three-dimensional displacements under both gravity and maximum design loading. Analyses of the Frigg CDP1 platform have been performed for:

(a) the three-dimensional idealization shown in Figure 14.2 and,
(b) the plane strain idealization shown in Figure 14.6.

In contrast to the previous example, the plane strain discretization included a full structural model composed of membrane and bar elements. Also, as noted previously, this platform has an annular base, and this is apparent in the displacement patterns (Figure 14.10). Loads for the plane strain analysis were calculated on the basis of an $(My/I + P/A)$ diagram for the base. Vertical displacements for the two calculations differ somewhat for maximum design loading, but horizontal displacements are remarkably similar. The difference in vertical deformations is partly explained by different stiffnesses of the structural models, the global one being too flexible. Displacements are larger in the plane strain case, presumably because of the load spreading noted above. Even so, in this example, both models predict stress levels which indicate the onset of plastic yielding at approximately design environmental loads ($K_0 = 1$).

The magnitude of the differences between the results of two- and three-dimensional analyses clearly depend upon a number of factors, which include base configuration and stiffness, and soil stratigraphy.

14.3.7 Soil-structure interface

Generally, the soil-structure interface does not allow separation of the base of the structure from the soil. Consequently, unrealistic tensile stresses can develop, in some cases, under the heel of the platform.

Any solution to this problem will involve iteration. A rigorous treatment would require the use of special interface elements,[13] but a simpler solution is to re-distribute tensile stresses using a 'no tension'[14] or limited tension criterion. An elasto-plastic no tension formulation, which allows for the simultaneous re-distribution of initial stresses due to tension and plasticity, has been presented by Hobbs[3] and is incorporated in the elasto-plastic program described in Section 14.3.2. In analyses of centrifuged dam embankments,[3,5] these no tension analyses have been found to increase computing time by 25–45 per cent over straightforward elasto-plastic solutions, while not significantly affecting shear stresses. In practice, such analyses are not considered to be a worthwhile investment of computing time.

14.3.8 Importance of soil stiffness

The deformation properties of the soil, used in the foundation analyses, are based upon empirical relationships with strengths derived from laboratory test data.

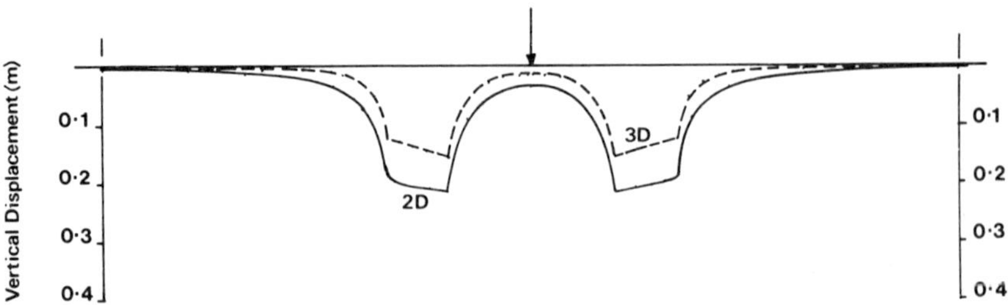

(a) Vertical displacements under gravity load

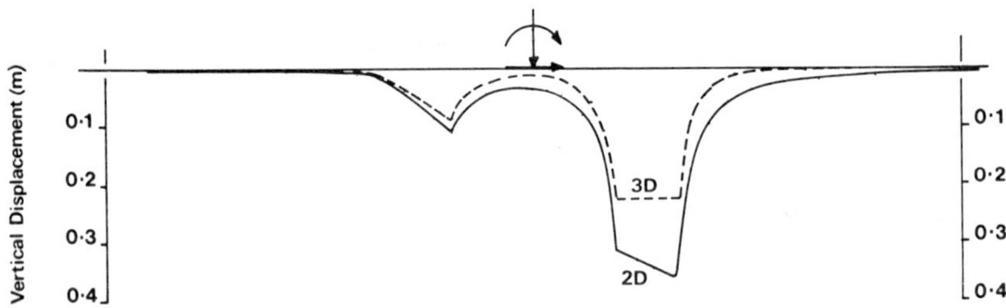

(b) Vertical displacements under gravity and environmental loads

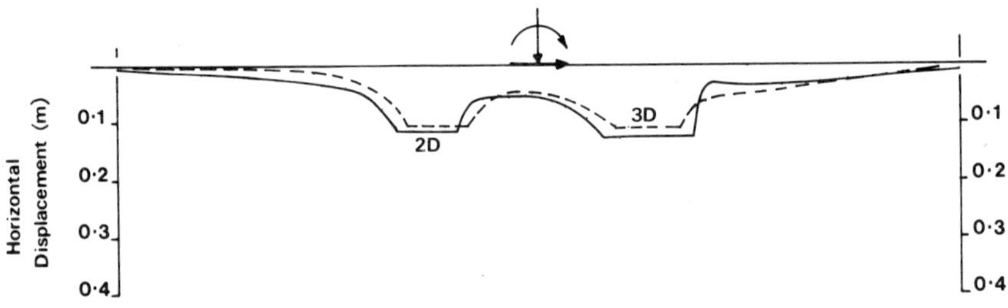

(c) Horizontal displacements under gravity and environmental loads

Figure 14.10 Comparison of deformations under gravity and environmental loading for plane strain and three-dimensional idealizations

For example, for clays, E values ranging from 100 to 400 times the undrained shear strength, Cu, are typically used, while for sands relationships such as those given by Terzaghi and Peck[15] are employed. Thus, caution must be exercised in the interpretation of results based on these properties.

The importance of soil stiffness with respect to calculated natural frequencies has already been demonstrated (Section 14.3.4).

To investigate the importance of soil stiffness on static stresses, three-dimensional studies have been carried out in which the magnitude and distribution of the soil spring stiffnesses in the structural model of a gravity platform were varied, within a range considered to be applicable to the actual site. Figure 14.11 is a sketch of the vertical displacement patterns of the base of the platform under gravity load with three different assemblages of horizontal and vertical springs. The base consists of a relatively stiff central portion surrounded by a more flexible collar, and the dashed line in Figure 14.11 indicates the point at which the stiffness changes. In case 2 the springs are about seven times stiffer than in 1, while case 3 is identical to 2 except that the stiffnesses of the edge springs are reduced to one-half of those in 2. The base displacements (approximated by straight line segments) show the effect of soil spring stiffnesses on overall flexure of the base. Increasing the soil stiffness by a factor of seven reduced the calculated stresses in the base slab by 20 per cent; halving the edge spring stiffnesses as well reduced these stresses by 35 to 45 per cent.

While there may be some doubt as to whether springs can correctly model soil behaviour, even for a hypothetical linearly elastic situation, the above example does illustrate the importance of soil stiffness on the behaviour of the platform. Conversely, it is important also that the stiffness of the base be correctly modelled in the foundation analyses.

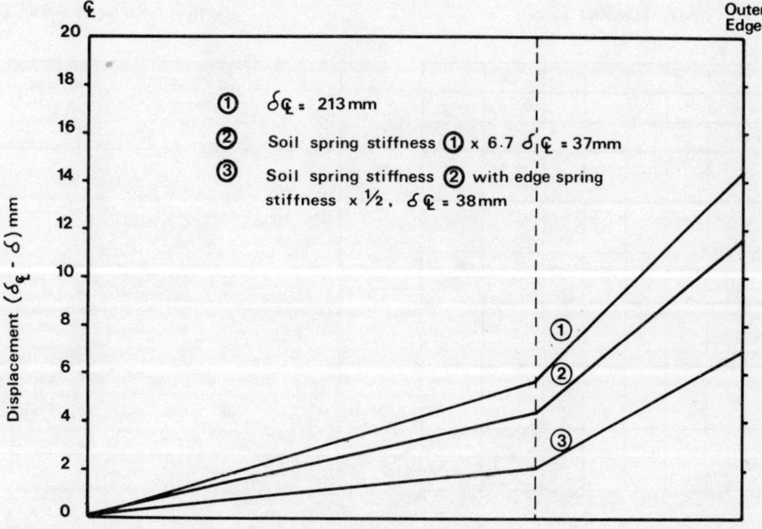

Figure 14.11   Effect of soil spring stiffness on deformation of base slab

14.4 BOUNDARY ELEMENT STUDIES

14.4.1 General

While finite element calculations constitute the major part of the computer studies, boundary element methods are currently being examined as complementary or alternative systems of analysis, with a view to reducing program running costs and enabling larger problems to be tackled.

The boundary element method is a boundary integral equation technique.[16,17,18,19,20,21,22,23] The problem boundaries are discretized so that a singular solution of the governing differential equation can be integrated around them, yielding an appropriate source distribution which will generate the specified boundary conditions. Alternatively, the method may be regarded as a 'super' finite element technique, where each element models a homogeneous zone. These 'finite elements' may be of any shape and can have a very large number of degrees of freedom. Figure 14.12 compares finite element and boundary element idealizations of a problem containing three homogeneous layers.

The incorporation of elasto-plasticity is achieved in a similar manner to that for finite elements (Section 14.3.2). An initial stress algorithm is used to re-distribute internal body forces which equilibrate the stress differences between elastic and plastic solutions of a load increment. The body forces require the introduction to the discretization of 'cells' in zones in which yielding

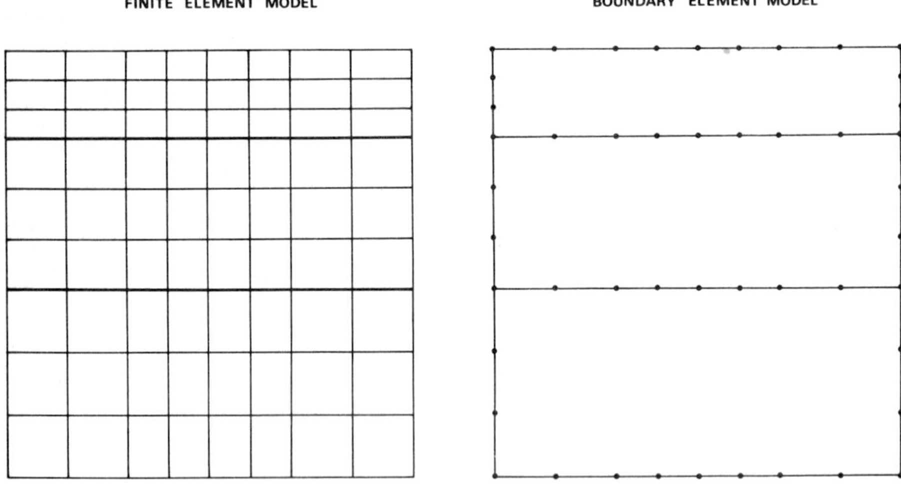

Figure 14.12 Comparison of finite element and boundary element discretizations of layered site

is expected to occur. These do not result, however, in an increase in the order of the final system of equations to be solved.

Some advantages of the boundary element method are summarized below:

(a) The systems of equations formulated are smaller than in corresponding finite element idealizations, except where the volume to surface ratio is low.

(b) A boundary discretization will usually be easier to generate than an internal division.

(c) Stresses and displacements (for example) need only be found where required.

(d) Stresses and displacements are found directly without resorting to numerical differentiation.

(e) There is a continuous variation of stress and displacement within a given elastic zone.

(f) Boundary conditions at 'infinity' can be modelled exactly.

(g) Errors are mainly confined to the boundary since the governing differential equation is solved exactly within the problem region.

(h) In plasticity solutions, the internal body cells need only be introduced in regions of yielding.

The disadvantages of the method include:

(a) Even though the systems of equations may be smaller they are, for single zoned problems at least, fully populated and not banded as in finite element formulations.

(b) If stresses and displacements are required at a large number of points, then the solution may be expensive.

(c) Problems must have a high volume to surface ratio so that the boundary discretization is simpler than a volume division. This means that the boundary element method does not readily lend itself to the analysis of structural members.

A possible answer to the final point might be a hybrid model with finite elements representing the structure and boundary elements the soil.

14.4.2 The boundary element technique

Theory
In this section an indirect formulation for two-dimensional elasticity problems is briefly discussed.

The technique involves using a convenient analytical singular solution for a point load in an infinite plane (see Appendix). The solution of a problem is achieved by distributing multiples of the singular solution so that prescribed boundary conditions are satisfied. Using this technique with the appropriate singular solution it is possible to solve both two- and three-dimensional problems with equal facility. However, due to the logarithmic behaviour of the two-dimensional singular solutions these present a small additional complexity as discussed below.

Two-dimensional elasticity problems (without body force)
In Figure 14.13(b) I^∞ represents an infinite plane within which is inscribed a body R, bounded by a curve S. R and S are identical to the real domain of the problem R^* and its boundary S^* shown in Figure 14.13(a). The infinite plane I^∞ is comprised of isotropic material identical to that of R^*. In the following text R will be called the fictitious body, and R^* the real body.

The curve S is divided into n straight line segments Δs_i which are called boundary elements (see Figure 14.13(c)). If the singular solution is uniformly distributed over each of the Δs_i of strength $\boldsymbol{\phi}_i$ per unit length along S, it is possible to express the boundary displacements $\boldsymbol{u}$ and the tractions $\boldsymbol{p}$ at the

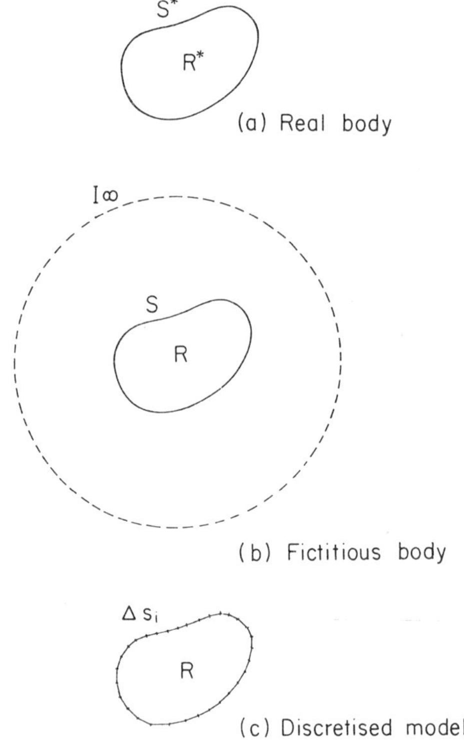

Figure 14.13 Boundary element discretization for two-dimensional elasticity

mid-points of the boundary elements due to the fictitious tractions ϕ_i in the matrix equations[25]

$$u = \mathbf{K}\phi \qquad \text{boundary displacements} \qquad (14.16)$$

$$p = \mathbf{T}\phi \qquad \text{boundary tractions} \qquad (14.17)$$

ϕ is a $2n \times 1$ vector containing the ϕ_i, $= 1, 2 \ldots n$. $\mathbf{K}$ and $\mathbf{T}$ are $2n \times 2n$ matrices which are obtained by integrating the singular solution (see Appendix) around the boundary S as described by Reference 17.

Due to the logarithmic behaviour of the singular solution for displacements (see Appendix), the matrix Equation (14.16) only provides relative values of displacements as this displacement field has an unknown rigid body displacement contribution.[24,25]

This complexity is overcome by applying the constraint

$$\sum_{i=1}^{n} \phi_i = 0,$$

and adding a rigid body displacement contribution c to the displacement field. This results in the matrix Equation (14.16) being replaced by

$$\left[\begin{array}{c|c} \mathbf{K} & \mathbf{I} \\ \hline \mathbf{L} & \mathbf{0} \end{array}\right] \left[\begin{array}{c} \phi \\ \hline c \end{array}\right] = \left[\begin{array}{c} u \\ \hline 0 \end{array}\right] \qquad (14.18)$$

$\mathbf{I}$ is a $2n \times 2$ matrix comprised of 2×2 unit matrices.
$\mathbf{L}$ is a $2 \times 2n$ matrix comprised of 2×2 diagonal matrices where the diagonals contain the lengths of the boundary elements.
$\mathbf{0}$ is a 2×2 null matrix.
0 is a 2×1 null vector.
The problem is solved by making use of the prescribed boundary displacements and tractions to obtain ϕ and c, and hence displacements and stresses anywhere within R^* by numerical integration of the singular solution. The procedure is easily extended to multi-zoned problems as is shown by Butterfield and Tomlin,[25] by simply incorporating the following interface compatibility conditions:

  (i) interface displacements are continuous
 (ii) interface tractions are equal and opposite.

Alternatively multi-zoned problems may be solved by using a stiffness approach as described by Banerjee and Butterfield,[17] where each homogeneous zone may be regarded as a very large 'finite element' with many degrees of freedom. The assembly of these 'finite elements' is indentical to the standard assembly technique in the finite element method.

14.4.3   Comparison of boundary element and finite element solutions

In order to give some indication of differences between boundary element and finite element solutions, a very simple plane strain problem (a flexible footing under vertical load on a uniform elastic soil) has been analysed. It is not intended that this should represent a gravity platform—as noted previously, the stiffness of the structure must be properly modelled and, further, uniform sites are a theoretician's dream. Advantage of symmetry has not been taken as the idealizations were generated for both symmetric and non-symmetric loading.

The finite element model discretizes the problem with 420 soil elements, as shown in Figure 14.14(a). The boundary conditions are that there is no

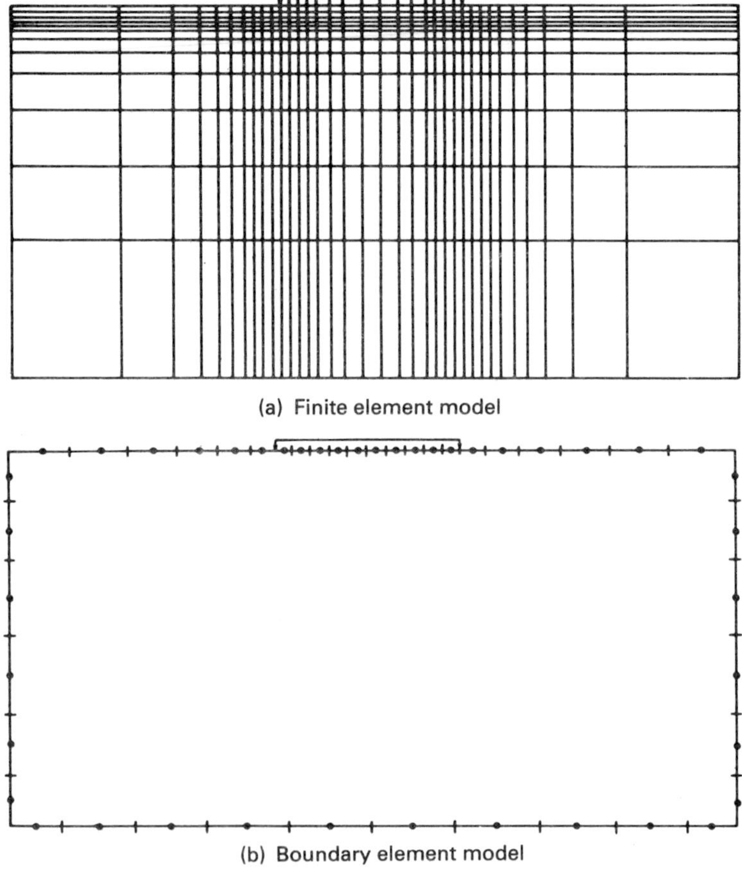

(a) Finite element model

(b) Boundary element model

Figure 14.14   Comparison of finite element and boundary element models
for flexible footing example

movement at the bottom of the soil mesh and that vertical movements only are permitted on the side boundaries. These conditions are satisfied at the nodes and, because of the nature of the linear strain elements, also along the entire length of the boundaries. A uniformly distributed vertical load is applied at seabed level, and the finite elements representing the base have a negligible stiffness.

The boundary element idealization is shown in Figure 14.14(b) and employs 44 elements. Identical boundary conditions are applied, but these are satisfied at the midpoints of the elements only. 'Infinite' boundaries have not been used because, in practice, soil stiffnesses increase with depth and the bottom boundary in the finite element mesh generally has very little influence on the computed static behaviour of the structure.

Computed vertical displacements for the two methods are compared in Figure 14.15 and it can be seen that there is reasonable agreement, although the boundary element results are approximately 7 per cent higher. This is largely explained by differences in fixity conditions. For example, the bottom boundary of the boundary element model is fixed at only ten points, allowing small movements at intermediate points.

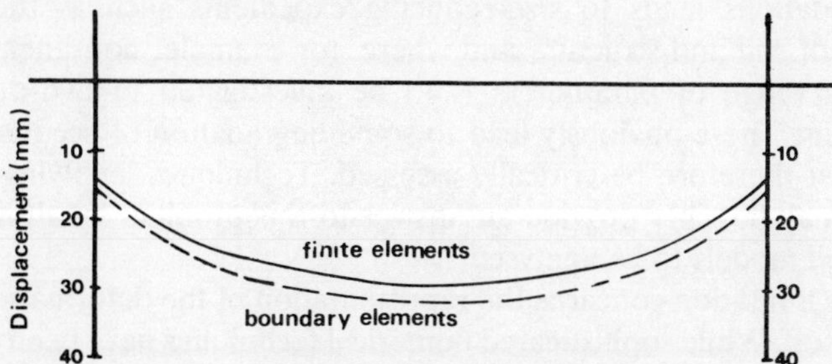

Figure 14.15 Comparison of displacements of flexible base under gravity loading computed using finite element and boundary element idealizations

The solution time for the boundary element method was approximately one-tenth of that for the finite element method. However, it should be noted that the finite element method could solve a layered problem for no increase in cost while the boundary element method would become more expensive, involving modelling the boundary between each stratum.

14.5 PILED STRUCTURES

This chapter has been concerned with some of the numerical techniques used in the assessment of offshore gravity structure foundations. However, the Certifying Authority's rôle also incorporates the assessment of piled structures.

Probably the most important calculation performed is that for axial (compressive and tensile) capacity, since this designs the pile length. Generally, the criteria laid down by API[26] are used.

Structural analyses which include the piles and represent the soil by '$P-y$' springs[27,28] are performed using NASTRAN. Complementary studies of lateral response are made using the boundary element program PGROUP.[29,30]

Calculations of pile drivability and driving stresses are made using a formulation based on the stress wave equation.[31] This is programmed to run on a main frame computer and also, in an interactive mode, on a PDP computer with visual display unit.

14.6 CONCLUSIONS

In this Chapter, numerical techniques used in practice for the analysis of the foundations of offshore structures have been described.

In any assessment based on numerical methods, the limitations of the methods should be borne in mind when interpretating the results. The cost of detailed three-dimensional finite element analyses of gravity structures and their foundations leads to size-reducing expedients such as the separate modelling of soil and structure and where, for example, non-linearity or the effect of a range of parameters is to be investigated to two-dimensional idealizations. These obviously lead to some degradation of accuracy and the results must therefore be critically assessed. Techniques involving boundary elements may possibly provide an answer to this problem by allowing more detailed soil models to be analysed.

Another limitation concerns the representation of the deformation properties of the soil. While sophisticated numerical techniques have been developed to model soil behaviour, the input data is reliant upon empirical relationships and relatively crude offshore sampling and laboratory testing procedures. Nor, unfortunately, do field measurements taken from the structure necessarily yield any further insight into soil behaviour since they tend to give only an overall picture.

The advantages of running centrifugal and 1g model tests to simulate gravity structure foundation behaviour are well known.[7,32] Centrifuge tests are expensive, however, both in cost and time, and where the effect of various parameters is to be studied this can only be achieved economically by numerical means. Moreover, the use of numerical techniques as a means of defining the limitations of model tests is an acknowledged procedure, and is an essential part of design studies where physical models are employed.[3]

Numerical tools such as the finite element method are clearly useful indicators of both small and large scale performance of the foundations of offshore structures. However, neither they nor physical models can point to all the problems which may be encountered. This requires the application of engineering judgement.

Acknowledgements

The Authors wish to thank the Committee of Lloyd's Register of Shipping for permission to publish this text. They also thank Total Oil Marine, Elf and C. G. Doris for their previous permission to publish data relating to the Frigg CDP1 platform.

APPENDIX: THE TWO-DIMENSIONAL POINT LOAD SOLUTION (PLANE STRAIN)

The displacement u_i at a point x_i due to a point load ϕ_j acting at a point y_i is given by

$$u_i = c_1 \left\{ c_2 \delta_{ij} \log r - \frac{\xi_i \xi_j}{r^2} \right\} \phi_j \qquad i, j = 1, 2$$

where

$$\xi_i = y_i - x_i, \quad r^2 = \xi_i \xi_i, \quad c_1 = -\frac{1}{8\pi G(1-\nu)}, \quad c_2 = 3 - 4\nu, \quad \delta_{ij} = \begin{cases} 0 & i \neq j \\ 1 & i = j \end{cases}$$

The strains in tensor form are

$$\varepsilon_{ij} = \frac{-c_1}{2r^2} \left[(c_2 - 1)\{\delta_{ik}\xi_j + \delta_{jk}\xi_i\} + \frac{4\xi_i \xi_j \xi_k}{r^2} - 2\delta_{ij}\xi_k \right] \phi_k \qquad i, j, k = 1, 2, 3$$

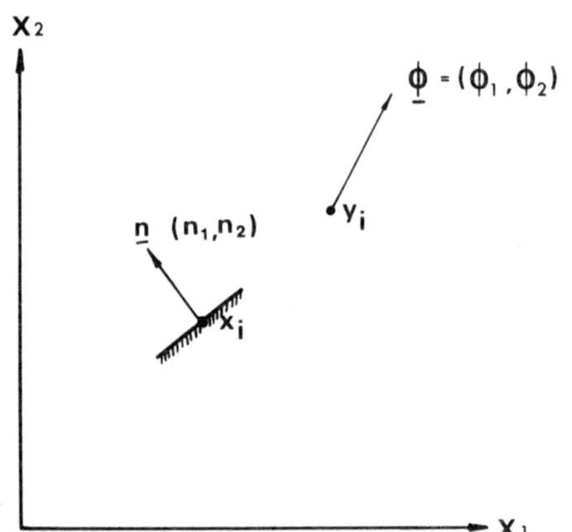

Figure 14.16   2-D Point load solution

The stresses σ_{ij} at the point x_i are calculated via

$$\sigma_{ij} = 2G\left[\varepsilon_{ij} + \frac{\nu}{1-2\nu} \delta_{ij}\varepsilon_{kk} \right] \qquad i = 1, 2, 3$$

with the plane strain condition $\varepsilon_{ij} = 0$, $i = 1, 2, 3$.

Finally the tractions p_i at a point x_i across a line plane with outward normal n_j are given by

$$p_i = \sigma_{ij}n_j$$

N.B. Repeated indices in the above expressions imply a summation with respect to the repeated index. (See Figure 14.15.)

REFERENCES

1. Brinch Hansen, J. (1961). 'A general formula for bearing capacity', Danish Geotechnical Institute, *Bulletin 11*.
2. Hansen, J. Brinch (1970). 'A revised and extended formula for bearing capacity', Danish Geotechnical Institute, *Bulletin 28*.
3. Hobbs, R. (1975). 'Finite element analyses of centrifuged soil slopes', *Ph.D. Thesis*, University of Manchester.
4. Zienkiewicz, O. C. (1971). *The Finite Element Method in Engineering Science*, McGraw-Hill.
5. Smith, I. M. and Hobbs, R. (1974). 'Finite element analysis of centrifuged and built-up slopes', *Geotechnique*, **24:4**, 531–559.
6. Smith, I. M. (1976). 'Aspects of the analysis of gravity offshore structures', *Numerical Methods in Geomechanics, Blacksburg*, **2**, 957–978.
7. George, P. J. and Shaw, P. G. H. (1976). 'Certification of concrete gravity platforms in the North Sea', Offshore Technology Conference, *Paper OTC 2436*, 131–148.
8. Chang, C-Y. and Duncan, J. M. (1970). 'Analysis of soil movement around a deep excavation', *Proc. ASCE*, **96:SM5**, 1655–1681.
9. Smith, I. M. (1976). 'Quasi-static design of foundations—including structure—foundation interaction'; Offshore Soil Mechanics Chapters 15–16, Lloyd's Register of Shipping/Cambridge University.
10. Smith, I. M. and Hobbs, R. (1976). 'Biot analysis of consolidation beneath embankments', *Geotechnique*, **26:1**, 149–171.
11. Hobbs, R. (1976). 'Finite element analysis of consolidation under Sale motorway embankment', *BGS Cooling Prize paper*.
12. Shaw, P. G. H., Coates, A. D., Hobbs, R., and Schumm, W. S. (1977). 'Analytical and field data studies of the dynamic behaviour of gravity offshore structures and foundations', Offshore Technology Conference, *Paper OTC 3008*.
13. Goodman, R. E., Taylor, R. L., and Brekke, T. L. (1968). 'A model for the mechanics of jointed rock', *Proc. ASCE*, **94:SM3**, 637–659.
14. Zienkiewicz, O. C., Valliappan, S., and King, I. P. (1968). 'Stress analysis of rock as a "no tension" material', *Geotechnique*, **18**, 56–66.
15. Terzaghi, K. and Peck, R. B. (1967). *Soil Mechanics in Engineering Practice*, 2nd edn., Wiley, London.
16. Banerjee, P. K. (1976). 'Integral equation methods for analysis of piecewise nonhomogeneous three-dimensional elastic solids of arbitrary shape', *Int. J. Mech. Sci.*, **18**, 293–303.
17. Banerjee, P. K. and Butterfield, R. (1976). 'Boundary element methods in geomechanics', *Numerical Methods in Soil and Rock Mechanics*, Ed. G. Gudehus, Wiley, London, Chapter 16 (in print).
18. Tomlin, G. R. (1972). 'Numerical analysis of continuum problems in zoned anisotropic media', *Ph.D. Thesis*, Southampton University.
19. Cruse, T. A. and Rizzo, F. J. (1975). Editors of the *Proceedings of A.S.M.E. Conference on Boundary Integral Equation Method A.S.M.E.*, New York.
20. Boissenot, J. M., Lachat, J. C., and Watson, J. O. (1974). 'Etude par equations Integrales d'une eprouvette C.T.15', Dept. D.T.E.—CETIM. Senlis, *France Revue de Physique Apliquee*, **9**, Juillet, 611–615.
21. Jaswon, M. A. (1963). 'Integral equation methods in potential theory; I', *Proc. Royal Soc.*, **A275**, V273, 237–246.

22. Watson, J. O. (1972). 'The analysis of 3-D problems of elasticity by integral representation of displacements', *Variational Methods in Eng.*, Vol. II. International Conference, Southampton.
23. Zienkiewicz, O. C. (1975). 'The finite element method and boundary solution procedures as general approximation methods for field problems'. *World Congress on Finite Element Methods in Structural Mechanics*, Bournemouth, Dorset.
24. Butterfield, R. (1971). 'The application of integral equation methods to continuum problems in soil mechanics', *Roscoe Memorial Symposium*, Cambridge, 573–587.
25. Butterfield, R. and Tomlin, G. R. (1972). 'Integral techniques for solving zoned anisotropic continuum problems', International Conference on Variational Methods in Engineering, Southampton University, *Paper 1*, Session, IX.
26. API. (1976). 'Recommended practice for planning, designing and constructing fixed offshore platforms', *API RP2A*, Seventh edn., American Petroleum Institute.
27. Matlock, H. (1970). 'Correlations for design of laterally loaded piles in soft clay', Offshore Technology Conference, *Paper OTC 1204*.
28. Reese, L. C., Cox, W. R., and Koop, F. D. (1975). 'Field testing and analysis of laterally loaded piles in stiff clay', Offshore Technology Conference, *Paper OTC 2312*.
29. Banerjee, P. K. and Driscoll, R. M. (1975). 'PGROUP: A program for the analysis of pile groups of any geometry subjected to horizontal and vertical loads and moments', *Report HECB/B/7*, Department of The Environment.
30. Banerjee, P. K. (1976). 'Three-dimensional analysis of raked pile groups', *Proc. I.C.E.*, 653–671.
31. Smith, E. A. L. (1960). 'Pile-driving analysis by the wave equation', *Proc. ASCE.*, **86**: EM4, 35–61.
32. Rowe, P. W. (1975). 'Displacement and failure modes of model offshore gravity platforms founded on clay', *Proc. Offshore Europe Conference*, Aberdeen, Spearhead Publications.

Chapter 15

Transient Phenomena of Offshore Foundations

I. M. Smith

SUMMARY

Two types of transient problem associated with foundations for offshore structures are described and analysed by finite element idealizations in space. The first of these concerns generation and dissipation of excess porewater pressures which occur beneath gravity structures and close to driven piles. The second concerns dynamic transient oscillations both of gravity structures, which may lead to permanent foundation displacements, and of piles during driving. Numerical problems discussed include mass lumping, explicit and implicit integration techniques, and control of damping. Physical problems discussed include the nature of wave forces, and the behaviour of cyclically loaded soil in element and large model tests.

15.1 INTRODUCTION

Finite element analyses of transient problems in engineering, with which the present paper is concerned, were first used some ten years ago. A transient process may be defined as a time dependent variation in behaviour, of limited duration, spanning between two steady states of much longer duration. One or both of the steady states can be periodic, or in the limit constant, in time (infinite period). The duration of the transient may be of the order of seconds, as during an earthquake, or of hours or days, as during an ocean storm.

Although all transient processes are fundamentally similar, and by suitable substitutions governed by the same types of partial differential equation,[1] it proves to be convenient to discuss two types of problem which arise in offshore engineering practice separately. These are the so-called 'field' problem, which in the present application concerns the variation in time of the excess fluid pressure field in the pores of soil. This process is governed by a differential equation involving the first derivative of the time only. Secondly there is the dynamic oscillation problem, governed by a differential equation which can involve both first and second derivatives of the time.

15.2 TRANSIENT POREPRESSURE ANALYSIS

In the analysis of field problems it is necessary to make a further subdivision, namely into uncoupled and coupled problems. In the former case the field variable is assumed to be independent of any other mechanical properties of the medium. For example in consolidation analysis in soil mechanics, excess porepressures are assumed to be calculable from a knowledge of initial and boundary porepressures alone, together with a coefficient of consolidation. Conversely in a coupled consolidation analysis, excess porepressures are a consequence of equilibrium being enforced between external loads and the solid and fluid phases of the soil, the solid phase usually being approximated by simple elasticity.

Analyses of uncoupled problems have far exceeded those of coupled ones. Visser[2] presented an early but limited finite element formulation of an uncoupled heat flow problem (entirely analogous to consolidation) while the work of Wilson and Nickell[3] contained a more general treatment. Since then, contributions to the study of such transient problems have concentrated on optimum methods for integrating the differential equations, which are often non-linear, with respect to time.

Numerical solution of such transient problems when discretized in space by other methods (usually finite differences) pre-dates the finite element work by some twenty years, for example Crank and Nicolson.[4] Over these years, many devices were developed for improving the stability, accuracy and efficiency of the numerical integration in time. The survey by Mitchell[5] of methods applicable to parabolic equations like the consolidation equation includes explicit, implicit, ADI, LOD, SOR and multi-level methods, all in association with finite difference discretizations in space. Recently, for example Neuman *et al.*,[6] Baker,[7] more of these methods have begun to be used in association with finite element discretizations in space.

The uncoupled, spatially discretized, ordinary differential equations to be solved take the form

$$\mathbf{A}\frac{d\mathbf{p}(t)}{dt} + \mathbf{B}\mathbf{p}(t) = \mathbf{s}(t) \tag{15.1}$$

where $\mathbf{A}$ and $\mathbf{B}$ are symmetric, positive definite matrices because of the parabolic nature of the problem, $\mathbf{p}$ is a vector of excess porewater pressures and $\mathbf{s}$ a vector of inflows or outflows. When finite difference discretizations are used, $\mathbf{A}$ reduces to the unit matrix $\mathbf{I}$ and even when finite element discretizations are used, it may be computationally convenient to 'lump' $\mathbf{A}$ into an equivalent diagonal matrix. The consequences of this will be discussed later.

Since for initial conditions $\mathbf{p}(0) = \mathbf{p}_0$, Equation (15.1) has the solution:

$$\mathbf{p}(t) = \exp\left(-t\mathbf{A}^{-1}\mathbf{B}\right)(\mathbf{p}_0 - \mathbf{B}^{-1}\mathbf{s}) + \mathbf{B}^{-1}\mathbf{s} \tag{15.2}$$

or in incremental form:

$$\mathbf{p}(t+\Delta t) = \exp\left(-\Delta t\,\mathrm{A}^{-1}\mathbf{B}\right)\mathbf{p}(t) - \mathbf{B}^{-1}\mathbf{s} + \mathbf{B}^{-1}\mathbf{s} \tag{15.3}$$

the nature of integrated solutions in time can be shown[1,8] to be governed by the nature of the approximation made to $\exp\left(-\Delta t\,\mathbf{A}^{-1}\mathbf{B}\right)$ and in particular by the approximation made to $\exp\left(-\Delta t\,\lambda_i\right)$ where λ_i are the eigenvalues of the matrix $\mathbf{A}^{-1}\mathbf{B}$.

In general the λ_i may be complex, but in the present instance, for real, symmetric, positive definite $\mathbf{A}$ and $\mathbf{B}$, they are all real. Difficulties in integrating with respect to time stem only from the 'stiffness' of the ordinary differential equations, a measurement of which for an n degree of freedom spatial discretization is the 'stiffness ratio' λ_n/λ_1. If this ratio is small, the integration should present no difficulties, and an Euler explicit method, with timestep limited to a fraction of $2\pi/\lambda_n$ or a Runge–Kutta explicit method with extended stability properties would form the basis of the most efficient algorithms.

Difficulties only arise when the stiffness ratio is large, say 100 or 1000. Under these circumstances the very high frequency components of the solution are usually physically unimportant 'noise', but they control the stability of explicit algorithms, rendering them inefficient. Conversely when implicit integration is resorted to, spurious oscillations or excessive damping are introduced into solutions for economic values of Δt.[1] These questions of stability and accuracy of integration in time will be raised again later, in connection with the more difficult transient dynamic problems. The present purpose is to show how a set of implicit methods due to Nørsett,[9,10] which have previously proved to be economical in the solution of uncoupled consolidation problems,[8] can be extended to solve coupled problems as well.

15.3 COUPLED PROBLEMS

The differential equations of equilibrium and continuity for laminar fluid flow through the pores of a porus elastic solid are due to Biot and were first treated in finite element terms by Sandhu and Wilson.[11] The advantages of such a coupled formulation lie in the ease with which interaction between a consolidating foundation and a flexible superstructure imposing load on that foundation can be analysed. The effects of superstructure flexibility on foundation porepressures have been found to be substantial in practical problems.[12] If an uncoupled analysis is chosen, the initial distribution of excess porepressure has to be estimated independently and the Mandel–Cryer and load redistribution effects are not modelled.

The spatially discretized coupled equations are:

$$\mathbf{Kr}(t) + \mathbf{Cp}(t) = \mathbf{T}(t) \tag{15.3a}$$

$$\mathbf{C}^T \frac{\mathrm{d}\mathbf{r}(t)}{\mathrm{d}t} - \mathbf{P}\mathbf{p}(t) = \mathbf{0} \tag{15.3b}$$

where $\mathbf{r}$ are the solid skeleton displacements, $\mathbf{p}$ the excess porewater pressures, $\mathbf{T}$ the external loads and there is no impressed inflow or outflow of fluid. Matrices $\mathbf{K}$, $\mathbf{P}$ and $\mathbf{C}$ are the system elastic stiffness, laminar fluid stiffness and connection components respectively, and for the present are assumed to be constant.

The simplest way of integrating these equations in time is to use an 'α method' of linear interpolation on all variables between times 0 and 1. For example,

$$\mathbf{r}(t) \simeq \alpha \mathbf{r}_1 + (1-\alpha)\mathbf{r}_0 \tag{15.4}$$

and so on where $\alpha = 0$ is an explicit, forward-looking method, $\alpha = 1$ is implicit or backward-looking, while $\alpha = \frac{1}{2}$ is centred (corresponding to the method of Reference 4 for simple uncoupled parabolic equations).

The resulting algorithm equivalent to Equations (15.3) for $\mathbf{T}$ constant is:

$$\alpha \mathbf{K}\mathbf{r}_1 + \alpha \mathbf{C}\mathbf{p}_1 = (\alpha - 1)\mathbf{K}\mathbf{r}_0 + (\alpha - 1)\mathbf{C}\mathbf{p}_0 + \mathbf{T} \tag{15.5a}$$

$$\alpha \mathbf{C}^T \mathbf{r}_1 + \alpha^2 \Delta t\, \mathbf{P}\mathbf{p}_1 = \alpha \mathbf{C}^T \mathbf{r}_0 - \alpha(\alpha - 1)\, \Delta t\, \mathbf{P}\mathbf{p}_0 \tag{15.5b}$$

or

$$\begin{bmatrix} \alpha \mathbf{K} & \alpha \mathbf{C} \\ \alpha \mathbf{C}^T & \alpha^2 \Delta t\, \mathbf{P} \end{bmatrix} \begin{Bmatrix} \mathbf{r}_1 \\ \mathbf{p}_1 \end{Bmatrix} = \begin{bmatrix} (\alpha - 1)\mathbf{K} & (\alpha - 1)\mathbf{C} \\ \alpha \mathbf{C}^T & -\alpha(\alpha - 1)\, \Delta t\, \mathbf{P} \end{bmatrix} \begin{Bmatrix} \mathbf{r}_0 \\ \mathbf{p}_0 \end{Bmatrix} + \begin{Bmatrix} \mathbf{T} \\ \mathbf{0} \end{Bmatrix}$$

$$\tag{15.6}$$

Note that the matrix on the left of these equations is symmetric but that on the right is not.

An alternative algorithm based on the same principles can be devised by combining Equations (15.3) in terms of a single variable, say $\mathbf{r}$. If $\mathbf{P}$ in Equation (15.3b) is split by Cholesky's method,

$$\mathbf{C}^T \frac{\mathrm{d}\mathbf{r}}{\mathrm{d}t} - \mathbf{L}\mathbf{L}^T \mathbf{p} = \mathbf{0} \tag{15.7}$$

Letting

$$\mathbf{L}^T \mathbf{p} = \mathbf{v} \quad \text{or} \quad \mathbf{p} = \mathbf{L}^{-T} \mathbf{v}$$

we have

$$\mathbf{C}\mathbf{L}^{-T} \mathbf{L}^{-1} \mathbf{C}^T \frac{\mathrm{d}\mathbf{r}}{\mathrm{d}t} = -\mathbf{K}\mathbf{r} + \mathbf{T} \tag{15.8}$$

If

$$\mathbf{CL}^{-T} = \mathbf{E} \quad \text{or} \quad \mathbf{E}^{T} = \mathbf{L}^{-1}\mathbf{C}^{T}$$

$$\mathbf{EE}^{T}\frac{\mathrm{d}\mathbf{r}}{\mathrm{d}t} = -\mathbf{Kr} + \mathbf{T} \tag{15.9}$$

demonstrating the parabolic nature of the combined equation in $\mathbf{r}$. Although such a method has some attractions since fewer unknowns need to be evaluated at each timestep, $\mathbf{EE}^{T}$ is in general a full matrix and the cost of forming and reducing it is high in linear problems and prohibitive in non-linear ones. Interest in the α-algorithm:

$$[\mathbf{EE}^{T} + \alpha\,\Delta t\,\mathbf{K}]\mathbf{r}_1 = [\mathbf{EE}^{T} + (\alpha-1)\,\Delta t\,\mathbf{K}]\mathbf{r}_0 + \Delta t\,\mathbf{T} \tag{15.10}$$

rests solely in the fact that it leads to precisely the same results as Equations (15.6), and it can be shown algebraically that it should. This encourages the belief that more efficient algorithms of the Nørsett type, whose formulation is appropriate to Equations (15.9), can be devised in terms of the uncombined Equations (15.3), and this turns out to be the case.

For example, a Nørsett method with good properties[1,8,13] is the third order L-stable method which, when applied to Equations (15.9) gives the algorithm:

$$\left[\mathbf{CP}^{-1}\mathbf{C}^{T} + \frac{\Delta t}{\gamma}\mathbf{K}\right]\mathbf{v}_1 = \frac{\Delta t}{\gamma}\mathbf{Kr}_0 - \frac{\Delta t}{\gamma}\mathbf{T} \tag{15.11a}$$

$$\left[\mathbf{CP}^{-1}\mathbf{C}^{T} + \frac{\Delta t}{\gamma}\mathbf{K}\right]\mathbf{v}_2 = \frac{\Delta t}{\gamma}\mathbf{Kv}_1 \tag{15.11b}$$

$$\left[\mathbf{CP}^{-1}\mathbf{C}^{T} + \frac{\Delta t}{\gamma}\mathbf{K}\right]\mathbf{r}_1 = \mathbf{CP}^{-1}\mathbf{C}^{T}(\mathbf{r}_0 + L_1\mathbf{v}_1 + L_2\mathbf{v}_2) + \frac{\Delta t}{\gamma}\mathbf{T} \tag{15.11c}$$

where $\mathbf{v}_1$ and $\mathbf{v}_2$ are intermediate dummy vectors and L_1, L_2 and γ are pure numbers based on the roots of Laguerre polynomials.[10]

After some algebra, the equivalent three-stage algorithm in terms of the uncombined variables $\mathbf{r}$ and $\mathbf{p}$ can be shown to be:

$$\begin{bmatrix} \mathbf{K} & \mathbf{C} \\ \mathbf{C}^{T} & -(\Delta t/\gamma)\mathbf{P} \end{bmatrix}\begin{Bmatrix} \mathbf{v}_1 \\ \mathbf{w}_1 \end{Bmatrix} = \begin{bmatrix} \mathbf{0} & \mathbf{0} \\ \mathbf{0} & -(\Delta t/\gamma)\mathbf{P} \end{bmatrix}\begin{Bmatrix} \mathbf{r}_0 \\ \mathbf{p}_0 \end{Bmatrix} \tag{15.12a}$$

$$\begin{bmatrix} \mathbf{K} & \mathbf{C} \\ \mathbf{C}^{T} & -(\Delta t/\gamma)\mathbf{P} \end{bmatrix}\begin{Bmatrix} \mathbf{v}_2 \\ \mathbf{w}_2 \end{Bmatrix} = \begin{bmatrix} \mathbf{0} & \mathbf{0} \\ \mathbf{0} & -(\Delta t/\gamma)\mathbf{P} \end{bmatrix}\begin{Bmatrix} \mathbf{v}_1 \\ \mathbf{w}_1 \end{Bmatrix} \tag{15.12b}$$

$$\begin{bmatrix} \mathbf{K} & \mathbf{C} \\ \mathbf{C}^{T} - (\Delta t/\gamma)\mathbf{P} \end{bmatrix}\begin{Bmatrix} \mathbf{r}_1 \\ \mathbf{p}_1 \end{Bmatrix} = \begin{bmatrix} \mathbf{0} & \mathbf{0} \\ \mathbf{C}^{T} & \mathbf{0} \end{bmatrix}\begin{Bmatrix} \mathbf{r}_0 \\ \mathbf{p}_0 \end{Bmatrix} + \begin{bmatrix} \mathbf{0} & \mathbf{0} \\ L_1\mathbf{C}^{T} & \mathbf{0} \end{bmatrix}\begin{Bmatrix} \mathbf{v}_1 \\ \mathbf{w}_1 \end{Bmatrix}$$

$$+ \begin{bmatrix} \mathbf{0} & \mathbf{0} \\ L_2\mathbf{C}^{T} & \mathbf{0} \end{bmatrix}\begin{Bmatrix} \mathbf{v}_2 \\ \mathbf{w}_2 \end{Bmatrix} + \begin{Bmatrix} \mathbf{T} \\ \mathbf{0} \end{Bmatrix} \tag{15.12c}$$

Note the constant symmetric left hand side matrix and the sparseness of the right hand side matrices in this algorithm. It is much quicker and more economical of storage than Equations (15.11) but gives precisely the same results.

In the next section, coupled transient problems of interest in offshore engineering will be analysed by the above methods.

15.4 RESULTS OF COUPLED TRANSIENT ANALYSES

Since it will be shown that spatial oscillations can be introduced by the presence of shear strains and near-incompressibility, consider first the simplest one-dimensional consolidation problem for a step loading. Computed and analytic mid-plane porepressures are shown in Figure 15.1, from which it can be seen that the completely centred α-algorithm gives violently oscillatory results, for the reasons discussed in Reference 8, although an averaging technique would clearly give a more accurate solution. There is some damping inaccuracy introduced by the $\alpha = 1$ algorithm, but it is without doubt the optimum α in this case. An alternative course[12] is to use $\alpha = \frac{1}{2}$ in Equation (15.3b) while writing Equation (15.3a) at the new time level only ($\alpha = 1$). This method leads to very similar results to the consistent $\alpha = 1$ method. The Nørsett result is, to plotting accuracy, indistinguishable from the analytic one, and this method will be used in subsequent solutions.

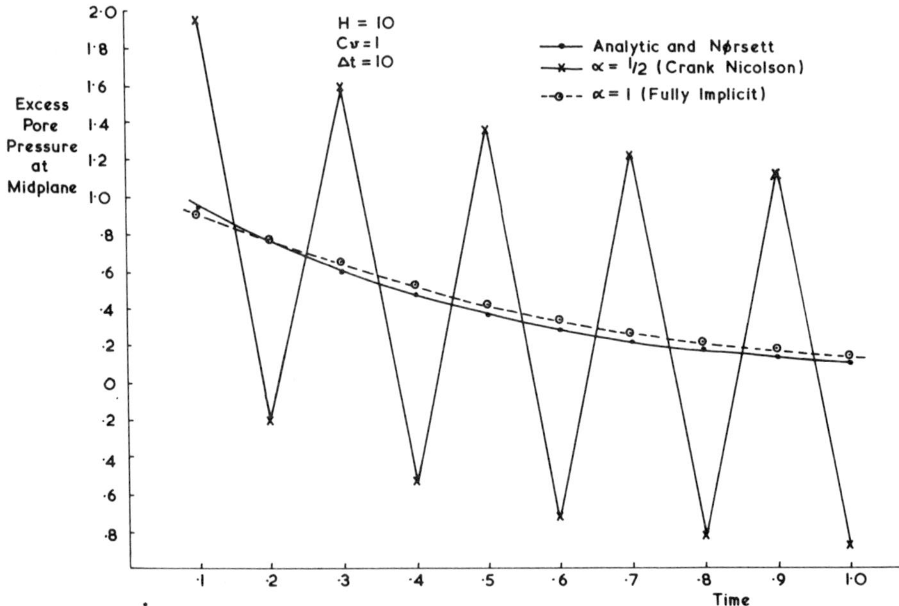

Figure 15.1　Effect of integration method on computed porepressures

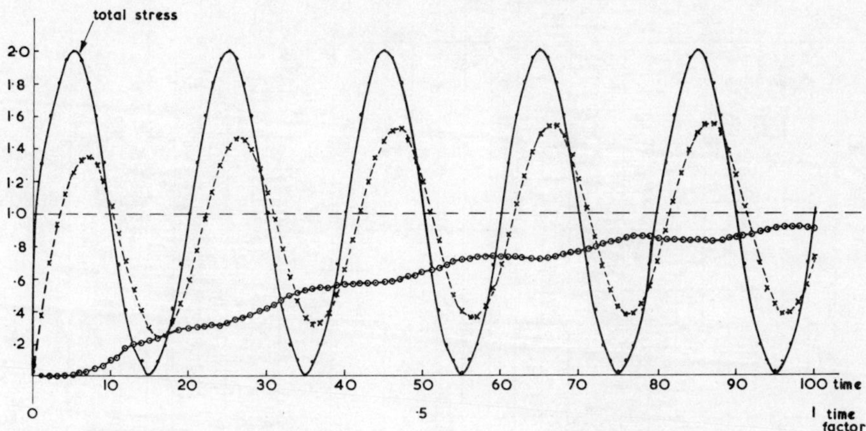

Figure 15.2   Effective stress variations in cyclically loaded foundation

The nature of consolidation under alternating loads has previously been discussed in Reference 14. The critical ratio governing porefluid behaviour is that between the frequency of loading, ω, and the drainage rate of the ground, c_v. Analytically this appears as the ratio $\sqrt{(\omega/2c_v)} = p$. For one-dimensional flow, the amplitude of steady periodic porepressures should decay according to $\exp(-zp)$ and lag behind the imposed wave according to zp. Results of a finite element computation illustrating these features are shown in Figure 15.2. The foundation layer analysed has $H = 10$, $c_v = 1$ and drains freely from the surface, on which the external pressure $1·0 + \sin(\pi t/10)$ acts. At $z = 1·25$ the phase lag is some 30° with amplitude decay 40 per cent while at $z = 10·0$ the phase lag is some 230° with amplitude decay 95 per cent. Effective stresses rather than porepressures are plotted.

Actual wave loadings are in fact essentially random as shown by the typical trace in Figure (15.3). The variations in wave spectra during the passage of a storm are shown in Figure (15.4), indicating narrow band characteristics at the height of the storm, and broader band characteristics during build up and decay. The consequences of phase lag and depth decay for this range of ω are clearly that an extremely complicated variation of excess porewater pressures can be anticipated in the uppermost layers of the foundations of gravity platforms. The writer has previously suggested[14] that this may be a cause of near-surface weakening of clays in physical model tests.

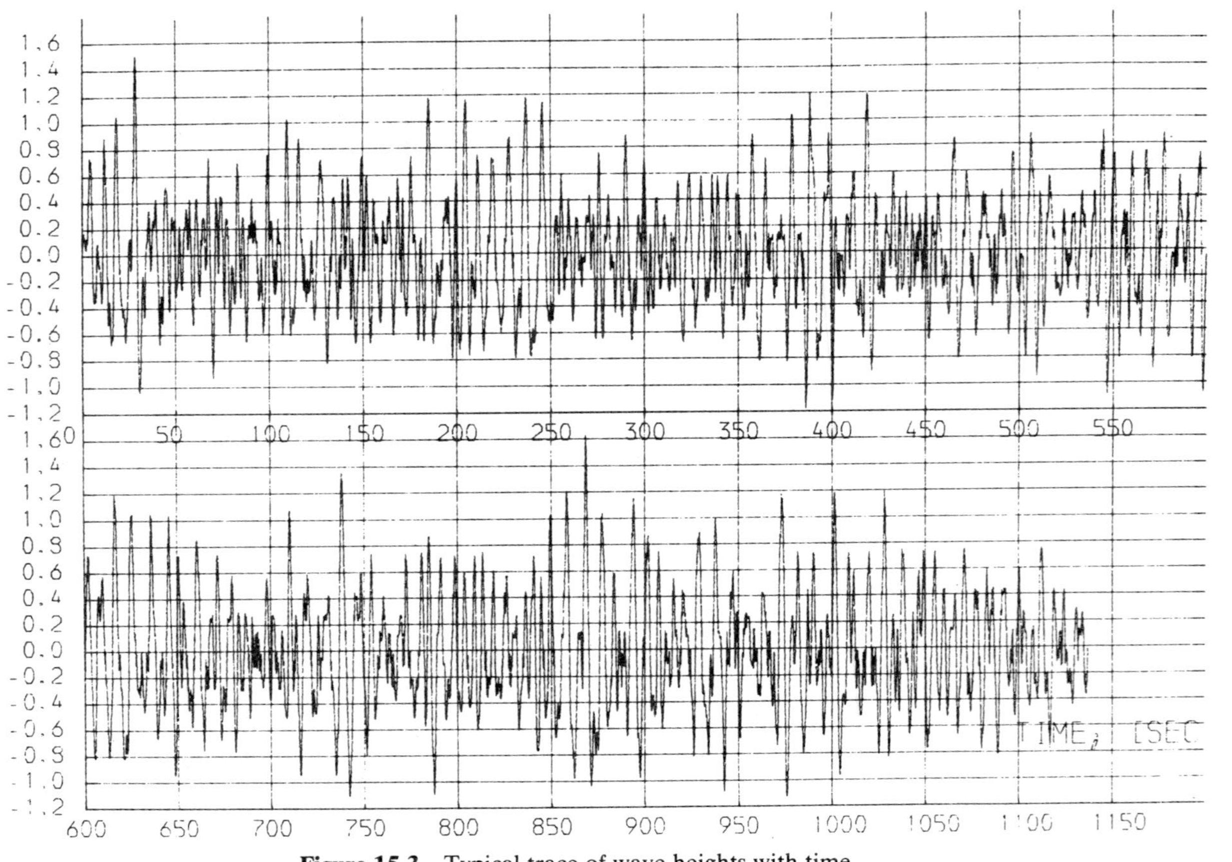

Figure 15.3   Typical trace of wave heights with time

The above treatment deals with vertical migration of water pressure in the sea bed. A second phenomenon is horizontal migration of porepressure, usually referred to in soil mechanics as 'porepressure spread'. An analysis of this process is given in Figure (15.5), and in this case the foundation surface boundary conditions are those appropriate to wave pressure loading, namely total sinusoidal pressure equals porewater pressure at the sea bed. The previous boundary conditions were more appropriate to a freely draining gravity platform base.

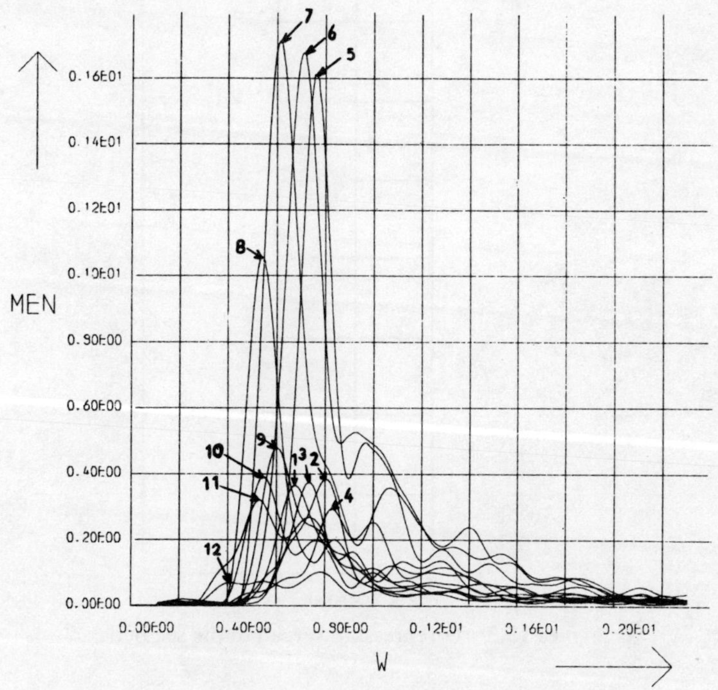

Figure 15.4 Wave height spectra during passage of a storm

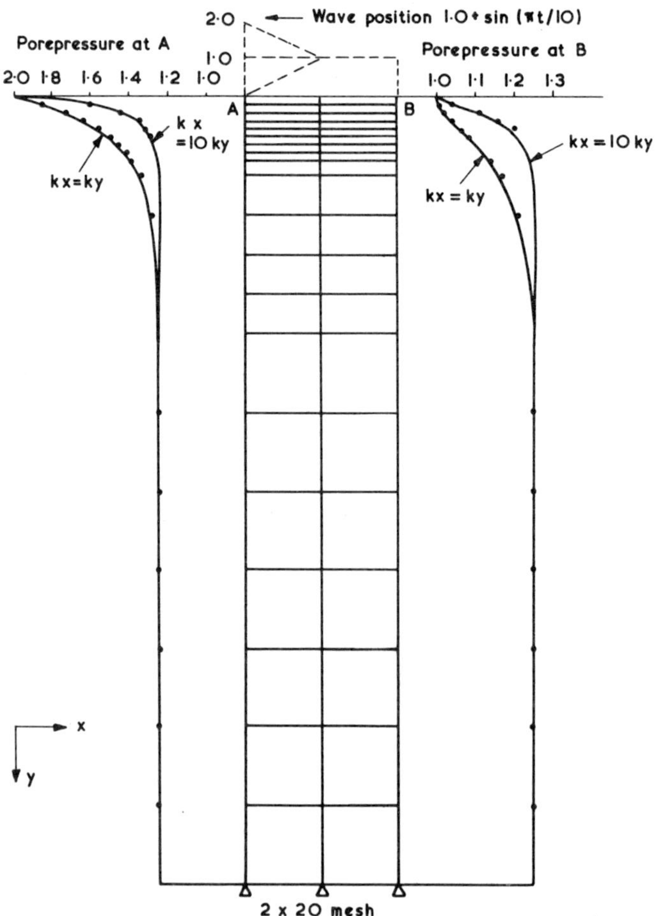

Figure 15.5 Porepressure spread in the sea bed

Again the rapid decay of excess porepressures with depth is apparent. Porepressure spread is much accentuated by high horizontal permeability.

Although the present paper is concerned with transient problems and with their integration with respect to time it should never be forgotten that serious spatial oscillations can arise in solutions to coupled problems at small times due to incompressibility restrictions.[15] Typical results for a suddenly loaded foundation (plane strain conditions) are shown in Figure (15.6). At small times, nodal porepressures are badly in error but averages (or values at 'reduced' integration points) are much more reasonable. These difficulties are similar to those encountered in other types of analysis where perfect conservation (for example

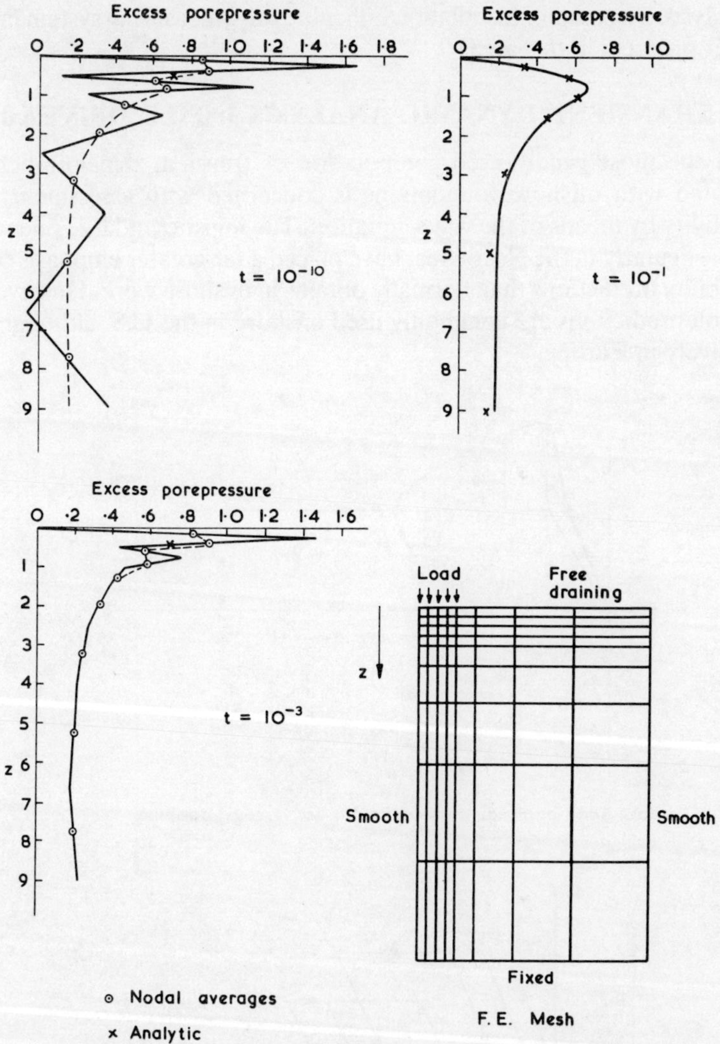

Figure 15.6 Porepressures in suddenly loaded foundation (plane strain case)

of mass in convective problems[13]) is sought. As time proceeds the computed results come into much better agreement with the truth.

This section of the paper has dealt with transient porepressure prediction. The present state of the art seems to be that computational capabilities exceed those of laboratory researchers or field engineers to produce appropriate data for analysis. Piezometer installation difficulties and measuring system lag times hamper progress in this area.

15.5 TRANSIENT DYNAMIC ANALYSIS I: PILE DRIVEABILITY

By far the most widely used computation of transient dynamic behaviour associated with offshore foundations is concerned with assessment of pile driveability by means of the wave equation. The logistics of large pile installation, particularly in the North Sea, have placed a far greater emphasis on good driveability predictions than normally obtains in onshore work. However wave equation predictions are commonly used onshore in the U.S. although not so extensively in Europe.

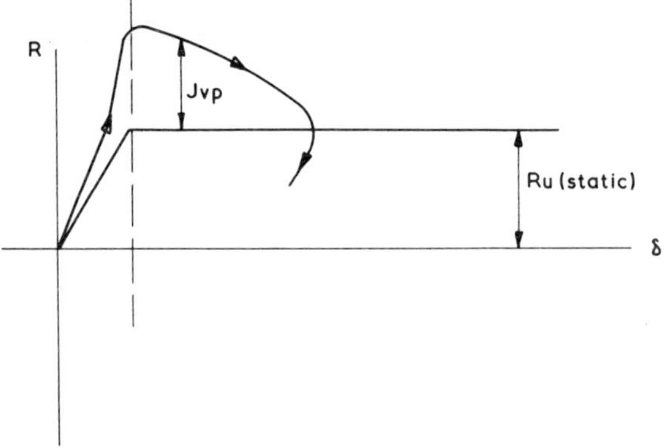

(a) Additional soil resistance due to viscous damping.

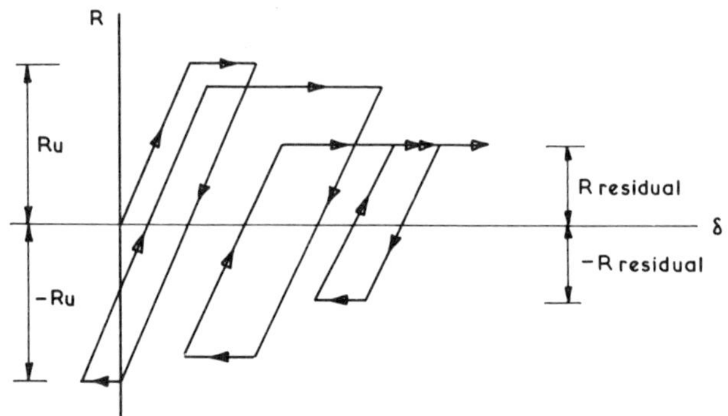

(b) Weakening due to a residual strength with cycling.

Figure 15.7 Soil spring properties used in wave equation solutions

In the 1930's the wave propagation nature of the pile driving process was first recognized but it was not until the advent of computers in the early 1950's that realistic calculations could be attempted. The pioneer in this work was E. A. L. Smith[16] and through his efforts foundation engineers were the first to solve transient wave propagation problems incorporating elasto-plastic material properties. The method employed finite difference approximation in (one dimensional) space, lumped masses and explicit integration in time, still popular choices in many wave propagation codes twenty five years on. Soil resistance was approximated by massless springs with the elasto-plastic force-deformation relationships shown in Figure 15.7(a). Viscous resistance, proportional to pile velocity relative to the soil, v_p, was incorporated through the empirical parameter J. If this model is used in subsequent transient analysis of the *in situ* pile, the complementary effect of weakening with cycling should be remembered, Figure 15.7(b).

The writer[17] has analysed pile driveability by alternative numerical methods employing finite element approximation in space, distributed masses and implicit integration in time, all in contrast with Smith's original techniques. The typical element arrangement is shown in Figure 15.8. It was not anticipated that this would lead to any radical change in established practice for one

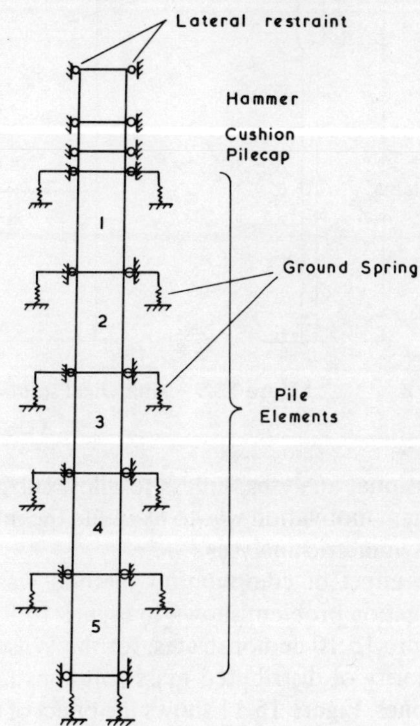

Figure 15.8 Finite element idealization of pile and soil

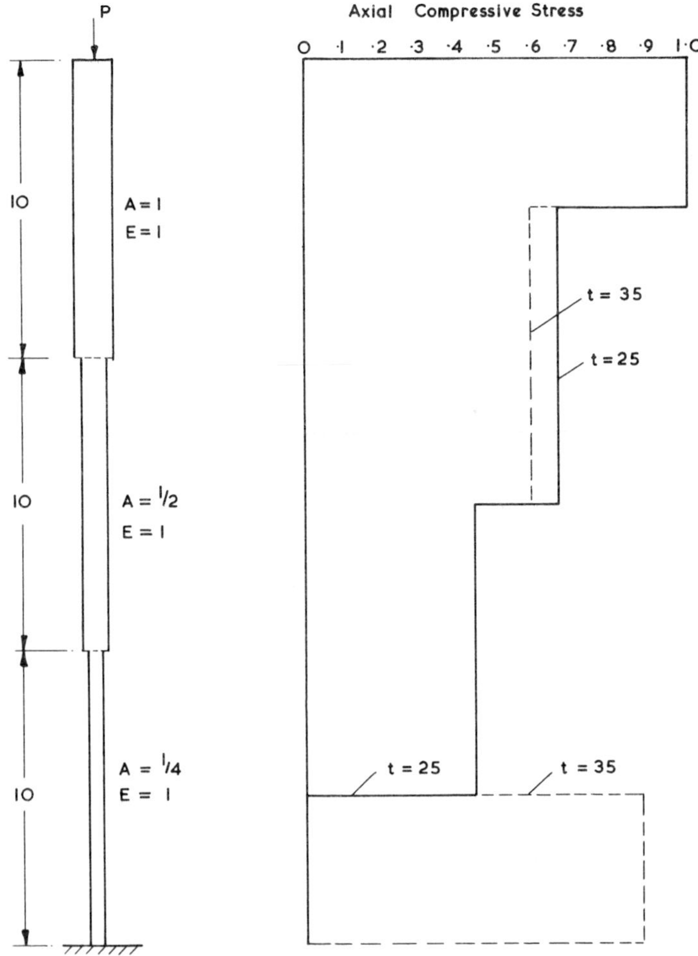

Figure 15.9   Analytical solutions for a stepped pile

dimensional analyses, unless to allow substantial relaxation of timestep size. The main motivation was to examine the influence of the Poisson's ratio effect in axisymmetric analyses.

The effect of computation method was first studied for the axial wave propagation problem shown in Figure 15.9, for a suddenly applied axial force *P*. Figure 15.10 demonstrates, for the Wilson implicit integration method, the superiority of distributed mass solutions for stresses compared with lumped mass ones. Figure 15.11 shows the effect of doubling the timestep while Figure

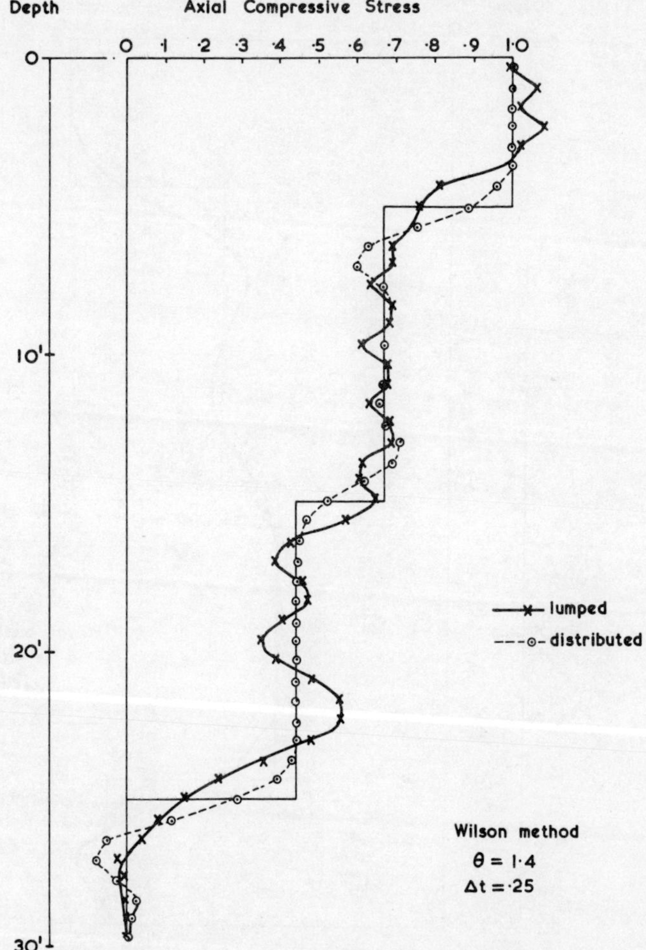

Figure 15.10 Comparison of lumped and distributed mass solutions, $t = 25$

15.12 compares results from explicit and implicit calculation methods. The explicit results are somewhat inferior.

When the various methods are applied to the pile driveability problem given in Smith's original publication[16] the results shown in Table 15.1 are obtained. It is clear that the computation method employed, with the exception of the Wilson method, does not affect the driveability calculation very seriously. However other aspects of the solution, such as velocities and accelerations in the pile during driving, which would be recorded in the course of a field test on a well instrumented pile, are quite strongly affected by algorithm choice.

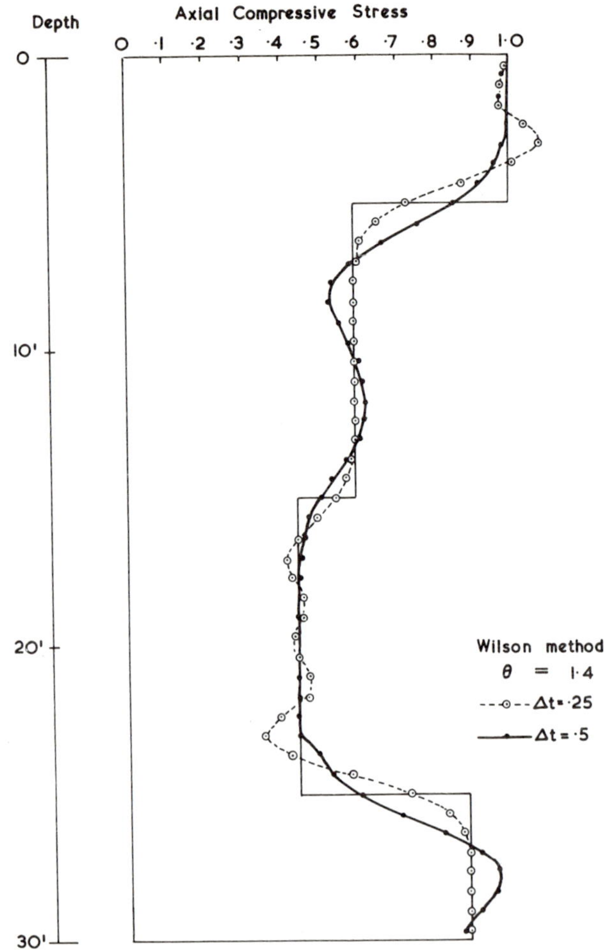

Figure 15.11 Effect of timestep variations, $t = 35$

Table 15.1 Parametric study of pile driveability

Method	Max. tip displacement	Time to reach		Tip velocity	
		Quake	Full pen.	Maximum	At quake
Explicit[16]	0·303	0·0080	0·0140	79·9	79·9
Newmark, L	0·328	0·0080	0·0115	120·1	111·1
Newmark, D	0·291	0·0080	0·0098	122·8	122·4
Wilson, L	0·207	0·0082	0·0112	68·2	68·2
Wilson, D	0·221	0·0082	0·0110	69·3	69·0
Nørsett, L	0·360	0·0079	0·0126	115·3	100·0
Nørsett, D	0·347	0·0077	0·0118	115·7	103·2

L = lumped mass D = distributed mass.

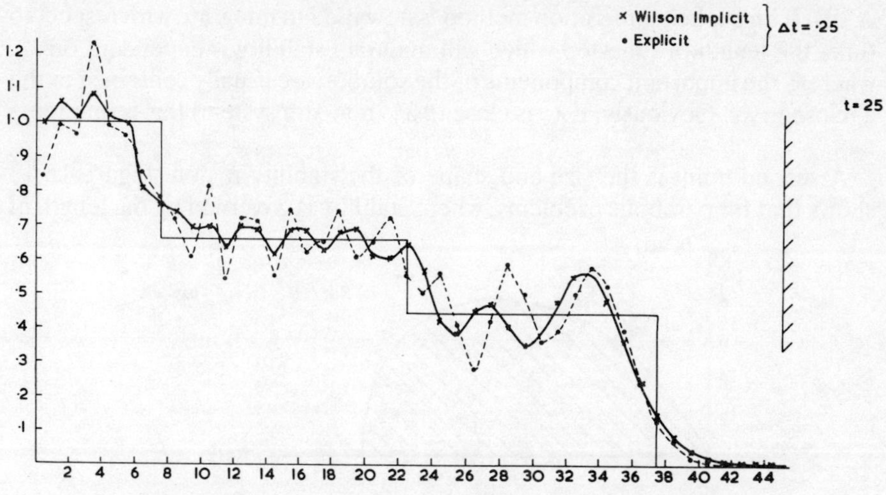

Figure 15.12 Comparison of explicit and implicit solutions

In the next section the mathematical properties of solutions will be briefly discussed. Thereafter the question of algorithm performance in general non-linear transient dynamic analyses will be re-examined.

15.6 STIFFNESS AND STABILITY OF NUMERICAL SOLUTIONS

In the previous section contrasting solution methods were used for the partial differential equation governing a transient dynamic foundation problem. These contrasts reflect three current controversies associated with solution of equations of this kind. The controversies are:

1. Should explicit or implicit methods be used for advancing the solutions in time?

2. Should the system mass be lumped at the nodal points or not? (This is associated with 1 because the full advantages of explicit methods cannot be realized unless lumped mass is assumed.)

3. Is it better to employ integration methods exhibiting underdamping or overdamping?

The reason that controversy has arisen is that the answers to these questions depend upon the type of problem being solved, whereas authors tend to make generalized statements on the basis of experience with a particular class of problem. This can be illustrated by reference to two mathematical properties of spatially discretized partial differential equations, namely their *stiffness ratio* and the *region of stability* of the solution algorithm in the complex plane.

Stiffness ratio has already been defined in relation to transient parabolic problems, Equation (15.1), as λ_n/λ_1 where the λ_i are the real eigenvalues of

$A^{-1}B$. If an explicit integration method is now used to integrate with respect to time, the length of timestep which will maintain stability is dependent on λ_n, whereas the important components of the solution are usually contained in the λ_i close to λ_1. Obviously, if λ_n is close to λ_1 (non-stiff system) this is unimportant.

A second point is the size and shape of the stability region. Figure 15.13 shows that for parabolic problems, where stability is governed by the length of

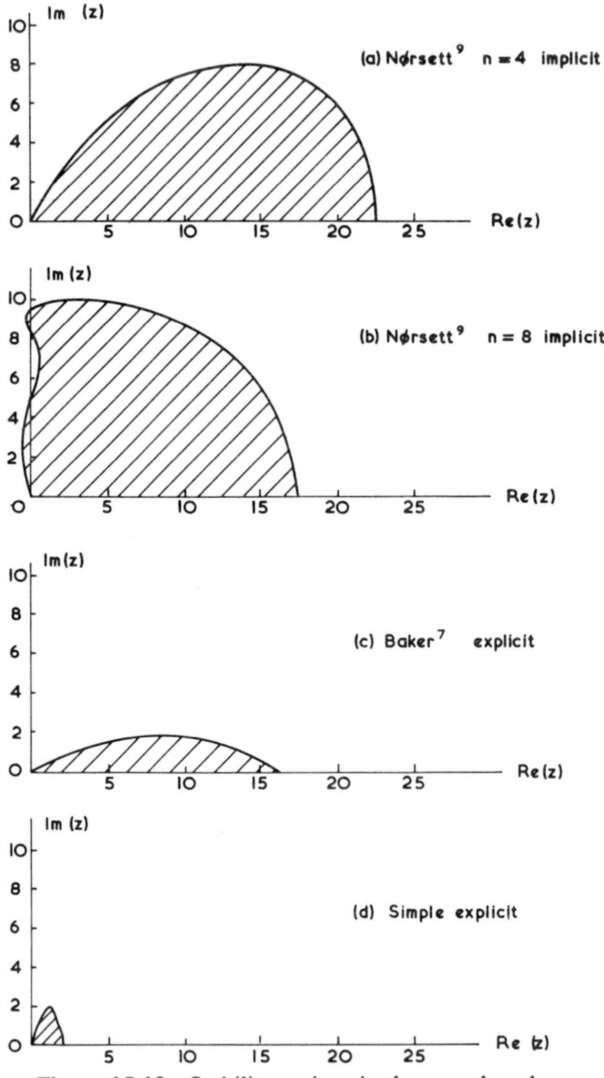

Figure 15.13   Stability regions in the complex plane

the intercept on the real axis, the explicit Runge–Kutta method advocated by Baker[7] has eight times the stability intercept of the simple (forward Euler) explicit method. Some conditionally stable Nørsett methods[9,10] are also shown for comparison purposes. Unconditionally stable methods like the implicit Crank–Nicolson procedure have an infinite intercept on the real axis.

Consider now, however, cases in which the λ_i are complex, and in the limit purely imaginary. Clearly the method represented by Figure 15.13(b) has a significant range of stability close to the imaginary axis, whereas the method used by Baker, Figure 15.13(c), has none at all. Both the hyperbolic transport problem

$$-u_x \frac{\partial c}{\partial x} = \frac{\partial c}{\partial t} \tag{15.13}$$

and free oscillation problems discretize into a form of Equation (15.1) such that the eigenvalues of $\mathbf{A}^{-1}\mathbf{B}$ are purely imaginary. Clearly the more expensive Runge–Kutta method would not preserve any advantage in these cases.

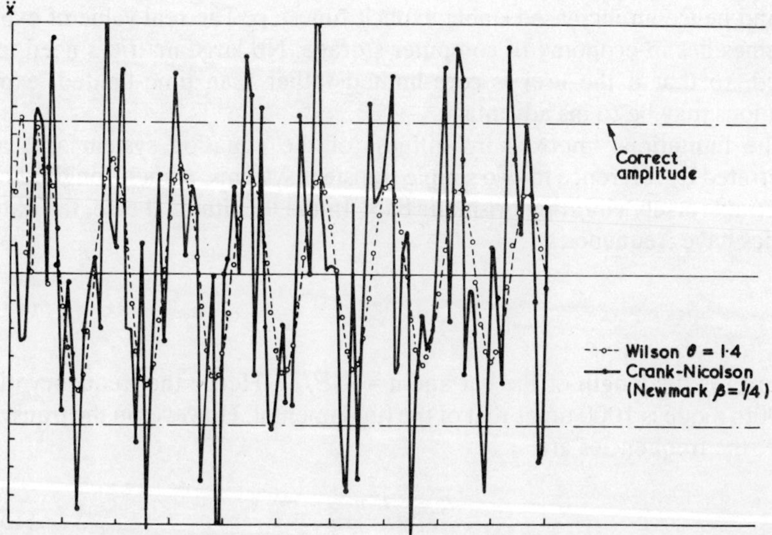

Figure 15.14 Inaccuracies introduced by solution algorithms: under-damping and overdamping

Difficulties associated with the stiffness ratio $|\lambda_n|/|\lambda_i|$ are as before, but the essential absence of stability in many explicit algorithms can now lead to very minute timesteps indeed being necessary. Components will in fact be amplified but presumably within acceptable limits for a transient of limited duration. Alternatively, artificial damping is introduced. The implicit methods described in this paper are stable over the complete half plane and these particular difficulties do not arise in their case.

The concept of stability regions is of particular value in transient diffusion-convection problems[13] after the range of eigenvalues of $\mathbf{A}^{-1}\mathbf{B}$ in the complex plane has been determined. The essential point is that if a method with limited stability is used, it should be chosen with particular reference to the type of problem being solved. For example some explicit Runge–Kutta methods do have a significant stability region along the imaginary axis. However it should also be remembered that additional function evaluations increase the cost per timestep despite the extra length of step. No one method is likely to be optimum for all types of problem.

Explicit algorithms are made to be competitive in computer time with implicit ones by means of various devices, one of which is nodal mass lumping. It was originally argued,[18] and the belief is still current,[19] that this is actually *beneficial* in propagation problems. There are no theoretical grounds for this assertion and indeed it is disproved by the results in Figures 15.10 and 15.12. An apparent benefit does arise in that the lumping leads to an underestimate of λ_n and hence an increased stable explicit timestep. The real value of explicit schemes lies in economy of computer storage. No large matrices need to be saved, so that if the user is core-limited rather than time-limited, explicit solutions may be to his advantage.

The limitations imposed by stiffness of the equation system are easily illustrated by reference to two simple transient systems, namely longitudinally and transversely vibrating prismatic bars. In the longitudinal case, the natural modes have frequencies

$$\lambda_i = \frac{i\pi a}{l} \tag{15.14}$$

where l is the length of the bar and $a = \sqrt{(E/\rho)}$. Hence the frequency of the 1000th mode is 1000 times that of the fundamental. However in the transverse case, the frequencies are

$$\lambda_i = \frac{i^2 \pi^2 a}{l^2} \tag{15.15}$$

so that the frequency of the 1000th mode is 10^6 times that of the fundamental. In general terms this means that in a 1000 degree of freedom discretization, stable explicit timesteps will be orders of magnitude smaller for transverse, as opposed to longitudinal, vibration problems. It is significant that much of the evidence for competitiveness of explicit algorithms is derived from problems concerned with compressional wave transmission. Of course these difficulties tend not to arise at all if one is content with 10 degree of freedom idealizations.

Lest it should seem that implicit methods, with their usual unconditional stability, have all the advantages except increased storage requirements, consider the third current controversy regarding underdamped versus over-damped solutions. The roots of this difficulty lie also in the stiffness of the discretized equations when the solution is oscillatory.[1] At least when explicit methods are used, if a solution is found at all, it will probably be accurate. However some very inaccurate solutions to stiff problems can result from implicit algorithms as shown in Figure 15.14. The underdamped Newmark $\beta = \frac{1}{4}$ method[20] for a timestep a reasonable fraction of $2\pi/\lambda_1$ exhibits spurious oscillations, while the overdamped Wilson method[21] introduces spurious damping. Improved implicit methods which alleviate these difficulties have been discussed at length elsewhere.[1,8] Of course if very small timesteps are used, or if non-stiff systems are analysed, the errors will not manifest them-selves. In the former case, a suitable explicit method would almost certainly be cheaper.

15.7 TRANSIENT DYNAMIC BEHAVIOUR II: PLATFORM BEHAVIOUR

Non-linear transient wave propagation in solids was studied in the early 1960's by Ang[22] using the same discretization techniques as Smith had used in his one-dimensional piling analyses, that is lumped masses and spring stiffnesses. However Ang used implicit Newmark integration with respect to time. An early finite element analysis (of dam vibrations) was presented by Clough and Chopra,[23] but used real modal superposition, a technique not popular nowa-days for structure-soil interaction problems because of its restriction to linearized systems and to frequency-independent damping. In 1967 Constan-tino[18] used finite element discretization in space, together with a step-by-step integration method in time.

During the intervening 10 years, solutions have been presented for structural vibration problems involving plasticity and large strains[24–27] and kinematic hardening.[28] Cundall[29] has applied this class of solution in soil mechanics, using finite difference discretizations in space. However few of these publications address themselves to the numerical problems raised in the previous section of the present paper; hence the resulting controversies because each presentation is an apparent story of success, achieved by conflicting methods.

For example, because of the essential instability of simple explicit integration in oscillatory problems, and in order to enable one-point integration, programs which use explicit methods incorporate artificial damping: 'Programs such as PISCES use several forms of damping: artificial quadratic viscosity for smoo-thing shocks; linear viscosity for damping mesh noise; anisotropic viscosity for reducing "hour glass" distortions and velocity damping for static problems'.[29]

'The explicit temporal integration of the discretized space equations often introduces annoying spurious oscillations. These can be effectively controlled by damping, but . . .' (Belytschko *et al.*[26]). This kind of intervention in the progress of the solution has been termed a 'black art'[30] and it would certainly be surprising if it were problem-independent or easily guessed at by the average user.

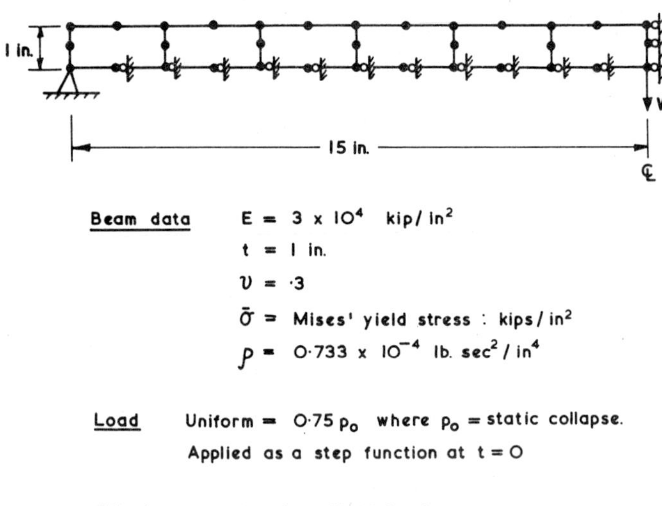

Figure 15.15   Finite element discretization of simply supported
elastic-plastic beam

As an example of a non-linear dynamic analysis incorporating plasticity consider the finite element discretization of a simply supported beam shown in Figure 15.15. This example is taken from Bathe *et al.*[25] and employs 8-node isoparametric elements to model bending stresses adequately. In order to lump the mass and use an explicit method, an assumption has to be made. Alternative nodal area weighting assumptions are shown in Figure 15.16 and lead to

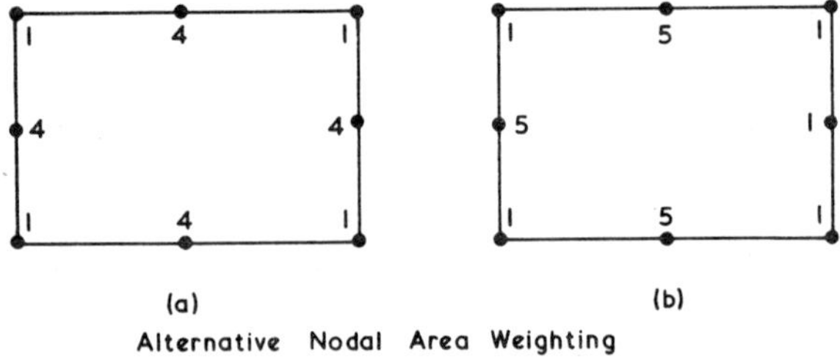

Figure 15.16   Alternative nodal lumping

the eigenvalues for the 50 degree of freedom system listed in Table 15.2. Method (a) is somewhat better than method (b) which is close to that recommended elsewhere.[27] The stiffness ratio for the distributed system is $1\cdot875\times10^6$ and for the lumped system $1\cdot25\times10^6$.

Table 15.2 Eigenvalues for structure in Figure 15

	Distributed mass	Lumped mass	
		Assumption (a)	Assumption (b)
λ_1	$1\cdot78\times10^6$	$1\cdot78\times10^6$	$1\cdot72\times10^6$
λ_2	$1\cdot29\times10^8$	$1\cdot28\times10^8$	$1\cdot24\times10^8$
λ_3	$8\cdot39\times10^8$	$8\cdot23\times10^8$	$7\cdot96\times10^8$
λ_4	$2\cdot67\times10^9$	$2\cdot54\times10^9$	$2\cdot46\times10^9$
—	—	—	—
—	—	—	—
λ_{50}	$3\cdot34\times10^{12}$	$2\cdot23\times10^{12}$	$2\cdot23\times10^{12}$

Explicitly and implicitly integrated solutions to this problem are shown in Figure 15.17. Although this is by no means a particularly stiff system, the implicit timestep used is 50 times greater than the stable explicit limit. As discussed earlier, the Wilson method tends to introduce some damping and phase shift which can be lessened by using a Nørsett method for example.[1] The effects of algorithmic damping and of relaxing the order of numerical integration governing spread of plasticity through the Gauss points are further illustrated in Figure 15.18.

An alternative method of transient dynamic analysis which has found popularity in earthquake engineering is the 'complex response' or 'complex transfer function' method which operates from the undamped equations of motion

$$\mathbf{M}\frac{d^2\mathbf{r}}{dt^2}+\mathbf{Kr}=\mathbf{F}(t) \tag{15.16}$$

by transforming to the frequency domain. Although this approach implies harmonic input and output, random wave forces can be split into harmonic components using a discrete Fourier transform. An advantage of the method is that frequency-dependent damping can be introduced in the form of a complex stiffness $\mathbf{K}$ which can be varied from element to element.[31] The random output is synthesized from its harmonic components by inverse Fourier transform.

Examples of the application of this method to analysis of an offshore structure, discretized into finite elements as shown in Figure 15.19, are presented in Figures 15.20 to 15.23.

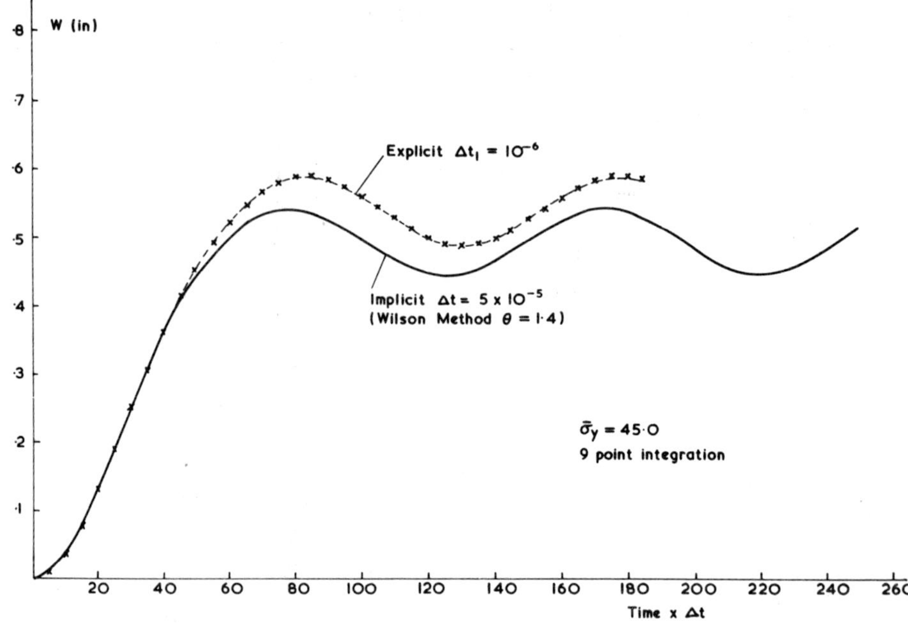

Figure 15.17　Comparison of explicit and implicit solutions

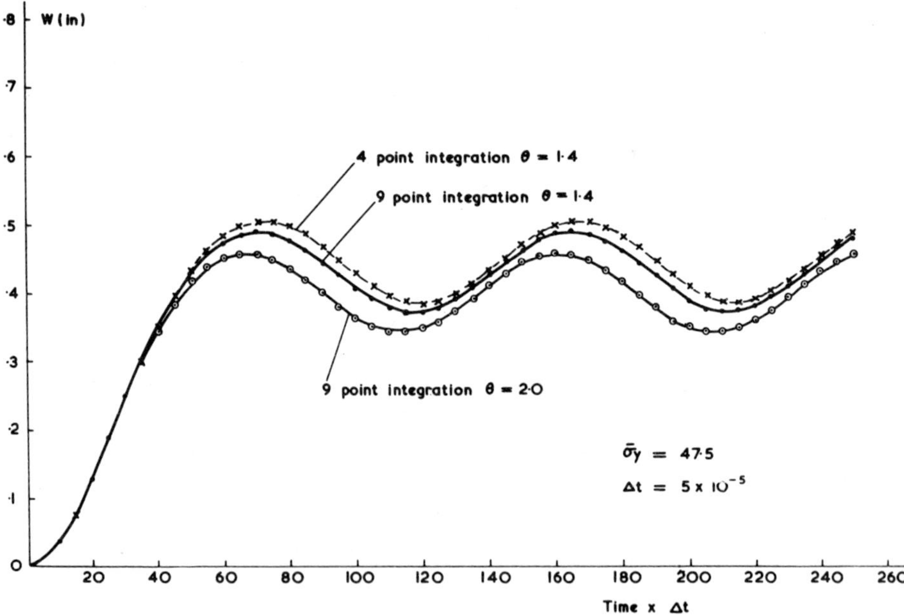

Figure 15.18　Effects of integration order and timestep on implicit solutions

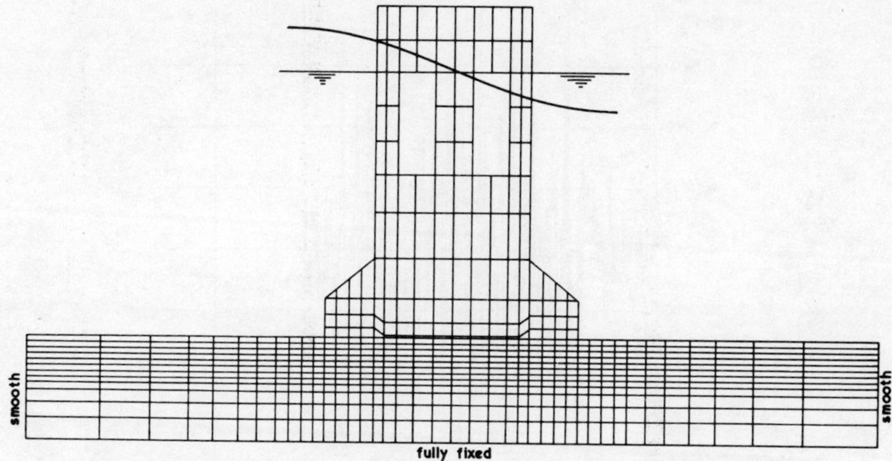

Figure 15.19 Finite element discretization of gravity offshore platform

Since the wave force is essentially inertial for such a gravity structure it is assumed to be proportional to waveheight for the random record plotted in Figure 15.20. The displacement response of the structure deck when subjected to this force is shown in Figure 15.21 and two acceleration records for increasing soil damping are shown in Figures 15.22 and 15.23.

These analyses are linearized in that repeated solutions are carried out until strain-compatible damping is achieved. Convergence difficulties have been reported[29] when using this approach and truly non-linear analyses are to be preferred because they adhere more faithfully to the actual physical nature of the deformation process. The dynamic analysis can then be directly linked to, and compared with, the equivalent static 'failure' computation. A consequence of quasi-linearization and harmonic input is that no permanent strains can be predicted by these methods. Nevertheless, due to the uncertainties in all analytical procedures, a good approach at present[14] is to use both linearized and non-linear methods and compare results.

15.8 ROLE OF PHYSICAL MODELS

In computer analysis of quasi-static platform deformations and failure modes,[14] comparisons with results obtained from physical model tests have proved to be invaluable, pending more detailed reports of field performance. Results of tests in a large centrifuge have been reported in the literature for quasi-static cyclic loading conditions[32,33] and consideration has been given to the difficulties of properly satisfying the modelling laws for dynamically loaded centrifuged oil platforms.[34]

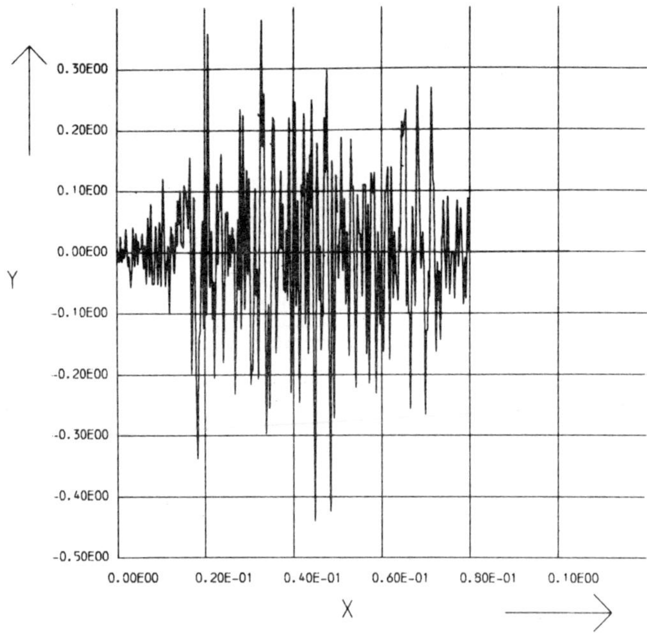

Figure 15.20   Random wave force on the platform

Broadly speaking the difficulties are these:

1. The transient phenomena of porepressure migration, platform oscillation, and soil creep all obey different scaling laws, namely N^2, N and 1 where N is the model dimension scale.

2. It has not yet been possible to model the full structure as well as its soil foundation, let alone the wave-structure interaction. Instead, stiff foundation plates have been subjected to forced oscillations by hydraulic jacks.

3. There are excellent reasons for using large foundation plates[33] and typical model bases have been 0·8 m diameter on foundations 1·0 m × 1·0 m × 0·6 m deep. However this does mean that relatively rigid boundaries are present quite close to the model, and this will influence its dynamic response.

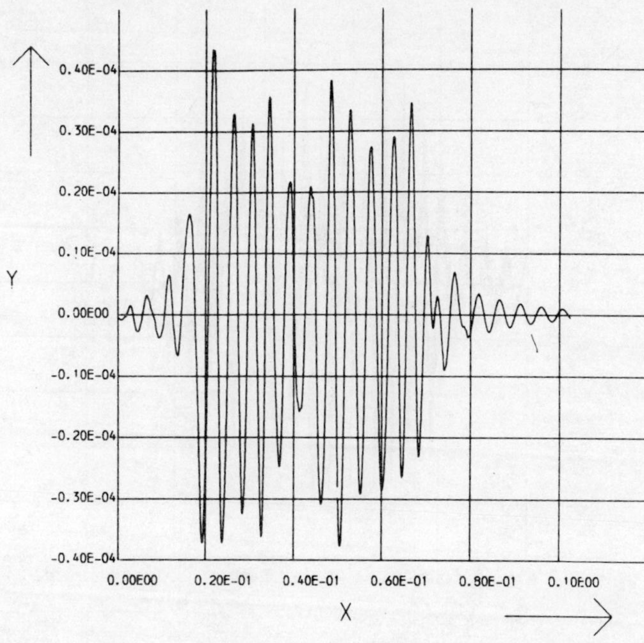

X: TIME. [SEC]

Y: DISPLACEMENT. [M]

Figure 15.21   Deck displacement due to random force

For these reasons, although equipment has now been installed which will enable dynamic loading of centrifuged model bases at up to the correctly scaled frequency of 10 Hz, it is believed that there will be a need for the computation methods described in the last section in order to gain as good an interpretation of the model test data as possible.

15.9 DISCUSSION

A survey of numerical methods for use in transient foundation analysis would be incomplete without a mention of the method of characteristics. This has been applied very successfully to non-linear one-dimensional problems[35] but the usual difficulties with characteristics are encountered when two and three

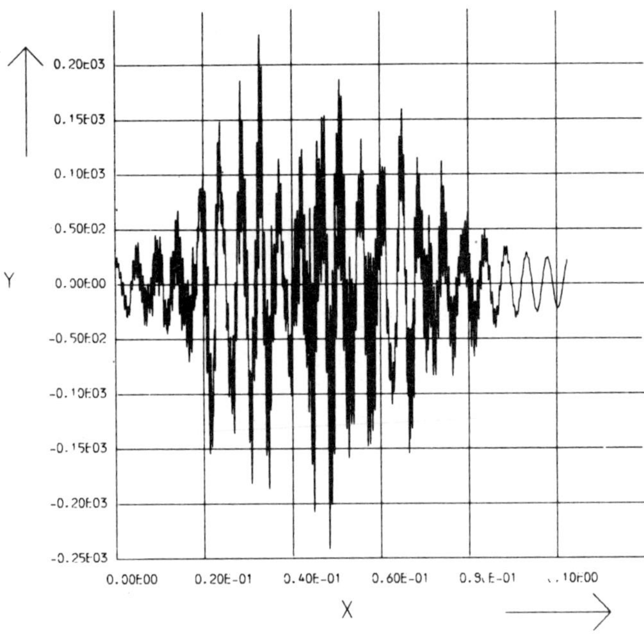

Figure 15.22 Deck accelerations for low soil damping

space dimensions are required. As with finite difference procedures, there do not seem to be sufficiently simple generalized coding procedures available to compete with finite elements.

Seed *et al.*[36] have treated simultaneous generation and dissipation of excess porewater pressures during earthquakes in an approximate manner by uncoupling the transient dynamic and 'field' aspects of the problem. The methods described in the present paper could be used to analyse offshore foundations in this way.

Lack of space has precluded a discussion of the importance of boundary conditions in erroneously reflecting waves back into the foundation. Fortunately this difficulty is much less serious than when shear waves are being propagated vertically through a stratum from a base rock shaking mechanism.

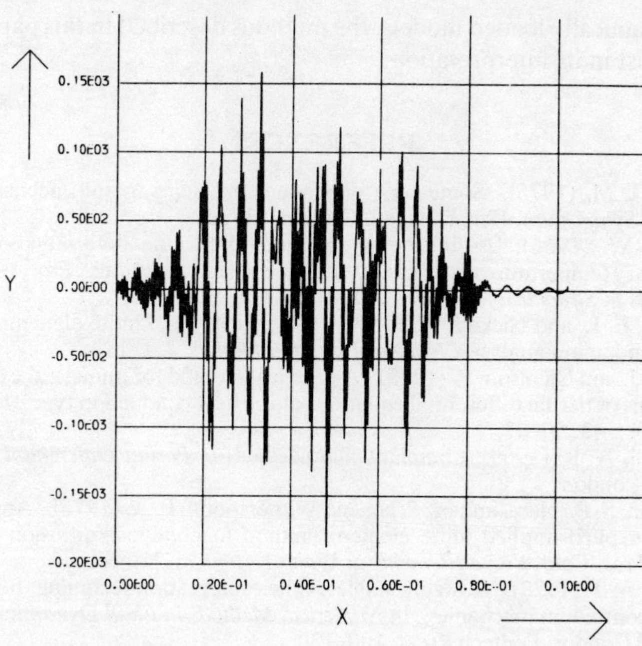

Figure 15.23 Effect on deck acceleration of increased soil damping

The reader is referred to the work of Roesset, for example Reference 37, for a treatment of consistent energy transmission boundaries.

The view taken of time-stepping algorithms for transient problems has been a fairly restricted one. For example attention has only been directed to methods which advance by one step at a time. This is still an area of research interest in which valuable new methods may still arise[38] involving multi-steps and so on. However any new approach needs to be fully analysed in terms of its stability and its damping and shift of component eigenmodes before it can be used with confidence for stiff differential equations with oscillatory solutions.

Numerical analysis has played a major role in the understanding of offshore foundation failure modes, and in the assessment of pile driveability. As information becomes available from presently installed structures and from future dynamically loaded models, the methods described in this paper should greatly assist in its interpretation.

REFERENCES

1. Smith, I. M. (1975). 'Some time dependent problems in soil mechanics', *Proc. NMSR Symposium*, Karlsruhe, ed. Gudehus, Wiley.
2. Visser, W. (1965). 'A finite element method for the determination of non-stationary temperature distributions and thermal deformations', *Proc. Conf. Matrix Methods in Structural Mechanics*, Dayton, Ohio, 925–944.
3. Wilson, E. L. and Nickell, R. E. (1966). 'Application of finite element method to heat conduction analysis', *Nuclear Eng. and Design*, **4**, 1–11.
4. Crank, J. and Nicolson, P. (1947). 'A practical method for numerical evaluation of solutions of partial differential equations of the heat conduction type', *Proc. Camb. Phil. Soc.*, **43**, 50–67.
5. Mitchell, A. R. (1969). *Computational Methods in Partial Differential Equations*, Wiley, London.
6. Neuman, S. P., Narasimhan, T. N., and Witherspoon, P. A. (1976). 'Application of mixed explicit-implicit finite element method to nonlinear diffusion-type problems', *Proc. Conf. Finite Elements in Water Resources*, Princeton.
7. Baker, A. J. (1973). 'A highly stable explicit integration technique for computational continuum mechanics', in *Numerical Methods in Fluid Dynamics*, ed. Brebbia and Connor, Pentech Press, 100–120.
8. Smith, I. M., Siemienieuch, J. L., and Gladwell, I. (1977). 'Evaluation of Nørsett methods for integrating differential equations in time', *Int. J. Num. Anal. Meth. in Geomechanics*, **1**, 1.
9. Nørsett, S. P. (1974). 'One-step methods of Hermite type for numerical integration of stiff systems', *BIT*, **14**, 63–77.
10. Siemienieuch, J. L. (1976). 'Properties of certain rational approximations to e^{-z}', *BIT*, **16**, 172–191.
11 Sandhu, R. S. and Wilson, E. L. (1969). 'Finite element analysis of seepage in elastic media', *Proc. Am. Soc. Civ. Eng.*, **95**, EM3, 641–51.
12. Smith, I. M. and Hobbs, R. (1976). 'Biot analysis of consolidation beneath embankments', *Geotechnique*, **26**, 1, 149–171.
13. Smith, I. M. (1976). 'Integration in time of diffusion and diffusion–convection equations', *Proc. Conf. Finite Elements in Water Resources*, Princeton.
14. Smith, I. M. (1976). 'Aspects of the analysis of gravity offshore structures', *Numerical Methods in Geomechanics*, Blacksburg, vol. 2, 957–978.
15. Naylor, D. J. (1974). 'Stresses in nearly incompressible materials by finite elements with application to the calculation of excess pore-pressures', *Int. J. Num. Meth. Eng.*, **8**, 443–460.
16. Smith, E. A. L. (1960). 'Pile driving analysis by the wave equation', *Proc. Am. Soc. Civ. Eng.*, **86**, SM4, 35–61.
17. Smith, I. M. (1976). 'Finite element analysis of axially loaded pile capacity and driveability', *Report to Fugro-Cesco B.V.*, March 1976.
18. Constantino, C. J. (1967). 'Finite element approach to stress wave problems', *Proc. Am. Soc. Civ. Eng.*, **93**, EM2, 153–176.
19. Bazant, Z. P., Glazik, J. L., and Achenbach, J. D. (1976). 'Finite element analysis of wave diffraction by a crack', *Proc. Am. Soc. Civ. Eng.*, **102**, EM3, 479–496.

20. Newmark, N. M. (1959). 'A method of computation for structural dynamics', *Proc. Am. Soc. Civ. Eng.*, **85**, EM3, 67–94.
21. Wilson, E. L. and Clough, R. W. (1962). 'Dynamic response by step by step matrix analysis', *Symp. on Use of Computers in Civ. Eng.*, Lisbon.
22. Ang, A. H. S. (1965). 'Numerical approach for wave motions in nonlinear solid continua', *Proc. Conf. Matrix Methods in Structural Mechanics*, Dayton, Ohio, 753–778.
23. Clough, R. W. and Chopra, A. K. (1966). 'Earthquake stress analysis in earth dams', *Proc. Am. Soc. Civ. Eng.*, **92**, EM2, 197–212.
24. Heifitz, J. H. and Constantino, C. L. (1972). 'Dynamic response of nonlinear media at large strains', *Proc. Am. Soc. Civ. Eng.*, **98**, EM6, 1511–1528.
25. Bathe, K. J., Ozdemir, H., and Wilson, E. L. (1974). 'Static and dynamic geometric and material nonlinear analysis', *SESM Report 74–4*, University of California, Berkeley, February 1974.
26. Belytschko, T., Chiapetta, R. L., and Bartel, H. D. (1976). 'Efficient large scale nonlinear transient analysis by finite elements', *Int. J. Num. Meth. Eng.*, **10**, 579–596.
27. Shantaram, D., Owen, D. R. J., and Zienkiewicz, O. C. (1976). 'Dynamic transient behaviour of two- and three-dimensional structures including plasticity, large deformation effects and fluid interaction', *Earthquake Eng. and Structural Dynamics*, **4**, 561–578.
28. Zudans, Z. (1975). 'Elastic-plastic creep response of structures under composite time history of loadings', *Nucl. Eng. and Design*, **35**, 2, 213–245.
29. Cundall, P. (1976). 'Explicit finite difference methods in geomechanics', *Numerical Methods in Geomechanics*, vol. 1, Blacksburg, 132–150.
30. Goudreau, G. L. and Taylor, R. L. (1972). 'Evaluation of numerical integration methods in elastodynamics', *Comp. Meth. App. Mech. Eng.*, **2**, 65.
31. Lysmer, J., Udaka, T., Seed, H. B., and Hwang, R. (1974). 'LUSH: A computer program for complex response analysis of soil-structure systems', *Report EERC 74–4*, University of California, Berkeley.
32. Rowe, P. W. (1975). 'Displacement and failure modes of model offshore platforms founded on clay', *Proc. Offshore Europe Conference, Aberdeen*, Spearhead Publications.
33. Rowe, P. W., Craig, W. H., and Procter, D. C. (1976). 'Model studies of offshore gravity structures founded on clay', *BOSS Conference*, Trondheim.
34. Rowe, P. W., Craig, W. H., and Procter, D. C. (1977). 'Dynamically loaded centrifugal model foundations', *Proc. Int. Conf. Soil Mech. and Found. Eng.*, Tokyo.
35. Papadakis, C. N. (1973). 'Soil transients by characteristics method', *Report UMEE-73R10*, University of Michigan.
36. Seed, H. B., Martin, P. P., and Lysmer, J. (1976). 'Pore water pressure changes during soil liquefaction', *Proc. Am. Soc. Civ. Eng.*, **102**, GT4, 323–346.
37. Roesset, J. M. and Kausel, E. (1976). 'Dynamic soil-structure interaction', *Numerical Methods in Geomechanics, Blacksburg*, vol. 1, 3–19.
38. Zienkiewicz, O. C. (1976). 'A new look at Newmark, Houbolt and other time stepping formulae. A weighted residual approach', University of Wales, Swansea, *Civil Engineering Report C/R/273/76*.

Chapter 16

Response of Saturated Sands to Earthquake and Wave Induced Forces

W. D. L. Finn, Kwok W. Lee, P. M. Byrne, and G. R. Martin

16.1 INTRODUCTION

The response of a saturated sand deposit to cyclic loading induced by earthquake motions or by wave loading of gravity structures is a very important and difficult problem of non-linear soil dynamics and a completely satisfactory generalized solution is not yet available. The resistance to deformation at any point in the sand deposit is a function of effective stress which, in turn, depends on the contemporaneous rates of generation and dissipation of porewater pressure. Therefore the dynamic response, at least for loose to medium dense sands, is dominated by the effects of the progressive increases in porewater pressure that develop during cyclic loading.[1]

Current methods of determining the dynamic response of a horizontal saturated sand layer are based on total stress procedures. The significant ground motions are assumed to be shear waves propagating vertically and the appropriate shear modulus, G, for use in the analysis, may be determined from an equation of the form[2]:

$$G = 1000K_2(\sigma'_m)^{1/2} \qquad (16.1)$$

in which K_2 is a parameter that varies with shear strain and σ'_m is the mean normal effective stress. The initial effective stresses are used in the computation of the initial value of G and thereafter G is modified to take into account the dependence of G on shear strains. Although the porewater pressure increases during shaking and decreases the effective stresses in the layer, the effect of decreasing mean normal effective stress on the shear modulus, G, is not taken into account in total stress methods of analysis.

In certain circumstances the porewater pressure at a given elevation in the horizontal deposit may reach a value equal to the overburden pressure and the sand layer may liquefy. Since the problem is one-dimensional, liquefaction occurs, in theory, at every point on the same horizontal plane simultaneously. In two-dimensional structures, the cyclic shear stress conditions on horizontal planes vary and liquefaction of sand layers in such structures develops progressively. For such cases Seed, Lee, and Idriss have developed an approximate iterative approach. In their analysis of the Sheffield dam[3] they took into consideration the possibility that certain zones of the dam would liquefy during earthquake motions. They examined the stress histories on horizontal planes within the dam after a period of shaking such as 10 s and for subsequent analysis assigned zero shear modulus to those zones which they deemed to have liquefied. By proceeding in this way, it was possible to predict the progressive development of liquefaction.

This method of analysis is a considerable improvement over the total stress method although it does not consider the effect of porewater pressures on modulus prior to liquefaction. However, the method cannot be applied to the one-dimensional problem of the dynamic response of horizontal strata of saturated sand in which liquefaction occurs at all points at a given level at the same time.

Seed, Martin, and Lysmer[4] have developed an approximate procedure for determining the rate of generation and dissipation of porewater pressures in horizontal sand strata. At various elevations in the horizontal sand deposit the time histories of dynamic shear stresses are determined by a total stress method of analysis; the effect of porewater pressure on dynamic response is not taken into account. At each elevation of interest, the actual time history of shear stresses is converted into an equivalent number of uniform stress cycles. The number of cycles of this uniform stress required to cause liquefaction is determined by laboratory tests.

It is now assumed that the rise in porewater pressure in saturated sands is given by a single characteristic curve which describes the functional relationship between the ratio of the porewater pressure to the initial effective overburden pressure and the ratio of the cycles of uniform stress to the cycles required to cause liquefaction. The generation of porewater pressure with cycles of uniform stress is determined easily using this curve. The combined effects of generation and dissipation of porewater pressure may be computed by solving the equation of diffusion with internal generation of porewater pressure.[5]

Streeter *et al.*[6] in 1973 presented a numerical method of dynamic response which takes into account the effects of porewater pressures. The responses of the porewater and the granular skeleton were treated separately as uncoupled problems. Porewater pressure effects were introduced by specifying pseudocyclic and permanent volume changes. However, a method for assigning values to such volume changes was not established.

Recently Liou, Streeter, and Richart,[7] in 1976, introduced a different model for the development of porewater pressure which depends on assuming a fixed relationship between the strain dependent shear modulus and the constrained rebound modulus. Comparisons between results predicted by the new model and laboratory experimental data have not yet been carried out so the adequacy of this model cannot yet be assessed.

It would seem that the best approximation to true site response to an earthquake would be obtained by an effective stress method of analysis which coupled the effect of the progressive development of porewater pressure with the usual procedures for dynamic analysis.

Two such methods of effective stress analysis are presented here for the dynamic response analysis of a horizontally layered saturated sand deposit under simple shear cyclic loading conditions and the extension of these methods to more complicated circumstances is indicated. The methods are based on constitutive laws which take into account important factors which are known to affect the response characteristics of saturated sands to cyclic loading including the simultaneous generation and dissipation of porewater pressures. The first method solves the non-linear equations of motion directly[8,9]; the second uses the technique of iterative elastic analysis incorporating strain compatible moduli and damping.[1,10] For consistency all equations are formulated for earthquake loading only.

16.2 GENERAL RESPONSE CHARACTERISTICS

Any general stress–strain relations which are proposed to describe the response of sands to cyclic loading must predict most or all of the important response characteristics if they are to be of practical value. The dynamic response of a saturated sand layer to cyclic loading is a very complicated process but nevertheless, the important response characteristics are readily identifiable. Dynamic shear stresses and shear strains are generated which cause slip at grain to grain contacts. This intergranular slip, in dry sands, would lead to volumetric compaction at the typical shear strain levels that are usually generated in sand by earthquakes[11,12] or by wave loading on stable gravity structures. In saturated sands, the volumetric compaction is retarded because the water cannot drain instantaneously to accommodate the volume change. Consequently the relaxing sand skeleton transfers some of its intergranular or effective stresses to the porewater and the porewater pressure increases.

The volume changes and hence the porewater pressures that develop during cyclic loading depend on the dynamic shear strains[11,12] and the strains in turn depend on the stiffness and damping characteristics of the sand layers. In the special case under consideration, the sand undergoes simple shear deformations and so the stiffness at any time may be expressed as a function of the shear modulus. In fundamental studies of effective stress–strain relations for sands, Seed and Idriss,[2] and Hardin and Drnevich[13] showed the shear modulus, G, to be a function of the mean normal effective stress and the shear strain. In saturated sands the progressive development of porewater pressure during cyclic loading is continuously diminishing the level of effective stress and hence the shear modulus and the resistance to deformation.

During cyclic loading the slips at grain contacts result in volumetric compaction[14,11,12] and increased values of K_0, the coefficient of effective lateral stress.[15] Both effects stiffen the sand against further deformation. It is also probable that the slips at grain contacts result in a more stable sand structure under the existing effective stress regime.[14] The processes leading to increased resistance are referred to collectively as hardening. The effect has been noted in dry sands[11,16] and in saturated undrained sand[14] at the strain levels typical of ground shaking.[17]

In summary, the important factors which must be considered when computing the response of saturated sand layers to a given earthquake are (a) the initial shear modulus *in situ*; (b) the variation of shear modulus with shear strain; (c) contemporaneous generation and dissipation of porewater pressures; (d) changes in effective mean normal stress; (e) damping; and (f) hardening. All of these factors are taken into account in the constitutive relations formulated below.

16.3   DIRECT NON-LINEAR ANALYSIS

16.3.1   Constitutive relations

Constitutive relations for dry or saturated sands in simple shear were presented by Finn, Lee and Martin.[18] These relations, which covered the previously cited factors affecting dynamic response were used to determine the response of undrained saturated sands in cyclic shear laboratory tests. Subsequently, analytical procedures were presented for the analysis of the dynamic response of saturated sands under field loading conditions including the dissipation of porewater pressure.[8,9]

During an earthquake a sand deposit is subjected to an irregular loading pattern which consists of intervals of loading, unloading and reloading. The sand has different behaviour characteristics in each of the different loading phases. Each phase is dealt with separately below.

(a)   Initial loading

The response of the sand initially to cyclic loading is controlled by its state *in situ* as specified by its initial shear modulus, G_{m0}. This initial maximum value of the shear modulus may be determined in a variety of ways, such as by geophysical methods[17] or by resonant column tests.[19] Values of G_{m0} cannot be reliably obtained from simple shear tests because the levels of shear strain associated with G_{m0} are beyond the sensitivity of the apparatus.[18] Alternatively, G_{m0} may be computed using equations by Hardin and Drnevich[13] or

Seed and Idriss.[2] For the work which follows we used the Hardin–Drnevich equation. This equation gives an average representation for the initial shear modulus based on test data from a wide variety of sands.

Following the presentation by Finn *et al.* in Reference 9, it is assumed that, up to the point of the first reversal in loading, the response of the sand follows the hyperbolic stress–strain relationship formulated by Konder and Zelasko[20] and also adopted by Hardin and Drnevich.[13] Therefore, the initial loading phase of the sand is assumed to be described by the following equation:

$$\tau = G_{m0}\gamma \Big/ \left(1 + \frac{G_{m0}}{\tau_{m0}}\gamma\right) \tag{16.2}$$

in which τ is the shear stress at strain amplitude γ, G_{m0} is the initial maximum tangent modulus and τ_{m0} is the maximum shear stress that can be applied to the sand in its initial state without failure (Figure 16.1).

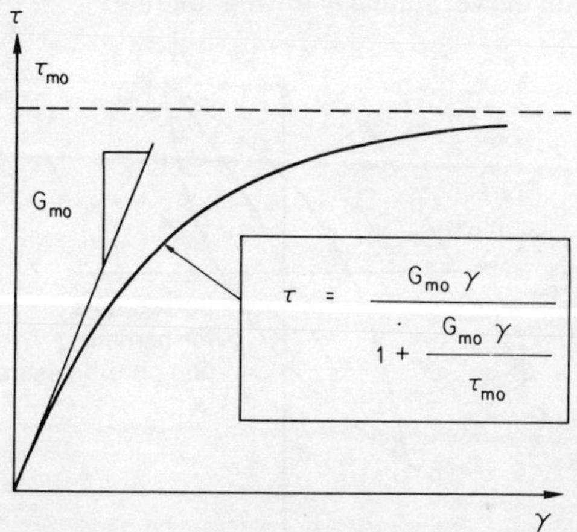

Figure 16.1 Hyperbolic stress–strain curve

The quantities G_{m0} and τ_{m0} are given in units of psf by the equations proposed in Reference 13, modified for horizontal sand layers;

$$G_{m0} = 14760 \frac{(2 \cdot 973 - e)^2}{1 + e} \left(\frac{1 + 2K_0}{3}\right)^{1/2} . \sqrt{\sigma'_v} \tag{16.3}$$

$$\tau_{m0} = \left\{ \left(\frac{1 + K_0}{2}\sin\phi'\right)^2 - \left(\frac{1 - K_0}{2}\right)^2 \right\}^{1/2} \sigma'_v \tag{16.4}$$

in which e = void ratio and is limited to values less than 2, σ'_v = vertical effective stress in psf, K_0 = coefficient of earth pressure at rest and ϕ' = effective angle of shearing resistance.

(b)   Unloading and reloading

The stress–strain curve for initial loading, called the skeleton curve [curve 1 of Figure 16.2(a)], is described by Equation (16.2) which may be written in the form

$$\tau = f(\gamma) \qquad (16.5)$$

If loading reversal occurs at (γ_r, τ_r) then the equation of the stress–strain curve [curve 2 in Figure 16.2(a)] during subsequent unloading or reloading from the reversal point is assumed to be given by

$$\frac{\tau - \tau_r}{2} = f\left(\frac{\gamma - \gamma_r}{2}\right) \qquad (16.6)$$

If the curve defined by Equation (16.6) crosses a curve described in a previous load cycle, the stress–strain curve follows that of the previous cycle. If during unloading the extension of the skeleton curve is intersected further unloading follows the skeleton curve; similarly during loading.

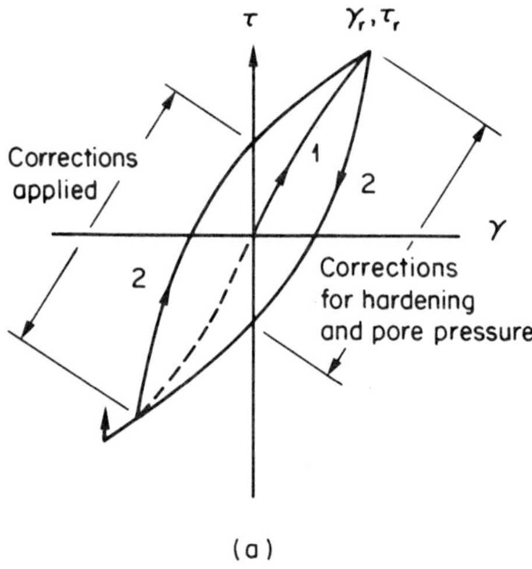

Figure 16.2(a)   First loading cycle

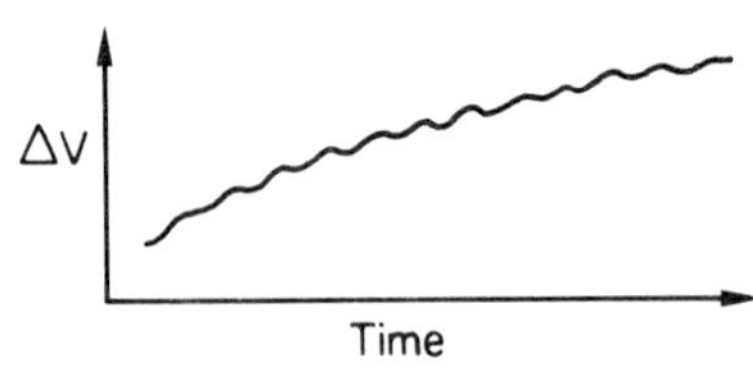

Figure 16.2(b)   Volume change with time

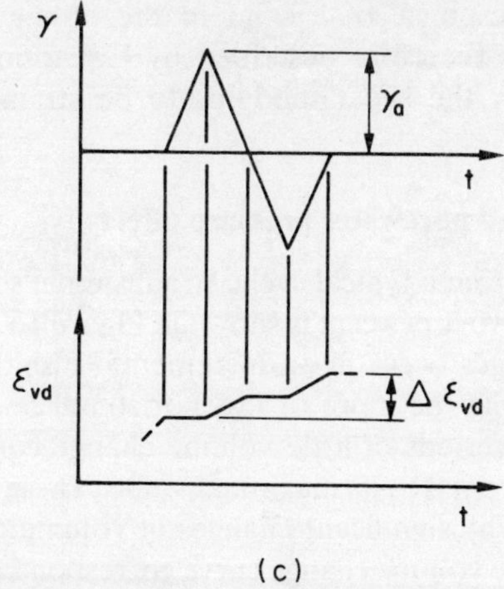

Figure 16.2(c)   Occurrence of volume change

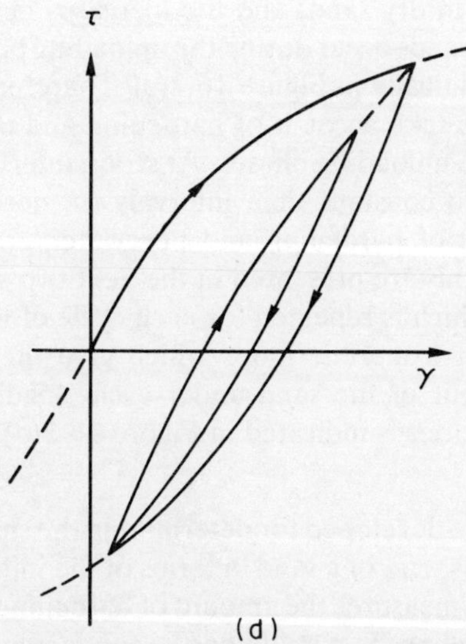

Figure 16.2(d)   General loading pattern

This type of response is referred to as Masing behavioúr.[21] Herrera[22] has shown that combinations of Coulomb friction and linear elastic elements exhibit this type of behaviour. Newmark and Rosenblueth[23] suggest the Masing model to be a good representation of soil behaviour and some experimental support for the suggestion is provided by Marsal.[24] Further corroborative data is given in References 18 and 25 by Finn *et al.*

The stress–strain response of a sand in the absence of hardening and porewater pressure is therefore described by Equations (16.2) and (16.6). These equations show the shear modulus to be strain-dependent and the damping hysteretic.[18]

16.3.2 Hardening and porewater pressure effects

Volume change data from a typical cyclic strain simple shear test with a shear strain amplitude of $\gamma = 0\cdot1$ per cent is shown in Figure 16.2(b). The data shows that the volume changes occur in an incremental fashion with little volume change occurring during the more or less horizontal portions of the volume change curve. These periods of little volume change correspond to intervals when γ is increasing from 0 to its maximum value. These intervals are defined as loading intervals. The significant changes in volume corresponding to the sloping segments of the volume change curve correspond to intervals when γ is reducing from its maximum amplitude to zero or when unloading is occurring. In general, we have observed in laboratory cyclic simple shear tests that most of the volume changes in dry sands and the increases in porewater pressure in undrained saturated sands occur during the unloading portion of the load cycle. This is shown schematically in Figure 16.2(c). Therefore modifications to the stress–strain curve to take account of hardening and porewater pressure are made only during the unloading phases. At strain intervals $\Delta\gamma$ during unloading, corresponding to constant time intervals Δt, necessary corrections are made for the effects of hardening and porewater pressure. The necessary equations for doing this are presented in the next two sections.

This procedure, which is repeated for each cycle of loading and unloading, results in a series of hysteretic loops which give the continually changing stress–strain behaviour of the sand under cyclic loading. The generality of possible loading histories is indicated in Figure 16.2(d).

(a) Hardening

Equations will now be developed for determining the shear modulus, G_{mn}, and maximum shear stress, τ_{mn}, of a sand in terms of the initial values G_{m0} and τ_{m0} and a parameter that measures the amount of hardening that has occurred due to previous cyclic loading. In a study on the fundamental behaviour of sands under cyclic loading in simple shear, Martin, Finn and Seed[11] suggested a stress–strain relationship which included the effects of hardening in the form

$$\tau_{hv} = \frac{\gamma\sqrt{\sigma'_v}}{a + b\gamma} \tag{16.7}$$

in which τ_{hv} and γ are associated horizontal shear stress and shear strain respectively, and $\sigma'_v =$ current effective normal stress. The parameters a and b

are constants for a given load cycle but in general they are functions of the volumetric compaction strain, ε_{vd}, and are given by

$$a = A_1 - \frac{\varepsilon_{vd}}{A_2 + A_3 \varepsilon_{vd}} \qquad (16.8)$$

$$b = B_1 - \frac{\varepsilon_{vd}}{B_2 + B_3 \varepsilon_{vd}} \qquad (16.9)$$

in which A_i and B_i are constants with $i = 1, 2, 3$. The accumulated volumetric strain is therefore a measure of the hardening that has occurred.

The maximum value of the shear modulus, G_{mn}, in any cycle of loading n is given by

$$G_{mn} = (d\tau_{hv}/d\gamma) \quad \text{at } \gamma = 0 \qquad (16.10)$$

From Equations (16.7) through (16.10) expressions can be derived for the maximum shear modulus of dry sand G_{mn}, and the maximum shear stress, τ_{mn}, after volumetric strains ε_{vd} have occurred as follows;

$$G_{mn} = G_{m0}\left[1 + \frac{\varepsilon_{vd}}{H_1 + H_2 \varepsilon_{vd}}\right] \qquad (16.11)$$

$$\tau_{mn} = \tau_{m0}\left[1 + \frac{\varepsilon_{vd}}{H_3 + H_4 \varepsilon_{vd}}\right] \qquad (16.12)$$

where ε_{vd} is the accumulated volumetric strain and H_1, H_2, H_3 and H_4 are constants. The constants are determined by fitting Equations (16.11) and (16.12) to the results of constant strain cyclic loading tests using simple shear apparatus as described in Reference 18. A typical example of data fitting is shown in Figure 16.3(a) which also shows the effect of hardening on the modulus.

The stress–strain behaviour of dry sand is now completely described by Equations (16.2), (16.6), (16.11) and (16.12). Corroboration of these equations (Fig. 16.3(b)) was offered by Finn *et al.*[18] who compared theoretical predictions based on these equations with experimental data by Pyke.[26] More recently an experimental program was conducted by Finn *et al.*[25] to allow a 'point by point' comparison between experimental stress–strain relationships in simple shear and those predicted by the theory. An example of comparisons between the experimental and theoretical data are shown in Figure 16.4, for the second and fourth cycles of loading. The linear segment near the apex of the loops on the unloading curves is due to friction in the apparatus. This was measured and incorporated in the theoretical analysis to allow a direct comparison between the experimental and theoretical loops. The theoretical

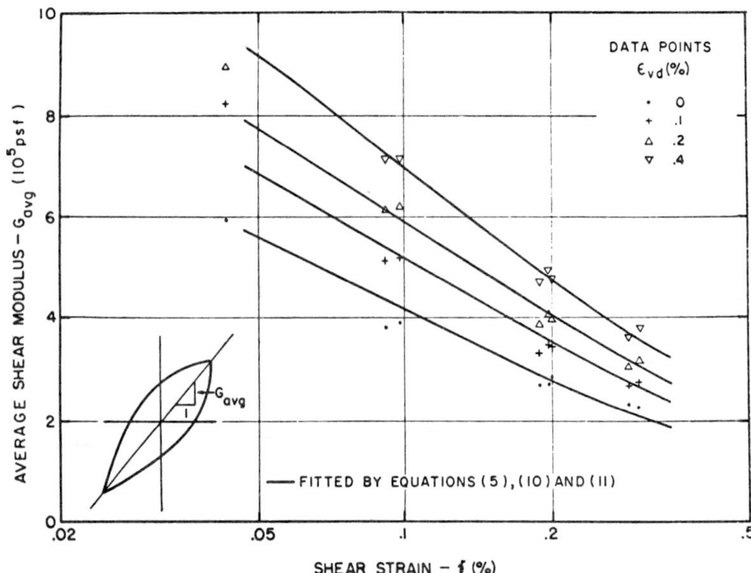

Figure 16.3(a)   Average shear modulus at various values of volumetric strain

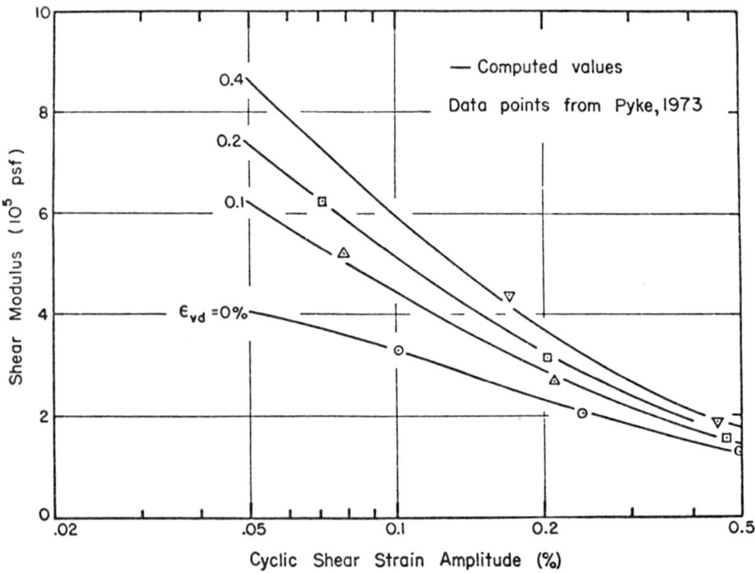

Figure 16.3(b)   Comparison of measured and computed shear moduli

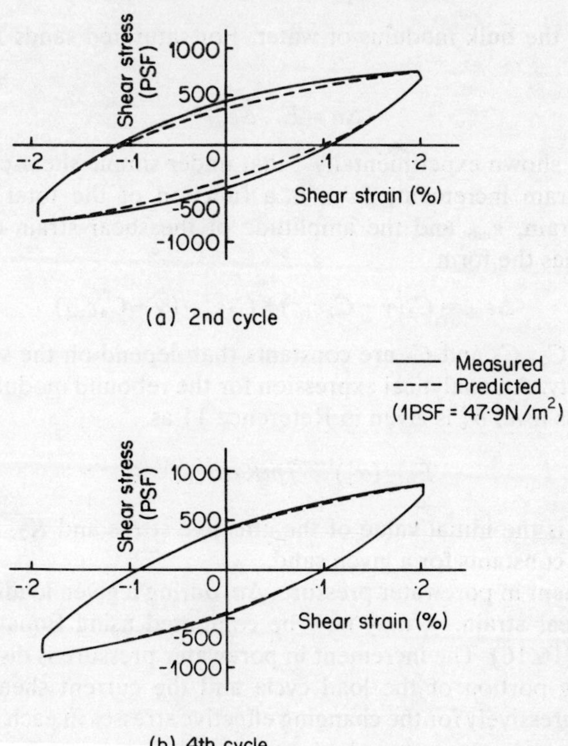

Figure 16.4 Comparison between predicted and measured stress–strain curves

stress–strain loops appear to be good approximations to the measured stress–strain curves up to five cycles of load, thereafter the curves are practically identical.

(b) *Generation of porewater pressure*

Consider a cubic element of saturated sand of unit volume and porosity n_p. Let the element be under a vertical effective stress, σ'_V, and horizontal stresses, $K_0\sigma'_V$. During a drained simple shear test a cycle of shear strain, γ, causes an increment in volumetric compaction strain, $\Delta\varepsilon_{vd}$, due to grain slip. During an undrained shear test starting with the same effective stress system, the cycle of shear strain, γ, causes an increase in porewater pressure, Δu. It was shown in Reference 11 that equating volume changes in water and sand gives

$$\Delta u = \Delta\varepsilon_{vd}\Big/\left[\frac{1}{\bar{E}_r}+\frac{n_p}{K_w}\right] \tag{16.13}$$

in which $\bar{E}_r$ = one-dimensional rebound modulus of sand at an effective stress

σ'_v and K_w is the bulk modulus of water. For saturated sands $K_w \gg \bar{E}_r$, and, therefore,

$$\Delta u = \bar{E}_r \, . \, \Delta \varepsilon_{vd} \qquad (16.14)$$

It has been shown experimentally[11] that under simple shear conditions the volumetric strain increment, $\Delta \varepsilon_{vd}$, is a function of the total accumulated volumetric strain, ε_{vd}, and the amplitude of the shear strain cycle, γ. The relationship has the form

$$\Delta \varepsilon_{vd} = C_1(\gamma - C_2 \varepsilon_{vd}) + C_3 \varepsilon_{vd}^2 / (\gamma + C_4 \varepsilon_{vd}) \qquad (16.15)$$

in which C_1, C_2, C_3 and C_4 are constants that depend on the sand type and relative density. An analytical expression for the rebound modulus, $\bar{E}_r$, at any effective stress level σ'_v is given in Reference 11 as

$$\bar{E}_r = (\sigma'_v)^{1-m} / m K_2 (\sigma'_{v0})^{n-m} \qquad (16.16)$$

in which σ'_{v0} is the initial value of the effective stress and K_2, m, and n are experimental constants for a given sand.

The increment in porewater pressure, Δu, during a given loading cycle with maximum shear strain, γ, may now be computed using Equations (16.14), (16.15), and (16.16). The increment in porewater pressure is distributed over the unloading portion of the load cycle and the current shear modulus is modified progressively for the changing effective stresses in each time interval, Δt.

The new level of effective stress will affect the initial modulus, G_{mn}, and the maximum shear stress, τ_{mn}, applicable to the next cycle of loading. For saturated sands, therefore, the maximum shear moduli and maximum allowable shear stresses for the nth loading cycle are related to the initial values by the following equations:

$$G_{mn} = G_{m0} \left[1 + \frac{\varepsilon_{vd}}{H_1 + H_2 \varepsilon_{vd}} \right] \left(\frac{\sigma'_v}{\sigma'_{v0}} \right)^{1/2} \qquad (16.17)$$

and

$$\tau_{mn} = \tau_{m0} \left[1 + \frac{\varepsilon_{vd}}{H_3 + H_4 \varepsilon_{vd}} \right] \frac{\sigma'_v}{\sigma'_{v0}} \qquad (16.18)$$

in which σ'_{v0} is the initial vertical effective stress, and σ'_v is the vertical effective stress at the beginning of the nth cycle.

In more general terms, Equations (16.17) and (16.18) allow us to compute the maximum shear modulus and shear stress compatible with the amount of hardening that has taken place and the current magnitude of the porewater pressure. Therefore, we can continually update Equation (16.6) and determine at any time the current tangent shear modulus for use in dynamic analysis. The updated explicit form of Equation (16.6) is given by

$$\frac{\tau - \tau_r}{2} = \frac{G_{mn}(\gamma - \gamma_r)}{2} \Bigg/ \left[1 + \frac{G_{mn}|(\gamma - \gamma_r)|}{2\tau_{mn}} \right] \tag{16.19}$$

The complete stress–strain response of undrained saturated sands is therefore described by Equations (16.2), (16.17), (16.18) and (16.19). We have used these equations together with Equation (16.13) to predict the number of cycles of the shear stress ratio τ/σ'_{v0} to cause liquefaction in a saturated crystal silica sand in constant volume cyclic simple shear tests. The theoretical predictions and the measured data are given in Figure 16.5 and compare quite well with each other.

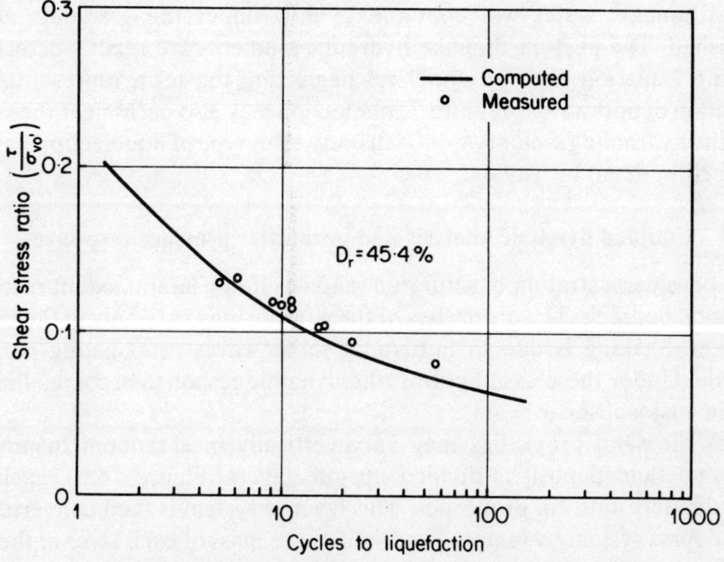

Figure 16.5 Liquefaction data for crystal silica sand

(*c*) *Dissipation of porewater pressure*
If the saturated sand layer can drain during shaking there will be simultaneous generation and dissipation of porewater pressure. Thus, the rate of increase of porewater pressure will be less than for completely undrained sand. The distribution of porewater pressure at time t is given by the equation

$$\frac{\partial u}{\partial t} = \bar{E}_r \frac{\partial}{\partial z}\left(\frac{k}{\gamma_w} \cdot \frac{\partial u}{\partial z} \right) + \bar{E}_r \frac{\partial \varepsilon_{vd}}{\partial t} \tag{16.20}$$

in which u is the porewater pressure, k the permeability and γ_w the unit weight of water. The term containing ε_{vd} represents the internal generation of porewater pressure. Equation (16.20) is identical in form to that for the conduction of heat in a bar with distributed heat sources.[5]

This equation also applies even when external drainage does not take place. In an undrained layer of sand the porewater pressures generated at various locations by an earthquake will not be in instantaneous equilibrium with each other and a continuous redistribution takes place under the instantaneous gradients established by the cyclic motions. The porewater pressure established at any time is in accordance with Equation (16.20) and reflects the net effects of contemporaneous generation and redistribution.

Equation (16.20) must be solved numerically in conjunction with the equations of motion of the sand layer in order to update continually the values of porewater pressure that are being developed during earthquake shaking. After the earthquake, water will continue to flow under the gradients already established. The post-earthquake hydraulic gradients are readily determined by using Equation (16.20) alone but neglecting the term representing the generation of porewater pressure. Liquefaction may also occur near the surface when the hydraulic gradients approach unity. This type of liquefaction has been described in detail by Taylor.[27]

16.3.3 Coupled dynamic analysis and porewater pressure response

Let us consider a stratum of saturated sand of infinite lateral extent resting on horizontal bedrock. The properties of the stratum may vary only in the vertical direction. Shaking is due to horizontal shear waves propagating vertically upwards. Under these assumptions, the dynamic response becomes that of a one-dimensional shear beam.

Since the sand properties may vary vertically in a random manner the horizontal sand deposit is divided up into layers (Figure 16.6), each with approximately uniform properties. This layered system is then converted to a lumped mass system by lumping one-half of the mass of each layer at the layer boundaries.[28] The masses are connected by non-linear springs with stress–strain properties given by Equation (16.2) for initial loading and Equation (16.19) for subsequent unloading and reloading. These equations reflect the non-linear, strain-dependent, and hysteretic behaviour of sands. In addition to the inherent hysteretic damping, viscous damping can be added independently if desired and we include it here for completeness of the formulation.

The differential equations of motion are, therefore,

$$[M]\{\ddot{x}\} + [C]\{\dot{x}\} + [K]\{x\} = -[M]\{\ddot{u}_g(t)\} \tag{16.21}$$

in which $[M]$ is the diagonal mass matrix, $[C]$ the viscous damping matrix, $[K]$ the non-linear stiffness matrix, $\ddot{u}_g(t)$ the earthquake accelerations at the base of

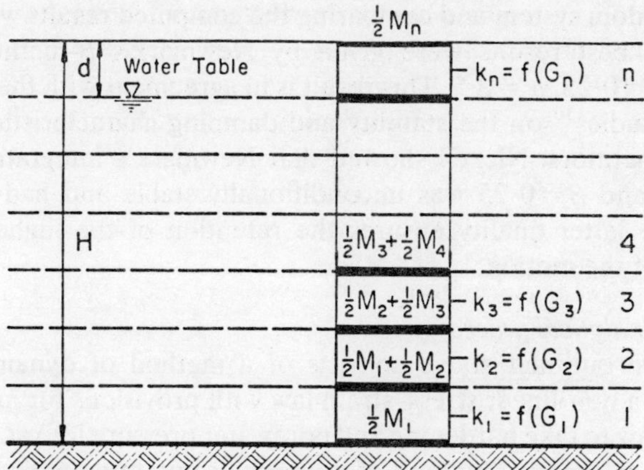

Figure 16.6 Discrete mass model for sand stratum

the layer and $\{x\}$, $\{\dot{x}\}$, and $\{\ddot{x}\}$ are the displacements, velocities, and accelerations of the masses $[M]$ relative to the base. Equation (16.22) is the typical equation of the system for the ith mass.

$$m_i \cdot \ddot{x}_i + (-c_i,\ c_i + c_{i+1},\ -c_{i+1}) \begin{Bmatrix} \dot{x}_{i+1} \\ \dot{x}_i \\ \dot{x}_{i+1} \end{Bmatrix}$$

$$+ (-k_i,\ k_i + k_{i+1},\ -k_{i+1}) \begin{Bmatrix} x_{i-1} \\ x_i \\ x_{i+1} \end{Bmatrix} = -m_i \ddot{u}_g(t) \qquad (16.22)$$

The individual stiffnesses k_i are defined as

$$k_i = \frac{f(\gamma_i)}{h_i \gamma_i} \qquad (16.23)$$

in which h_i = layer thickness, γ_i = shear strain in the ith layer, and $f(\gamma_i)$ is given by Equation (16.2) for initial loading and Equation (16.19) for subsequent loading and unloading. The maximum shear modulus and shear stress appropriate to conditions at any time in the centre of the ith layer are used in Equation (16.23). In the examples which follow stiffness proportional viscous damping was used equivalent to 2 per cent of critical damping as determined for the initial state of the sand to account for any nonhysteretic damping that may be present in saturated sands.

Equations (16.21) are solved numerically using Newmark's unconditionally stable integration operator.[29] The behaviour of several unconditionally stable integration operators was studied by investigating the behaviour of a single degree of freedom system and comparing the computed results with the exact solution. The best results were given by Newmark's β-method with the parameters $\beta = 0\cdot25$, $\alpha = 0\cdot5$. This result is in agreement with the conclusions of Nickell's studies[30] on the stability and damping characteristics of several integration operators. Nickell showed that Newmark's integration operator with $\alpha = 0\cdot5$ and $\beta = 0\cdot25$ was unconditionally stable and had no artificial damping. The latter quality ensures the retention of the higher frequency components of the motion.

(a)   Test of analytical procedure

The procedure outlined above consists of a method of dynamic response analysis using a non-linear stress–strain law with provisions for modifying the stress–strain law to take hardening and porewater pressure into account. In the absence of an alternative method of doing the same thing a direct analytical check is not available. However, an independent method of analysis exists for dry sands which ignores hardening and is embodied in the SHAKE computer program.[31] This method is based on obtaining the solution of a non-linear problem by iterative equivalent linear elastic analysis rather than by following the actual stress–strain curve. The sand deposit in Figure 16.7(a), 16.7(b) was analysed using SHAKE and the method proposed above by neglecting both porewater pressure and hardening. Sinusoidal base motion was used with a wide range of frequencies and a maximum acceleration of $0\cdot065$ g. Results showing the amplification of base accelerations at the surface for various frequencies computed by both methods are given in Figure 16.7(c). It is clear that for dry sands, neglecting hardening, both methods give similar response curves in practical terms. The resonance peak in the SHAKE response curve is to be expected since the results are based on elastic analysis.

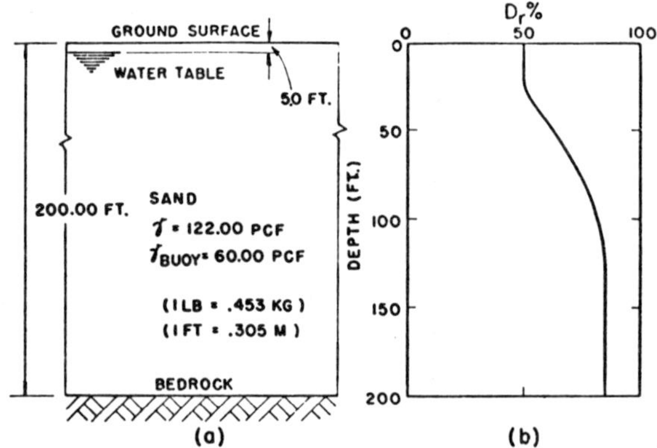

Figure 16.7(a & b)   Properties of a soil deposit

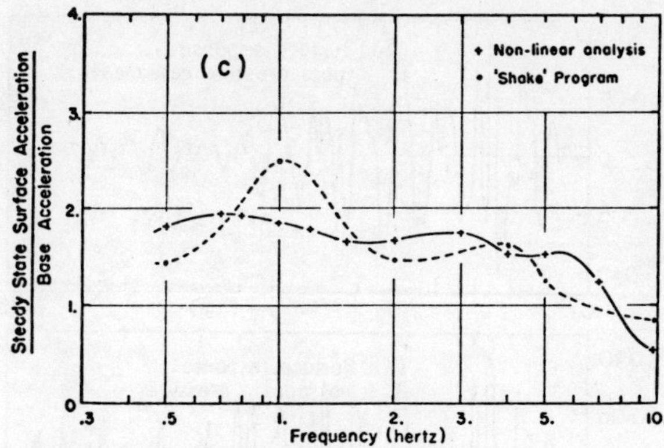

Figure 16.7(c) Results from 'shake' and proposed method (dry sand)

(*b*) *Response to earthquake motions*

The random acceleration function used by Parmalee *et al.*[32] scaled to a maximum acceleration of 0·065 g was used as input earthquake motion at the base of the sand deposit [Figure 16.7(a)]. The response analysis was carried out with and without external drainage. Initially no external drainage was allowed and there was no internal redistribution of porewater pressure. In effect, the permeability of the layer, $k = 0$. For this case, the computed time histories of surface accelerations with and without porewater pressure generation are shown in Figures 16.8(a) and 16.8(b). The input motion is given in Figure 16.8(c) for comparison. The marked changes in the characters of these motions after about 6–7 s of shaking are seen in Figure 16.9(c) to occur at the time of liquefaction of the sand at a depth of 15 ft (4·6 m). Here the steadily increasing porewater pressure has reached the level of the effective overburden pressure as indicated by the arrow. The shear strain and shear stress response at this depth are shown in Figure 16.9(a) and 16.9(b). As has been noted in cyclic simple shear tests,[14] the analysis predicts increasing shear strains as the porewater pressure approaches values associated with liquefaction. After liquefaction only very small shear stresses are transmitted by the liquefied layer. These result from the presence of viscous damping and a residual value of the shear modulus equal to three per cent of G_{m0} which is retained to maintain stability in the calculations.

Similar results may be computed at various depths in the deposit. From such data the distributions of porewater pressures at various times may be plotted as shown in Figure 16.10(a) in which the distributions of porewater pressure at $t = 4$ s and $t = 8$ s are given. The solid line in Figure 16.10(a) represents the initial effective stress distribution, and it is evident that liquefaction is about to occur under continued shaking.

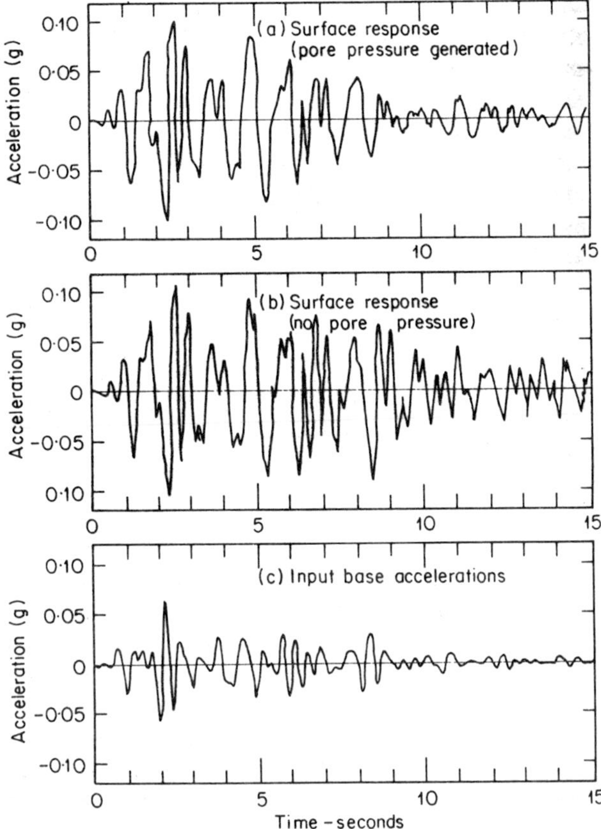

Figure 16.8 Surface acceleration response to base motions

Porewater pressure distributions that develop at various times when drainage occurs during the earthquake are given in Figures 16.10(b) and 16.10(c) for uniform permeabilities $k = 0\cdot0003$ ft/s (10^{-2} cm/s) and $k = 0\cdot03$ ft/s (1 cm/s) respectively. It can be observed that drainage increased the time to liquefaction and raised the level at which liquefaction first occurred for $k = 0\cdot0003$ ft/s (10^{-2} cm/s) whereas the more permeable sand did not liquefy at all under the given earthquake loading. It is clear from Figure 16.10(c) that at about $t = 15$ s or so, the earthquake motions are such that porewater pressure dissipation is occurring faster than porewater pressure generation.

It is interesting to consider what might happen in a saturated sand stratum sealed on both sides by impermeable surfaces. In this case, no external drainage can occur but an internal redistribution of porewater pressure will take place in

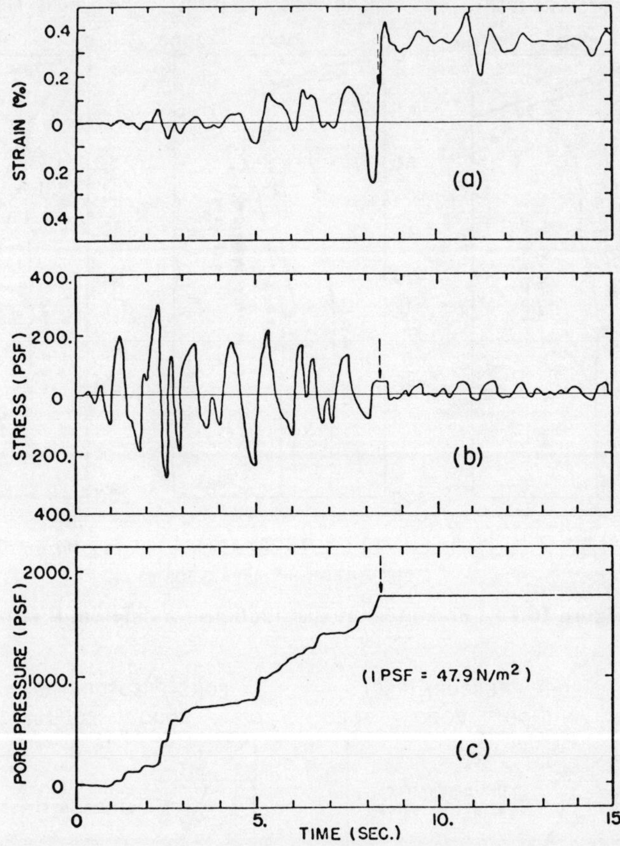

Figure 16.9 Strain, stress and pore pressure response at
depth of 15 ft (4·6 m)

accordance with Equation (16.20). Some results from a parametric study are
given in Figures 16.11(a) and 16.11(b) for a 50 ft (15·3 m) layer from which
drainage is prevented. The layer has a uniform relative density of $D_r = 45$ per
cent and there is an impermeable boundary at the water table (W.T.) at a depth
of 5 ft (1·5 m). The base of the layer was shaken by the N-S acceleration
component of the El Centro earthquake of 1940. The results in Figure 16.11
indicate that redistribution of porewater pressure in a sealed saturated stratum
of sand of uniform relative density raises the level of initial liquefaction and
decreases the time to liquefaction.

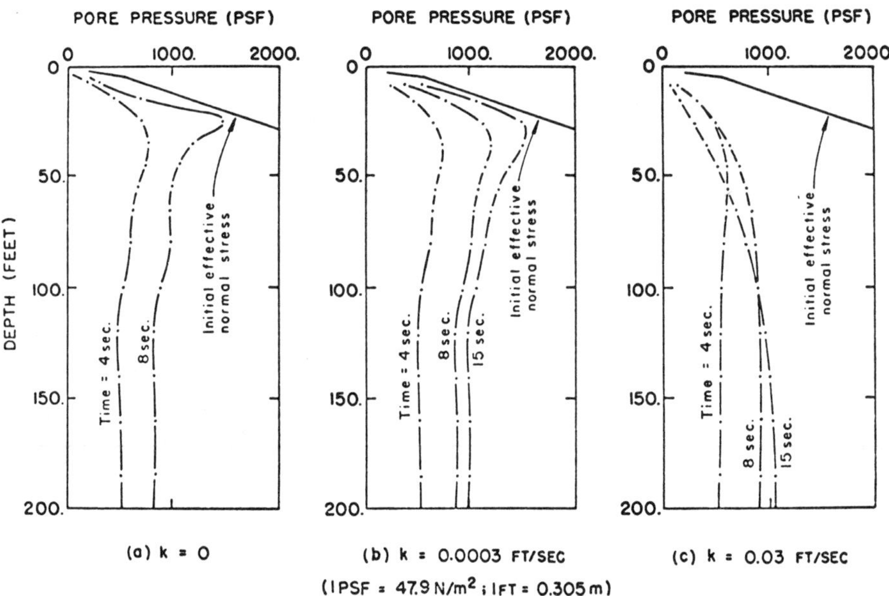

Figure 16.10   Pore pressure distributions for different *k* values

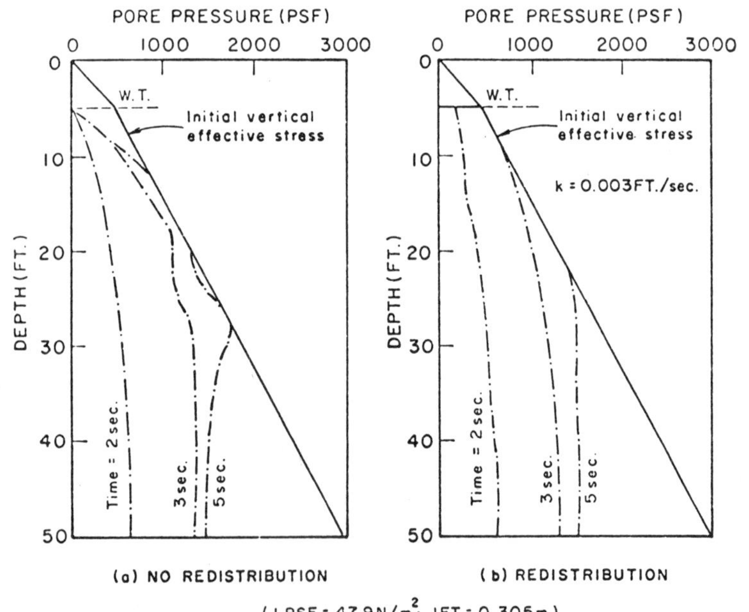

Figure 16.11   Effect of internal redistribution of porewater pressure

16.3.4 Comments

The method of analysis developed and applied above is a non-linear effective stress analysis for the dynamic response of dry or saturated sands. The method consists of a procedure for dynamic analysis, a specific stress–strain law, and a method for computing volume changes and porewater pressures concurrently with the dynamic response. Each of the elements of the method has been tested independently to ensure a reasonable correlation with experimental results.

To use the method in practice the constants describing the volume change characteristics of the sands and the associated porewater pressures must be measured for each sand at the present stage of development. More extensive testing of sands may allow generalizations to be made about the values of these constants which would be adequate for practical applications.

The method predicts the phenomenological features of the dynamic response of saturated sand layers that commonly occur as the porewater pressure rises in the sand during earthquake shaking. It allows the distribution of porewater pressure and the effects that drainage and internal flow have on the location and time of liquefaction to be determined quantitatively.

The method is efficient computationally. The analysis of the 200 ft (61 m) layer took 18 s of central processing time which includes recording data on tape every $0 \cdot 01$ s for plotting and giving hard copy every second. The corresponding time for the 50 ft ($15 \cdot 3$ m) layer was 10 s.

16.4 ITERATIVE ELASTIC ANALYSIS

A stratum of saturated sand of infinite lateral extent resting on horizontal bedrock is shown in Figure 16.6. The layer is of thickness, H, and the water table is at a depth, d, below the surface. The properties of the sand layer vary only in the vertical direction. Shaking is due to shear waves propagating vertically upwards. Under these assumptions, the dynamic response becomes that of a one-dimensional shear beam.

Since the layer properties may vary vertically in a random manner, the shear beam is represented by a discrete mass model shown in Figure 16.6. The masses are connected by appropriate springs and dampers. The stress–strain characteristics of sands are non-linear, hysteretic and strain-dependent. These characteristics are introduced into the model by using the equivalent linear strain dependent shear moduli and viscous damping ratios defined by Seed and Idriss,[2] and shown in Figure 16.12. The elemental spring stiffnesses, K_i, are based on the current moduli, G_i, at the middle of each layer of thickness, h_i; $K_i = (G_i/h_i)$.

The equations of motion of the discrete mass system are given by

$$[M]\{\ddot{x}\} + [C]\{\dot{x}\} + [K]\{x\} = -[M]\{\ddot{u}_g(t)\} \tag{16.24}$$

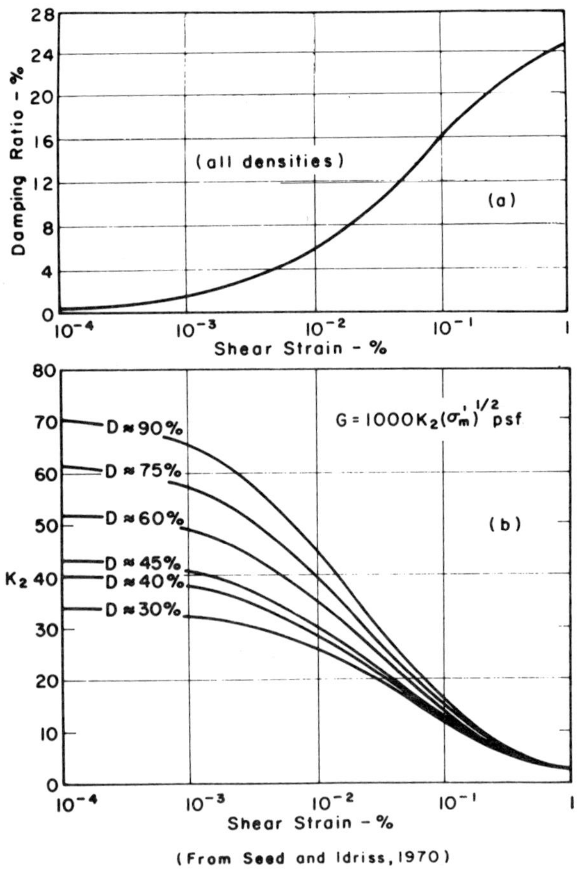

Figure 16.12 Moduli and damping ratios for sand

in which $[M]$ is the diagonal mass matrix, $[C]$ the damping matrix, $[K]$ the stiffness matrix, $\ddot{u}_g(t)$ the base rock acceleration and $\{x\}$, $\{\dot{x}\}$, and $\{\ddot{x}\}$ are the displacement, velocity and accelerations of the masses $[M]$ relative to the bedrock. Since an equivalent linear method of analysis will be used, the equations are uncoupled by the transformation

$$\{y\} = [\phi]^T [M] \{x\} \tag{16.25}$$

in which $[\phi]$ is such that

$$[\phi]^T [M][\phi] = [I] \tag{16.26}$$

$$\{\phi_i\}^T [C]\{\phi_i\} = 2\lambda_i \omega_i \tag{16.27}$$

and

$$\{\phi_i\}^T[K]\{\phi_i\} = \omega_i^2 \tag{16.28}$$

where $\lambda_i = $ damping in the ith mode and $\omega_i = $ the ith natural frequency. The equations then reduce to normal mode equations

$$\ddot{y}_i + 2\lambda_i\omega_i\dot{y}_i + \omega_i^2 y_i = -\{\phi_i\}^T[M]\{\ddot{u}_g(t)\} \tag{16.29}$$

An iterative procedure for solving Equation (16.29) has been described by Seed and Idriss.[28] An initial strain distribution is assumed and corresponding damping ratios and shear moduli selected from Figure 16.12. The response of the system to the *entire* prescribed base motion record is then computed assuming elastic behaviour and the average strains within the soil profile determined. If these strains differ significantly from the assumed strains, new values for the moduli and damping are selected and the system of equations again solved. The procedure is repeated until two consecutive sets of effective strains are obtained which agree within prescribed tolerances. In this procedure, the modifications in the current moduli and damping ratios due to strain are made for the effective strain response to the entire earthquake record and the effect of porewater pressure is neglected. In this paper it is desired to make adjustments to the soil properties as the porewater pressure is generated and consequently the above procedure must be modified.

The analysis incorporating the effects of porewater pressure generation proceeds as follows. Initial values for shear modulus and damping ratio are assumed to be those for a strain, $\gamma = 10^{-4}\%$. The analysis is then carried out for a short time interval, Δt, as described above until a consistent set of average strains are obtained. At this stage the corresponding increments in porewater pressure, Δu, occurring during the time interval, Δt, are computed as described earlier for the volumetric strain increment.

The increment of porewater pressure, Δu, reduces the effective stress to $(\sigma_v' - \Delta u)$ and so reduces the shear modulus according to Equation (16.1). There is some data on the variation of damping ratio with effective stress but considering its sparseness and the many uncertainties connected with damping itself it did not seem worthwhile to introduce a correction to the damping for changes in effective stress.

Thus, at the end of a time interval, Δt, the stiffness matrix $[K]$ is modified for the effects of strain and porewater pressure and the damping matrix $[C]$ is modified for strain effects only. These instantaneous changes in stiffness and damping will not affect the instantaneous values of displacements and velocities but will alter the instantaneous accelerations to maintain dynamic equilibrium. Dynamic equilibrium requires that the forces before and after the changes in stiffness and damping must be equal or

$$[M]\{\ddot{x}\}_1 + [C]_1\{\dot{x}\}_1 + [K]_1\{x\}_1 = [M]\{\ddot{x}\}_2 + [C]_2\{\dot{x}\}_1 + [K]_2\{x\}_1$$

$$\tag{16.30}$$

where the subscript, 1, indicates 'before modification of properties,' and subscript, 2, indicates 'after modification.' Therefore:

$$\{\ddot{x}\}_2 = \{\ddot{x}\}_1 + [M]^{-1}[[[C]_1 - [C]_2]\{\dot{x}\}_1 + [[K]_1 - [K]_2]\{x\}_1] \qquad (16.31)$$

where $\{\ddot{x}\}_2$ is the acceleration after the change in stiffness in each iteration and $[K]_2$ has been modified not only for strains but also for porewater pressures. The damping matrix for use in Equation (16.27) is conveniently constructed as follows: transformation to normal co-ordinates requires that

$$[\phi]^T[C][\phi] = \begin{bmatrix} 2\lambda\omega_1 & 0 & 0 \\ 0 & 2\lambda\omega_2 & \\ 0 & & 2\lambda\omega_n \end{bmatrix} = [A] \qquad (16.32)$$

$$[C] = [\phi^T]^{-1}[A][\phi]^{-1} \qquad (16.33)$$

or

$$[C] = [M][\phi][A][\phi]^T[M] \qquad (16.34)$$

Thus, $\{x\}$, $\{\dot{x}\}$, and $\{\ddot{x}\}$ at the beginning of the next time interval are now known, and integration proceeds as before in normal co-ordinate space.

The time interval of integration used in the analysis was $0 \cdot 01$ s and the shear modulus was changed every $0 \cdot 5$ s. The time interval for changing the modulus was established by a parametric study using the forcing functions described below and time intervals ranging from $0 \cdot 1$ to $1 \cdot 0$ s.

16.5 RESULTS

The dynamic response of a horizontal stratum of saturated sand subjected to a variety of base accelerations was determined by the total stress procedure and the effective stress method described above. The object of the analyses is to investigate the differences in ground accelerations, dynamic stresses and strains determined by the two methods and to consider the implications of these differences for engineering practice. All the results presented below are for a saturated sand stratum 50 ft ($15 \cdot 24$ m) deep at a relative density, $D_r = 45$ per cent with the water table 5 ft ($1 \cdot 52$ m) below the surface except as noted.

16.5.1 Harmonic base input

A sinusoidal acceleration with a frequency of 2 cycles per second and with a maximum acceleration of $0 \cdot 805$ ft/s^2 ($0 \cdot 245$ m/s^2) was applied to the base of the stratum. When analysed by the total stress method the motion in any layer quickly stabilized to a sinusoidal response at 2 cycles/s. The acceleration history for the surface of the layer is presented in Figure 16.13(a) which shows a

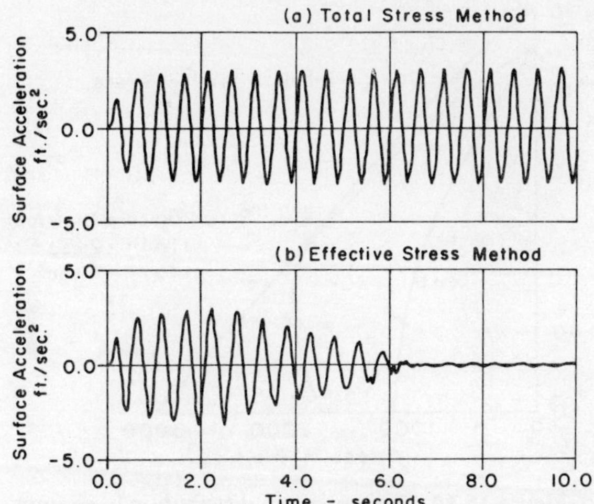

Figure 16.13 Surface accelerations computed by (a)
total stress method and (b) proposed effective stress
method for harmonic input

$(1 \text{ ft/s}^2 = 0\cdot3048 \text{ m/s}^2)$

steady sinusoidal response of 2 cycles/s with a magnitude of approximately $3\cdot0 \text{ ft/s}^2$ $(0\cdot91 \text{ m/s}^2)$ indicating an amplification of about four.

The surface accelerations determined by the effective stress analysis show a different response history [Figure 16.13(b)]. The response is again about 2 cycles/s and the maximum value is also about $3\cdot0 \text{ ft/s}^2$ $(0\cdot91 \text{ m/s}^2)$ which occurs at about 2 s. It can be seen that up to this time the acceleration histories by both methods are quite comparable. This is to be expected because in the early stages of the shaking the porewater pressures have not developed to the point where they have a significant effect on the dynamic response characteristics. After about 2 s the accelerations determined by the effective stress method steadily decrease and are effectively zero after about 6 s even though shaking continues. The accelerations determined by the total stress method persist at their maximum value of $3\cdot0 \text{ ft/s}^2$ $(0\cdot91 \text{ m/s}^2)$ at 2 cycles/s.

The reason for the difference in response becomes clear when the development of porewater pressure in the sand stratum is examined. The distribution of the porewater pressure after 2 s and 6 s is shown in Figure 16.14 in which the distribution of initial effective stress is also shown. It is clear that the porewater pressures after 2 s are far from causing liquefaction and are low enough not to affect significantly the stiffness of the stratum which varies as the square root of the effective stress. After 6 s, however, the porewater pressure at a depth of $27\cdot5$ ft $(8\cdot38$ m), corresponding to the middle layer 5, is equal to the effective overburden pressure and so layer 5 is at the point of liquefaction. At this stage

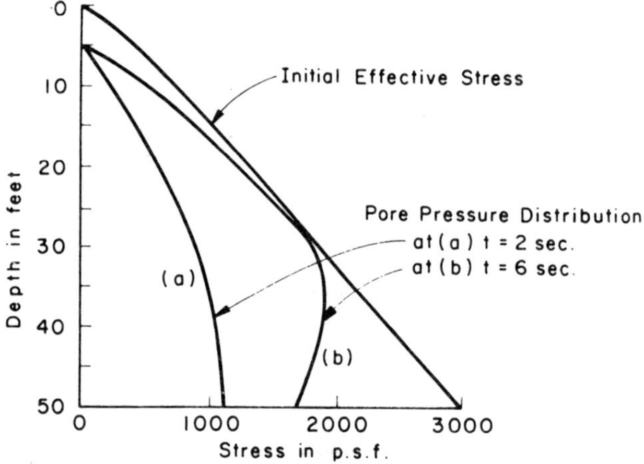

Figure 16.14   Pore pressure distribution in stratum
$(1\ \text{p.s.f.} = 47{\cdot}9\ \text{N/m}^2)$

no shear stresses can be transmitted across layer 5 and so the accelerations at the surface drop to zero. In practice, due to the viscosity of the liquefied soil layer very low, long period shear stresses and accelerations will be transmitted. In the model to control the stability of the calculation some residual stiffness is also retained by specifying that the vertical effective stress, σ'_v, be not less than 1/10,000 of the initial vertical effective stress.

Prior to liquefaction between about 2 s and 6 s the porewater pressures have developed to a significant level as may be seen in Figure 16.15 which shows the development of porewater pressure with time in the critical layer 5. At the

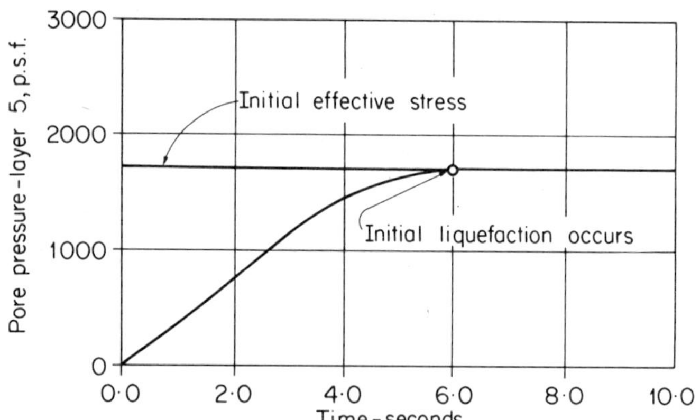

Figure 16.15   Pore pressure development in layer 5
$(1\ \text{p.s.f.} = 47{\cdot}9\ \text{N/m}^2)$

levels of porewater pressure developed between about 2 s and 5 s the sand stratum has softened considerably leading to longer instantaneous natural periods and so a reduction in response to the 2 cycles/s input. Thus the values of surface accelerations decrease as shown in Figure 16.13(b).

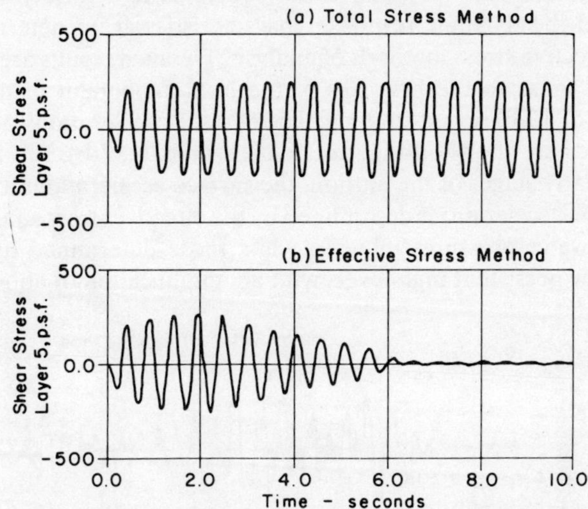

Figure 16.16 Shear stresses computed by (a) total stress method and (b) proposed effective stress method for harmonic input

$$(1 \text{ p.s.f.} = 47 \cdot 9 \text{ N/m}^2)$$

It is also interesting to examine the shear stresses in layer 5 which is the first layer to liquefy. The stresses computed by the total stress method are shown in Figure 16.16(a). The response is again sinusoidal in form at 2 cycles/s. The steady state shear stress amplitude is about 250 psf (12,000 N/m^2). The corresponding results from the proposed effective stress analysis are shown in Figure 16.16(b). The initial parts of both records are again very similar indicating that up to about 2 s of shaking the effect of the porewater pressure does not significantly modify the motion. However, after about 2 s and continuing up to 6 s when liquefaction occurs, the shear stresses computed by the effective stress analysis reduce steadily to near zero.

The computed strains exhibit the type of behaviour already familiar from cyclic loading tests on sand; as the porewater pressure increases the strains increase and increase more rapidly as liquefaction occurs until stabilizing finally at values determined by the residual shear modulus.

The responses to sinusoidal base input show very clearly the differences in the dynamic responses of saturated layers determined by the total stress and effective stress methods. In practice, however, one is concerned with the response to actual earthquake motions.

16.5.2   Earthquake accelerations input

The sand stratum described above was subjected to a variety of earthquake motions and the dynamic response characteristics again determined by the total and effective stress methods of analysis. Detailed results are given for the dynamic response to the first 10 s of the N-S component of the El Centro earthquake of 1940 scaled to 0·1 g. The surface accelerations determined by both methods are shown in Figures 16.17(a) and 16.17(b). It is again evident that in the early stages of the motion, the surface accelerations are similar but thereafter the accelerations determined by the effective stress analysis decrease as the porewater pressures increase while those determined by the current method show persistent high levels with an amplification of about 2.

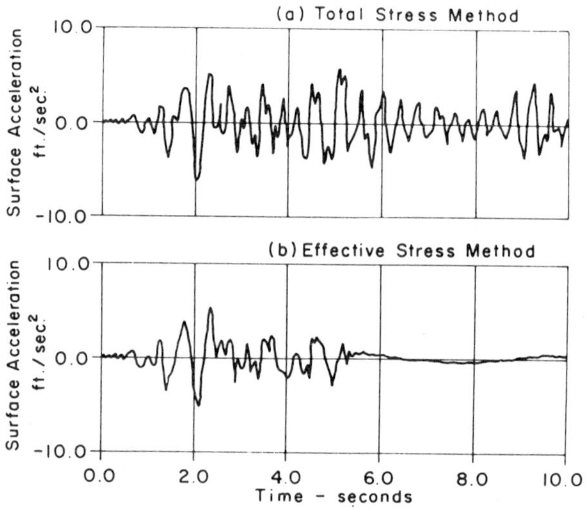

Figure 16.17   Surface accelerations computed by (a) total stress method and (b) proposed effective stress method for El Centro (1940) scaled to 0·1 g

$$(1 \text{ ft/s}^2 = 0·3048 \text{ m/s}^2)$$

The shear stresses in layer 6, which is the layer that liquefies first, are shown in Figures 16.18(a) and 16.18(b). Again both methods give similar results in the early stages of motion. Thereafter, as the porewater pressure increases the shear stresses computed by the effective stress method decrease until ultimately near liquefaction only very low stresses are transmitted.

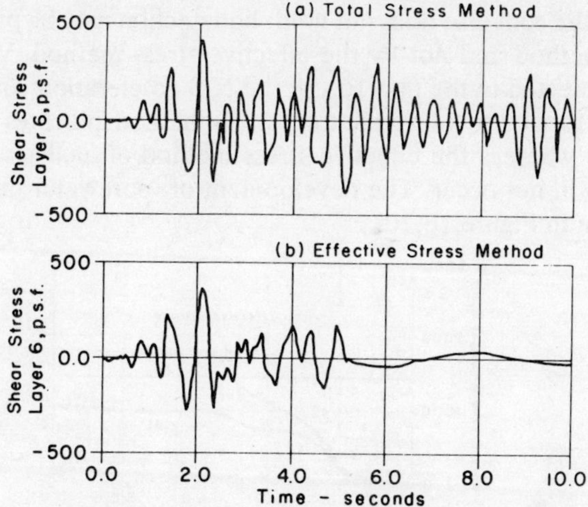

Figure 16.18 Shear stresses computed by (a) total stress method and (b) proposed effective stress method for El Centro (1940) scaled to 0·1 g

$(1 \text{ p.s.f.} = 47 \cdot 9 \text{ N/m}^2)$

The porewater pressures that these sets of shear stresses would develop over time in a simple shear specimen of the same sand under the initial stress conditions in layer 6 were computed and the results are shown in Figure 16.19.

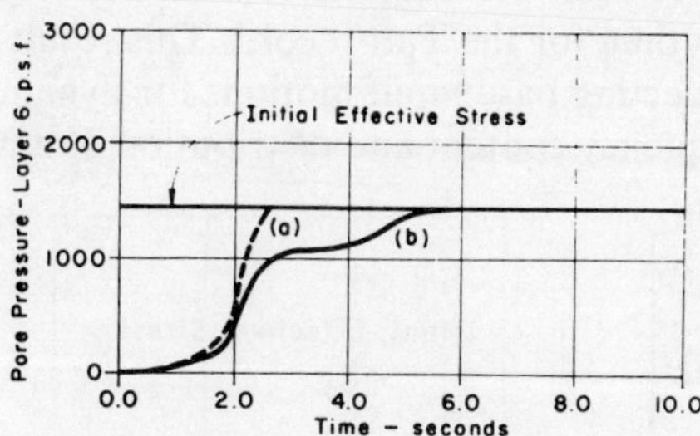

Figure 16.19 Comparison of pore pressure development in layer 6 by (a) current and (b) proposed methods under El Centro (1940) scaled to 0·1 g

$(1 \text{ p.s.f.} = 47 \cdot 9 \text{ N/m}^2)$

It may be noted that both methods predict that liquefaction will occur but the shear stresses computed by the total stress method lead to liquefaction much sooner. For all analyses performed in this study, whenever liquefaction was predicted by the effective stress analysis it was also predicted by the total stress method. But the converse does not hold; liquefaction may be predicted by the total stress method and not by the effective stress method. When the sand stratum is subjected to the first 10 s of the N-S acceleration component of El Centro (1940) scaled to 0.05 g the total stress method indicates liquefaction in layer 6 in 5 s whereas the effective stress method of analysis indicated that liquefaction will not occur. The development of porewater pressure in both cases is shown in Figure 16.20.

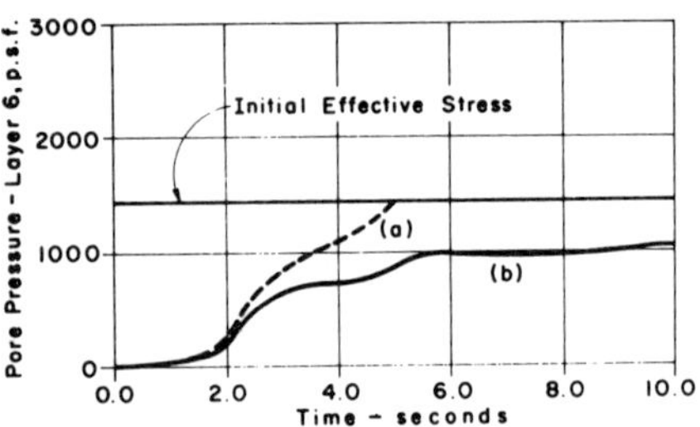

Figure 16.20 Comparison of pore-pressure development in layer 6 by (a) current and (b) proposed methods under El Centro (1940) scaled to 0·05 g

The development of porewater pressures with time due to shaking by the first 10 s of the accelerations of El Centro (1940) and Taft (1952) both scaled to 0·1 g are shown in Figure 16.21 for layer 6. It may be seen that for the El Centro record the porewater pressures increase at a faster rate and liquefaction occurs much sooner than for the Taft record. This result confirms the importance attached to selecting base input motions for dynamic response analyses with the proper frequency content and distribution of amplitudes.[33,28]

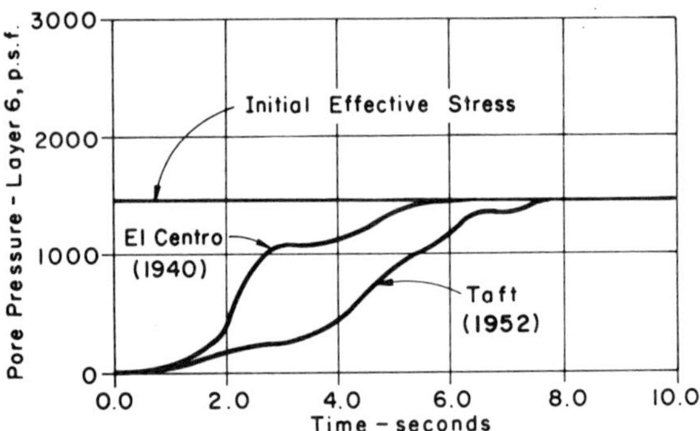

Figure 16.21 Development of porepressure in sand layer 6 for two different earthquakes both scaled to 0·1 g

(1 p.s.f. = 47·9 N/m²)

In general, from the studies completed to date it seems that there may be significant differences between the dynamic responses to earthquake shaking of a saturated sand stratum computed by the total stress and the proposed effective stress methods whenever the porewater pressure builds up to more than about 30 per cent of the over-burden pressure.

Many examples of liquefaction occurring in the field have been reported in the literature but in no case is the information available that would allow a rigorous check on the capability of the effective stress methods. The required information consists of (a) the *in situ* properties of the soil, (b) the earthquake accelerations at bedrock, and (c) the ground accelerations at the surface. If this information were available it would be possible to compare in detail the computed and measured accelerations at the ground surface.

During the Niigata earthquake of 1964 an acceleration record was obtained from a recorder in the basement of an apartment building founded on sand which liquefied.[33] This record is shown in Figure 16.22(a). The characteristics of the record may be described in general terms as follows: Initially there is an interval of high frequency motion in which the accelerations generally are less than half the peak value. This is followed by an interval in which the motion has significantly longer periods and in which the peak acceleration is reached. At the end of this interval the recorded motion shows very long predominant periods and rather low amplitudes.

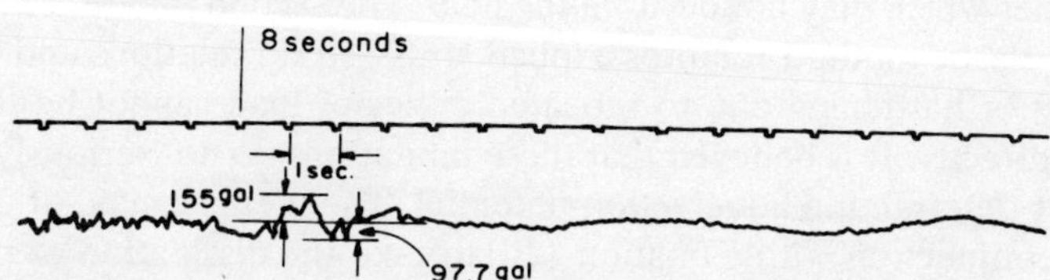

Figure 16.22(a) (*After Seed and Idriss*[33]) Earthquake accelerogram at Basement of No. 2 Apartment Building, Kawagishi, Niigata, Japan

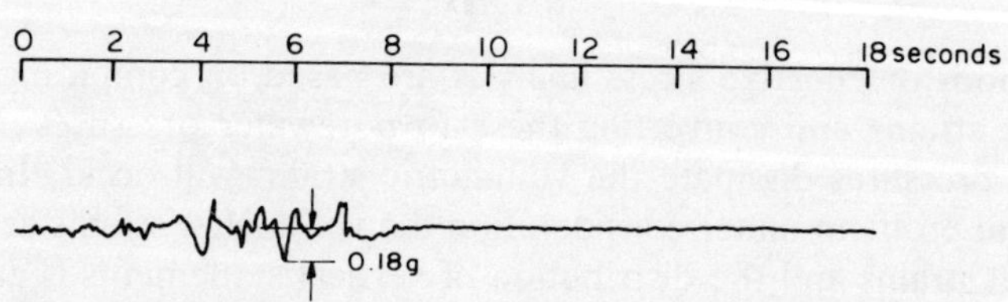

Figure 16.22(b) Computed surface acceleration at Niigata Site using the 1952 Taft Earthquake scaled to 0·125 g as base input

The Niigata site was analysed using the data on *in-situ* densities and the soil profile given by Seed and Idriss[33] and using the stress–strain relations in Figure 16.12. The base accelerations at the Niigata site are not known so we adopted the base motions used by Seed and Idriss[33] in the original study of the Niigata site, i.e., the earthquake accelerations recorded during the Taft earthquake in 1952 scaled to a maximum acceleration of 0.125 g. The computed surface accelerations are shown in Figure 16.22(b). The results indicate liquefaction after about 8 s of shaking. As in the Niigata record there is an initial period of high frequency motion followed by longer period motions in which the acceleration reaches its maximum value, and thereafter motions occur of much longer periods and reduced amplitudes consistent with the residual resistance of the layer. The peak acceleration is 0.18 g compared with the measured 0.16 g and the distribution of amplitudes in the record is different as might be expected since we are not using the true base accelerations.

16.6 COMMENTS

There are two important differences between the non-linear and the iterative elastic methods of dynamic effective stress analysis. The first is that the iterative elastic method cannot give permanent deformations in the directions of motion and, since it is elastic, it may give pseudo-resonant response at certain frequencies which may not occur in the field. The second difference is that the iterative elastic method requires explicit stress–strain relations and therefore the effect of hardening due to repeated cycles of load cannot be taken into account directly. It is believed that these limitations do not seriously affect its utility for determining liquefaction potential. The method may, of course, be used in conjunction with Equation (20) to take the dissipation of porewater pressure into account.

16.7 DYNAMICALLY INDUCED SETTLEMENTS IN SATURATED SAND

Both methods of effective stress analysis are based on computing potential volumetric strains and converting these to porewater pressures. When the porewater pressures dissipate the volumetric strains will occur. In the one-dimensional problem under consideration the volumetric strains are equal to the vertical strains and the distribution of vertical settlements is easily computed from the distribution of vertical strains. Settlements under cyclic loading have been analysed by Finn and Byrne[10] using the iterative elastic method of analysis and by Finn and Lee,[34] using the non-linear method of analysis.

Settlements in a sand layer 50 ft (15·2 m) thick were computed for a range of earthquake accelerations using the iterative elastic method. The relative density of the sand varied from $D_r = 45$ per cent to $D_r = 80$ per cent. The first 10 s of the N-S acceleration component of the El Centro (1940) earthquake was used as input motion at the base of the sand layer. This earthquake record was scaled by various factors to provide different acceleration histories.

For the purpose of analysis, the layer was divided into 10 slices each 5 ft (1·52 m) thick. The dynamic shear stresses and associated shear strains for a typical slice are shown in Figures 16.23(a) and 16.23(b). The volumetric strains generated by the strain history in Figure 16.23(b) were computed using Equation (16.15) and are shown in Figure 16.23(c). In a horizontal layer of

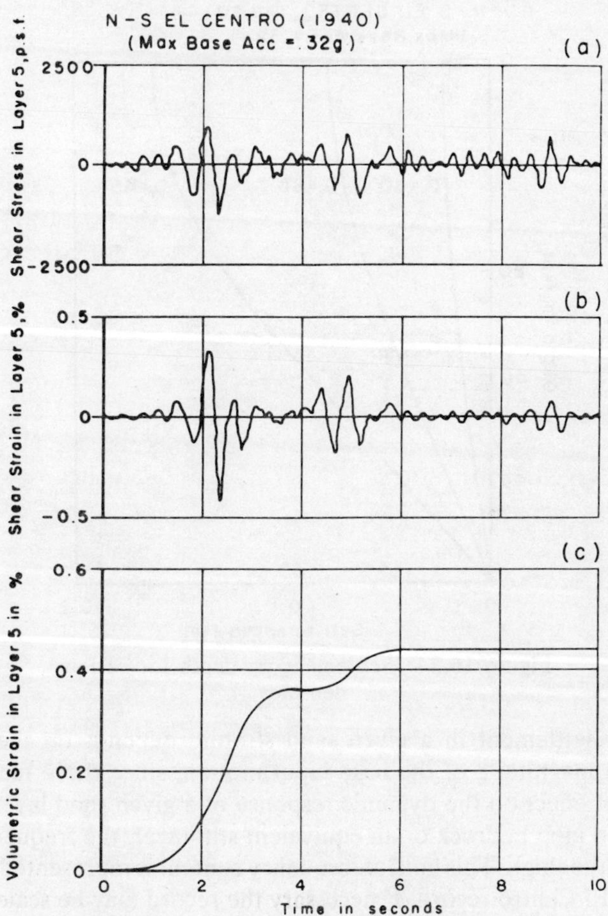

Figure 16.23 Dynamic stresses and associated strains

sand these strains are equal to the vertical strains and when multiplied by the thickness of the slice give the total settlement of this slice as a function of time. When similar results for other slices are summed, the distribution of settlements throughout the depth of the layer are obtained as a function of time.

The distribution of total settlements under shaking by the N-S component of El Centro (1940) with maximum acceleration of 0.32 g is shown in Figure 16.24 for relative densities $D_r = 45$ per cent, 60 per cent, and 80 per cent. The dependence of settlement on relative density is obvious. It should be noted that even very dense sands ($D_r = 80$ per cent) can undergo appreciable settlement under strong shaking. In the example shown, the settlement was of the order of 0.25 in (0.64 cm) when the relative density $D_r = 80$ per cent.

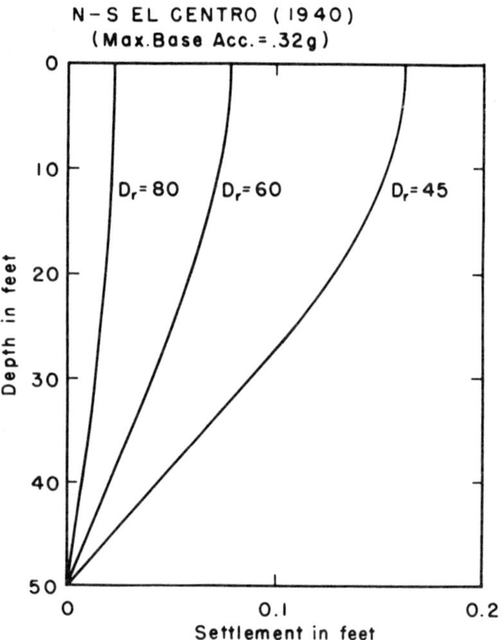

Figure 16.24 Settlement vs. depth for various densities

The total settlement in a given sand stratum depends on the frequency content and magnitude of the base input motion, since these factors have a predominant effect on the dynamic response of a given sand layer. Since the motion is fed in at bedrock or an equivalent stiff layer, the frequency content should be rather high. This kind of frequency content is represented reasonably well by the El Centro record. If necessary the record may be scaled to higher frequencies.

The effect of variations of the maximum accelerations are shown as functions of the relative density in Figure 16.25. The non-linear nature of the response of the sand is clearly evident.

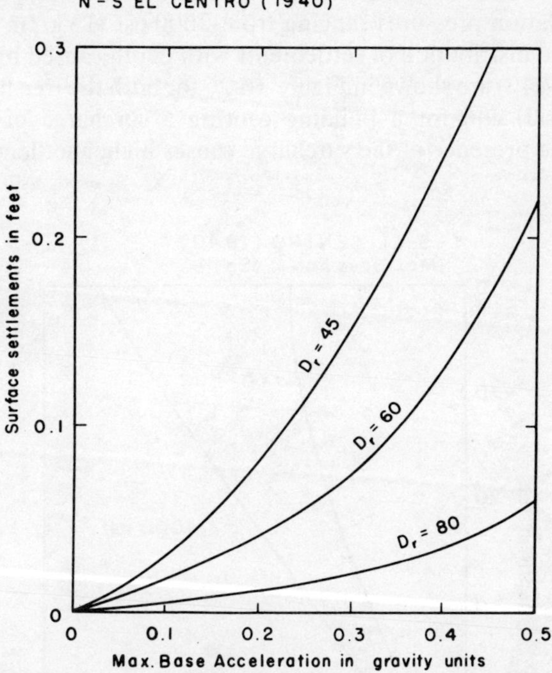

Figure 16.25 Surface settlements vs. maximum base acceleration

The settlement of a sand will be affected by the existence of a structure founded on the sand. The weight of the structure increases the mean normal effective stress in the sand and so increases the shear modulus G, in accordance with Equation (16.1). The increased values of G lead to increased resistance to shearing strain and hence to volume change. This effect tends to reduce the total settlement. However, the mass of the building has an opposite effect. During an earthquake, the base shear generated in the building by inertia forces generates additional shear stresses and strains in the sand and tends to increase settlements.

A very approximate model of the action of a structure was used to investigate these effects. This model took into account only the increased effective pressures between the sand grains due to the weight of the structure and the increased shear stresses generated by the mass of the structure. The flexibility of the structure and such vibration modes as rocking were not taken into account as they would be incompatible with the simple one-dimensional analysis used.

The effect of a structure on settlements may be modelled by adding an extra slice to the discrete mass system with a very high stiffness and a mass sufficient to cause foundation pressures ranging from 2000 psf (1 kg/cm^2) to 8000 psf (4 kg/cm^2). The distribution of settlements with depth caused by the first 10 s of El Centro (1940) are shown in Figure 16.26 for both the free field case of no surcharge ($\sigma = 0$) and for a building causing a surcharge of $\sigma = 4000$ psf (2 kg/cm^2). The presence of the surcharge causes higher settlements.

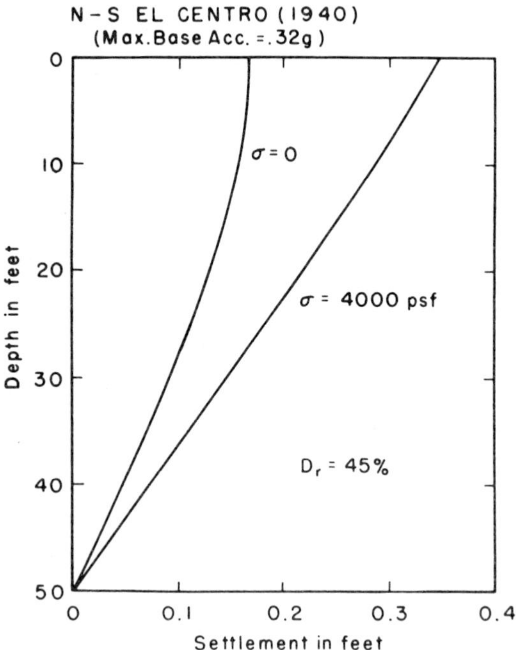

Figure 16.26   Effect of surface loads on settlements

16.8   APPLICATION AND EXTENSION OF METHODS

The proposed methods of effective stress analysis are one-dimensional and apply strictly only to a horizontal sand stratum in which there are no shear stresses on horizontal planes before the excitation occurs due to waves or earthquakes. Such stress conditions would exist under the centre of a large gravity structure. At the edges of such a structure however, shear stresses act on horizontal planes. If we assume that as far as the response of the saturated sand is concerned that shearing deformations are the most important factor then we can still treat the problem as one-dimensional provided we obtain the volume change data under conditions of various levels of initial static shear stress; the constants in Equation (16.15) must be determined for each relative density at various levels of static shear stress.

It is also possible to carry out two- and three-dimensional dynamic analyses using G and the bulk modulus K as parameters and to compute the shear strains and from these using the appropriate form of Equation (16.15) to compute the porewater pressures as described previously. Work is now in progress at the University of British Columbia to measure the volume change characteristics of sands in shear under various levels of initial static shear stress. This data will allow the feasibility of two- and three-dimensional dynamic effective stress analysis for saturated sands to be checked.

16.9 CONCLUSION

A review has been presented of the current state of knowledge of the dynamic effective stress analysis of saturated sands under cyclic loading. Two different methods of dynamic analysis have been presented: one is a non-linear method which integrates the equations of motion directly while taking into account progressively the effects of straining, hardening, and porewater generation and dissipation; the other solves the equations of motion using an iterative linear elastic analysis.

Each of the proposed methods consists of four elements: (i) a method of solving the equations of motion based on effective stresses, (ii) a stress–strain law for sands, (iii) a procedure for computing volume changes caused by shear strains, and (iv) a procedure for converting potential volume changes to porewater pressures to allow the determination of the effective stresses required in (i). Any one of these elements is open to improvement and almost certainly (ii), (iii) and (iv) will require further development. Work is going on at many universities and laboratories in these areas and rapid progress appears likely. The methods outlined above are first steps in the direction of realistic dynamic effective stress analysis.

However, even at their present stage of development both methods can predict reliably the development of porewater pressures in laboratory cyclic simple shear tests and give results that check reasonably well with available data from the field. The extension of these methods to two- and three-dimensional problems will require the development of much more general stress–strain relations.

Acknowledgements

The financial support of the National Research Council of Canada under grant No. 1498 is gratefully acknowledged. The authors are grateful to Desirée Cheung for her skilled typing of the manuscript and to K. Wah Lee and Richard Brun for their expeditious and careful preparation of the sketches.

REFERENCES

1. Finn, W. D. L., Byrne, P. M., and Martin, G. R. (1976). 'Seismic response and liquefaction of sands', Department of Civil Engineering, University of British Columbia, Vancouver, Soil Mechanics Series No. 24, 1975. Journal of the *Geotechnical Engineering Division, ASCE*, Vol. 102, No. GT8, Proc. Paper 12323, August 1976, 841–856.
   University, Hamilton, Paper No. 8, 1–8.
2. Seed, H. B. and Idriss, I. M. (1970). 'Soil moduli and damping factors for dynamic response analyses', *Report No. EERC 70-10*, University of California Earthquake Engineering Research Centre, Berkeley, December 1970.
3. Seed, H. B., Lee, Kenneth K., and Idriss, I. M. (1969). 'Analysis of Sheffield dam failure', *Journal of the Soil Mechanics and Foundations Division, ASCE*, **95**, No. SM6, November 1969, 1453–1490.
4. Seed, H. B., Martin, P. P., and Lysmer, J. (1975). 'The generation and dissipation of porewater pressures during soil liquefaction', *Report No. EERC 75-26*, Earthquake Engineering Research Centre, University of California, Berkeley, California, August 1975.
5. Sneddon, I. N. (1957). *Elements of Partial Differential Equations*, McGraw-Hill. 274–275.
6. Streeter, V. L., Wylie, E. B., and Richart, F. E. (1974). 'Soil motion computations by characteristics method', *Journal of the Geotechnical Engineering Division, ASCE*, **100**, No. GT3, March 1974, 247–263.
7. Liou, C. P., Streeter, V. L., and Richart, F. E. (1976). 'A numerical model for liquefaction', ASCE National Convention, Philadelphia, Specialty Session, Liquefaction Problems in Geotechnical Engineering, *Preprint No. 2752*, 169–199, September 1976.
8. Finn, W. D. L., Lee, K. W., and Martin, G. R. (1976). 'Constitutive laws for sand in dynamic shear', Proceedings of the Second International Conference on Numerical Methods in Geomechanics, Blacksburg, Virginia, June 20–25, 1976, Vol. I, 270–281.
9. Finn, W. D. L., Lee, K. W., and Martin, G. R. (1976). 'An effective stress model for liquefaction', ASCE National Convention, Philadelphia, Specialty Session, Liquefaction Problems in Geotechnical Engineering, *Preprint No. 2752*, 169–199, September 1976. Also in *Journal of the Geotechnical Engineering Division, ASCE*, **103**, No. GT6, Proc. Paper 13008, June, 1977, pp. 517–533.
10. Finn, W. D. L. and Byrne, P. M. (1975). 'Settlements in sands during earthquakes', Proceedings, Second Canadian Conference on Earthquake Engineering, McMaster University, Hamilton, Paper No. 8, 1–8. Also in *Canadian Geotechnical Journal*, **13**, 4, 1976, pp. 355–363.

11. Martin, G. R., Finn, W. D. L., and Seed, H. B. (1974). 'Fundamentals of liquefaction under cyclic loading', *Journal of the Geotechnical Division, ASCE*, **101**, No. GT5, May 1975, 423–438. Also in *Soil Mechanics Series No. 23*, University of British Columbia, Vancouver.

12. Silver, M. L. and Seed, H. B. (1971). 'Volume changes in sands during cyclic loading', *Journal of the Soil Mechanics and Foundations Division, ASCE*, **97**, No. SM9, September 1971, 1171–1182.

13. Hardin, B. O. and Drnevich, V. P. (1972). 'Shear modulus and damping in soils: design equations and curves', *Journal of Soil Mechanics and Foundations Division, ASCE*, **98**, No. SM7, July 1972, 667–692.

14. Finn, W. D. L., Bransby, P. L., and Pickering, D. J. (1970). 'Effect of strain history on liquefaction of sand', *Journal of Soil Mechanics and Foundations Division, ASCE*, Vol. 97, No. 2M6, November 1970, 1917–1934.

15. Youd, T. L. and Craven, T. N. (1975). 'Lateral stress in sands during cyclic loading', *Journal of the Geotechnical Engineering Division, ASCE*, **101**, No. GT2, February 1975, 217–221.

16. Silver, M. L. and Seed, N. B. (1971). 'Deformation characteristics of sands under cyclic loading', *Journal of the Soil Mechanics and Foundations Division, ASCE*, **97**, No. SM8, August 1971, 1081–1098.

17. Shannon and Wilson, Inc. and Agbabian-Jacobsen Associates (1972). *Soil Behaviour Under Earthquake Loading Conditions*, Seattle, Los Angeles, January 1972, 78–86.

18. Finn, W. D. L., Lee, K. W., and Martin, G. R., 'Stress-strain Relations for Sand in Simple Shear', ASCE National Convention, Denver, Colorado, Nov. 1975, Meeting Preprint 2517. Also, in Soil Mechanics Series No. 26, Dept. of Civil Eng., Univ. of British Columbia, B.C. 1975.

19. Hardin, B. O. and Drnevich, V. P. (1972). 'Shear modulus and damping in soils; measurement and parameter effects', *Journal of Soil Mechanics and Foundations Division, ASCE*, **98**, No. SM6, June 1972, 603–624.

20. Kondner, R. L. and Zelasko, J. S. (1963). 'A hyperbolic stress–strain formulation for sands', *Proceedings, 2nd Pan American Conference on Soil Mechanics and Foundations Engineering*, 289–324.

21. Masing, G. (1926). 'Eigenspannungen and Verfestigung beim Messing', *Proceedings, 2nd International Congress of Applied Mechanics*, Zurich.

22. Herrera, I. (1964). 'Modelos Dinamicos para Materiales Estructuras del Tipo Masing', *Boletin Sociedad Mexicana de Ingenieria Sismica*, **3(1)**, 1–8.

23. Newmark, N. M. and Rosenblueth, E. (1971). *Fundamentals of Earthquake Engineering*, Prentice-Hall, Englewood Cliffs, N.J., 162–163.

24. Marsal, R. J. (1963). *Internal Report*, Institute of Engineering, National University of Mexico, Mexico.

25. Finn, W. D. L., Lee, K. W., and Vaid, Y. (1976). 'Experimental and analytical study of stress–strain relations in simple shear', Department of Civil Engineering, University of British Columbia, Vancouver, *Soil Mechanics Series No. 30.*

26. Pyke, R. M. (1973). 'Settlement and liquefaction of sands under multidirectional loading', *Ph.D. Thesis*, University of California, Berkeley, California.

27. Taylor, D. W. (1954). *Fundamentals of Soil Mechanics*, Wiley, New York, 7th printing, 130–133.

28. Seed, H. B. and Idriss, I. M. (1969). 'Influence of soil conditions on ground motions during earthquakes', *Journal of the Soil Mechanics and Foundations Division, ASCE*, **95**, No. SM1, January 1969, 99–137.

29. Newmark, N. M. (1959). 'A method of computation for structural dynamics', *Journal of the Engineering Mechanics Division, ASCE,* **85**, No. EM3, July 1959, 67–94.
30 Nickell, R. E. (1972). 'A survey of direct integration methods in structural dynamics', *Technical Report No. 9,* Division of Engineering, Brown University, Providence, R.I., April 1972.
31. Schnabel, P. B., Lysmer, J., and Seed, H. B. (1972). 'SHAKE: A computer program for earthquake response analysis of horizontally layered sites', *Report No. EERC 72-12,* Earthquake Engineering Research Centre, University of California, Berkeley, California, December 1972.
32. Parmalee, R. A., Perelman, D. S., Lee, S. L., and Keer, L. M. (1968). 'Seismic response of structure foundation system', *Journal of the Engineering Mechanics Division, ASCE,* **94**, No. EM6, December 1968, 1295–1315.
33. Seed, H. B., and Idriss, I. M. (1967). 'Analysis of soil liquefaction: Niigata earthquake', *Journal of the Soil Mechanics and Foundations Division, ASCE,* **93**, No. SM3, May 1967, 83–108.
34. Finn, W. D. and Lee, K. W. (1977). 'Earthquake generated settlements in saturated sands', *Proceedings Sixth International Conference on Earthquake Engineering,* Theme 6, Dynamics of Soils and Soil Structures, New Delhi, 1977.

Chapter 17

Some Aspects of Marine Pipeline Analysis

R. Nielsen and J. W. Pendered

17.1 INTRODUCTION

First to be described will be the matrix techniques used in producing a computer program to analyse pipe laying. This program is called OPLS2D and 3D.

Secondly, the dynamic lay stress question is discussed. The final section which is riser analysis by transfer matrices discusses a riser analysis where a connector is used between the pipeline and riser. In addition the question of pipeline spanning on the sea bed is discussed. This analysis makes use of transfer matrices. Bibliography attached to this chapter lists current papers of special significance to the general pipe problem.[1-26]

17.2 A GENERAL DESCRIPTION OF THE PIPELINE ANALYSIS

Figure 17.1 indicates the finite element model of the total system used to lay pipe from the lay barge, including a stinger. The tangent stiffness method is used in the analysis described by Oran.[15-16] Oran has completely described the applicable matrix required to solve what is essentially a linear beam problem with large rotations and displacements.

It is amusing to note that in his paper, he indicates that lack of fit and support settlements would not be difficult to incorporate into the analysis. Whereas this may be true from a theoretical point of view, the production of a computer program is such that these matters require the most attention and introduce multiple complexities into the analysis. This can be seen by observing Figures 17.2, 17.3 and 17.4 which give some idea of the constraints on the system such as elastic foundations on the sea bed and elastic supports by the stinger, as well as fixed supports on the lay barge. In addition, the boundary conditions are particularly troublesome because the pipe is able to lift off supports and off the bottom depending on the tensions applied.

The beam element analysis is based on the beam column theory which refers to the initially straight structural members subjected to axial and lateral loads.

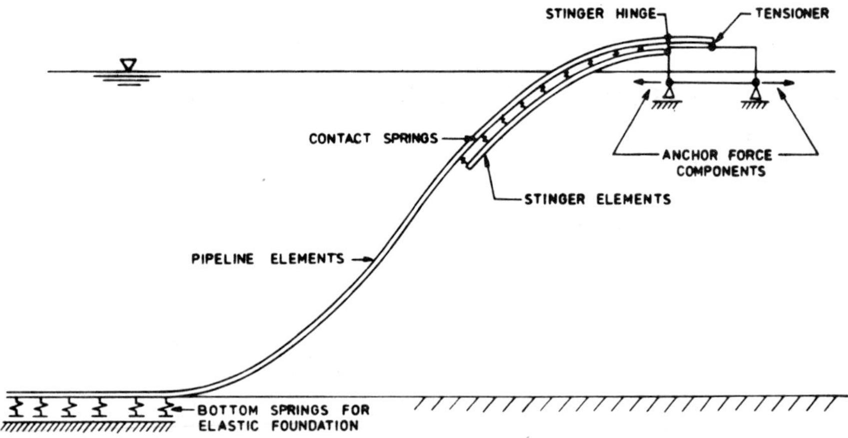

Figure 17.1　Finite element model of total system

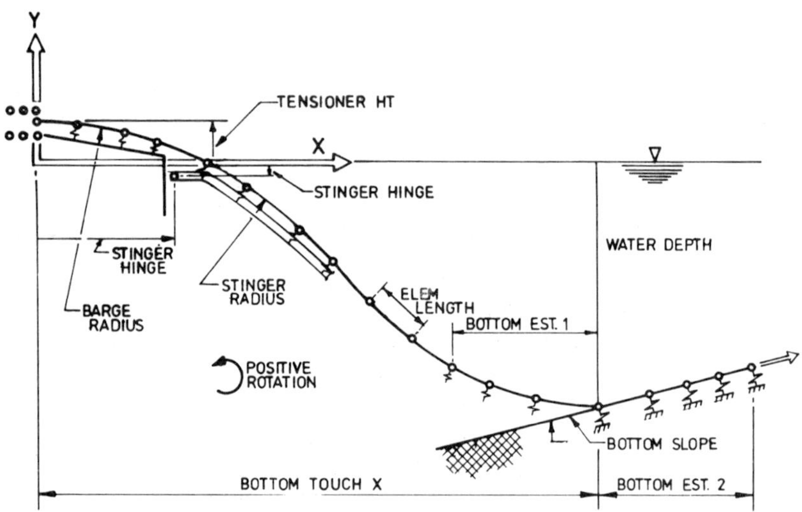

Figure 17.2　OPLS geometry in the vertical plane

In this theory, the axial force is considered to be larger than the effects caused by flexural deformations of the member. Hence axial force is finite where all displacements are infinitesimal. This results in the formulation which is linear; i.e. curvatures are linear functions of displacements. Whereas it is possible to include the effect of the tension on the beam element, the analysis described herein set the axial force as zero. The contribution of the axial force is therefore taken into account at nodes only.

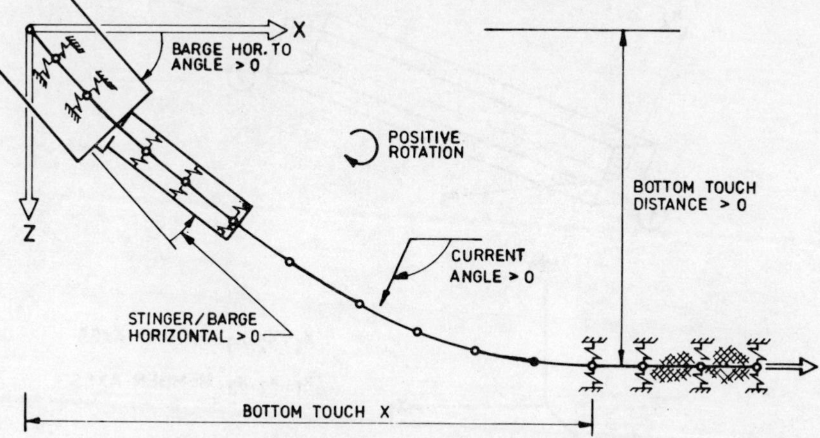

Figure 17.3　OPLS geometry in the horizontal plane

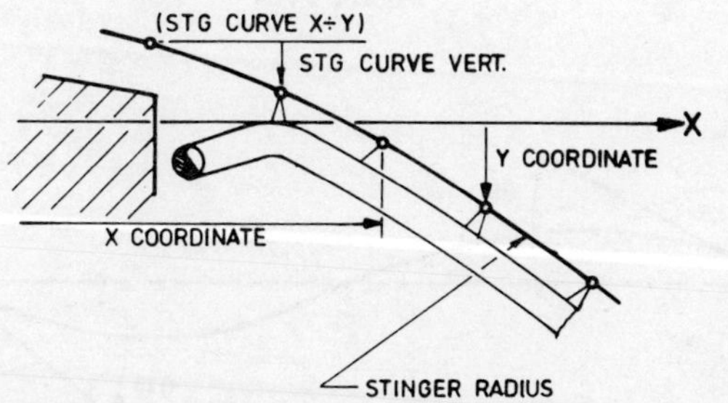

GENERATION OF THE STINGER PIPE SUPPORT COORDINATE

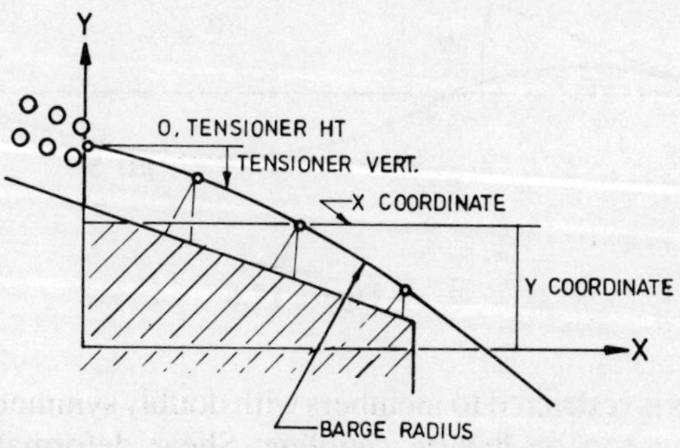

GENERATION OF THE LAYBARGE PIPE SUPPORT COORDINATE

Figure 17.4　Generation of the stinger and laybarge pipe support
co-ordinate

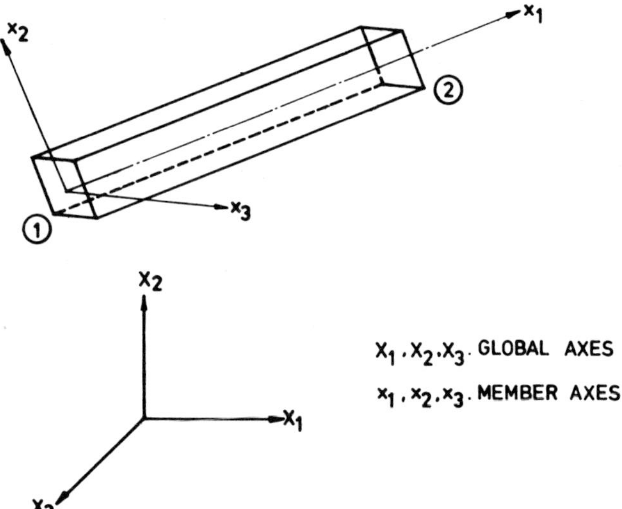

Figure 17.5

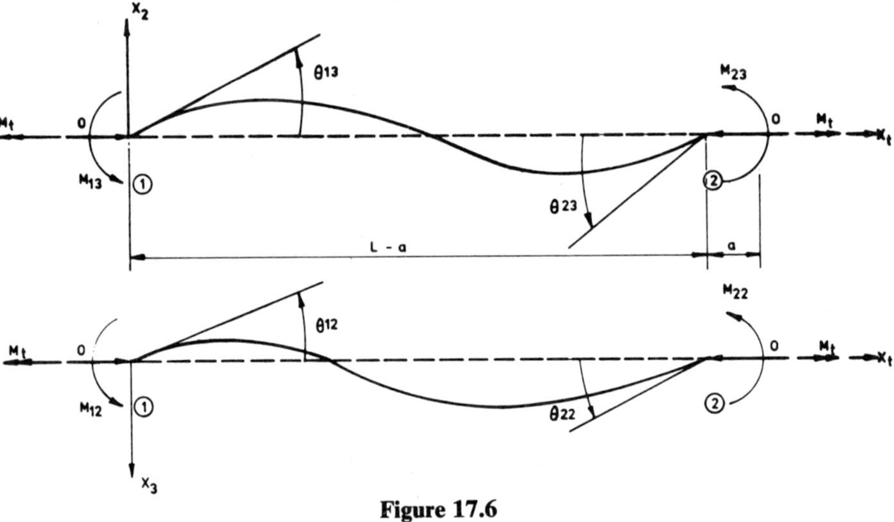

Figure 17.6

The analysis is restricted to members with doubly symmetrical cross sections, thus ruling out torsion-flexure coupling. Shear deformations and restraint against warping are neglected.

The sections that follow describe the equilibrium and co-ordinate system used in the analysis and the notation is shown in Figures 17.5 and 17.6.

This is then followed by discussion of the solution of the equilibrium equation and a section on calculating the derivative matrix.

The section on pipe laying is then concluded with a discussion of dynamic laying stresses.

EXTERNAL FORCES

Some of the external forces, pipe support reactions and concentrated loads such as buoys, are modelled as being applied directly to the pipe nodes. Other forces, the pipe weight and current drag, are applied as distributed loads along the length of the pipe. The simplifying assumptions used to derive equations included *neglecting* the distributed loading acting along the beam element. In the finite element method these distributed loadings, which obviously cannot be ignored for the overall problem, are represented by 'equivalent end loads' or concentrated forces (moment, tension, and shear) applied to the nodes at the ends of each element.

EQUILIBRIUM AND CO-ORDINATE SYSTEMS

The pipe is modelled by a 'chain' of beam elements connected to each other at points called nodes. The FEM model is in *equilibrium* when all of the 'internal' and 'external' forces described in the preceding section sum to zero at the nodes.

The 'internal' forces and 'equivalent end loads' defined by equations elsewhere are expressed in terms of *element* co-ordinate systems which are different for each element. In order for the different forces to be summed their components must be expressed in the *same* co-ordinate system. The program OPLS accomplishes this by *rotating* the element forces from the different element co-ordinate systems to the 'global' co-ordinate system.

OPLS uses three distinct co-ordinate systems:

(1) A 'global' co-ordinate system $(X, Y, Z, \psi, \alpha, \beta)$ is used to describe the overall problem. The independent variables of the overall problem are the global co-ordinates of the nodes.

(2) A unique 'local' element co-ordinate system (x_e, y_e, z_e) is defined for each element. The x axis is aligned with the longitudinal axis of the element. The element tensions and shears are expressed in the directions of the element x and y axes; moments are expressed as rotating about the z axis.

(3) A unique 'local' nodal co-ordinate system (x_η, y_η, z_η) is defined for each node. For pipe nodes the x axis is tangent to (coincident with) the longitudinal axis of the pipe. The nodal y_η and z_η axes are the 'vertical' and 'horizontal' principle axes of the pipe cross-section and are perpendicular to the pipe and to each other.

SOLVING THE EQUILIBRIUM EQUATIONS

The forces calculated in the global co-ordinate system are obtained by summing the components of the element end forces and external forces. A non-zero value indicates that the forces at that node and direction are not balanced and not in equilibrium. Each unbalanced force represents an equilibrium equation of the form:

$$F_i \text{ (node } j) = F_i \text{ (element } a, \text{ node } j) + F_i \text{ (element } b, \text{ node } j)$$

$$+ \cdots \text{ (all elements at node } j) \tag{17.1}$$

From Equation (17.1) it is obvious that the element end forces, and thus the equilibrium equations are non-linear functions of the independent variables X, Y and β. The overall pipelaying problem is represented by a system of simultaneous non-linear equations written for each node in each global direction in which the pipe is free to move. Equilibrium exists when all of these equations (unbalanced forces) are equal to zero. Thus finding the equilibrium configuration of the pipe means finding the 'zero root' of a system of non-linear equations. There are a number of methods for solving non-linear equations. OPLS uses Newton method which consists of the following steps:

(1) Given a close initial approximation to the equilibrium of the problem, calculate the unbalanced force vector (F) by summing the components of all the element end forces and external forces in each 'global' direction at each node.

(2) Given the unbalanced force vector F, calculate the matrix of the partial derivatives of F with respect to all the independent global coordinates X, and use Equation (17.2) to calculate a new improved solution to the equilibrium equations.

$$\{X\}_{i+1} = \{X\}_i - \left[\frac{\partial\{F\}}{\partial\{X\}}\right]^{-1} \{F\} \quad \text{or} \quad -\left[\frac{\partial\{F\}}{\partial\{X\}}\right]^{-1} \{Si\} = \{F\} \tag{17.2}$$

(3) Calculate the unbalanced forces for solution X_{i+1} of step (2) and if the forces are not sufficiently close to zero, repeat step (2) as many times as necessary.

Newton's method has several important characteristics:

(a) It is an iterative procedure, each new solution calculated by Equation (17.2) should be better than the previous one.

(b) When it works, Newton's method is the fastest converging method for solving non-linear equations.

(c) Newton's method will not work without an approximate initial solution which is 'close' to the final equilibrium solution.

(d) Newton's method requires the calculation of the matrix of partial derivatives of the non-linear equilibrium equations with respect to each global co-ordinate involved.

(e) The solution (zero root) of the non-linear unbalanced force equations is obtained by a series of linear steps [Equation (17.2)]. Each linear step involves evaluating the matrix of partial derivatives $\partial\{F\}/\partial\{X\}$ and then solving the resulting system of linear equations for the displacement vector $\{\Delta X\}$ by Gaussian elimination or some similar means.

CALCULATING THE DERIVATIVE MATRIX

The unbalanced forces at a node are evaluated by computer by calculating the element end forces and equivalent end loads for all the elements attached to the node and summing these along with any other external forces that may exist. Because the different elements are mathematically similar the calculation of the unbalanced forces consists of repeating a single set of equations for each element.

The partial derivatives of the unbalanced forces can also be calculated by a repetitive procedure. From Equation (17.1) it can be seen that the partial derivative of a single unbalanced force F_i with respect to a particular co-ordinate X_j will be equal to the summation of the partial derivatives of the element force components that contributed to the unbalanced force. For example:

$$\frac{\partial F_i}{\partial X_j} = \frac{\partial F_i \,(\text{element } a)}{\partial X_j} + \frac{\partial F_i \,(\text{element } b)}{\partial X_j} + \cdots \tag{17.3}$$

The element forces are only functions of the co-ordinates (x, y, β) of the two end nodes of the element. Thus the unbalanced forces at a node are only functions of the global co-ordinates $(X, Y \text{ and } \beta)$ of the node itself and of the nodes connected to it by a finite element.

In OPLS the derivatives of the external force terms $\{F_e\}$ and $\{f_e\}$ are neglected in calculating the *derivative* matrices because:

(1) The concentrated external forces $\{F_e\}$ applied in global co-ordinates are constant and independent of $\{x\}$.

(2) The equivalent end loads $\{f_e\}$ are much 'smaller' and thus less important than the internal forces.

(3) Including the equivalent end load terms $\{f_e\}$ would destroy the symmetry and positive definiteness of the derivative matrix in two dimensions and greatly complicate the calculation of the matrices in both two and three dimensions.

Taking derivatives of the element internal force equations with respect to the co-ordinates x we have:

$$\left[\frac{\mathrm{d}\{F\}}{\mathrm{d}\{X\}}\right] = \left[\left[\frac{\mathrm{d}[R\cdot T]}{\mathrm{d}\{X\}}\right]\right]\cdot\{f\} + [R\cdot T]\cdot\left[\frac{\mathrm{d}\{f\}}{\mathrm{d}\{e\}}\right]\left[\frac{\mathrm{d}\{e\}}{\mathrm{d}\{X\}}\right]$$

where $\qquad\qquad\qquad\qquad\qquad\qquad\qquad\qquad\qquad\qquad\qquad\qquad\qquad$ (17.4)

$$\left[\frac{\mathrm{d}\{f\}}{\mathrm{d}\{e\}}\right] = [K]\frac{\partial\{e\}}{\partial\{x\}}$$

Note that matrices $[R]$ and $[T]$ have already been multiplied together and that $[[D[K]/\mathrm{d}\{e\}]] = 0$. Thus equation (17.4) becomes:

$$\left[\frac{\partial\{F_i\}}{\partial\{X\}}\right] = \left[\left[\frac{\partial[RT]}{\partial\{X\}}\right]\right]\cdot\{f\} + [RT][K]\left[\frac{\partial\{e\}}{\partial\{X\}}\right]$$
$$\qquad\qquad\qquad\qquad\qquad\qquad\qquad\qquad\qquad\qquad\qquad (17.5)$$

For a beam element the term $[\mathrm{d}\{F\}/\mathrm{d}\{X\}]$ is a two-dimensional (square) matrix. In two dimensional problems it has six rows and columns because there are six end forces $(M_1, M_2, S_1, S_2, T_1$ and $T_2)$ and six end node coordinates $(x_1, y_1, \beta_1, x_2, y_2, \beta_2)$ for each element.

The beam element derivative matrix has two parts: the first part represents the change in the rotation matrix used to calculate the global components of the element end forces; the second part represents the change in the element end forces themselves. For the two-dimensional FEM the matrix $[\partial\{e\}/\partial\{X\}]$ is equal to $[RT]^T$ and the corresponding matrix product $[RT][K][RT]^T$ is the 'stiffness' matrix used in linear FEM analysis. In three dimensions however, this is not the case.

Each term of the matrix of partial derivatives for the overall problem is calculated by summing the appropriate derivative terms from the derivative matrices of all the elements connected to the node. In OPLS the overall derivative matrix is formed by first, calculating the terms of an individual element derivative matrix, and then adding these terms to the appropriate locations in the overall derivative matrix. The resulting overall derivative matrix is narrowly banded because the terms corresponding to co-ordinates which are not connected by elements are equal to zero.

In two dimensions the position and orientation of a point are uniquely defined by exactly three independent co-ordinates (x, y, β). Similarly the forces at a point are uniquely defined by their components in two mutually perpendicular directions and a moment about an axes perpendicular to the two-dimensional plane. (F_x, F_y, M_z). Thus there are three possible equilibrium equations and three possible independent co-ordinate variables for each node. When the components of all the 'internal' and 'external' forces are summed in one 'global' direction at one node, this makes up one equilibrium equation.

The independent variables are the co-ordinates $(X, Y, Z, \Psi, \alpha, \beta)$ whose values are not fixed. They are called 'free displacements' because they represent the directions in which the FEM model is free to move to reach equilibrium. If a node is not free to move in one of the $(X, Y, Z, \Psi, \alpha, \beta)$ directions then there is no equilibrium in that direction unless it is assumed that a corresponding 'support reaction' acts on the nodes.

DYNAMIC LAYING STRESSES

In the true environment the limitation to laying stresses and the advantage to be gained from later generation lay barges is the dynamic stress level within the pipe under the action of wave induced effects. Firstly, the lay barge will be in a state of dynamic response to the wave environment and this motion will be applied to the stinger and hence to the pipe catenary. Secondly, wave forces are applied directly to the submerged portion of the pipe. Thus the pipe experiences an additional distributed loading due to viscous drag and added inertia contributions arising from both structural and wave orbital motions. The problem is more complicated by the fact that the stiffness of a large diameter pipe will have a measurable effect upon the motion of the lay barge in some of its modes—certainly more effect than the barge anchors.

The motion analysis of floating structures can be carried out by several means according to the type (i.e. shape) of vessel concerned. Lay barges having ship forms will be more readily analysed by means of a ship motion program such as SCORES. Semi-submersible forms comprising tubular members and/or rectangular pontoon sections can be analysed by means of accumulating the equation of motion over each member with forces derived on the assumption of Morrison's equation, and techniques are now available to solve this equation in either a linearized form in terms of a modal or spectral technique or to incorporate non-linear effects and to solve by means of Newmark time stepping technique or a Newton–Raphson iteration (see for example Reference 21). The non-linear capability is particularly necessary for consideration of the restraining effect of the pipe catenary. Alternatively the larger later generation barges will be more appropriately analysed using the now available dynamic capabilities of source distribution loading programs.

In order to fully assess dynamic laying stresses and to place a rational limit to the sea state in which laying is possible with a particular barge it is thus seen that two complex computer programs have now to be combined—a laying program such as OPLS on the one hand with a non-linear motion analysis program on the other. Work directed towards this objective is at the present time being constructed.

RISER ANALYSIS BY TRANSFER MATRIX

Many pipeline problems are more akin to simple (straight) beam configurations. In the discussion of pipelaying analysis it was seen that 'finite element' methods applied to such single beam problems led to an extremely narrow banded matrix and the only true advantage of finite element methods to that problem was their capability of extension to the non-linear geometry problem. Without such considerations it is natural to look to a more versatile and simple analytical tool for the routine pipe stress analysis.

One important criterion for such a method is that it shall free the analyst from concern for deriving an approximate discretized model of the system under analysis and should permit a direct input of the physical and engineering parameters. Just such a philosophy and ready application to 'line based' problems is provided by the transfer matrix initial parameter technique which is based upon the Laplace Transform method of solution. Whilst this is not a new technique—having its origins in the German elasticity school of more than a century ago—its potential has been enhanced by modern computing capabilities and it is now available in serveral popular 'beam analysis' computer codes.

LINK is one such code which is available and is based on Reference 20. It is equally applicable to beam, plate, cylinder or open section beam members and permits static, buckling and free or forced dynamic analyses. It is particularly suited to pipeline problems by virtue of its wide range of distributed and discrete load conditions, its ability to account for changes of section and material properties, and its range of offset or in-line support conditions. All input corresponds directly to the engineering formulation of the problem and output is in terms of shear, moment, deflection and slope at the chosen increments 'along' the problem together with reactions, natural frequencies and mode shapes where appropriate.

One of the most common problems to which this type of analysis can be applied is that of the assessment of stresses in a submarine pipeline when laid on a defined sea bed profile—so-called spanning analysis. LINK provides the ability to model such a profile with considerable realism as to the heights of individual rock outcrops and levels of sediment foundations and then to lay the pipe over such supports. The alternative permitted by simple beam analysis is usually restricted to simple supports and over-idealized end fixity assumptions which may be useful in indicating the height of obstructions which could cause problems but cannot cope with the 'fine-tuning' demanded by the high operating stresses in such pipelines. The natural frequency of such spans has also to be determined to assess the likelihood of vortex shedding resonance and this again is easily provided by LINK from the same problem formulation. An example of such a pipeline spanning analysis is given in Figure 17.7.

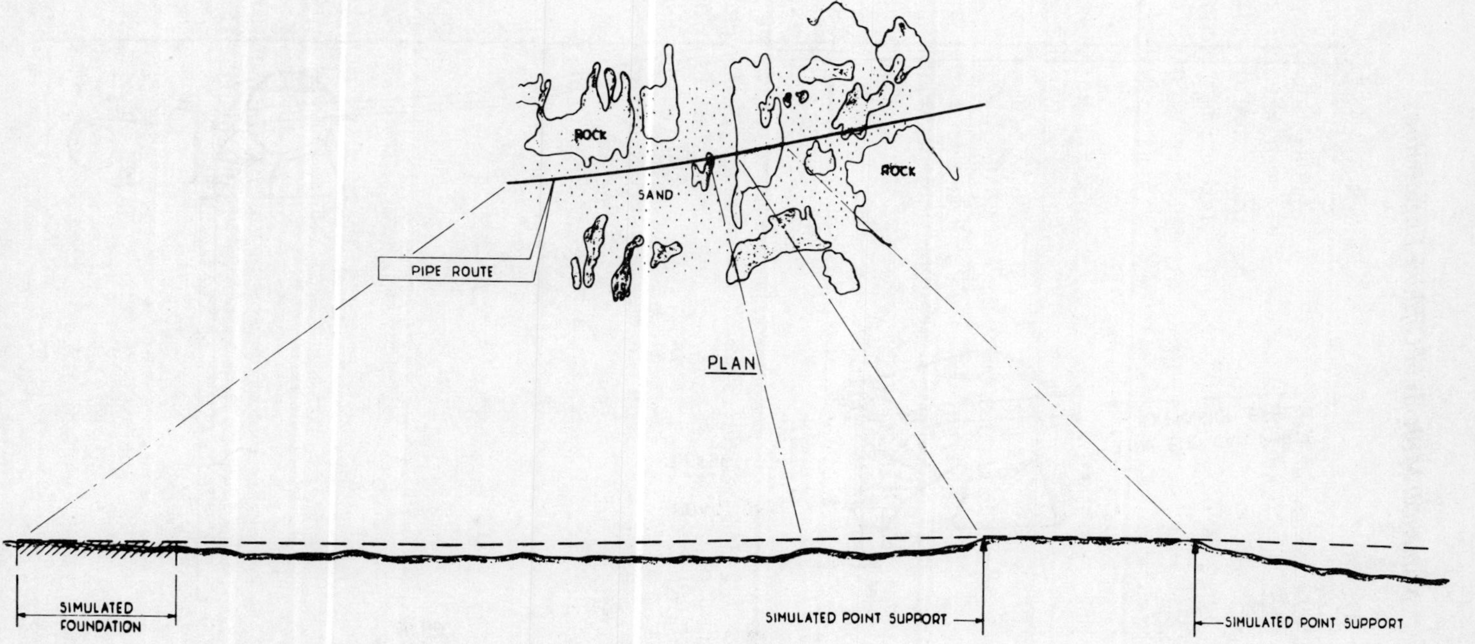

Figure 17.7 Calculated pipeline profile (broken line) superimposed on sea-bed profile

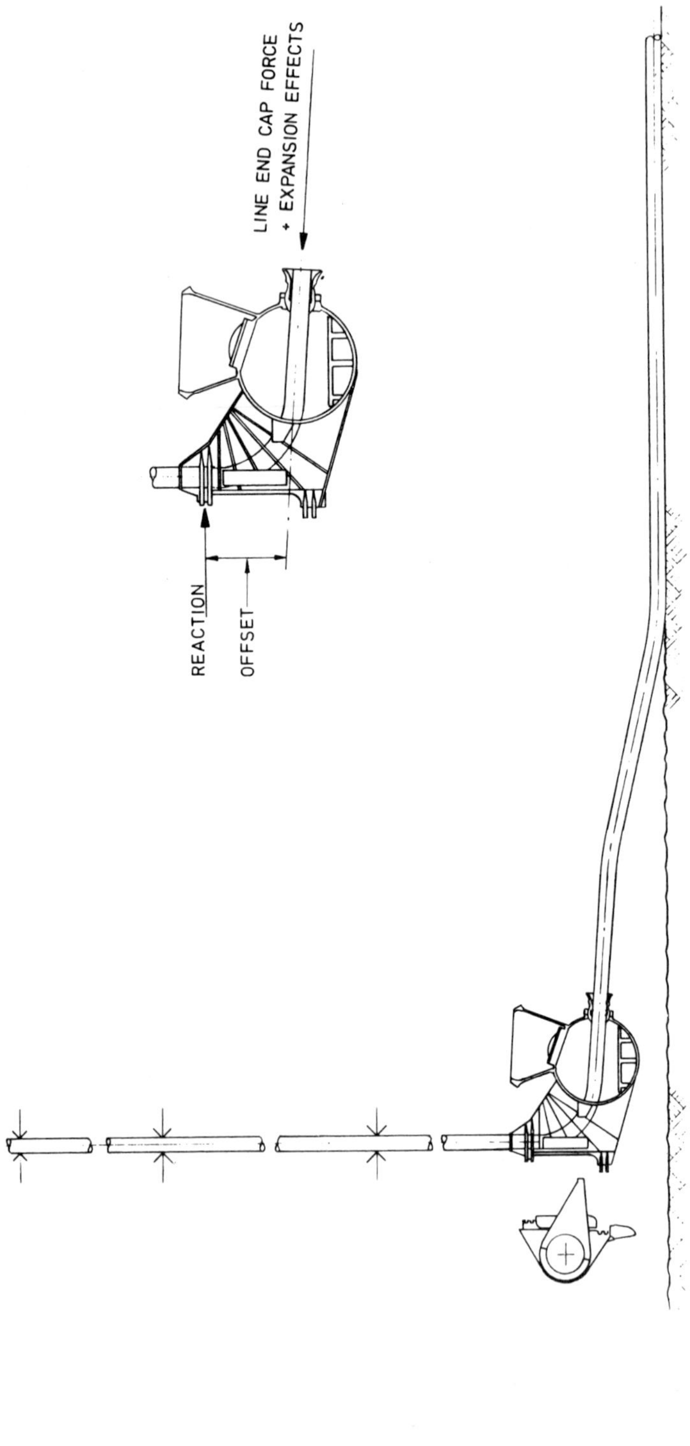

Figure 17.8

The range of features of LINK—and indeed those which are demanded of any similar code—permit its application to more complex problems—again with advantages over an equivalent finite element analysis. Figure 17.8 shows some of the features (in schematic form) of a system for pipeline/riser connection in which the pipeline is pulled into a chamber attached to the bottom of the riser and the welded connection is then made within the chamber. Such an arrangement makes for a complex analysis. The chamber is permitted a finite deflection until it abuts against the platform bracing and such deflection affects the riser and pipeline stresses. The chamber has vertical and rotational freedom. Both riser and pipeline are subject to thermal and pressure expansion. The riser is clamped at discrete levels. The pipeline is pulled up the chamber and thus has a non-horizontal entry angle and an indetermined free span before being supported on an elastic foundation. And the problem is finally complicated by the fact that the pipeline has to be analysed for the possibility of out-of-plane buckling over its unburied length.*

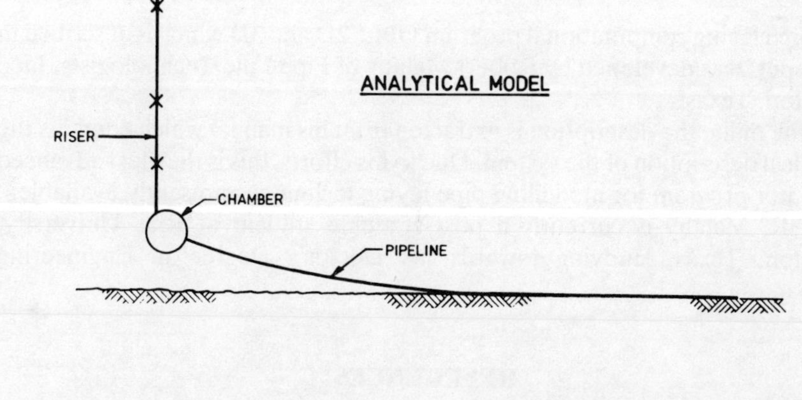

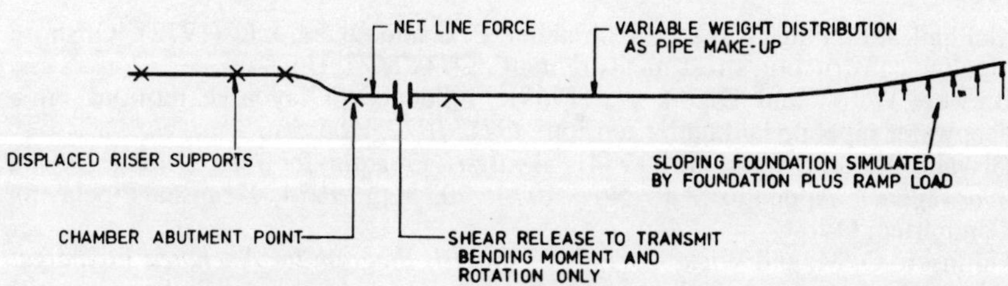

Figure 17.9 Axial tensions added on each section

* The buckle mechanism for a pipeline under axial loading is complicated by the contribution of the weight associated potential energy when buckling in a vertical or inclined plane. A snap-through type of instability results followed by a complex post buckling phase which can alleviate the axial load within a constrained length. This matter is discussed in Reference 19.

Figure 17.9 indicates how such a problem was idealized and modelled for a transfer matrix solution. The bend at the chamber is accommodated by a shear release with axial/transverse loads transferred mutually across (as a manual iteration outside the program). The angle of entry of the pipeline to the chamber introduces the requirement to model a sloping foundation and this is readily achieved in LINK by deforming a normal elastic foundation with a superimposed ramp load. The point of touchdown onto the foundation is again a matter of iteration.

LINK and other similar programs are written as general purpose transfer matrix codes and, although readily applicable to pipeline analysis situations, there are some additional features which could be added to improve this application. Automatic iteration of touch-down points and displaced foundations are two examples of such modifications. These and other modifications are currently being incorporated into a specialized transfer matrix code.

ACKNOWLEDGEMENTS

The pipe laying computational program OPLS2D and 3D which is described in the paper, was developed by Robert Malahy of Pipe Line Technologists, Inc., Houston, Texas.

In the main, the description is extracted from his manual which contains the technical description of the system. Due to his efforts, this is the most advanced computer program for modelling pipe laying techniques presently available.

Mr R. Malahy is currently a post-graduate student at Rice University, Houston, Texas, studying towards his Doctor's Degree in Engineering Mechanics.

REFERENCES

1. McPhail, J. F., Finn, W. D. L., Rohmaller, P. L. and Okers, J. C. (1972). 'Offshore pipeline construction stress measurement', *OTC 1573*, Houston.
2. Brewer, W. V. and Dixon, A. (1969). 'Influence of laybarge motions on a deepwater pipeline laid under tension', *OTC 1072*, Houston.
3. Gisvald, K. and Wick, T. (1974). *Analysis Program for Pipe During Laying* (*Norwegian*), Appendix 2 to Nov. 1974: 90, Sept. 1974, Deapsea Pipelaying Committee, Oslo.
4. Faltinsen, O. M. and Wick, T. (1974). *Analysis Program for Anchored Laybarges* (*Norwegian*), Appendix 4 to Nov. 1974; 40, April 1974, Deepsea Pipline Committee, Oslo.
5. *Preliminary Study Concerning Model Investigations of Pipe During Laying* (*Norwegian*), Appendix 6 to Nov. 1974. 40, Sept. 1974, Deepwater Pipelaying Committee, Oslo.
6. Clauss, G. and Kruppa, C. (1979). 'Model testing techniques in offshore pipelining', *OTC 1937*, Houston.

7. Ovunc, B. and Mallareddy, H. (1971). 'Stress analysis of offshore pipelines under dynamic loads', *OTC 1361*, Houston.

8. Finn, W. D. L. 'Dynamic stresses in offshore pipelines', *ASCE 1972*, National Structural Engineering Meeting.

9. Gnone, E., Signorelli, P., and Guiliano, V. (1975). 'Three-dimensional static and dynamic analyses of deepwater sealines and risers', *OTC 2326*, Houston.

10. Ovunc, B. and Mallareddy, H. (1970). 'Stress analysis of offshore pipelines', *OTC 1222*, Houston.

11. Brando, P. and Sebastiani, G. (1970). 'Determination of sealines elastic curves and stresses to be expected during laying operations', *OTC 1354*, Houston.

12. Dixon, D. A. and Rutledge, D. R. (1968). 'Stiffened catenary calculations in pipeline laying problem', *Journal of Engineering for Industry*, February 1968.

13. Wilkins, J. R. (1970). 'Offshore pipeline stress analysis', *OTC 1227*, Houston.

14. Powell, G. H. (1969). 'Theory of non-linear elastic structures', *Journal of the Structural Division, ASCE*, December 1969.

15. Oran, C. (1973). 'Tangent stiffness in plane frames', *Journal of the Structural Division, ASCE*, June 1973.

16. Oran, C. (1973). 'Tangent stiffness in space frame', *Journal of the Structural Division, ASCE*, June 1973.

17. Darcing, D. W. and Heathery, R. F. (1970). 'Marine pipeline analysis based on Newtons method with an arctic application', *Paper No. 70-PET*-16, Sept. 1970 ASME, Denver, Colorado.

18. Powers, J. T. and Finn, W. D. L. (1969). 'Stress analysis of offshore pipelines during installation', *OTC 1071*, Houston.

19. Nielsen, R. (1976). 'Some special problems of deepwater pipelaying and connections between platforms or facilities', *ASME*, Sept. 1976, Mexico City.

20. Pilkey, W. D. and Nielsen, R. (1968). 'Line solution technology as a general engineering approach to the static, stability, and dynamic response of structural members and mechanical elements', *38th Shock and Vibration Information Symposium*, Wash. D.C., May 1968.

21. Natvig, B. J. and Pendered, J. W. (1977). 'Non-linear response of floating structure to wave excitation', *OTC 2796*, Houston.

22. 'Static analysis of marine pipelines during laying'. *Det norske Veritas*, NV457.

23. Palmer, A. C., Hutchinson, G., and Wells, J. W. (1973). 'Configuration of submarine pipeline during laying operations', *ASME Paper No. 73-WA/OCT-4*.

24. Malahy, R. 'Ocean pipe laying simulation', *OPLS 2D and OPLS 3D, Vol. I—Users Manual*, Pipeline Technologists, Houston.

25. Malahy, R. 'Ocean pipe laying simulation', *OPLS 2D and OPLS 3D*, Vol. *Technical Description*, Pipeline Technologists, Houston.

26. Pedersen, P. T. (1975). 'Equilibrium of offshore cables and pipelines during laying', *Report No. 80*, Feb. 1975, The Technical University of Denmark, Danish Center for Applied Mathematics and Mechanics.

Subject Index